Die Design Fundamentals

Vukota Boljanovic
J.R. Paquin

Industrial Press Inc.

Library of Congress Cataloging-in-Publication Data

Boljanovic, Vukota.
Die design fundamentals/Vukota Boljanovic, J.R. Paquin, Robert E. Crowley.—3rd ed.
p. cm.
Rev. ed. of: Die design fundamentals/J.R. Paquin and R.E. Crowley. c1987.
Includes index.
ISBN 13: 978-0-8311-3119-7 ISBN 10: 0-8311-3119-5
1. Dies (Metal-working) I. Paquin, J.R. II. Crowley, R.E. III. Paquin, J.R. Die design fundamentals. IV. Title.

TS253.P3 2005
621.9′84—dc22 2005047463

Industrial Press Inc.
989 Avenue of the Americas
New York, New York 10018
Tel: 212-889-6330 Fax: 545-8327

Cover Design: Janet Romano
Managing Editor: John Carleo
Developmental Editor: Robert Weinstein
Copy Editor: Kathy McKenzie

10 9 8 7 6 5 4 3 2

TABLE OF CONTENTS

PREFACE TO THIRD EDITION

The field of tool engineering and die design, a complex and fascinating subject, continues to advance rapidly. This broad and challenging topic continues to incorporate new concepts at an increasing rate, making tool and die design a dynamic and exciting field of study. In preparing this third edition, my most important goal has been to provide a comprehensive state-of-the-art textbook on die design fundamentals, which also encompasses the additional aims of motivating and challenging students.

This new edition provides balanced coverage of relevant fundamentals and real-world practices so that the student can understand the important and often complex interrelationships between die design and the economic factors involved in manufacturing sheet metal-forming products.

A groundbreaking and comprehensive reference with many thousands of copies sold since it first debuted in 1962 as J.R. Paquin's *Die Design Fundamentals,* this new third edition of *Die Design Fundamentals* basically follows the same design philosophy: It is a step-by-step introduction to the design of stamping dies. However, the original book has been completely revised and updated, and the order of the chapters has been changed to follow the logical process of designing a die.

The plan of the book remains unique. After introductory material and a discussion of 20 types of dies, the design of a representative die is separated into 14 distinct steps. Each step is illustrated in two ways: first, as a portion of an engineering drawing, that is, as the component is actually drawn on the design; second, the die design is shown pictorially in order to improve the user's visualization. In successive sections of the book, each step is detailed as it is applied to the design of the various types of dies listed in Chapter 2. In many figures a punch shank is shown because it is still in everyday use in many small stamping shops. However, according to the OSHA Standard 1910.217(7) it cannot be used for clamping the punch holder to the slide (ram) of a press, but can be used for aligning the die in the press. Slide (ram) mounting holes or another clamping system must be provided in the punch holder for fastening. The final chapter deals with presses and quick die-changing (QDC) systems.

The intent of this new edition is to provide students, instructors, and working professionals with graphically detailed assistance in understanding the underlying principles of designing single-station dies as well as small progressive dies of a type generally used once for short runs of parts manually cut from strip sheared from sheets.

For the first time, a dual (English and metric) system is included. New methods of producing blanks widely acknowledged within the industry, such as waterjet cutting and laser cutting are included, as well. The chapter 20 "Presses and Quick Die-Change Systems" has been considerably revised, with the addition of a new subtopic, "Quick

Die-Change Systems." To this third edition of the book has also been added a Glossary of the terms used.

In response to comments and suggestions by numerous reviewers, several major and minor changes have also been made throughout the text. A page-by-page comparison with the second edition will show that literally hundred of changes have been made for improved clarity and completeness.

It is hoped that by reading and studying this third edition of the book, students and other users will come to appreciate the vital nature of tool and die engineering as an academic subject that is as exciting, challenging, and as important as any other engineering and technology discipline.

The author of the third edition owes much to many people. I am grateful to my son Sasha for valuable contributions in the preparation of this edition. Finally, I wish to thank Em Turner Chitty for her competent proofreading of this new edition.

Vukota Boljanovic
Knoxville, Tennessee

1 INTRODUCTION TO DIE DESIGN

Die design, a large division of tool engineering, is a complex, fascinating subject. It is one of the most exacting of all the areas of the general field of tool designing.

How then shall we enter into the study of die design? Obviously, we shall have to begin cautiously, learning each principle thoroughly before proceeding to the next one. Otherwise it is quite likely that we should soon become hopelessly involved in the complexities of the subject and in the bewildering number and variety of principles that must be understood. What, then, is a die?

The word "die" is a very general one and it may be well to define its meaning as it will be used in this text. It is used in two distinct ways. When employed in a general sense, it means an entire press tool with all components taken together. When used in a more limited manner, it refers to that component which is machined to receive the blank, as differentiated from the component called the "punch," which is its opposite member. The distinction will become clear as we proceed with the study.

The die designer originates designs of dies used to stamp and form parts from sheet metal, assemble parts together, and perform a variety of other operations.

In this introduction you will learn basic meanings and the names of various die components; then, operations that are performed in dies will be listed and illustrated. In other sections of the book, the design of dies and die components will be explained in a far more thorough manner, so that your understanding will be complete in every respect.

1.1 BASIC MEANINGS

1.1.1 Part Drawing

To begin our study of the various components that make up a complete die, let us consider the drawing of the link illustrated in Figure 1.1. This part is to be blanked from steel strip and a die is to be designed for producing it in quantity. The first step in designing any die is to make a careful study of the part print because the information given on it provides many clues for solving the design problem.

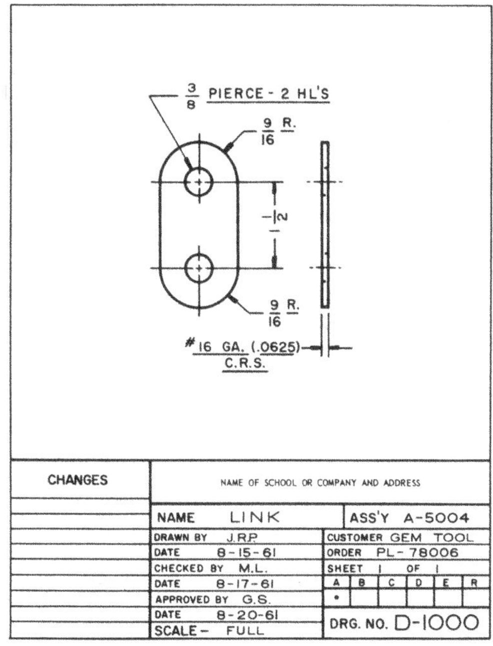

Figure 1.1 A typical part drawing.

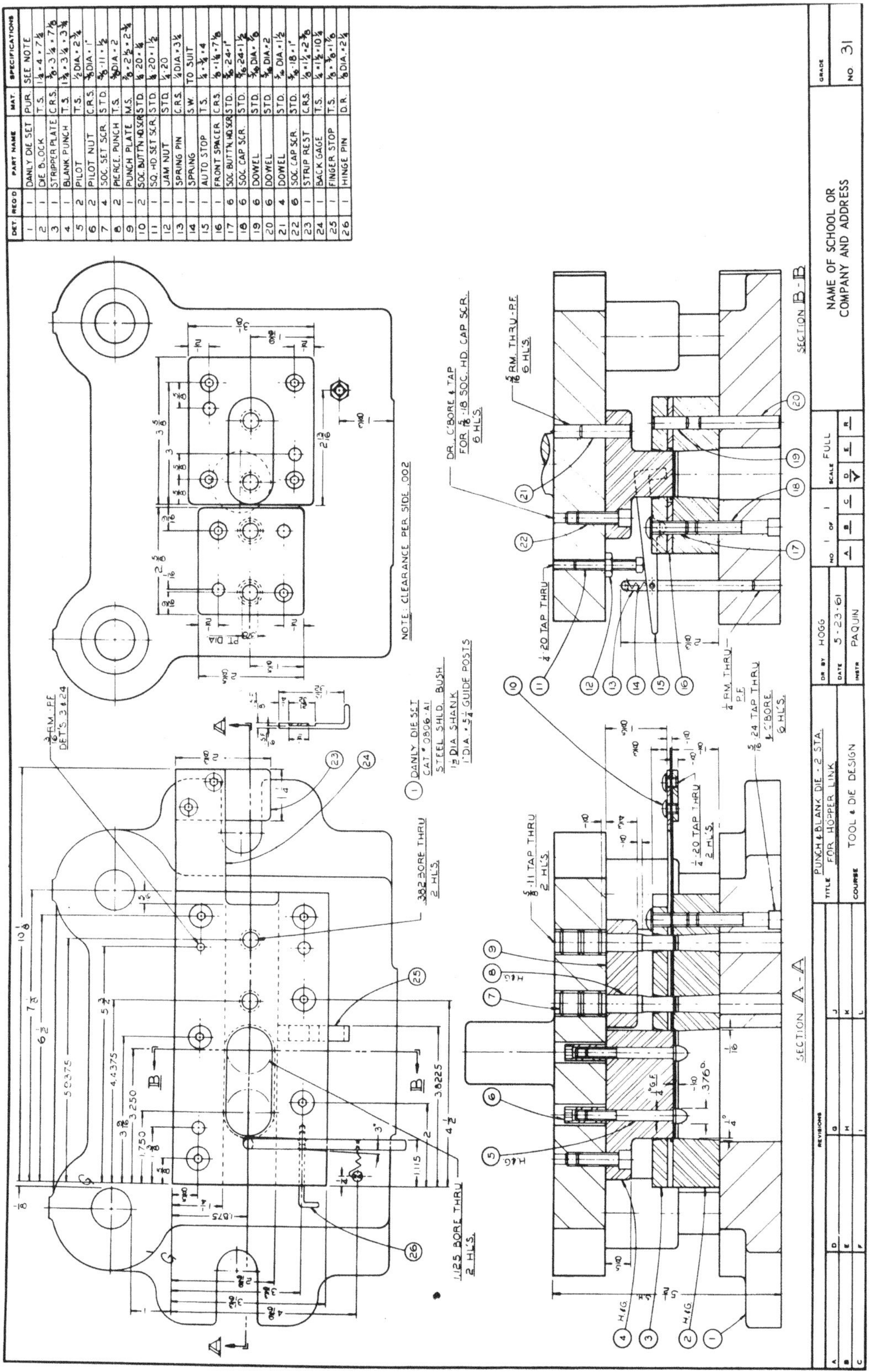

Figure 1.2 A complete die drawing.

1.1.2 Die Drawing

Figure 1.2 is a complete die drawing ready to be printed in blueprint form. To the uninitiated it might appear to be just a confusing maze of lines. Actually, however, each line represents important information that the die makers must have to build the die successfully. In illustrations to follow, we will remove the individual parts from this assembly and see how they appear both as three- and as two-view drawings, and as pictorial views, to help you to visualize their shapes. As you study further, keep coming back to this illustration to see how each component fits in. When you are through, you should have a good idea of how the various parts go together to make up a complete die.

1.1.3 Blueprints

After a die has been designed on tracing paper using traditional techniques or AutoCAD, blueprints are produced for use in the die shop where the dies are actually built by die makers. This is how a blueprint of a die drawing appears. From such prints, die makers build the die exactly as the designer designed it. The drawing must be complete with all required views, dimensions, notes, and specifications. If the die maker is obliged to ask numerous questions, the drawing was poorly done. Figure 1.3 shows a typical blueprint.

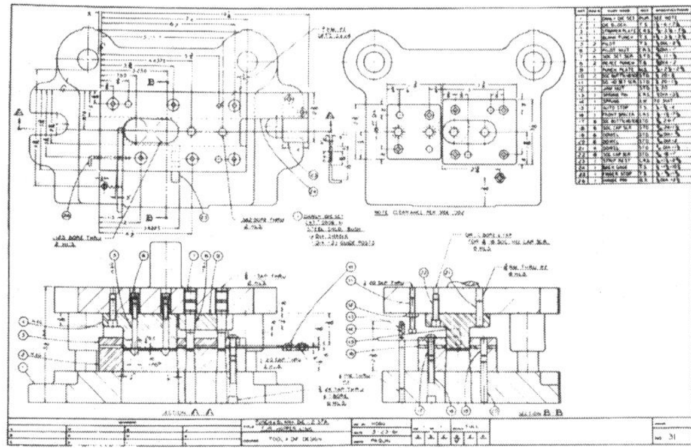

Figure 1.3 A typical blueprint.

1.1.4 Bill of Material

The bill of material (Figure 1.4) is filled in last. This gives required information and specifications for ordering standard parts and for cutting steel to the correct dimensions. This material is cut and assembled in the stock room, then placed in a pan, along with a print of the die drawing. When filled, the pan must contain everything the die maker will require for building the die, including all fasteners and the die set.

27	4	SOC. CAP SCR.	STD.	3/8 - 16 x 1 1/2
26	1	FRONT SPACER	C.R.S.	1/8 x 3/4 x 4 5/8
25	4	BUTTON HD. SOC. CAP SCR	STD.	3/8 x 16 x 7/8
24	2	DOWEL	STD.	3/8 DIA. x 1 3/4
23	2	DOWEL	STD.	3/8 DIA. x 7/8
22	1	JAM NUT	STD.	1/4 - 20
21	1	SQ. HD. SET SCR.	STD.	1/4 - 20 x 1 3/4
20	2	DOWEL	STD.	3/8 DIA. x 1 1/4
19	4	SOC. CAP SCR.	STD.	3/8 - 16 x 7/8
18	2	DOWEL	STD.	3/8 DIA. x 1 1/2
17	4	SOC. CAP. SCR.	STD.	3/8 - 16 x 1
16	2	RIVET	C.R.S.	1/8 DIA. x 3/8
15	1	STRIP REST	C.R.S.	1/8 x 3/4 x 1 1/2
14	1	BACK GAGE	T.S.	1/8 x 3/4 x 8 1/8
13	2	PIERCE PUNCH	T.S.	3/4 DIA. x 1 7/8
12	1	PUNCH PLATE	M.S.	1 x 2 x 2 1/8
11	1	BLANK-PUNCH	T.S.	1 1/2 x 2 x 2 1/2
10	1	STRIPPER	C.R.S.	3/8 x 2 1/2 x 4 5/8
9	1	DIE BLOCK	T.S.	1 1/4 x 2 3/4 x 4 5/8
8	1	DOWEL	STD.	3/16 x DIA. x 1/2
7	1	FINGER STOP	T.S.	1/8 x 3/8 x 2 1/8
6	1	SOC. CAP SCR.	STD.	#10-24 x 1/4
5	1	SPRING	S.W.	TO SUIT
4	1	SPRING PIN	C.R.S.	1/4 DIA. x 2 5/8
3	1	HINGE PIN	D.R.	1/8 DIA. x 1 7/8
2	1	AUTO STOP	T.S.	1/4 x 1/2 x 2 1/4
1	1	DIE SET	PUR.	SEE NOTE
DET.	**REQ'D**	**PART NAME**	**MAT.**	**SPECIFICATIONS**

NAME OF SCHOOL OR COMPANY AND ADDRESS

ASSEMBLY 2 STA. P. & D. FOR – "HOPPER LINK"	TYPE 40-BL
DRAWN BY J. R.P.	CUSTOMER TELECHRON
DATE APR. 25, 1961	ORDER 49268
CHECKED BY C. H. H.	SHEET 1 OF 2
DATE 4-27-61	A B C D E F (●under E)
APPROVED BY R.S.	
DATE 4-28-61	DRG. NO. D-1000
SCALE-FULL	

Figure 1.4 A typical bill of material.

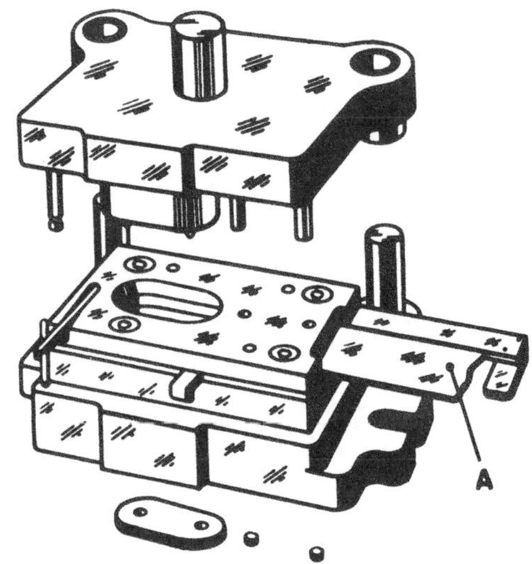

Figure 1.5 A pictorial view of an entire die.

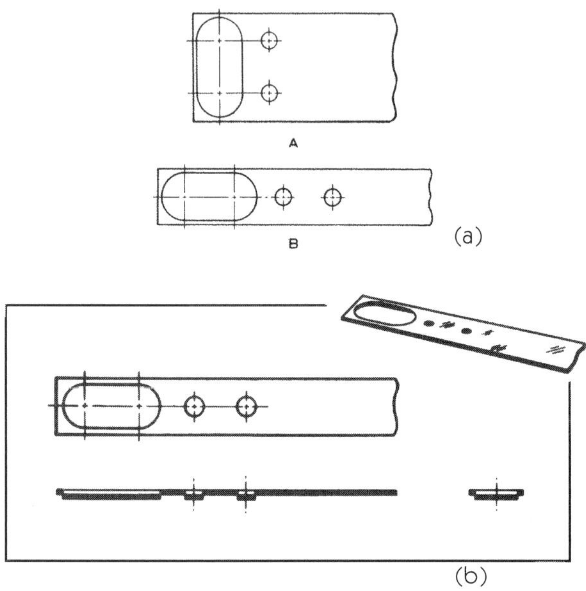

Figure 1.6 Scrap strips: (a) Typical scrap strip layouts and (b) Three views of the scrap strip.

1.1.5 Die Assembly

Figure 1.5 is a pictorial view of the entire die as shown in Figure 1.2. The die pierces two holes at the first station, and then the part is blanked out at the second station. The material from which the blanks are removed is a cold-rolled steel strip. Cold-rolled steel is a smooth, medium-hard steel, and it gets its name from the process by which it is produced. It is rolled, cold, between rollers under high pressure to provide a smooth surface. The strip **A** is shown entering the die at the right.

1.1.6 Scrap Strip

A scrap strip (Figure 1.6) is designed as a guide for laying out the views of the actual die. Figure 1.6a shows a typical scrap strip. This illustration shows the material strip as it will appear after holes have been pierced and the blank has been removed from it. We would first consider running the blank the wide way as shown at **A**. When blanks are positioned in this manner, the widest possible strip is employed and more blanks can be removed from each length of strip. In addition, the distance between blanks is short and little time is consumed in moving the strip from station to station. However, for this particular blank there is a serious disadvantage in this method of positioning. Because the grain in a metal strip runs along its length, the grain in each blank would run across the short width; the blanks would be weak and lacking in rigidity. This defect is important enough for the method to be discarded. Instead, the blanks should be positioned the long way in the strip as shown at **B**. The grain will then run along the length of each blank for maximum stiffness and strength.

Three views of the material strip are shown in Figure 1.6b exactly as they appear in the die drawing in Figure 1.2. In addition, a pictorial view is supplied at the upper right corner to help in visualizing the strip. In other words, this is the way you would imagine the strip if you were to draw it in three views. The top or plan view shows the strip outline, as well as all openings. This would be made actual size on the drawing. The holes are represented by circles at the first station, and the blanked opening is shown at the second station. At the lower left, a side view of the strip is drawn. It is shown exactly as it would appear at the bottom of the press stroke, with the pierced slugs pushed out of the strip at the first station and the blank pushed out of the strip at the second station. The narrow end view at the lower right corner is shown as a section through the blanking station, and the blank is shown pushed out of the strip. The strip in many instances is often drawn shaded to differentiate it from the numerous lines that will represent die members. In the upper plan view, shading lines would appear on the surface of the metal. In the two lower views, the lines are shown in solid black to further differentiate the strip from the die members.

1.1.7 Stampings

Stampings are parts cut and formed from sheet material. Look around you! Wherever you may be, you will find stampings. Many are worn on your own person; the ring on your finger is probably a stamping. Most of the parts in old-fashioned wrist watches are stampings, including the wristband. Your belt buckle, the metal grommets through which your shoe laces pass, eyeglass frame, the clip on your ball point pen, and zipper—all these are stampings.

Look around the room, any room, and you will find products of the pressed-metal industry. Most of the parts in the lighting fixture are stampings; so are threaded portions of light bulbs, door knobs, and the radiator cover. The list is a long one indeed. In the home we find stampings by the score: pots and pans, knives, forks, and spoons, coffee pot, canister set, pie plates and muffin pans, cabinet handles, kettle, can opener, and more.

The refrigerator is almost entirely made of stampings. So are the stove, toaster, and other appliances. And each single part in all these requires an average of three to six dies to produce.

Every automobile contains hundreds of stampings. The largest are the roofs, hoods, quarter-panels, doors, etc. Even the wheels are stampings. There are hundreds of smaller parts, many of which are covered and seldom seen. For example, even the points require very complex dies with multiple stations each, costing thousands of dollars to build, in addition to assembly dies for joining the components.

Office machines and computers provide another big stamping field. So do adding machines, calculators, and dictating machines. We could go on and on; the list is almost endless. Radio and television components require thousands of dies. So do streamlined trains, aircraft, and missiles. All of these are improved from year to year, so an enormous number of new dies is constantly required.

The foregoing should give you some idea of the great size and importance of the pressed-metal industry.

Minster Machine Co.

Figure 1.7 A typical punch press.

1.1.8 Punch Press

Figure 1.7 is a photograph of a typical mechanical punch press in which dies are operated to produce stampings. The bolster plate **A** is a thick steel plate fastened to the press frame. The complete die is clamped securely on this bolster plate. The upper portion of the die is clamped in ram **B**, which is reciprocated up and down by a crank. As the material strip is run through the die, the upper punches, which are fastened to the moving ram **B** of the press, remove blanks from it.

1.2 DIE COMPONENTS

Figure 1.8 is an exploded drawing of the die shown in Figure 1.5 with the names of various die components listed. These names should be memorized because we will refer to them many times in future work.

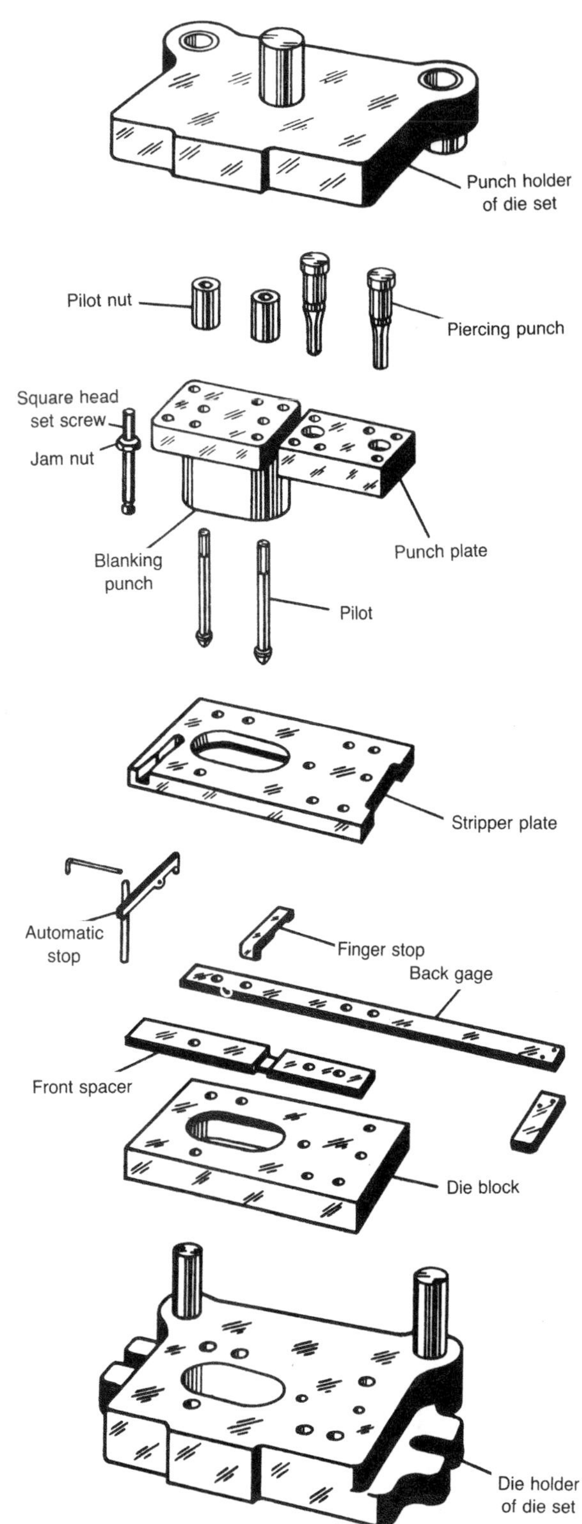

Figure 1.8 An exploded view of the die shown in Figure 1.5.

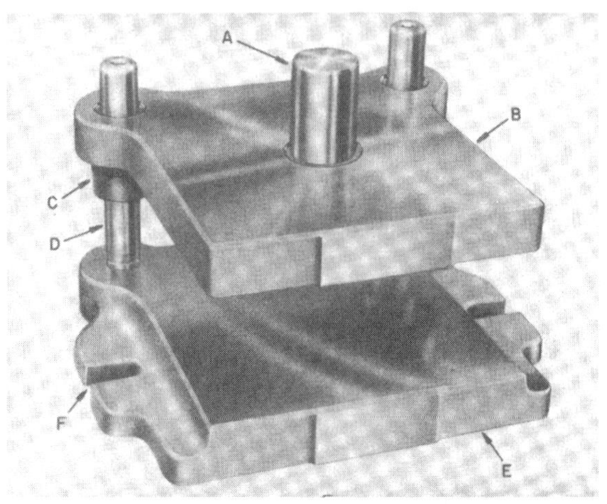

Danly Machine Specialties, Inc.

Figure 1.9 A typical die set.

1.2.1 Die Set

Figure 1.9 shows a die set, and all parts the die assembly comprises are built within it. Die sets are made by several manufacturers and they may be purchased in a great variety of shapes and sizes. The "center posts" **A** are called "punch shanks" in the die set manufacturers' jargon. And, no, they cannot be used for clamping the punch holder, but they can be used for aligning the die in the press. Ram mounting holes must be provided in the punch holder for mounting. In operation, the upper part of the die set **B**, called the "punch holder," moves up and down with the ram. Bushings **C**, pressed into the punch holder, slide on guide posts **D** to maintain precise alignment of cutting members of the die. The die holder **E** is clamped to the bolster plate of the press by bolts passing through slots **F**.

In Figure 1.10, the die set is drawn in four views. The lower left view shows a section through the entire die set. The side view, lower right, is a sectional view also, with a portion of the die set cut away to show internal holes more clearly. The upper left view is a plan view of the lower die holder with the punch holder removed from it.

The punch holder is shown at the upper right, and it is drawn inverted, or turned over, much like an opened book. In the complete die drawing, Figure 1.2, all punches are drawn with solid lines. If the punch holder were not inverted, most lines representing punches would be hidden and the drawing would contain a confusing maze of dotted lines.

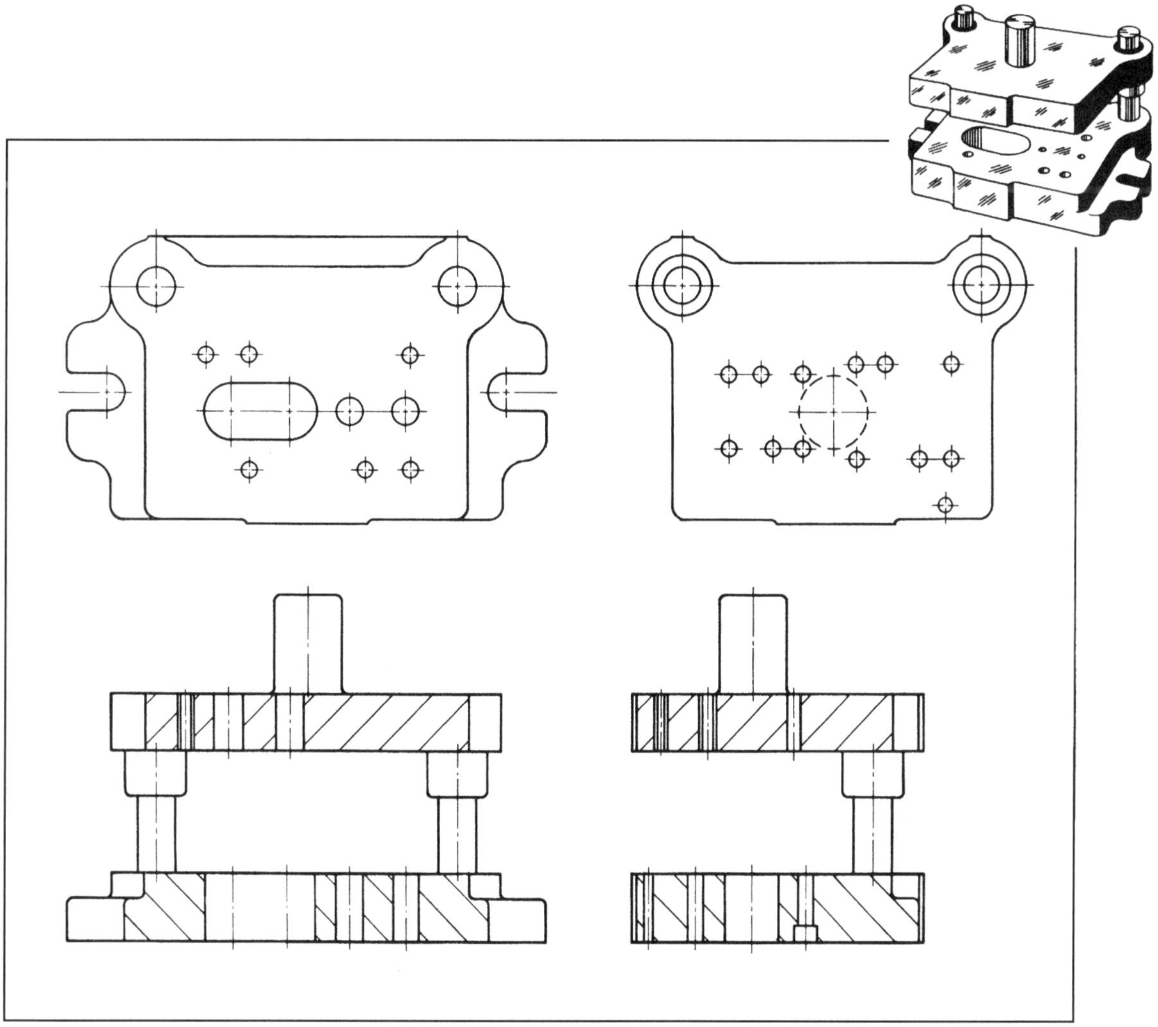

Figure 1.10 A die set.

Another reason for inverting the punch holder is that this is actually the position assumed by the die holders and punch holders on the bench as the die makers assemble the die, and it is easier for die makers to read the drawing when the views have been drawn in the same position as the die on which they perform assembly motions.

1.2.2 The Die Block

Figure 1.11 shows the die block of the die shown in Figure 1.2. The die block is made of hardened tool steel into which holes have been machined, before hardening, at the piercing station and also at the blanking station. These are the same size and shape as the blank holes and contour. Other

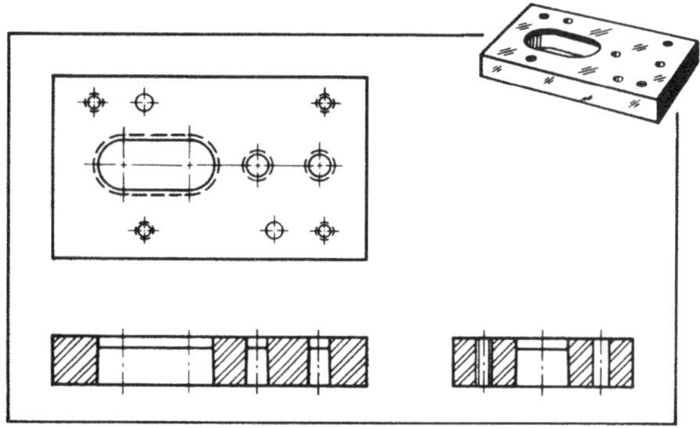

Figure 1.11 The die block.

holes are tapped holes used to fasten the die block to the die holder, and reamed holes into which dowels are pressed to fix the block's location relative to other die parts.

The top view is a plan view of the die block. The lower left view is a section through the holes machined for piercing and blanking. Lines drawn at a 45-degree angle, called "section lines," indicate that the die block has been cut through the center, the lines representing the cut portion. Similarly, the end view is a section cut through the die block at the blanking station. A tapped hole is shown at the left and a reamed hole at the right side. These are for the screws and dowels that hold the die block to other die components. Sectioning a die, that is, showing the die as if portions were cut away to reveal the inside contours of die openings, is a very common practice. In fact, practically all dies are sectioned in this manner. The die maker can then "read" the drawing far more easily than he could if outside views only were shown because these would contain many dotted or hidden lines.

Always remember that all drafting is, in a sense, a language. A die drawing is a sort of shorthand, which is used to convey a great deal of information to the die makers. Anything that can be done to make it easier for them to read the drawing will save considerable time in the shop.

Now refer back to Figure 1.2 and see how easily you can pick out the three views of the die block. That is exactly what the die maker has to do in order to make the die block.

1.2.3 The Blanking Punch

The blanking punch (Figure 1.12) removes the blank from the strip. The bottom, or cutting edge, is the shape and size of the part. A flange at the top provides metal for fastening the blanking punch to the punch holder of the die set with screws and dowels. Two holes are reamed all the way through the blanking punch for retaining the pilots, which locate the strip prior to the blanking operation. Locate the views of the blanking punch in the die drawing, Figure 1.2, to improve your ability to read a die drawing.

1.2.4 Piercing Punch

A piercing punch (Figure 1.13) pierces holes through the material strip or blank. It is usually round and provided with a shoulder to keep it in

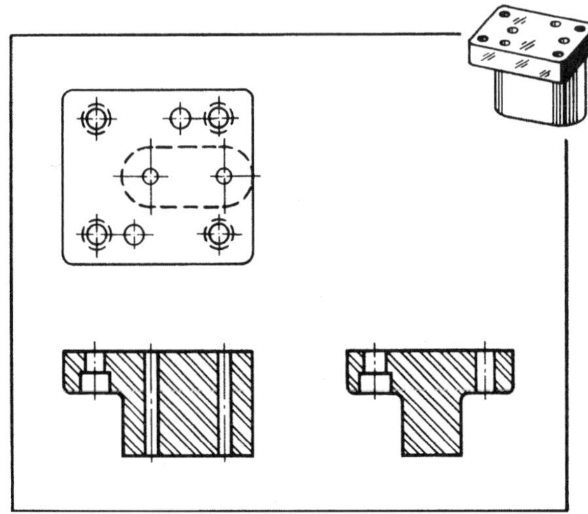

Figure 1.12 The blanking punch.

the punch plate. When a piercing punch penetrates the strip, the material clings very tightly around it. A way must be provided to strip or remove this material from around the punches. The means employed to remove such material is called a "stripper."

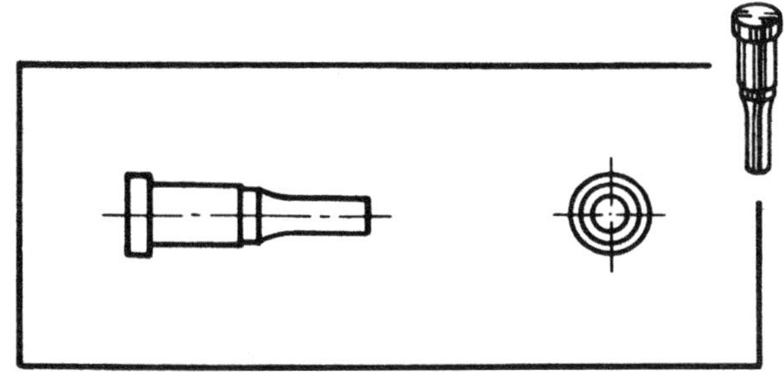

Figure 1.13 A piercing punch.

1.2.5 Punch Plate

The punch plate (Figure 1.14) is a block of machine steel that retains punches by keeping the punch heads against the punch holder of the die set. The punches are held in counterbored holes into which they are pressed. Four screws and two dowels retain the punch plate to the punch holder of the die set. The screws prevent it from being pulled away from the punch holder. Dowels, which are accurately ground round pins, are pressed through both the punch plate and punch holder to prevent shifting. Locate the

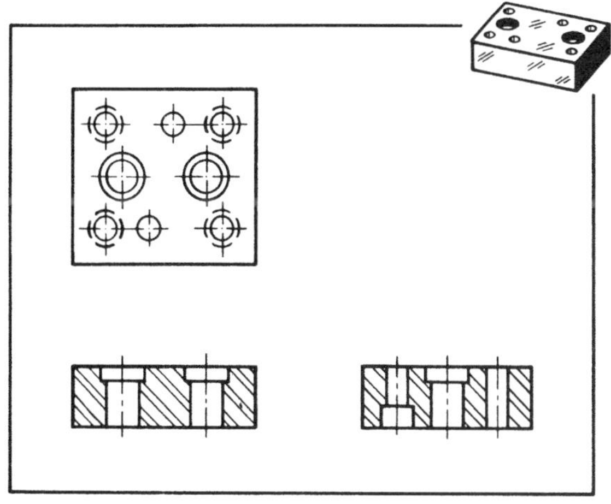

Figure 1.14 A punch plate.

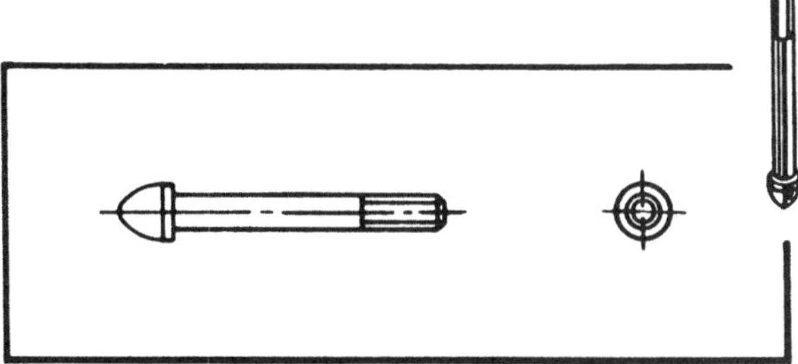

Figure 1.15 A pilot.

front view and plan view of the punch plate in the die drawing Figure 1.2.

1.2.6 Pilot

Pilots (Figure 1.15) are provided with acorn-shaped heads, which enter previously pierced holes in the strip. The acorn shape causes the strip to shift to correct register before blanking occurs.

1.2.7 The Back Gage

The back gage (Figure 1.16) is a relatively thin steel member against which the material

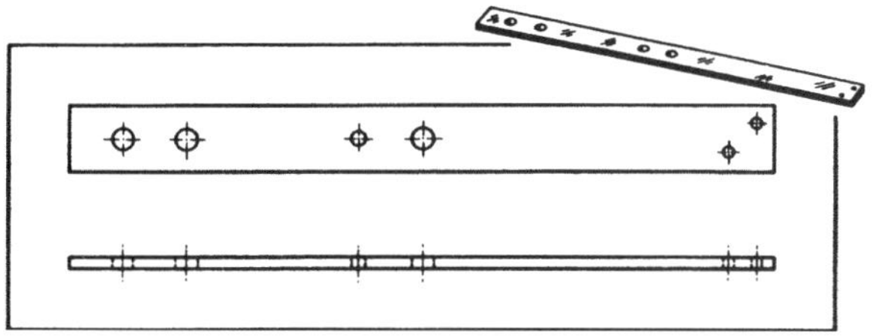

Figure 1.16 The back gage.

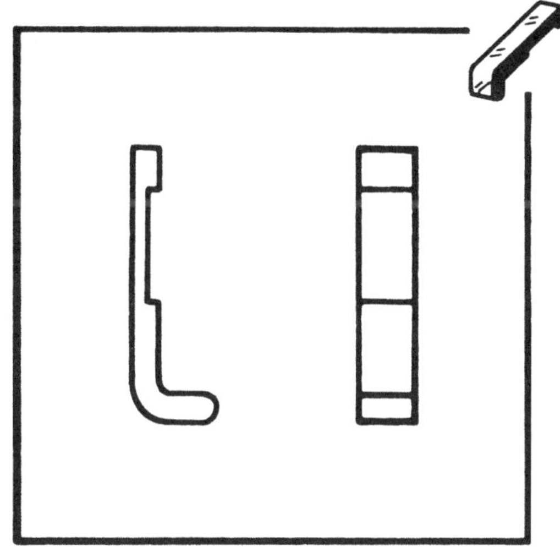

Figure 1.17 A finger stop.

strip is held by the operator in its travel through the die. The front spacer is a shorter component of the same thickness. The strip is fed from right to left. It rests on the die block and is guided between the back gage and front spacer. The distance between the back gage and front spacer is greater than the strip width to allow for possible slight variations in width.

1.2.8 The Finger Stops

The finger stop (Figure 1.17) locates the strip at the first station. In progressive dies having a number of stations, a finger stop may be applied at each station to register the strip before it contacts the automatic stop. Finger stops have slots machined in their lower surfaces to limit stop travel.

1.2.9 Automatic Stops

Automatic stops (Figure 1.18) locate the strip automatically while it is fed through the die. The operator simply keeps the strip pushed against the automatic stop toe, and the strip is stopped while the blank and pierced slugs are removed from it, then it is automatically allowed to move one station further and stopped again for the next cutting operation.

1.2.10 The Stripper Plate

The stripper plate (Figure 1.19) removes the material strip from around blanking and piercing punches. There are two types of stripper plates:

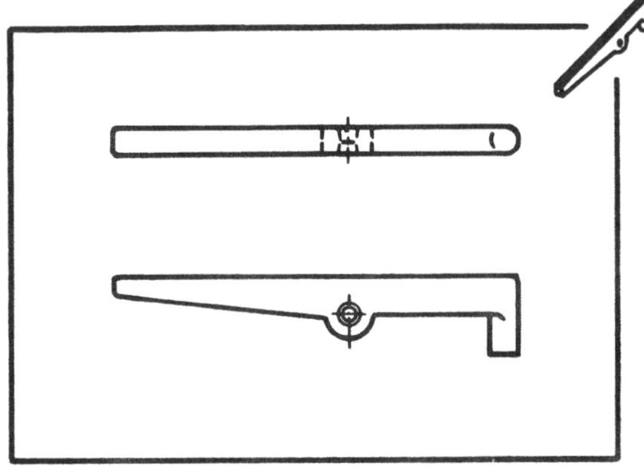

Figure 1.18 An automatic stop.

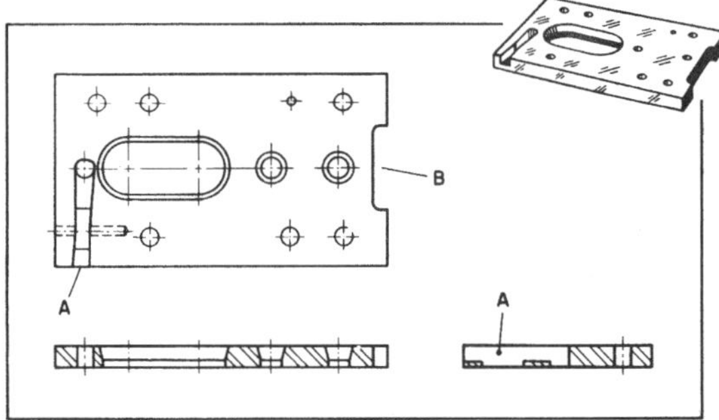

Figure 1.19 The stripper plate.

spring-operated and solid. The one illustrated is solid. The stripper plate has a slot **A** machined into it in which the automatic stop operates. Another slot **B** at the right provides a shelf for easy insertion of a new strip when it is started through the die.

1.2.11 Fasteners

Fasteners hold the various components of the die together. Figure 1.20 shows the commonly used socket cap screw. These fasteners are available from various suppliers, and all have a threaded portion and a larger round head provided with an internal hexagon for wrenching. As you have been doing for previous illustrations, pick out the fasteners shown in the die drawing, Figure 1.2. Note that in section views, screws are shown on one side and dowels on the other.

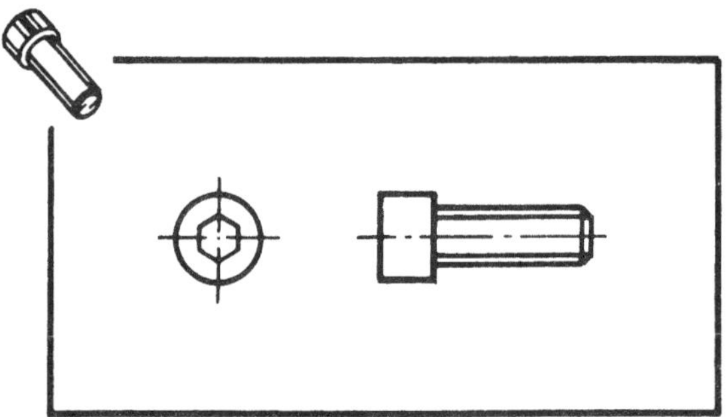

Figure 1.20 Socket head cap screw for use as a fastener.

1.3 PROCESSING A DIE

Let us now consider the steps taken in designing, building, and inspecting a representative die. At the same time, you will gain an insight into the operation of press shops, tool rooms, and manufacturing plants so that your understanding of tooling and manufacturing will be better than average.

1.3.1 The Product

First, we will consider the product to be manufactured. The product engineering department designs the product. In most plants, the work consists in improving the product from year to year to meet changing styles and changing requirements of customers.

After management has decided upon the final form of the new or improved product, a directive is sent to the process planning department to route the various parts through the appropriate manufacturing departments. The process or methods engineers then plan the order of manufacturing operations and decide what operations will be used. They request that the tool design department produce designs of all jigs, fixtures, cutting tools, and dies needed for efficient production of the parts.

After a product designer has prepared layouts and assembly drawings of the product to be manufactured, the engineering department prepares detail drawings of each component the shop has to produce. These drawings contain all required views, dimensions, and explanatory notes to represent all detail features of the objects.

The part which is to be machined, formed, pressed, or inspected is called by one of the following terms:

- Part
- Work
- Workpiece

Part is the preferred term, but *workpiece* or, simply, *work* are often employed as alternate names; all three terms will be used interchangeably throughout this book.

The print on which this part, work, or workpiece is represented is called a *part print*. In designing a die for producing a stamping, the die designer works from a part print.

1.3.2 Process Planning

Prints of detail drawings are sent to the process planning department. When stampings are required, it is the function of this department's employees to determine how the stampings are to be made. They decide how many operations will be required and what presses will be employed to make them. This department thus assumes the responsibility of determining the sequence of manufacturing operations. The information is noted on a series of forms:

a) Route Sheet

The route sheet (Figure 1.21) is designed to suit the requirements of the individual plant and, therefore, the information route sheets contain will vary. However, the following elements are usually included:

1. The heading. This is located at the top of the sheet and contains information such as:
   - Part name
   - Part number
   - Drawing number
   - Number of parts required
   - Name of product engineer
   - Date

   In addition, the product name and model number may be included.
2. The number of each operation required to make and inspect the part. Numbers are most frequently listed in increments of 5, such as 5, 10, 15, 20, etc., to provide numbers in sequence for additional operations which may be found necessary in manufacture or when changes are made in the design of the product.
3. The name of each operation.
4. The name and number of the machine on which the operation is to be performed.
5. Estimates of the number of parts that will be completed per hour for every operation. These estimates are altered after production rates have been measured accurately by the time study department. Route sheets are supplied to the following departments:
   - Tool design department
   - Production department
   - Inspection department

Of course, any machine or product will contain many components, which have been standardized and which can be purchased from outside suppliers or vendors. Such items would include screws and dowels, bearings, clutches, motors, and many others. The purchasing department would be supplied with a bill of material, and purchase orders would be issued for all parts to be bought.

NAME OF SCHOOL OR COMPANY AND ADDRESS

ROUTE SHEET

PART NAME: Housing Cover NO OF PARTS: 800,000 DATE: Feb. 15, 2005

PART NO: 10568 DRG. NO. 1225 PROD. ENG. J. White

SHEET NO. 1 OF 1

OPER. NO.	OPERATION	MACHINE	DEPT.	TIME (HOURS)	SET UP (HOURS)
5	Shear sheet into strips	Bliss	30	.0001	—
		Squaring			
		Shears			
		No. 37			
10	Blank	Federal Press	25	.0001	—
		No. 33			
		No. 442			
15	Form	Federal Press	25	.0002	—
		No. 44			
		No. 337			
20	Wash and ship to stores	Truck			

Figure 1.21 A typical route sheet.

NAME OF SCHOOL OR COMPANY AND ADDRESS						
TOOL OPERATION SHEET						
PART NAME: Housing Cover				DATE: Feb. 16, 2005		
PART NO: 10568				APPROVED BY: J. White		
OPER. NO.	OPERATION	TOOL NAME	TOOL NUMBER	INSTRUC-TIONS	DEPT. NO.	EQUIPMENT NAME & NO.
5	Shear sheet into strips				30	Bliss Squaring
						Shears Shop
						No. 37
10	Blank	Blanking Die	T-3073	Des.	25	Federal Press
						No. 33
						Shop No. 442
15	Form	Forming Die	T-3074	Des.	25	Federal Press
						No. 44
						Shop No. 337
20	Wash and ship to stores	Truck				

Figure 1.22 A typical tool operation sheet.

b) Tool Operation Sheet

The tool operation sheet (Figure 1.22) is prepared from the route sheet and it usually lists the following:

- Number of each operation
- Name of each operation
- Machine data
- List of all standard and special tools required for the job
- Names and numbers of all special tools that are to be designed and built. These numbers are marked on tool drawings and later stamped or marked on the actual tools for identification.

Tool operations sheets are helpful in planning and developing a tooling program. Copies go to the tool designers and to the tool purchasing department. Before proceeding further, study carefully the tool operation sheet illustrated.

c) Design Order

The design order (Figure 1.23) authorizes work on an actual design. An order is prepared for each die or special tool required and the information is

NAME OF SCHOOL OR COMPANY AND ADDRESS DESIGN ORDER NO. 102	
TO: Tool Design Department	DATE: Feb. 18, 2005
DESIGN: Forming Die	
FOR: Forming Lower Flange	
PART NAME: Housing Cover	
PART NO. 10567	TOOL NO. T-3074
USED IN No. 20 Bliss Press	SHOP NO. 406
DEPARTMENT Press	NO. 22
NUMBER OF PARTS REQUIRED 800,000	
COMPLETED	CANCELLED
REASON	
	SIGNED:

Figure 1.23 A typical design order.

taken from the route sheet. In addition, the design order may give instructions regarding the type of die preferred. The following lists the information usually given on a design order:

- Department name
- Tool name
- Date
- Tool number
- Part name
- Part number
- Operation
- Machine in which tool is to be used.

1.3.3 Designing the Die

Before designers begin to draw, they must seriously consider a number of things. It is now possible for them to list all the items that will be required so they can begin designing intelligently. These items are:

- The part print
- The tool operation sheet, or route sheet
- The design order
- A press data sheet.

In addition, designers may have either a reference drawing of a die similar to the one they are designing or a sketch of the proposed design prepared by the chief tool designer or group leader suggesting a possible approach to solving the problem. Let us consider further the information required:

Part print. The part drawing gives all necessary dimensions and notes. Any missing dimension must be obtained from the product design department before work can proceed.

Operation sheet. The operation sheet or route sheet must be studied to determine exactly what operations were previously performed upon the workpiece. This item is very important. When the views of the stamping are laid out, any cuts that were applied in a previous operation must be shown.

Design order. This item must be studied very carefully because it specifies the type of die to be designed. Consider particularly the operation to be performed, the press in which the die is to be installed, and the number of parts expected to be stamped by the die. The latter will establish the *class* of die to be designed.

Press (machine) data sheet. The die to be designed must fit into a particular press and it is important to know what space is available to receive it and what interferences may be present.

In time you will come to realize the importance of careful and repeated study of the part print, operation sheet, and design order because there can be no deviation from the specifications given.

a) Die Drawing

If the information on a drawing is complete, concise, and presented in the simplest possible manner, the die maker can work to best advantage. The first step in originating plans for a new die is to prepare a sketch or sketches of significant features of the proposed die. These are a guide for beginning the actual drawing of the full-size layout. However, it is a mistake to spend too much time in this phase of the work or to try to develop the entire design in sketch form because doing so can result in arbitrary and inflexible decisions.

Always keep your mind open to possible improvements as you develop the design in layout form. You will often find that the first or second idea sketched out can be considerably improved by alteration as work progresses. Often the first idea proves entirely impractical and another method of operation must be substituted.

Before beginning the sketch, gather before you the part print, operation sheet, and design order. The three must be studied together so that a complete and exact understanding of the problem will be realized. This study will form the basis for creating a mental picture of a tool suitable for performing the operations—one which will meet every requirement. The sketch you make may be a very simple one, for simple operations, or it may be more elaborate. In fact, a number of sketches may be required for more complex operations and intricate designs. In any event, the sketch will clarify your ideas before you attempt a formal layout. In addition, it will form the basis for a realistic estimate of the size of the finished die so that you may select the proper sheet size for the layout.

Layout. Laying out the die consists of drawing all views necessary for showing every component in its actual position. In the layout stage, no dimensions are applied and neither is the bill of material nor the record strip filled out. After the die has been laid out, the steps necessary for completing the set of working drawings are more or less routine.

Assembly drawing. A properly prepared assembly drawing contains six general features:

1. All views required for showing the contour of every component including the workpiece.
2. All assembly dimensions. These are dimensions that will be required for assembling the parts, as well as for machining operations to be performed after assembly.
3. All explanatory notes.
4. Finish marks and grind marks to indicate those surfaces to be machined after assembly.
5. A bill of material listing sizes, purchased components, materials, and number (quantity) required for all parts.
6. A title block and record strip with identifying information noted properly.

Detail drawing. After the assembly drawing of the die has been completed, detail drawings are prepared, unless all dimensions were previously placed on the assembly drawing (as is done for simple dies). Detail drawings are drawings of individual components. They contain all dimensions, notes, and supplementary information so that each part can be made without reference to the assembly drawing or to other detail drawings. Such information usually includes 10 distinct elements:

1. All views required for identifying every detail of the part must be drawn.
2. Every dimension needed for making the part must be given.
3. Suitable notes for furnishing the supplementary information that dimensions do not cover must be applied.
4. Finished surfaces must be identified.
5. The name of the part and its number must be given.
6. The material from which the part is to be made must be specified.
7. The number of each material required per assembly must be stated.
8. The scale to which the drawing has been laid out must be listed.
9. The draftsman's name or initials must be signed.
10. The date must be specified.

Dimension and notes. With the die design completed, all dimensions and notes are applied to the drawing. Figure 1.24 shows the die set note, which tells the die maker exactly what die set to order and gives required information about punch shank diameter, type of guide bushings, and diameter and length of guide posts.

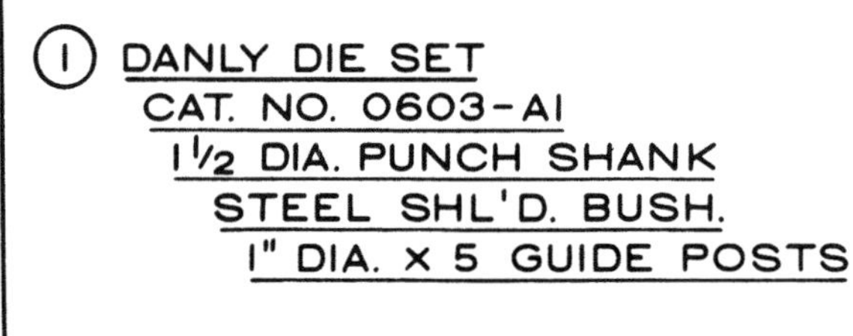

Figure 1.24 A typical die set note.

b) Checking the Die Drawings

After a set of drawings has been completed and the designer has reviewed them for possible omissions or errors, the set is turned over to the group leader, who will bring it to the checkers for further review. The design order, part print, and any notes or sketches that may have accompanied the drawings when the designer first received the job will now travel with the drawings. Checkers require all of these in order to do their work properly.

Checkers first study the operation of the die to make sure that it will function properly and that its cost will not be excessive for the work it is to perform. After they are satisfied that it has been designed properly, they will check every dimension, note, and specification for accuracy. They usually work from a *check print*. This is a blue and white print having blue lines and a white background. With a yellow crayon, they will cover every dimension they find to be accurate, and with a red crayon they will cover every dimension they find to be wrong. Above or to the side they will write the correct dimension in red.

The tracings, along with the check prints, are then returned to the designer for correction. Incorrect dimensions are carefully erased to remove all graphite from the paper. An erasing shield is ordinarily used to prevent smudging of other dimensions or lines. Correct dimensions are then lettered in place.

After the tracings have been corrected, they are returned to the checkers, who review the job again to make sure that no correction was overlooked. They then sign the drawing in the space provided and enter the date the drawing was checked.

After drawings have been completed and checked, they must be approved by the chief designer, chief tool engineer, and possibly the plant superintendent and others who are held responsible by the management for the cost and quality of dies used in the plant. Usually, these approvals are routine after drawings have been approved by the checker. However, it may sometimes happen that these personnel will refuse to sign because they believe that the die will not work as well as expected, will not deliver the number of parts required per day, will be too expensive to build, or for some other reason. If they convince others that their objections are valid, the drawings will have to be altered or a new design begun, depending upon the extent of the changes to be made.

c) Prints and Distribution

After drawings have been approved, blueprints are made from the tracings, or originals. A small print is taken of the bill of material only. This is sent to the stock cutting department where steel is stored and cut as required. The stock cutter goes over the list, selects bars of proper thickness and width, or diameter, and saws the bars to the lengths specified for each item listed. These cut blocks and plates are placed in a pan, along with screws, dowels, and other parts, which are kept in stock. When purchased components are delivered to the plant, they are also placed in the pan. Finally, the pan contains a set of die prints and a part print and it is delivered to the tool room, where the tool room foreman turns it over to the die maker, who will build the die.

One of the prints is sent to the purchasing department. There, orders are written to authorize purchase of all components needed to build the die. If the entire die is to be built by an outside tool shop, a purchase order is sent to it. If it is to be built within the plant, an order authorizing construction is sent to the tool room. In addition, requisitions are made out for the following:

- Standard parts or assemblies that are not kept in stock and which must be purchased.
- Castings, forgings, or weldments required for construction of the die.
- Steels of special analysis not carried in stock.
- Special sizes of steels or other materials not stocked.

The purchasing agent must plan for delivery of all these components before the date set for beginning construction of the die.

The files for the die drawings—whether electronic or card—are then made out by the die design department. The file contains the number of the drawings and the job by name and number. Each die has only one file.

1.3.4 Tool and Die Inspection Department

After a die has been designed, a set of prints is sent to the tool inspection department. Then after the die has been built, the department will inspect it to make certain that it was constructed to specifications given in the tool print.

When the die is built by an outside tool shop, it is inspected by the tool inspection department upon delivery. The same inspection procedures are followed to determine if the stampings it produces hold to tolerances specified on the part print.

1.3.5 Production

After the tool inspection department has approved its construction and accuracy, the die is delivered to the production department where it will be used. The set-up person for that department installs it in the press where it will be operated. A few sample parts are then produced under the same conditions in which the die will run in actual production. These parts are taken to the *production inspection department.* There, they are inspected to determine whether or not sizes hold to tolerances specified on the part print.

Once the production inspection department has determined that the samples are satisfactory, a form is issued and signed by the chief inspector authorizing production with the die. After receiving production orders from the production department, the production foreman will proceed to go into production of the stampings. Production orders specify how many parts are to be run, when they will be required, and where they are to be delivered.

After a new die has been in production for a few hours or so, and it is found to perform satisfactorily, the order that was issued to the tool room to build the die is closed. No more time may be charged against it. In this regard, it is worth noting that records are kept of all time devoted to designing, building, inspecting, and trying out the die in order to determine the actual tool cost, illustrating perfectly that "time is money."

1.4 DIE OPERATIONS

Just exactly what operations are performed in dies? This question is asked often and we have prepared the following illustrated list of the 20 types of operations.

1.4.1 Blanking

Stampings that have an irregular contour must be blanked from the coil or from the strip (Figure 1.25). Piercing, embossing, and various other operations may be performed on the strip prior to the blanking station.

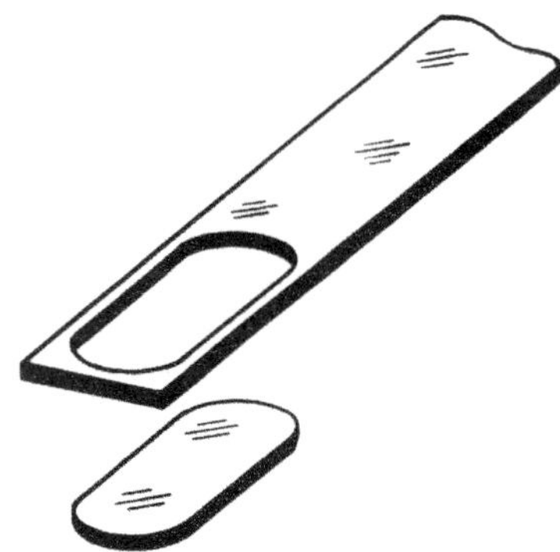

Figure 1.25 A blank and the strip from which it is been cut.

1.4.2 Cut-off

Cut-off operations (Figure 1.26) are those in which strip of suitable width is cut to length. Preliminary operations before cutting off include piercing, notching, and embossing. Although they are relatively simple, many parts can be produced by cut-off dies.

1.4.3 Piercing

Piercing dies pierce holes in previously blanked formed, or drawn, parts (Figure 1.27). It is often impractical to pierce holes while forming or before forming because the holes would become distorted in the forming operation. In such cases they are pierced in a piercing die after forming.

1.4.4 Piercing and Blanking

Compound dies pierce and blank simultaneously at the same station (Figure 1.28). They are more expensive to build and they are used where considerable accuracy is required in the part.

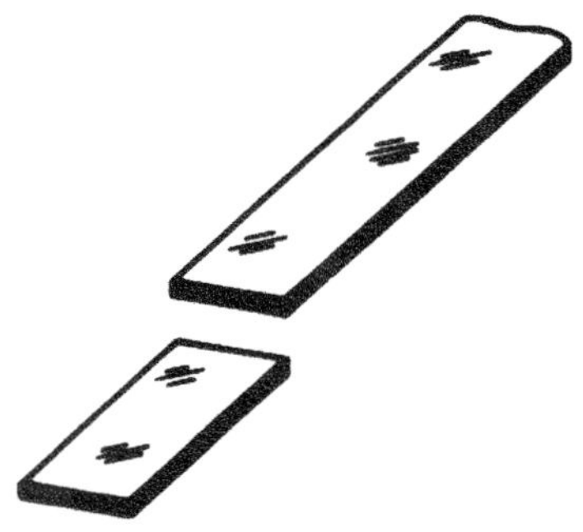

Figure 1.26 Part separated from strip in cut-off operation.

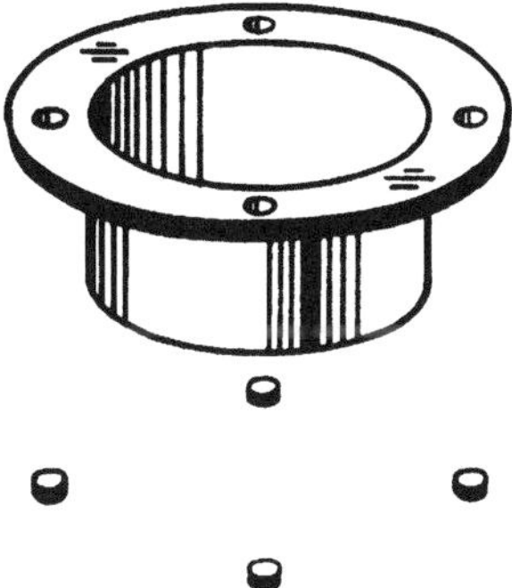

Figure 1.27 Holes pierced in a previously drawn part

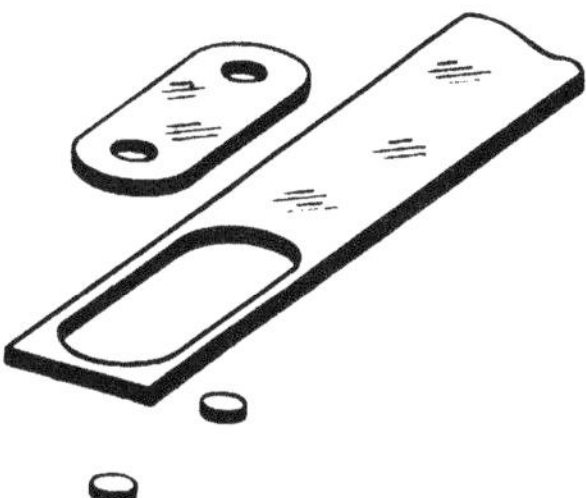

Figure 1.28 Part is blanked and pierced simultaneously in a compound die.

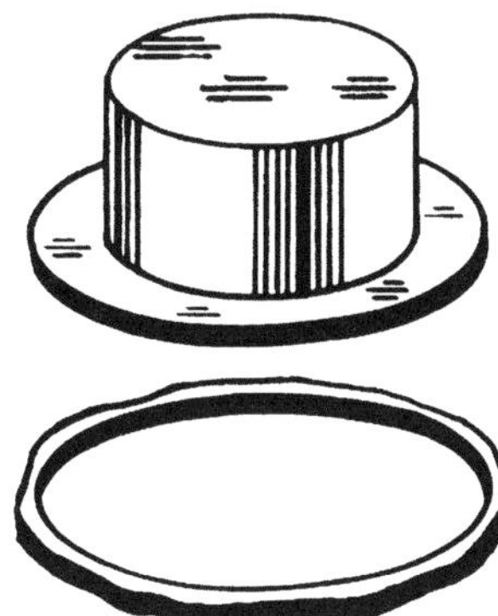

Figure 1.29 The result of trimming in a trimming die.

1.4.5 Trimming

When cups and shells are drawn from flat sheet metal the edge is left wavy and irregular due to uneven flow of metal. This irregular edge is trimmed in a trimming die. Figure 1.29 shows a flanged shell, as well as the trimmed ring removed from around the edge.

1.4.6 Shaving

Shaving consists of removing a chip from around the edges of a previously blanked stamping (Figure 1.30). A straight, smooth edge is provided. Therefore, shaving is frequently performed on instrument parts, watch and clock parts, and the like. Shaving is accomplished in shaving dies especially designed for the purpose.

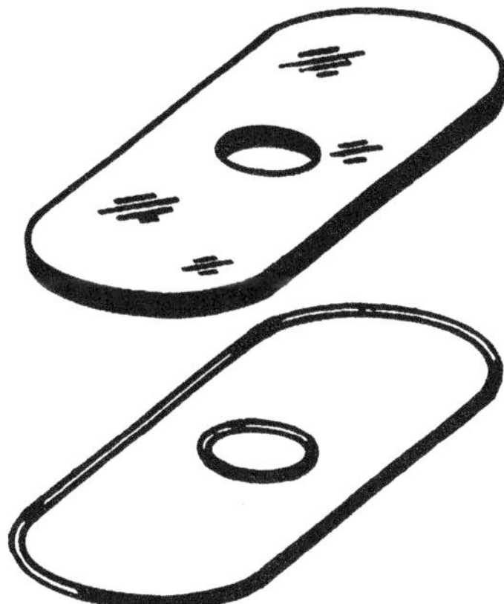

Figure 1.30 The result of shaving in a shaving die.

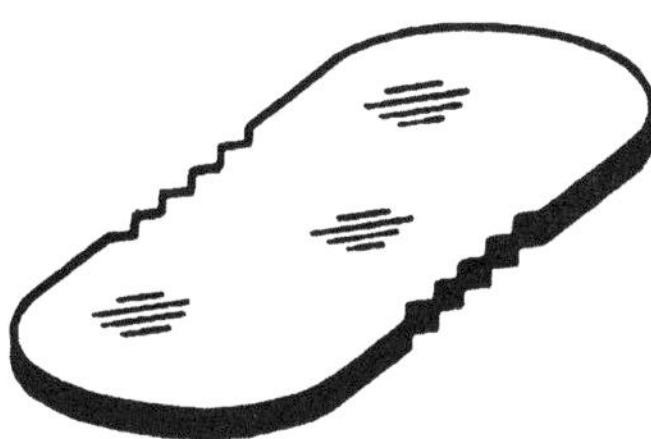

Figure 1.31 Serrations applied in a broaching die.

1.4.7 Broaching

Figure 1.31 shows serrations applied in the edges of a stamping. These would be broached in a broaching die. Broaching operations are similar to shaving operations. A series of teeth removes the metal instead of just one tooth, as in shaving. Broaching must be used when more material is to be removed than could effectively be done with one tooth.

1.4.8 Horning

Horn dies are provided with an arbor or horn over which the parts are placed for secondary operations such as seaming (Figure 1.32). Horn dies may also be used for piercing holes in the sides of shells.

1.4.9 Side Cam Operations

Piercing a number of holes simultaneously around a shell (Figure 1.33) is done in a side cam

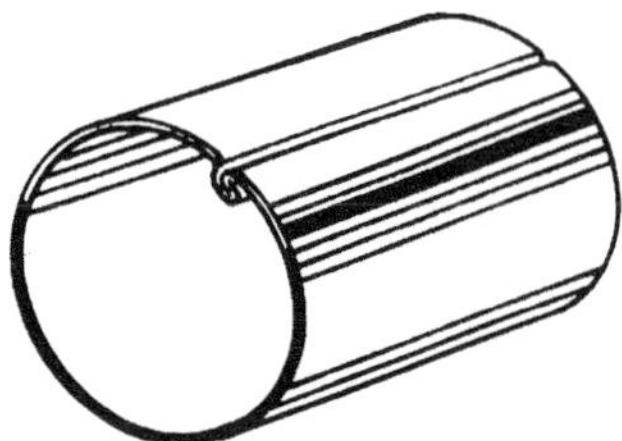

Figure 1.32 The seam on this part is done as a secondary operation in a horn die.

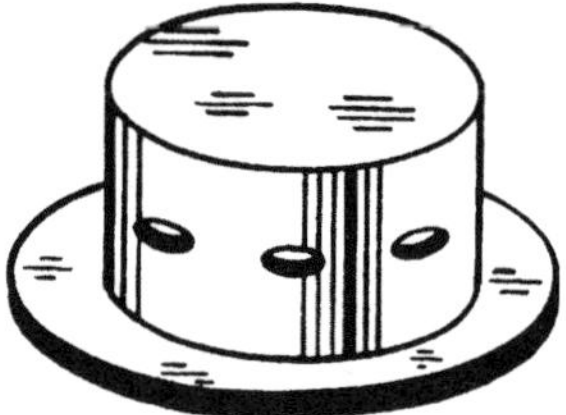

Figure 1.33 The holes are pierced simultaneously in a side cam die.

die. Side cams convert the up-and-down motion of the press ram into horizontal or angular motion when the nature of the work requires it.

1.4.10 Bending

Bending dies apply simple bends to stampings. A simple bend is one in which the line of bend is straight. One or more bends may be involved (Figure 1.34). Bending dies are a large and important class of press tool.

1.4.11 Forming

Forming dies apply more complex forms to workpieces. The line of bend is curved instead of straight and the metal is subjected to plastic flow or deformation (Figure 1.35).

Figure 1.34 Stamping bent in a bending die.

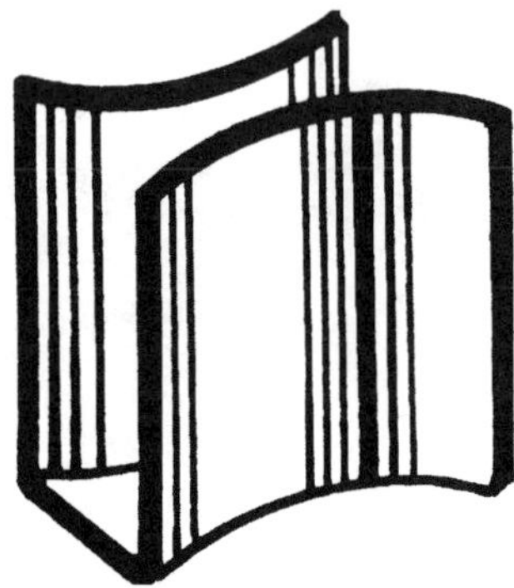

Figure 1.35 Stamping formed in a forming die.

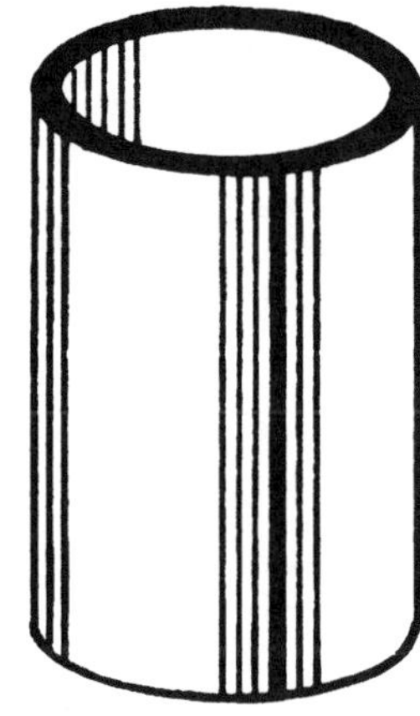

Figure 1.36 Shell drawn from a flat sheet.

1.4.12 Drawing

Drawing dies transform flat sheets of metal into cups, shells, or other drawn shapes by subjecting the material to severe plastic deformation. Figure 1.36 shows a rather deep shell that has been drawn from a flat sheet.

1.4.13 Curling

Curling dies curl the edges of drawn shells to provide strength and rigidity (Figure 1.37). The curl may be applied over a wire ring for increased strength. You may have seen the tops of sheet metal pails curled in this manner. Flat parts may be curled also. A good example is a hinge in which both members are curled to provide a hole for the hinge pin.

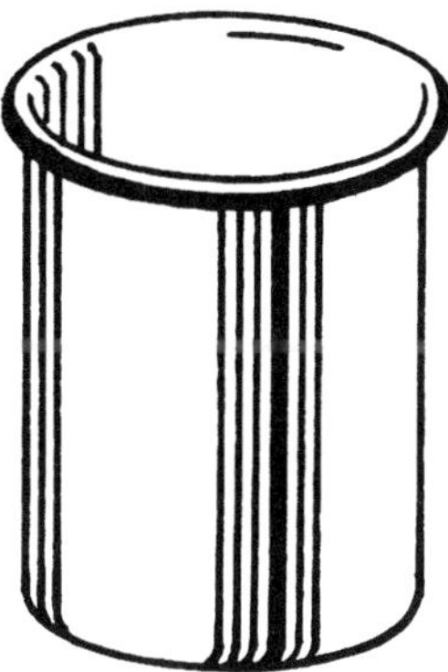

Figure 1.37 Lip on this drawn shell produced in curling die.

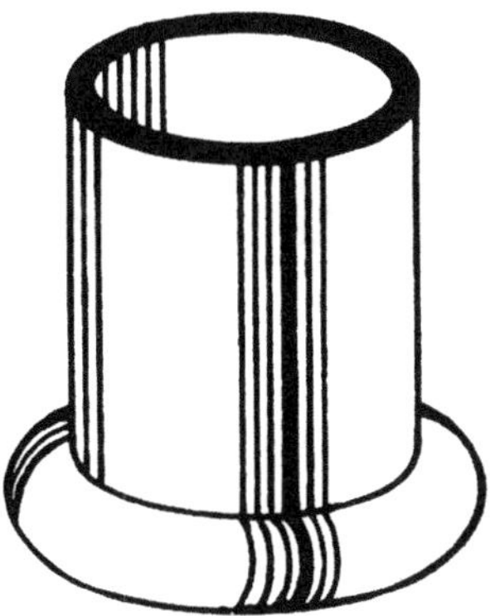

Figure 1.38 Bulge in this drawn shell produced in bulging die.

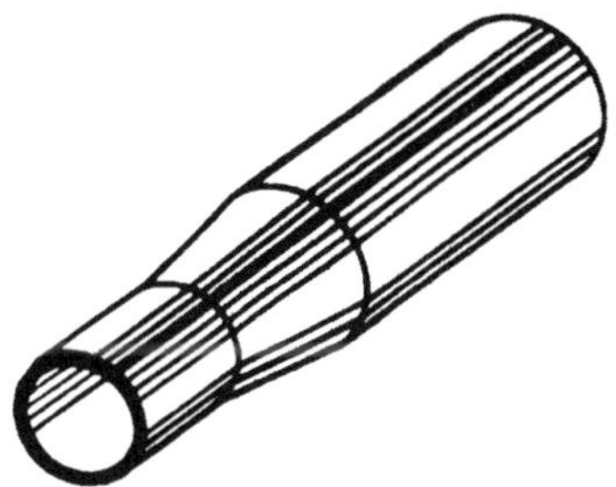

Figure 1.39 Drawn shell that has been swaged.

Figure 1.40 Drawn shell that has been extruded.

1.4.14 Bulging

Bulging dies expand the bottom of previously drawn shells (Figure 1.38). For example, the bulged bottoms of some types of coffee pots are formed in bulging dies.

1.4.15 Swaging

In swaging operations, drawn shells or tubes are reduced in diameter for a portion of their lengths (Figure 1.39). The operation is also called "necking."

1.4.16 Extruding

Extruding dies cause metal to be extruded or squeezed out, much as toothpaste is extruded from its tube when pressure is applied. Figure 1.40 shows a collapsible tube formed and extruded from a solid slug of metal.

1.4.17 Cold Coining

In cold forming operations, metal is subjected to high pressure and caused to flow into a predetermined form. In coining (Figure 1.41), the metal is caused to flow into the shape of the die cavity. Coins such as nickels, dimes, and quarters are produced in coining dies.

1.4.18 Progressive Operations

Progressive operations are those in which progressive dies perform work at a number of stations simultaneously. A complete part is cut off at the final station with each stroke of the press. Figure 1.42 shows part and strip produced in a progressive die.

1.4.19 Sub Press Operations

Sub press dies are used for producing tiny watch, clock, and instrument components, repre-

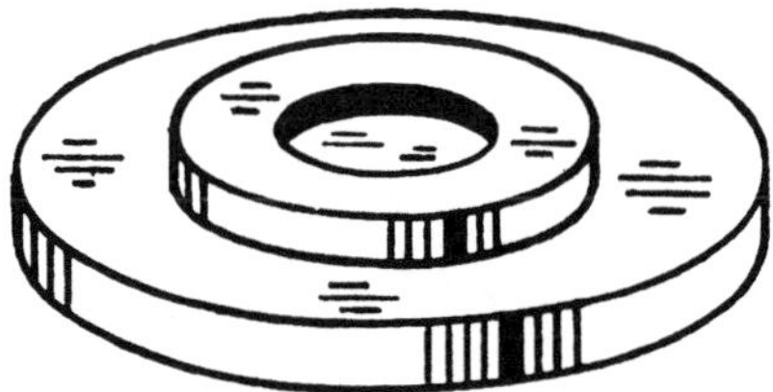

Figure 1.41 Cold-coining part in which metal flow is caused by high pressure.

Figure 1.43 Typical precision parts produced in sub press dies.

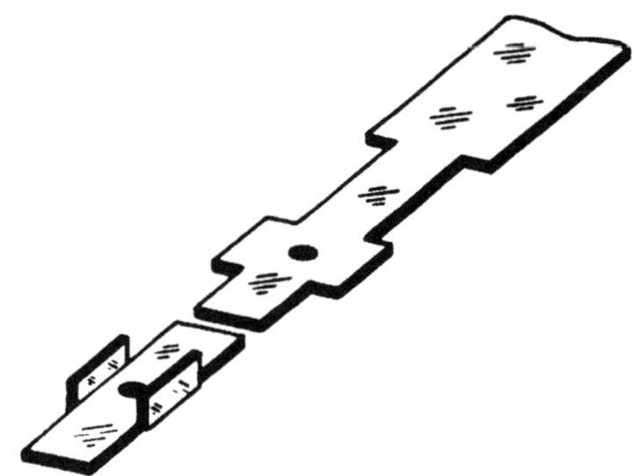

Figure 1.42 Part and strip produced in a progressive die.

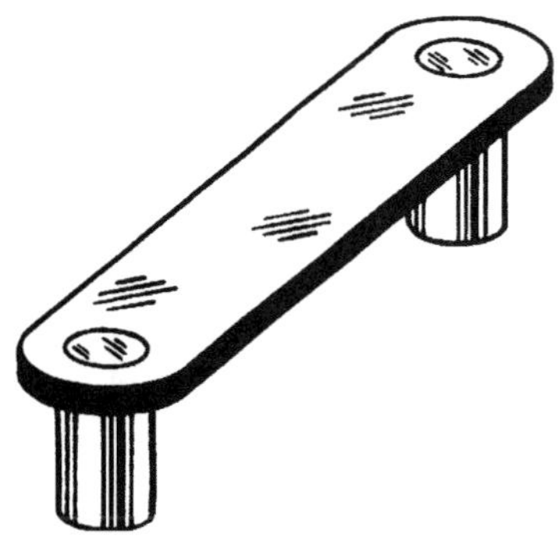

Figure 1.44 Part produced in an assembly die.

sented by the watch needles shown in Figure 1.43. Sub presses are special types of die sets used only for such precision work.

1.4.20 Assembly Operations

Assembly dies represent an assembly operation in which two studs are riveted at the ends of a link (Figure 1.44). Assembly dies assemble parts with great speed; they are being used more and more.

From the foregoing, you can perhaps appreciate what a wide field die design engineering really covers. You must have come to realize that it is indeed a pleasant and interesting occupation, one which will stimulate your mind in much the same manner as the working out of fascinating puzzles. In addition, you will come to find that is a very profitable one.

As you study the chapters that follow, you will be introduced, step by step, to the fundamental die components. You will learn the methods by which die designers assemble these components when they design dies. When you have completed the book you will know the elements of die design quite thoroughly. Knowledge such as this is well-compensated professionally. You will have acquired the foundation of a career that can benefit you for the rest of your life.

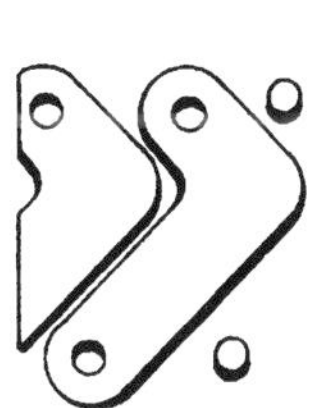

CLASSIFICATIONS AND TYPES OF DIES

2.1 DIE CLASSIFICATIONS

Dies can be classified according to a variety of elements and in keeping with the diversity of die designs. In this section, we will discuss primarily die classifications depending on the production quantities of stamping pieces (whether high, medium, or low) and the number of stations. In choosing these, we are not trying to downplay or ignore other classifications such as the number of operations, manufacturing processes, or guide methods.

2.1.1 Die Classifications Depending on the Production Quality of Parts

Depending on the production quality of pieces—high, medium, or low—stamping dies can be classified as follows:

Class A. These dies are used for high production only. The best of materials are used. All easily worn items or delicate sections are carefully designed for easy replacement. A combination of long die life, constant accuracy throughout the die life, and ease in maintenance are prime considerations, regardless of tool cost.

Class B. These dies are applicable to medium production quantities and are designed to produce the designated quantity only. Die cost as related to total production becomes an important consideration. Cheaper materials may be used, provided they are capable of producing the full quantity. Less consideration is given to the problem of ease of maintenance.

Class C. These dies represent the cheapest usable tools that can be built. They are suitable for low-volume production of parts.

2.1.2 Die Classifications According to Number of Stations

According to the number of stations, stamping dies may be classified as:

- Single-station dies
- Multiple-station dies

a) Single-Station Dies

Single-station dies may be either compound dies or combination dies.

Compound die. A die in which two or more cutting operations are accomplished to produce a part at every press stroke is called a compound die.

Combination die. A die in which both cutting and noncutting operations are accomplished to produce a part at one stroke of the press is called a combination die.

b) Multiple-Station Dies

Multiple station dies are arranged so that a series of sequential operations is accomplished with each press stroke. Two die types are used:

- Progressive dies
- Transfer dies

Progressive die. A progressive die is used to transform coil stock or strips into a completed part.

This transformation is performed incrementally, or progressively, by a series of stations that cut, form, and coin the material into the desired shape. The components that perform operations on the material are unique for every part. These components are located and guided in precision cut openings in plates, which are in turn located and guided by pins.

The entire die is actuated by a mechanical press that moves the die up and down. The press is also responsible for feeding the material through the die, progressing it from one station to the next with each stroke.

Transfer die. In transfer die operations, individual stock blanks are mechanically moved from die station to die station within a single die set. Large workpieces are done with tandem press lines where the stock is moved from press to press at which specific operations are performed.

2.2 TYPES OF DIES

There are 20 types of dies, and each is distinct and different from all the other types. However, as you study the descriptions to follow, observe how the elements are applied and reapplied with suitable modifications to adapt them for each particular job to be performed.

2.2.1 Blanking Dies

A blanking die (Figure 2.1) produces a blank by cutting the entire periphery in one simultaneous operation. Three advantages are realized when a part is blanked:

1. *Accuracy.* The edges of blanked parts are accurate in relation to each other.
2. *Appearance.* The burnished edge of each blank extends around its entire periphery on the same side.
3. *Flatness.* Blanked parts are flat because of the even compression of material between punch and die cutting edges.

The inset at **A** shows a material strip ready to be run through a blanking die. At **B** is shown the top view of the die with punches removed. The section view at **C** shows the die in open position with the upper punch raised to allow advance of the strip against the automatic stop. At **D**, the die

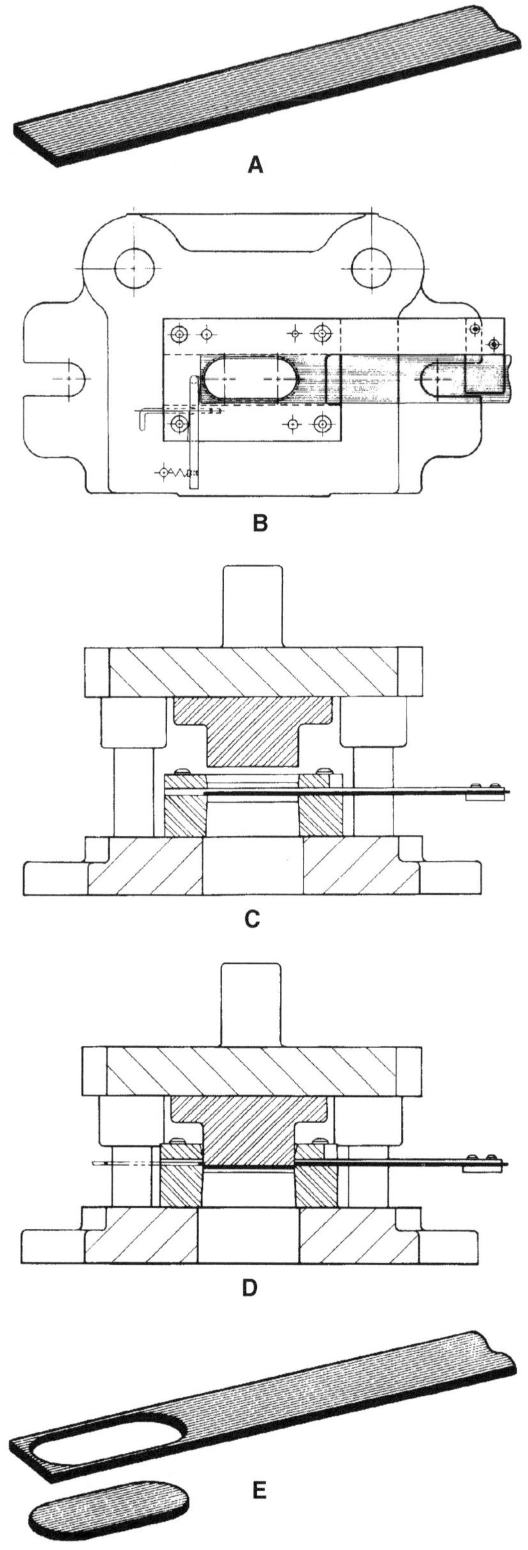

Figure 2.1 Blanking die.

is shown closed with a blank pushed out of the strip.

Blanking dies may produce plain blanks as shown in inset **E**, but more frequently holes are pierced at one station and the part is then blanked out at the second station. Such dies are called "pierce and blank" dies.

2.2.2 Cut-Off Dies

The basic operation of a cut-off die (Figure 2.2) consists of severing strips into short lengths to produce blanks. The line of cut may be straight or curved, and holes or notches or both may be applied in previous operations. Cut-off dies are used for producing blanks having straight, parallel sides because they are less expensive to build than blanking dies. In operation, the material strip **A** is registered against stop block **B**. Descent of the upper die causes the cut-off punch **C** to separate the blank from the strip. Stop block **B** also guides the punch while cutting occurs to prevent deflection and excessive wear on guide posts and bushings. A conventional solid stripper is employed.

2.2.3 Piercing Dies

Piercing dies (Figure 2.3) pierce holes in stampings. There are two principal reasons for piercing holes in a separate operation instead of combining piercing with other operations:

1. When a subsequent bending, forming, or drawing operation would distort the previously pierced hole or holes
2. When the edge of the pierced hole is too close to the edge of the blank for adequate strength in the die section. This occurs in compound and combination dies in which piercing and blanking are done simultaneously.

The inset at **A** shows a flanged shell requiring four holes to be pierced in the flange. If the holes were pierced before the drawing operation, they would become distorted because of the blank holder pressure applied to the flange in the drawing process.

The shell is located in an accurately ground hole in the die block. Piercing punches are retained in a punch plate fastened to the punch holder, and a knockout affects stripping after the holes have been pierced.

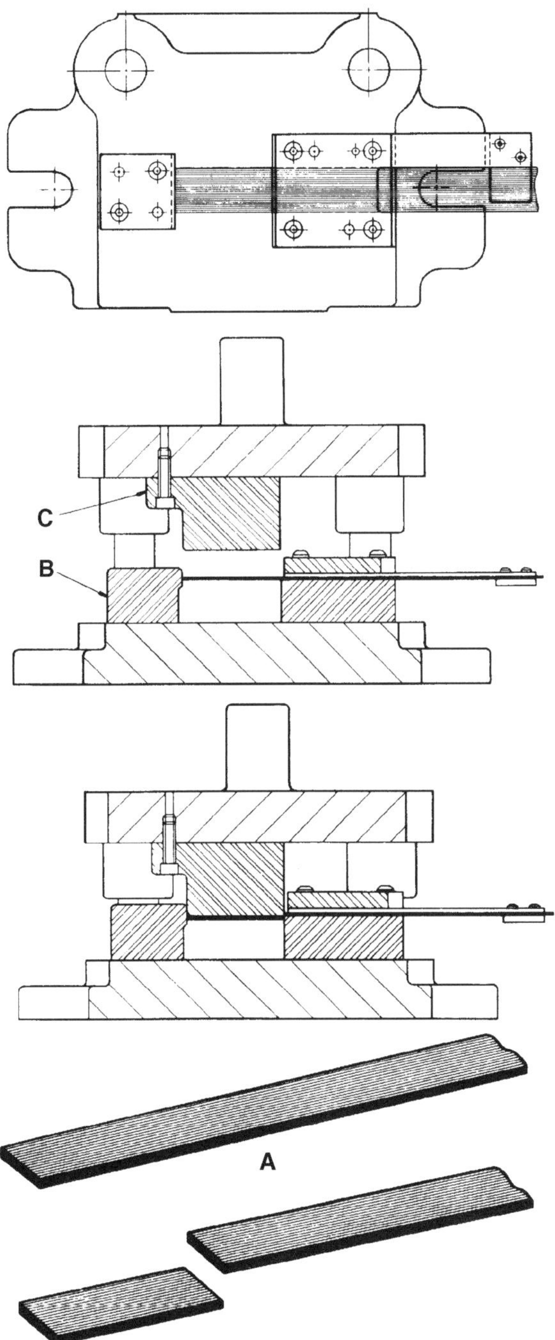

Figure 2.2 Cut-off die.

2.2.4 Compound Dies

In a compound die (Figure 2.4), holes are pierced at the same station where the part is blanked, instead of at a previous station, as is done in a

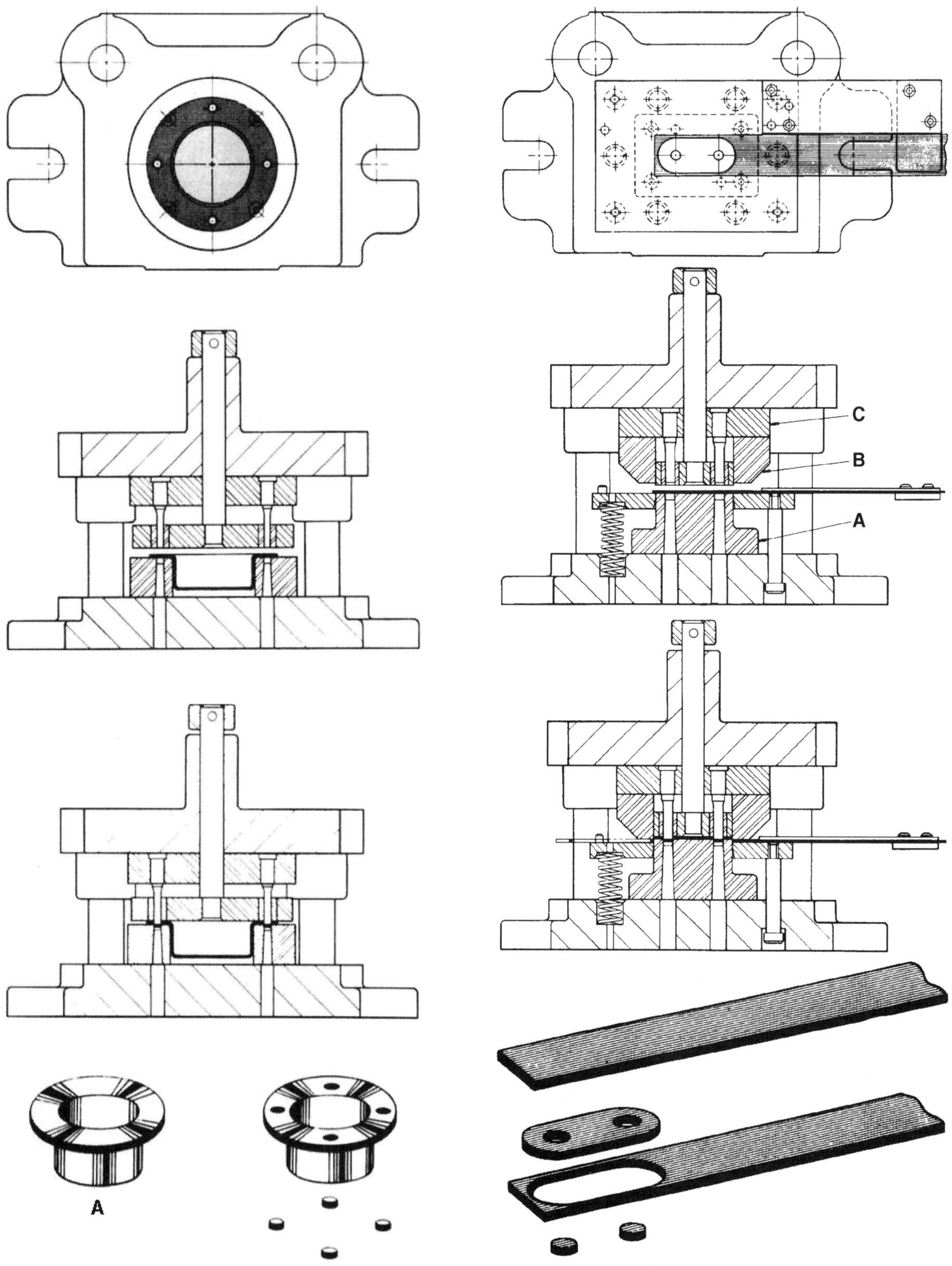

Figure 2.3 Piercing die.

Figure 2.4 Compound die.

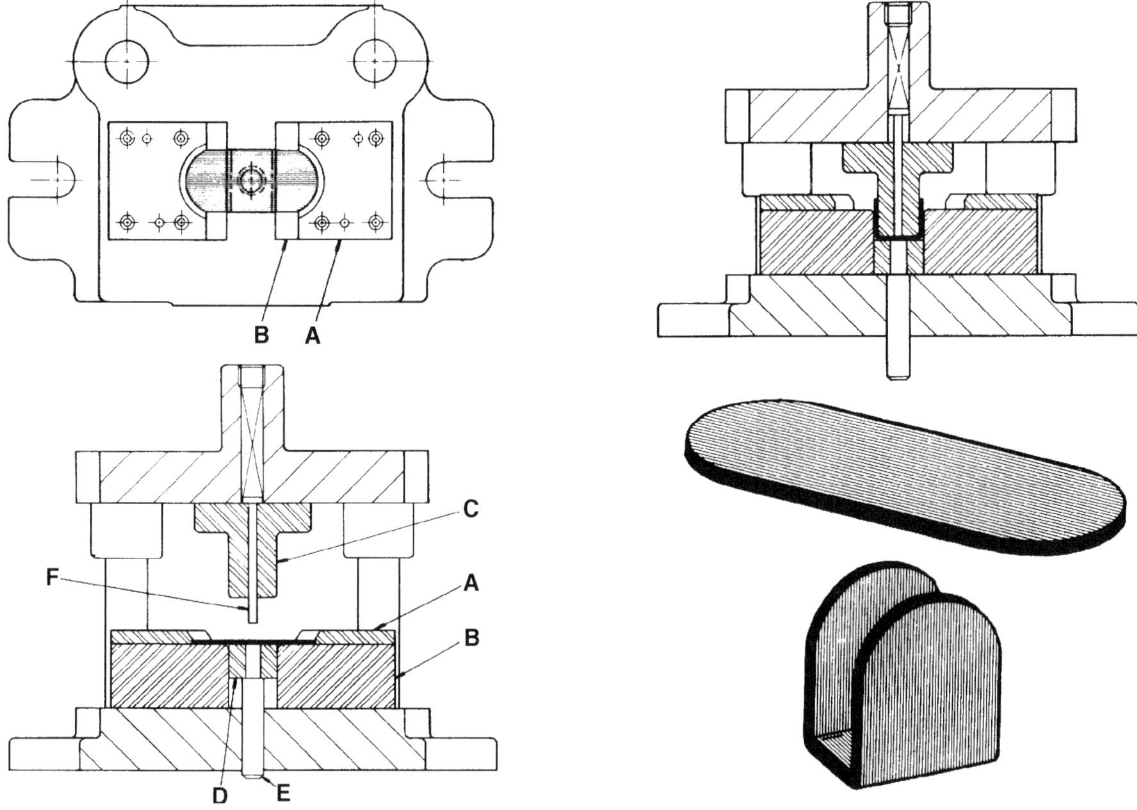

Figure 2.5 Bending die.

pierce and blank die. The result is greater accuracy in the blank. Whatever accuracy is built in will be duplicated in every blank produced by the die.

Compound dies are inverted dies. The blanking punch **A** is located on the die holder of the die set instead of being fastened to the punch holder as in conventional dies, and it is provided with tapered holes for disposal of slugs.

The die block **B** is fastened to the punch holder and it is backed up by a spacer **C,** which retains piercing punches. A positive knockout removes the blank from within the die cavity near the top of the press stroke. A spring stripper removes the material strip from around the blanking punch.

Although most compound dies are designed for producing accurate, flat blanks, they are occasionally used for producing blanks that are too large for production in more than one station. Because all operations are performed at the same station, compound dies are very compact and a smaller die set can be applied.

2.2.5 Bending Dies

A bending die (Figure 2.5) deforms portions of flat blanks to some angular position. The line of bend is straight along its entire length, as differentiated from a forming die, which produces workpieces having a curved line of bend. In the illustration, a flat blank is to be given a double bend to form a U shape. The blank is inserted in gages **A** fastened on bending blocks **B**. The bending blocks, in turn, are fastened to the die holder. Upon descent of the upper die, the bending punch **C** grips the blank between its lower face and pressure pad **D**. Pins **E** extend to the pressure attachment of the press. Shedder **F** strips the workpiece from the punch.

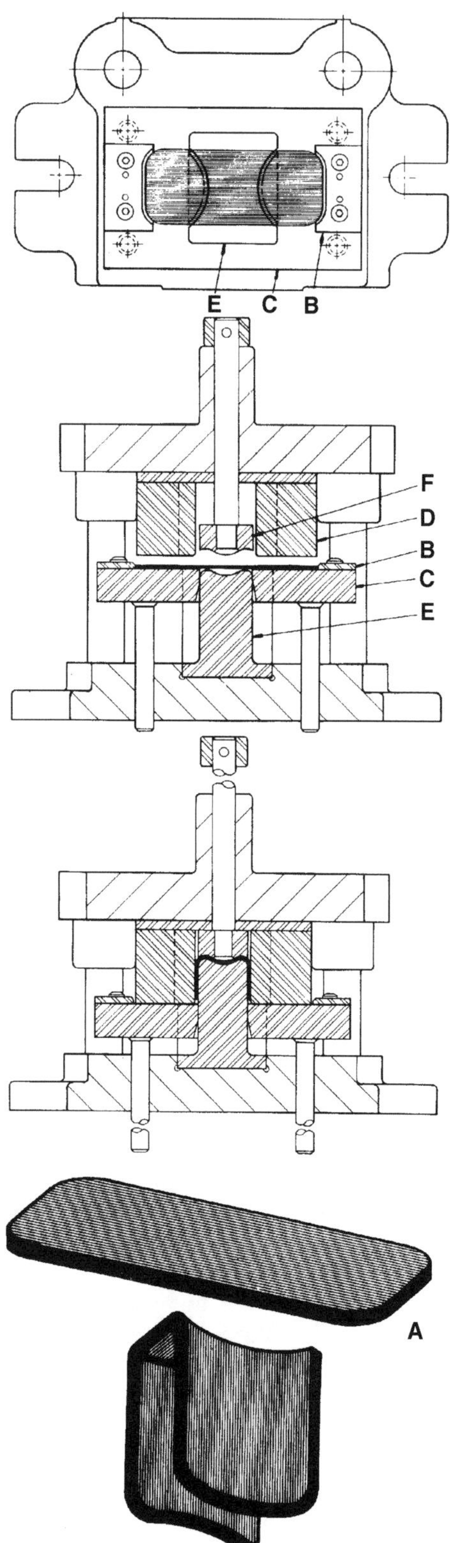

Figure 2.6 Forming die.

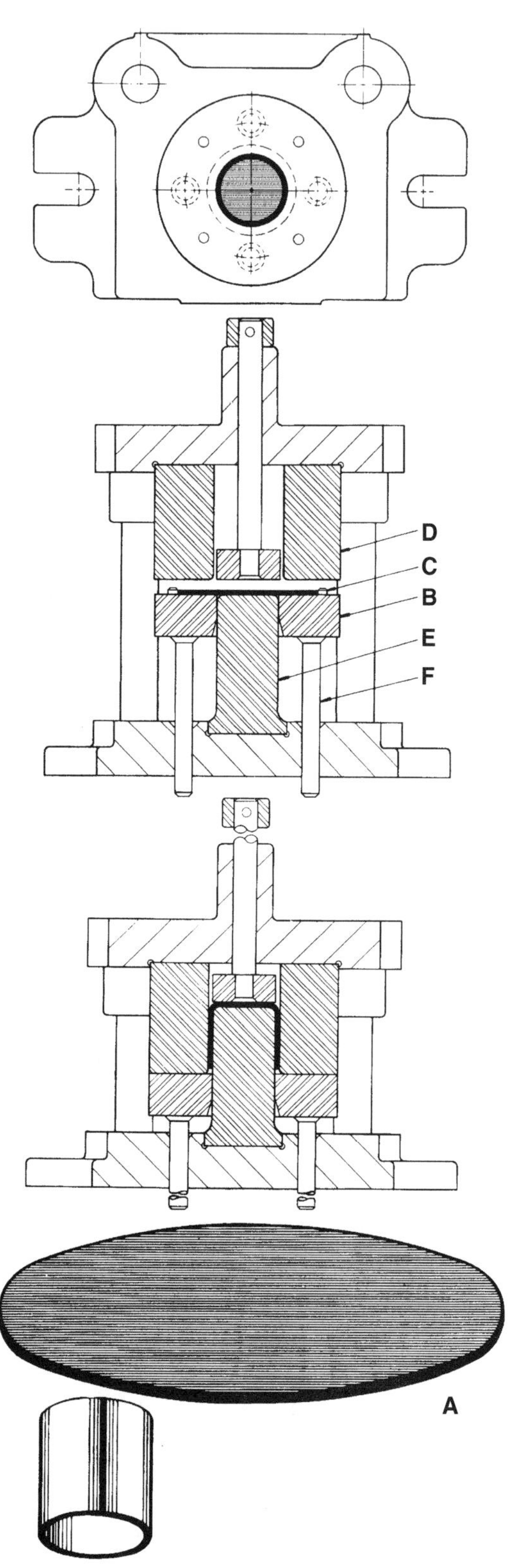

Figure 2.7 Drawing die.

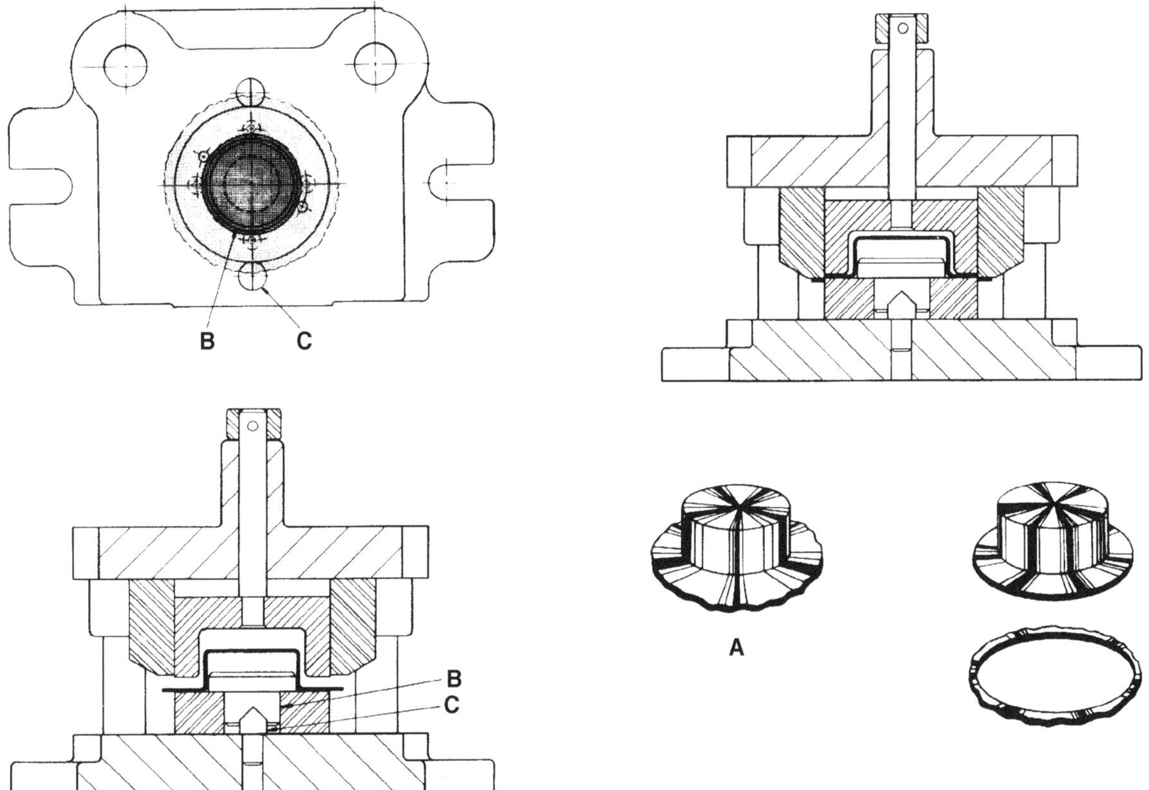

Figure 2.8 Trimming die.

2.2.6 Forming Dies

The operation of forming is similar to bending except that the line of bend is curved instead of straight and plastic deformation in the material is more severe. In Figure 2.6, the flat blank at **A** is to be formed into a part having a curved contour. The blank is positioned in nest **B** composed of two plates mounted on pressure pad **C**. When the ram descends, the blank is gripped between the bottoms of forming blocks **D** and the surface of pressure pad **C**. Further descent causes the sides of the blank to be formed to the curved shape of forming blocks **D** and forming punch **E**. At the bottom of the stroke, knockout block **F** applies the final form. It bottoms against a hardened spacer fastened to the punch holder, thus setting the form. When the die ascends, the part is carried up within form blocks **D**. Near the top of the stroke it is ejected by knockout **F**.

2.2.7 Drawing Dies

The drawing of metal, or deep-drawing manufacturing technology, is defined as the stretching of sheet metal stock, commonly referred to as a blank, around a punch. The edges of the metal blank are restrained by rings and the punch is deep drawn into a top die cavity to achieve the end shape that is desired. There are many shapes that can be made through deep drawing and stamping, such as cups, pans, cylinders, domes, and hemispheres, as well as irregularly shaped products.

In Figure 2.7 at **A**, a flat disk is to be drawn into a cup. The blank is placed on pressure pad **B** of the drawing die and is located by four spring-loaded pins **C**. Descent of the upper die causes the blank to be gripped securely between the surface of pressure pad **B** and the lower surface of draw ring **D**. Further descent of the ram causes the blank to be drawn over punch **E** until it has assumed the cup shape shown in the closed view at the right. Pressure pins **F** extend to the pressure attachment of the press.

The amount of pressure must be adjusted carefully. Excessive pressure would cause the bottom of the cup to be punched out. Insufficient pressure would allow wrinkles to form. With the proper amount of pressure, a smooth, wrinkle-free cup is produced. Drawing dies are extensively used for producing stampings ranging from tiny cups and ferrules to large shells for pressure vessels, ships, cars, aircraft, and missiles.

2.2.8 Trimming Dies

Trimming dies (Figure 2.8) cut away portions of formed or drawn workpieces that have become wavy and irregular. This condition occurs because of uneven flow of metal during forming operations. Trimming removes this unwanted portion to produce square edges and accurate contours.

The illustration at **A** shows a flanged shell after the drawing operation. A trimming die is required to trim the irregular edge of the flange. The shell is placed over a locating plug **B.** Descent of the upper die then causes the scrap ring to be cut from around the flange. After trimming, the shell is carried up in the upper die and a positive knockout ejects it near the top of the stroke. The scrap rings are forced down around the lower trimming punch until they are split in two by scrap cutters **C** applied at the front and back of the die. The scrap pieces fall to the sides, away from the operation of the press.

2.2.9 Shaving Dies

Shaving is the operation of removing a small amount of metal from around the edges of a blank or hole in order to improve the surface. A properly shaved blank has a straight, smooth edge and it is held to a very accurate size. Many instruments, business machines, and other parts are shaved to provide better functioning and longer wear.

In Figure 2.9, a blank **A** is to be shaved, both along outside edges and in the walls of the two holes. The shaving die for this workpiece consists of an inverted shaving punch **B** fastened to the die holder, and a shaving die block **C** fastened to the punch holder. A spacer **D** backs up the die block and it retains the shaving punches for the holes.

The blank is located in a nest **E**, beveled to provide clearance for the curled chip. The nest is mounted on a spring stripper plate guided on

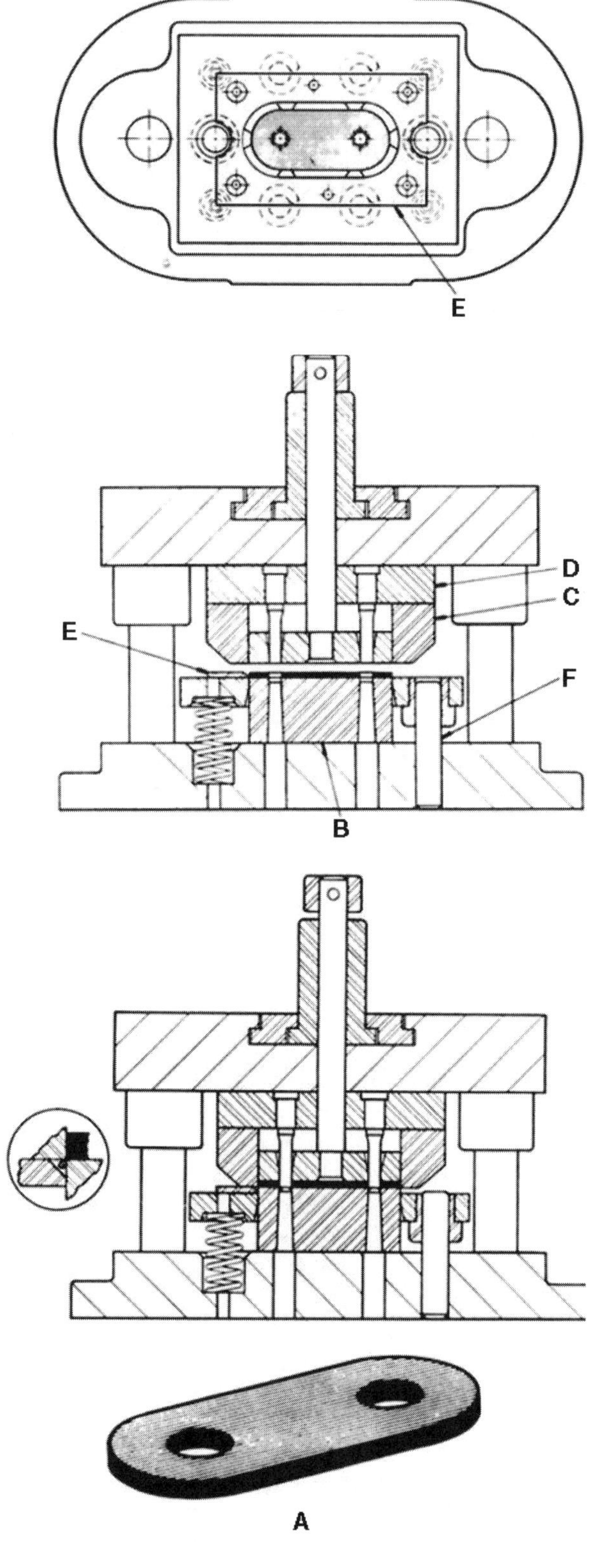

Figure 2.9 Shaving die.

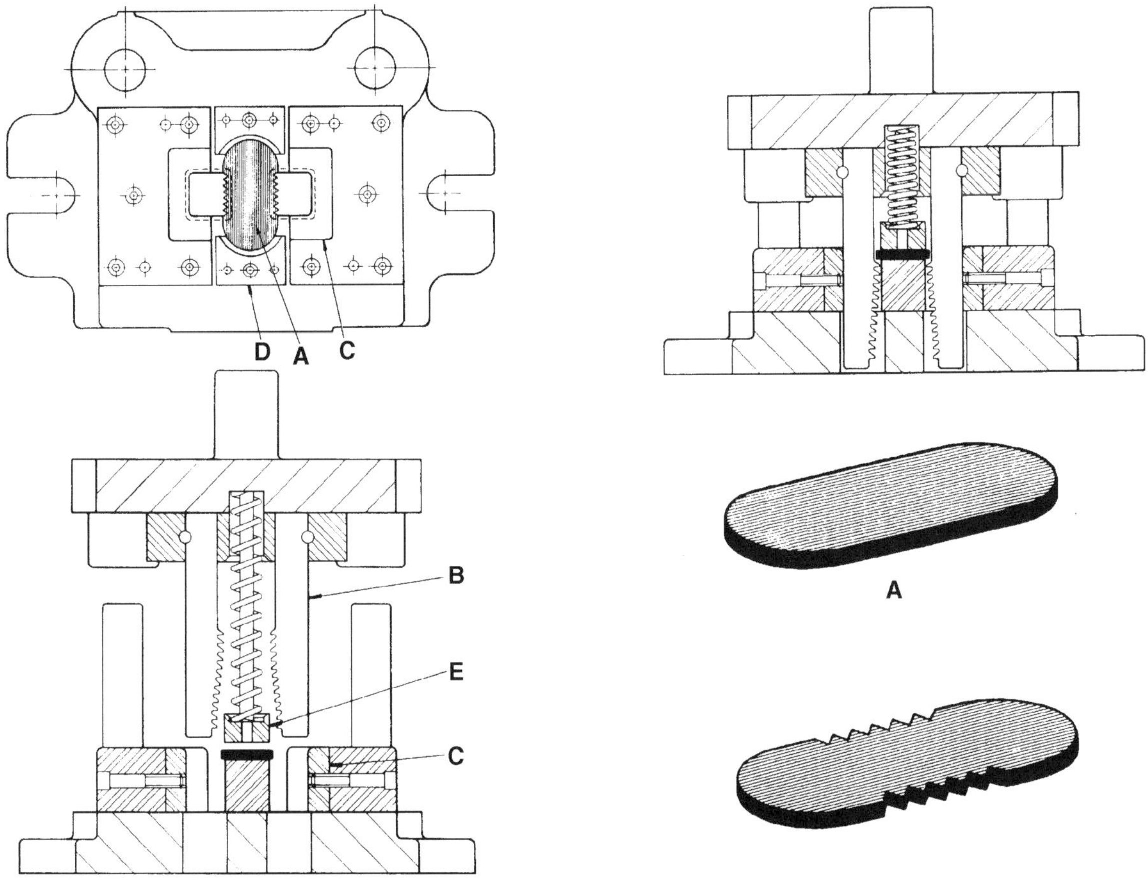

Figure 2.10 Broaching die.

two guide pins **F**. The shaved blank is carried up, held in the die block with considerable pressure, and ejected near the top of the stroke by a positive knockout. Shaving dies are ordinarily held in floating adapter die sets for better alignment. This is necessary because no clearance is applied between punches and die block.

2.2.10 Broaching Dies

Broaching may be considered to be a series of shaving operations performed one after the other by the same tool. A broach is provided with a number of teeth, each of which cuts a chip as the broach traverses the surface to be finished. Internal broaches finish holes; surface or slab broaches finish outside surfaces. Two conditions make broaching necessary:

1. Blanks are too thick for shaving: If considerable metal must be removed from the edges of thick blanks, a series of shaving dies would be required to produce a smooth finish. It would then be more economical to use a broaching die.
2. When considerable metal must be removed: This occurs when ridges or other shapes are required in the edges of the blank. It is often impractical to blank such shapes directly because the cutting edges would be weak and subject to breakage.

In Figure 2.10, a blank at **A** must have small pointed serrations machined in the sides. The die is provided with two broaches **B** supported during the cutting process by hardened backing blocks **C**. The blank is located in a nest **D** composed of two opposed plates machined to fit the contour. Pressure pad **E**, backed up by heavy springs, clamps the blank securely before cutting begins. The first three or four teeth of the broach

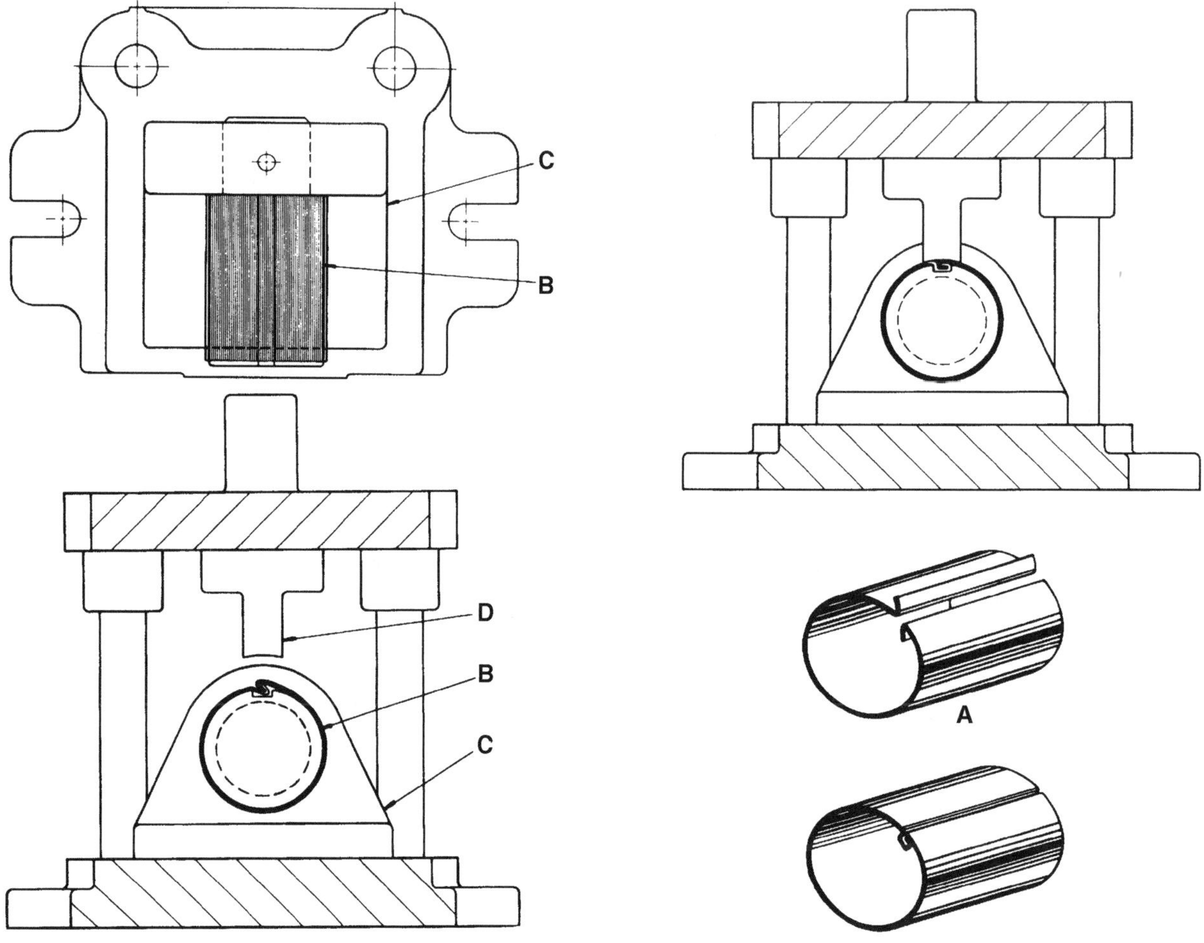

Figure 2.11 Horn die.

are made undersize; ordinarily they do no cutting unless an oversize blank is introduced into the die. The last three or four teeth are sizing teeth. Intermediate teeth are called working teeth and they take the successive chips to machine the serrations.

2.2.11 Horn Dies

A horn die (Figure 2.11) is provided with a projecting post called a horn. Bent, formed, or drawn workpieces are applied over the horn for performing secondary operations.

In the illustration at **A**, a blank has been reverse bent in a previous operation and the ends are to be hooked together and seamed in a horn die. The horn **B** is retained in a holder **C** fastened to the die holder. When the ram descends, seaming punch **D** strikes the workpiece to form the seam.

Many other operations, such as piercing and staking, are also performed in horn dies.

2.2.12 Side Cam Dies

Side cams transform vertical motion from the press ram into horizontal or angular motion and they make possible many ingenious operations. In Figure 2.12, at **A**, a flanged shell requires two holes pierced in its side. The shell is placed over die block **B** of the die. Descent of the upper die causes pressure pad **C** to seat the shell firmly over the block. Further descent causes side cams **D** to move the punch-carrying slides **E** for piercing the holes. Spring strippers **F** strip the shell from around the piercing punches as they are withdrawn.

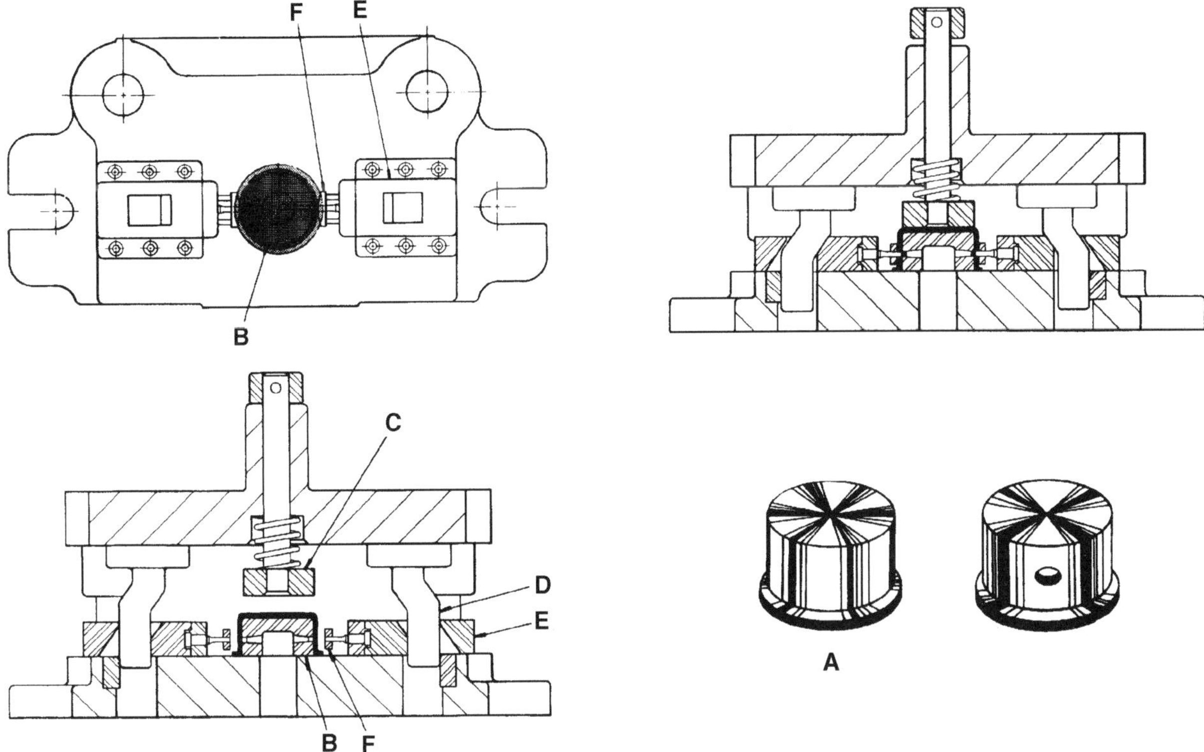

Figure 2.12 Side cam die.

2.2.13 Curling Dies

A curling die (Figure 2.13) forms the material at the edge of a workpiece into a circular shape or hollow ring. Flat blanks may be curled; a common application is a hinge formed of two plates each of which is curled at one side for engagement of the hinge pin. More often, curling is applied to edges of the open ends of cups and shells to provide stiffness and smooth, rounded edges. Most pans used for cooking and baking foods are curled.

In the illustration, a drawn shell shown at **A** is to be curled. The shell is placed in the curling die where it rests on knockout pad **B**. Descent of the upper die causes the knockout pad to be pushed down until it bottoms on the die holder. Further descent causes curling punch **C** to curl the edge of the shell. Near the bottom of the stroke, the lip of the material contacts an angular surface machined in curling ring **D** to complete the curl. When the punch goes up, the knockout raises the shell for easy removal.

2.2.14 Bulging Dies

A bulging die (Figure 2.14) expands a portion of a drawn shell causing it to bulge. There are two types: fluid dies and rubber dies. Fluid dies use water or oil as the expanding medium and a ram applies pressure to the fluid. In rubber dies, a pad or block of rubber under pressure moves the walls of the workpiece to the desired position. This is possible because rubber is virtually incompressible. Although it can be made to change its shape, the volume remains the same.

In the illustration at **A**, a drawn shell is to be bulged at its closed end. The shell is placed over punch **B** of the bulging die and its lower end is confined in lower die **C**. The upper end of punch **B** is a rubber ring within which is applied a spreader rod **D**. This rod is conical at its upper end and it helps the rubber to flow outward to the desired shape. When the press ram descends, the upper die applies a force to the shell bottom, and since the rubber cannot compress, it is forced outward bulging the walls of the shell. When the

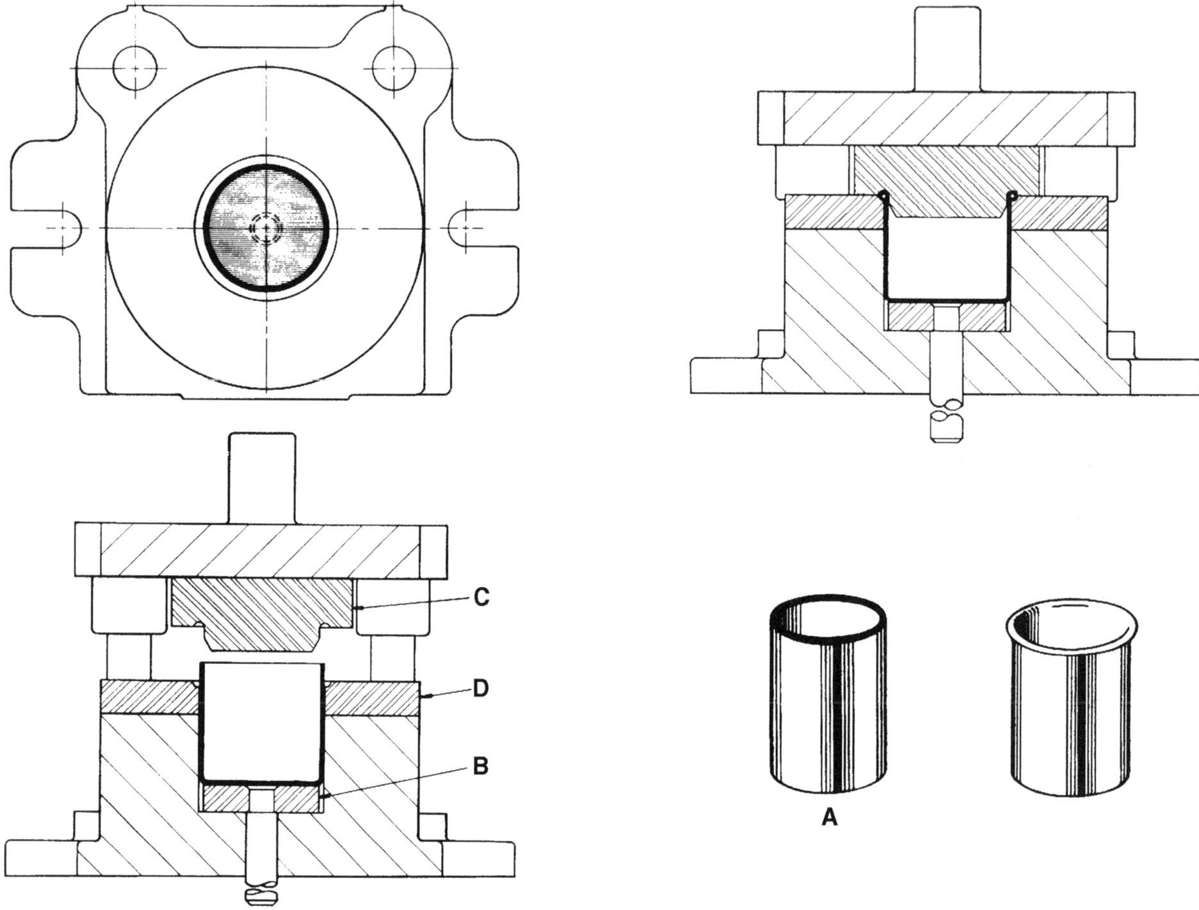

Figure 2.13 Curling die.

ram goes up, the rubber returns to its original shape and the bulged shell can be removed from the die. After bulging, a shell is shorter than it was previously.

2.2.15 Swaging Dies

The operation of swaging, sometimes called necking, is exactly the opposite of bulging. When a workpiece is swaged a portion is reduced in size. This process causes the part to become longer than it was before swaging. In Figure 2.15, at **A**, a shell is to be swaged at its open end. It is inserted in the swaging die where it rests on knockout pad **B** and its lower end is surrounded by the walls of block **C**. When the ram descends, swaging die **D** reduces a portion of the diameter of the shell and this portion becomes longer.

2.2.16 Extruding Dies

The function of all the dies discussed so far is to perform work on sheet material—to cut sheet material into blanks, to perform further operations upon the blanks, or to perform operations on workpieces bent, formed, or drawn from the blanks. We come now to some interesting classes of dies that perform secondary operations on small thick blanks called slugs. In these dies, the slugs are severely deformed to make parts having no resemblance to the slugs from which they were made.

The first class of dies are extruding dies. In this type of die, each slug is partly confined in a cavity. Then extremely high pressure is applied by a punch to cause the material in the slug to extrude or squirt out, much like toothpaste is extruded

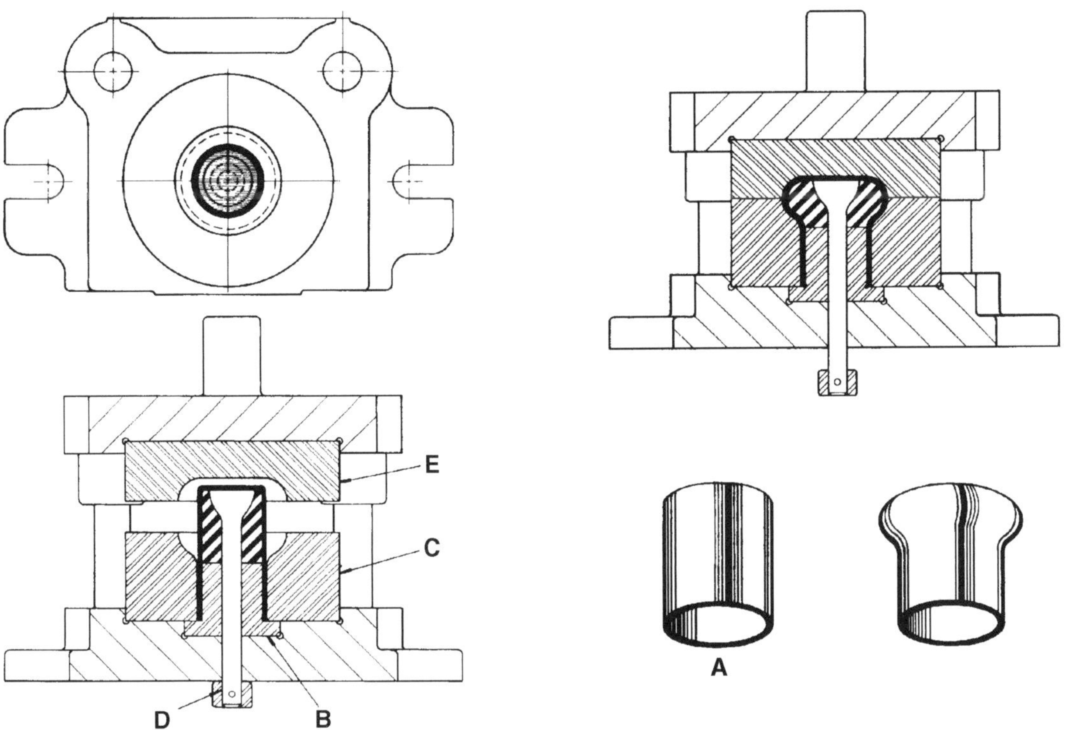

Figure 2.14 Bulging die that uses rubber as the bulging medium.

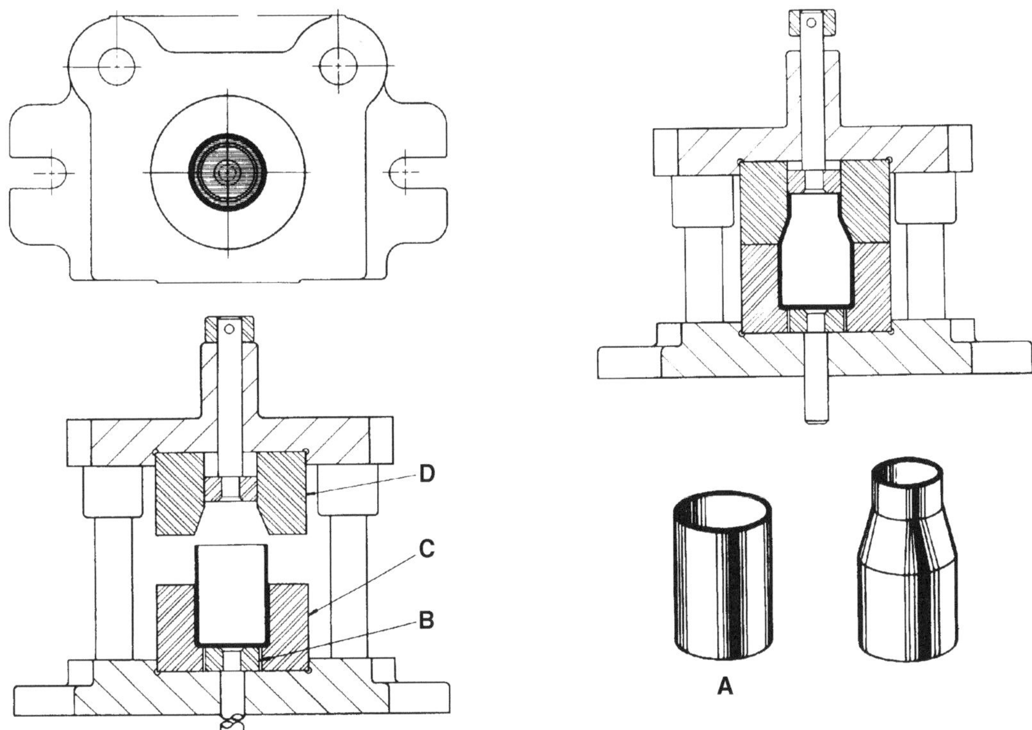

Figure 2.15 Swaging die.

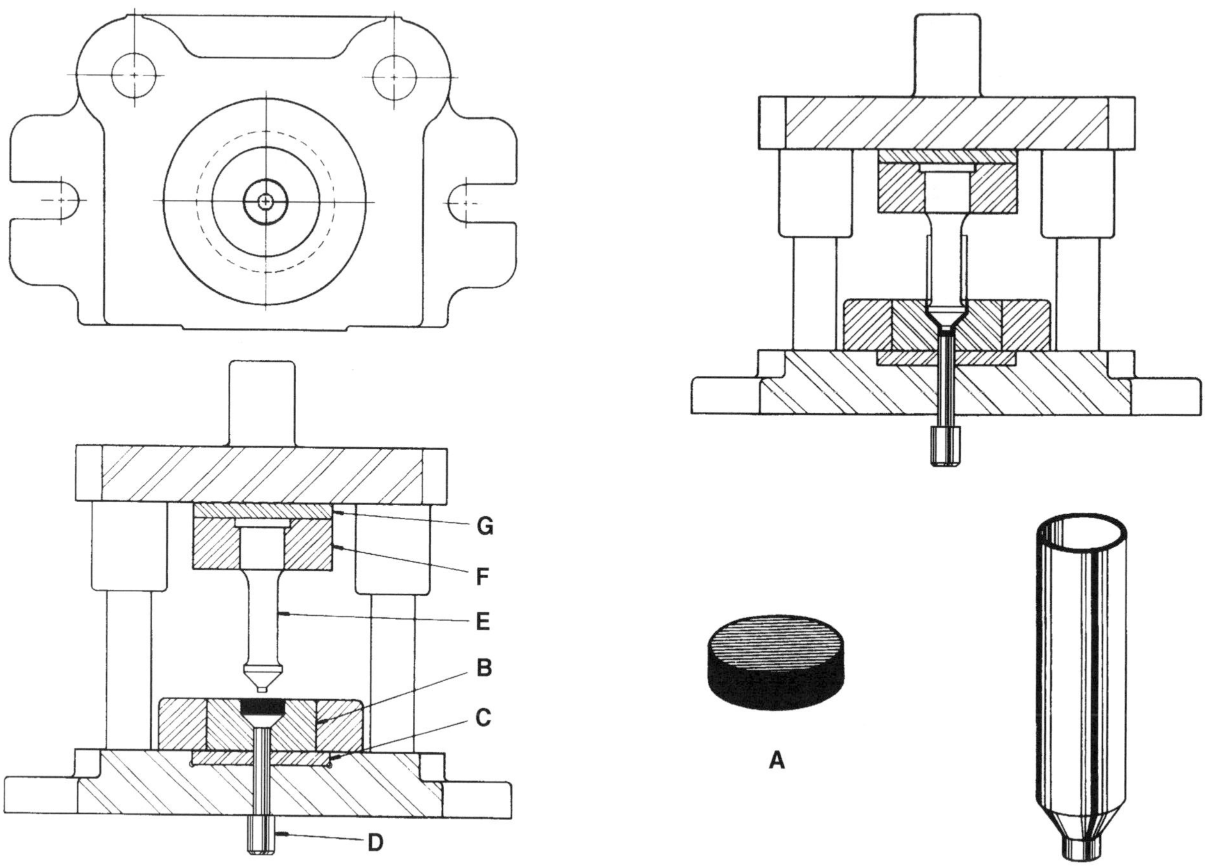

Figure 2.16 Extruding die.

when the tube is squeezed. In Figure 2.16, the slug **A** is to be extruded into a thin-walled shell having a conical closed end. The slug is placed in die block **B**, backed up by a hardened plate **C**. The bottom of the cavity in the die block is formed by the end of knockout rod **D**. When the press ram descends, extruding punch **E** first squeezes the slug until it assumes the shape of the die cavity and of the working end of the extruding punch. Continued descent causes the material to extrude upward between the wall of the punch and the wall of the die cavity. The amount of clearance between the two determines the thickness of the wall of the extruded shell. The extruding punch is retained in punch plate **F** and, because of the high pressure involved, it is backed up by backing plate **G**.

2.2.17 Cold-Coining Dies

Cold-coining dies (Figure 2.17) produce workpieces by applying pressure to blanks, squeezing and displacing the material until it assumes the shape of the punch and die. In the illustration at **A**, a slug is to be formed into a flanged part in a cold coining die. It is placed on punch **B** located within spring-loaded V gages **C**. Descent of the upper die causes the material under the upper die block to be displaced outward to form the flange. As the flange increases in diameter, the gages are pushed back as shown. When the die goes up, the part is carried upward within it and ejected near the top of the stroke by knockout plunger **D** actuated by knockout rod **E**.

The cylindrical part is the slug (blank); another illustration is the flanged part. This is only one simple example of cold coining dies. A basic postulate of plastic deformation of material states that "the shape (of course, dimensions, too) of blanks can be changed, but volume is constant."

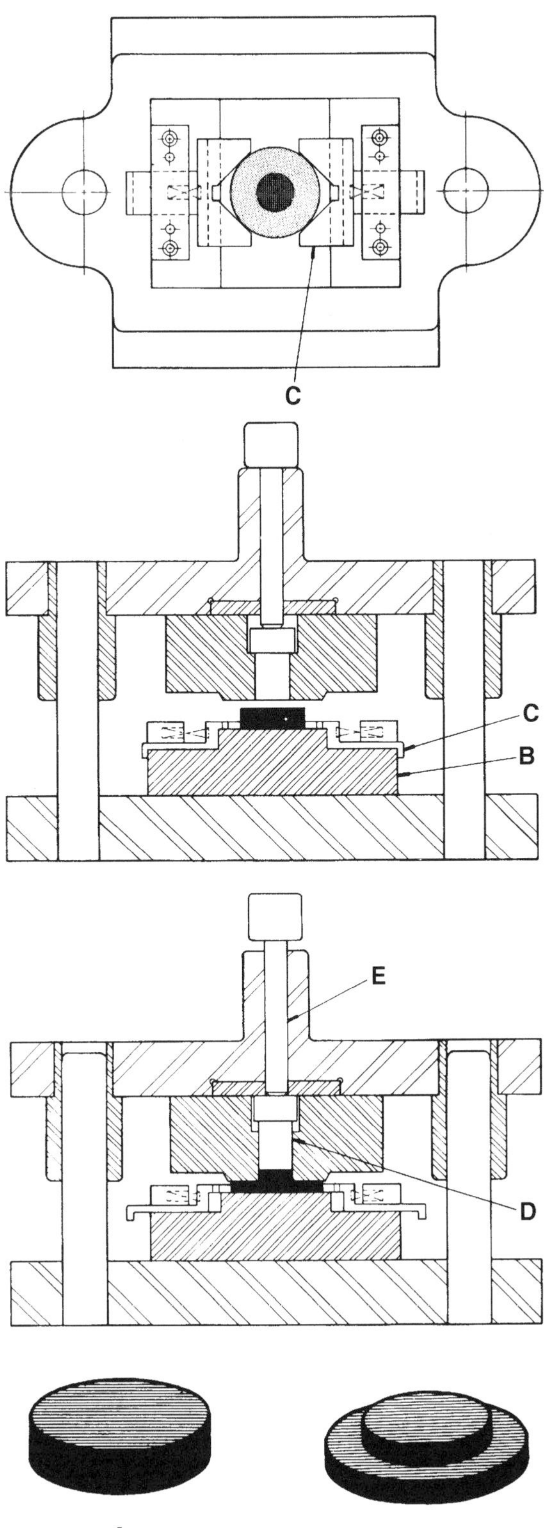

Figure 2.17 Cold coining die.

2.2.18 Progressive Dies

All of the operations described previously may be performed in progressive dies. For example, a single die of this type may do piercing at the first station, trimming at the second station, bending at the third, forming at the fourth, etc. A progressive die may thus be considered a series of different dies placed side by side with the strip passing through each successively. This analogy has some merit, although it does not give a true picture of the extremely close interrelationship between the various stations.

In Figure 2.18, at **A**, a pierced, trimmed, and bent part is to be produced complete in a simple progressive die. At the first station the strip is notched and pierced and at the second station the blank is cut off and bent. You should easily recognize all of the elements in this die—the die block, piercing punch, trimming punch, knockout, and stop block, along with all the others.

2.2.19 Sub-Press Dies

Sub-press dies (Figure 2.19) blank and form very small watch, clock, and instrument parts. An example is the small instrument cam shown at **A**. The die components are retained in a sub-press which is, as its name implies, actually a small press operated in a larger one. The sub-press is composed of base **C**, barrel **B**, and plunger **D**. A long, tapered babbit bearing **E** provided with longitudinal key slots guides the plunger and prevents rotation. Tightening spanner nut **F** against bearing **E** causes it to close around plunger **D** to remove all looseness. The top portion of plunger **D** is engaged by actuator **G** threaded into a central tapped hole. The slot of the actuator is engaged loosely by a yoke fastened to the press ram. Thus, the press ram does not guide the sub-press in any way. It simply applies the up-and-down motion. Sub-press dies are usually of the compound type because of the considerable accuracy required.

2.2.20 Assembly Dies

Assembly dies assemble two or more parts together by press-fitting, riveting, staking, or other means. Components are assembled very quickly and relationships between parts can be maintained closely. Figure 2.20 shows a link and two studs that are to be riveted together in an assembly die. The studs are positioned in die block **A** and they sit on plungers **B**. The link is

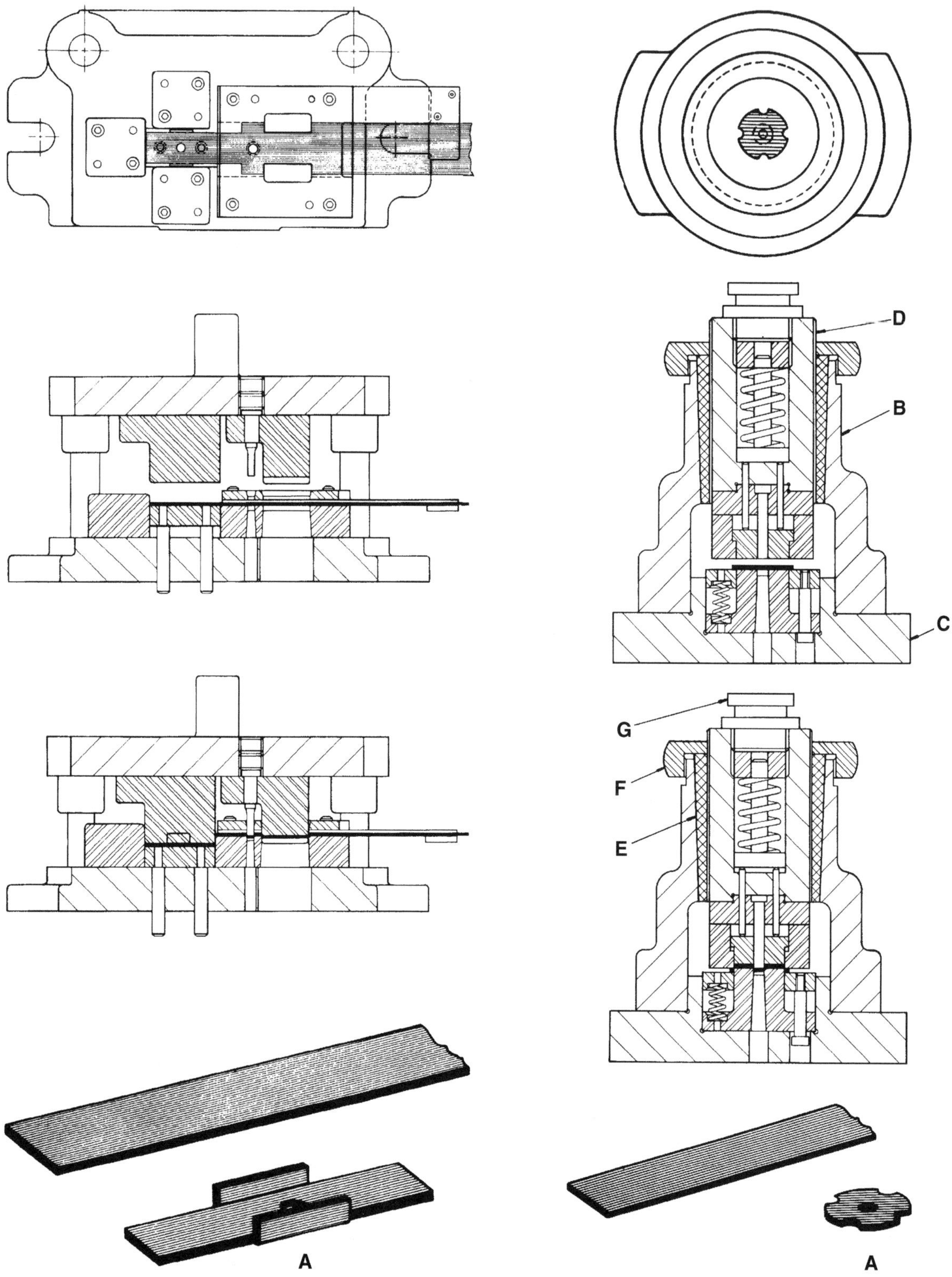

Figure 2.18 Progressive die.

Figure 2.19 Sub-press die.

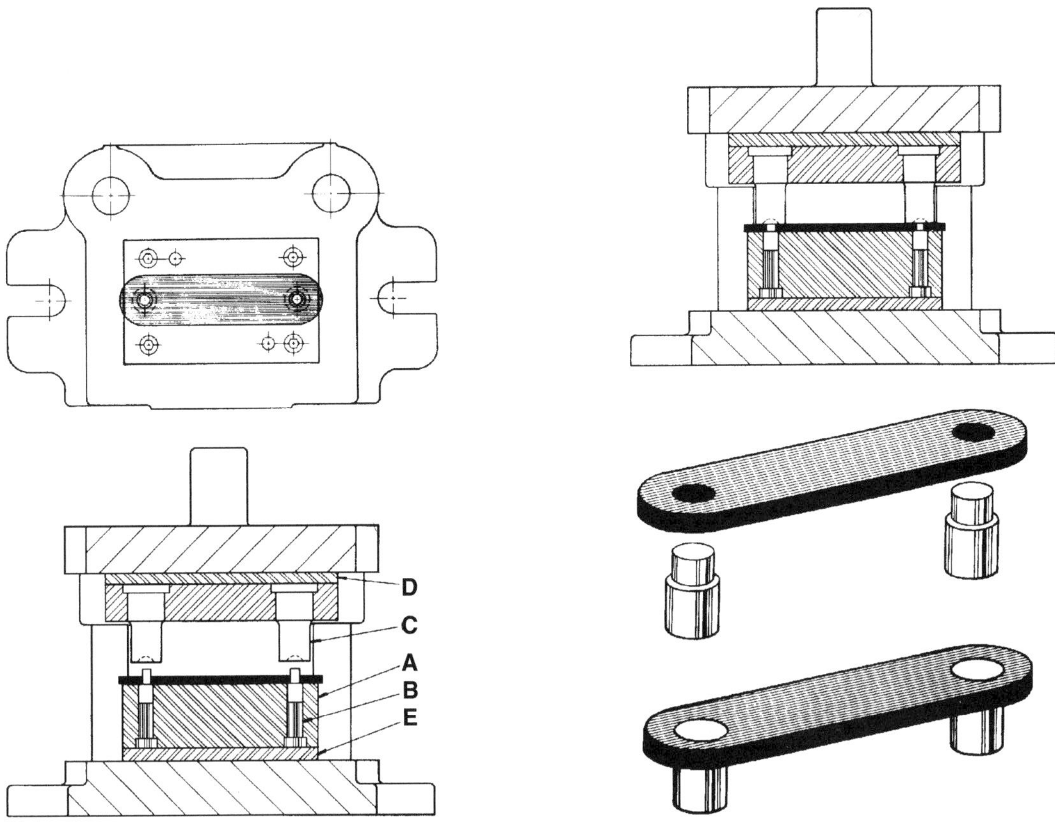

Figure 2.20 Assembly die.

then positioned over the studs, the turned-down ends of the studs engaging in holes in the link. Descent of the press ram causes riveting punches **C** to deform the ends of the studs into the shape of rivet heads. A hardened plate **D** backs up the punches to prevent the heads from sinking into the relatively soft material of the die set. Another hardened plate **E** backs up the plungers.

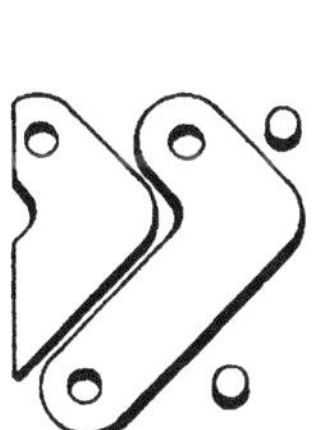

THE MATERIAL STRIPS

3.1 INTRODUCTION

Most stampings are made of steel. Carbon content varies from AISI-SAE 1010 to AISI-SAE 1030 and, therefore, most blanks are in the machine or cold-rolled steel range. Stampings are also made from these other materials:

1. Aluminum
2. Brass
3. Bronze
4. Copper
5. Stainless steel
6. Silicon steel
7. Fiber
8. Plastic sheet, etc.

3.2 STEEL

Steel is an alloy of iron and carbon. Carbon must be present to the extent of about 0.05 percent by weight in order for the material to be known as "steel" rather than commercial iron. The composition and processing of steels are controlled in a manner that makes them suitable for numerous applications. They are available in various basic product shapes: sheet, strip, sheet, and plate.

3.2.1 Hot-Rolled Steel

Hot-rolled sheets are formed easily. Low-carbon sheets are used for tanks, barrels, pails, farm implements, lockers, cabinets, truck bodies, and other applications where scale and discoloration are not objectionable because surfaces are painted after forming. Hot-rolled sheets are readily available in thicknesses ranging from #30 gage (0.012 in. or 0.3 mm) to #7 gage (0.1875 in. or 4.8 mm).

a) Pickled and Oiled Sheets

Pickling, or the immersing of hot-rolled sheets in acid solution, results in smooth, clean, scale-free surfaces having a uniform gray color. Oiling protects the surfaces against rust.

These sheets are readily stamped or welded. Long-lasting painting or enameling is possible because of the absence of scale. Pickled and oiled sheets are used for household appliances, automotive parts, toys, and the like.

b) Copper-Bearing Sheets

Copper-bearing sheets are hot-rolled sheets having 0.20 percent minimum copper content. They are used for parts designed for outdoor exposure, or for indoor use under corrosive conditions. These sheets have a service life from two to three times longer than can be expected from non-copper-bearing steels. They are used for roofing and siding, farm and industrial buildings, truck bodies, railroad cars, farm implements, signs, tanks, dryers, ventilators, washing machines, and other similar applications.

c) Medium-Carbon Sheets

Hot-rolled sheets having a 0.40 to 0.50 percent carbon content provide hardness, strength, and resistance to abrasion. They can be heat-treated to make the material even harder and stronger and are primarily used for scrapers, blades, hand tools, and the like.

3.2.2 Cold-Rolled Sheets

Cold-rolled sheets have a smooth, deoxidized satin finish, which provides an excellent base for paint, lacquer, and enamel coating. Thicknesses are held to a high degree of accuracy. Cold-rolled steel is produced by the cold rolling of hot-rolled sheets to improve size and finish. Refrigerators, ranges, panels, lockers, and electrical fixtures are among their many uses.

a) Possibility of Deformation

Six tempers of cold-rolled steel sheets and strips are available; it is important to know exactly what operations can be performed on each (Figure 3.1):

1. *Hard.* Hard sheets and strips will not bend in either direction of the grain without cracks or fracture. These tempers of steel are employed for flat blanks that require resistance to bending and wear. Direction of grain is shown along lines **A** in the illustration. Hardness is Rockwell B 90 to 100.

2. *Three-quarter hard.* This temper of steel will bend a total of 60 degrees from flat across the grain. This is shown as dimension **B** in the illustration. Hardness is Rockwell B 85 to 90.

3. *One-half hard.* This temper will bend to a sharp 90-degree angle across the grain, shown as dimension **C.** Hardness is between Rockwell B 70 and 85.

4. *One-quarter hard.* This commonly used temper of steel will bend over flat on itself across the grain and to a sharp right angle along the grain. Hardness is Rockwell B 60 to 70.

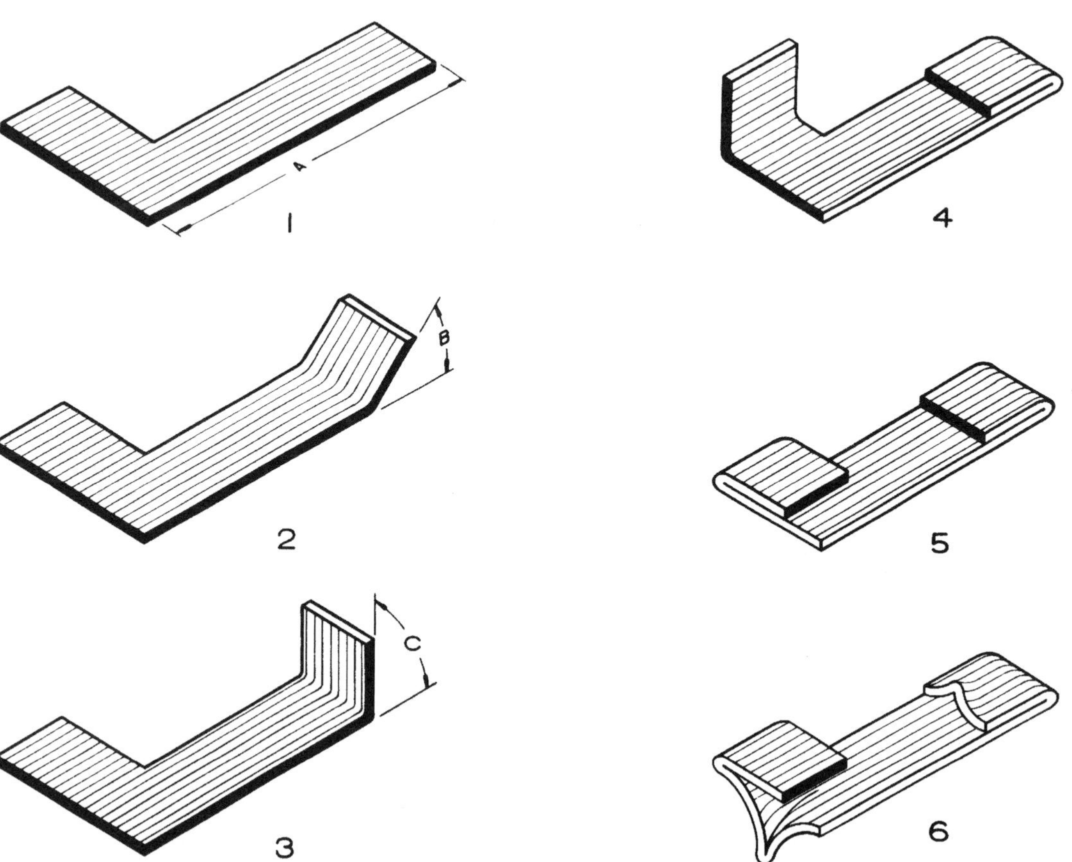

Figure 3.1 Various tempers of cold-rolled steel from hard (1) to dead soft (6) and kinds of deformation possible with each.

5. *Soft.* This temper will bend over flat upon itself both across the grain and along the grain. It is also used for moderate forming and drawing. Hardness is Rockwell B 50 to 60.

6. *Dead soft.* This temper of steel is used for deep drawing and for severe bending and forming operations. Hardness is Rockwell B 40 to 50.

b) Finish

Cold-rolled steel is available in three grades of finish:

1. *Dull finish.* This is a gray lusterless finish to which lacquer and paint bond well.

2. *Regular bright finish.* This is a moderately bright finish suitable for most work. It is not recommended for plating unless buffed first.

3. *Best bright finish.* This finish has a high lustre well suited for electroplating. It is the brightest finish obtainable.

c) Stretcher-Leveled Sheets

These are cold-rolled steel sheets that have been further processed by stretcher leveling and resquaring. They are used in the manufacture of metal furniture, table tops, truck body panels, partitions, and other equipment requiring perfectly flat material.

d) Deep-Drawing Sheets

Deep-drawing steel is prime quality cold-rolled steel having a low carbon content. Sheets are thoroughly annealed, highly finished to a deoxidized silver finish, and oiled. Deep-drawing sheets are used for difficult drawing, spinning, and stamping operations such as those which produce automobile bodies, fenders, electrical fixtures, and laboratory equipment.

e) Silicon Steel

Also called "electrical steel," silicon steel is extensively used for motors and generators. Lighter gages are suitable for transformers, reactors, relays, and other magnetic circuits.

3.3 MECHANICS OF SHEAR

The shearing process involves the cutting of flat material forms, such as sheets and plates. The cutting may be done by different types of blades or cutters in special machines driven by mechanical, hydraulic, or pneumatic power.

Figure 3.2 shows the mechanics of shear in 8 steps:

1. This illustration shows the cutting edges of a die with clearance **C** applied. The amount of this clearance is important, as will be shown.
2. A material strip is introduced between the cutting edges and is represented by phantom lines. Cutting a material strip occurs when it is sheared between cutting edges until the material between the edges has been compressed beyond its ultimate strength and fracture takes place.
3. The upper die begins its downward travel and the cutting edge of the punch penetrates the material by the amount **A**. The following stresses occur: The material in the radii at **B** is in tension; that is, it is *stretched*. The material between cutting edges **C** is *compressed,* or squeezed together. Stretching continues beyond the elastic limit of the material, then plastic deformation occurs. Observe that the same penetration and stretching is applied to both sides of the strip.
4. Continued descent of the upper cutting edge causes cracks to form in the material. These cleavage planes occur adjacent to the corner of each cutting edge.
5. Continued descent of the upper die causes the cracks to elongate until they meet. Here then is the reason for the importance of correct clearance. If the cracks fail to meet, a bad edge will be produced in the blank.
6. Further descent of the upper die causes the blank to separate from the strip. Separation occurs when the punch has penetrated approximately 1/3 of the strip.
7. Continued descent of the upper die causes the blank to be pushed into the die hole where it clings tightly because of the compressive stresses introduced prior to separation of the blank from the strip. In other words, the material at **C** in step 3 was compressed and it acts like a compressed spring. The blank, confined in the die hole, tends to swell, but it is prevented from doing so by the confining walls of the die block. Conversely, the material around the punch tends to close in and, therefore, the strip clings tightly around the punch.

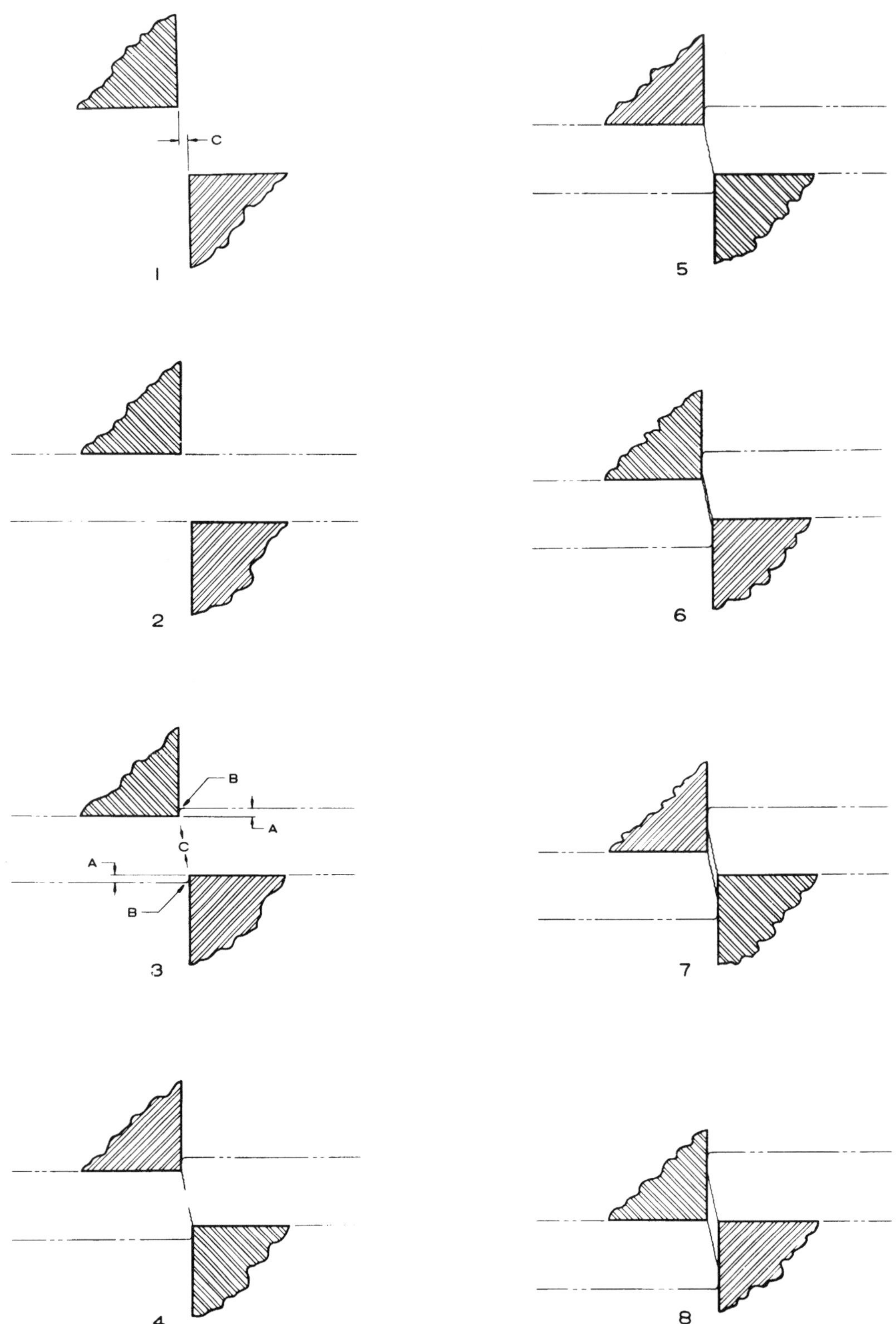

Figure 3.2 Mechanics of shear-enlarged views of clearance between cutting edges of a shearing die (step 1) and material undergoing shear (steps 2 to 8).

8. The punch has now penetrated entirely through the strip and the blank has been pushed entirely within the die hole. Observe that the edge of the blank and the edge of the strip have identical contours except that they are reversed. The strip will cling around the upper punch with approximately the same pressure as the blank clings within the die hole and a stripper will be required to remove it.

3.3.1 Sheared Edges

It now becomes necessary to understand exactly what occurs when sheet material is cut between the cutting edges of a punch and die. Figure 3.3 shows the cut edge of a blank with correct clearance **c** applied, enlarged many times to reveal its contour. Observe the following:

The top corner is defined by a small radius **R**. The size of this radius depends upon the thickness and hardness of the strip and on the sharpness of the punch and die members.

A smooth, straight, burnished band goes around the periphery of the blank. The extent of this band, distance **D**, is approximately 1/3 the thickness **T** of the blank when the die is properly sharpened and when the correct clearance has been applied.

The remaining 2/3 of the edge is called the breakoff. The surface is somewhat rough and tapers back slightly. The extent of the taper, distance **B**, is the amount of clearance between cutting edges. If burrs are produced in cutting the blanks, they occur on this breakoff side of the blank. Burrs are produced when improper clearance has been applied and also when cutting edges become dull. The other side of the blank, which has the radius and smooth, shiny band, is called the burnished side of the blank.

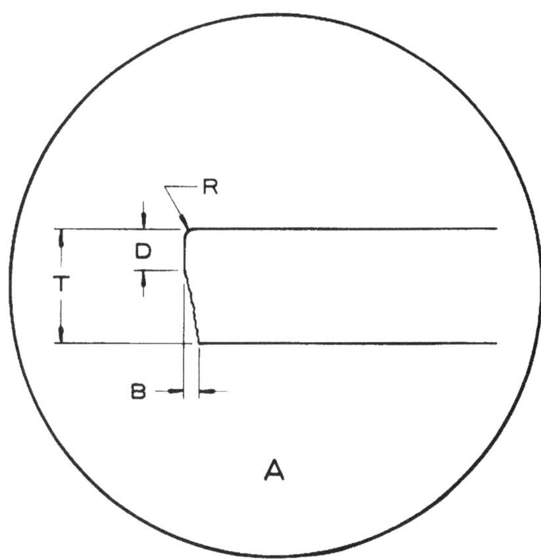

Figure 3.3 Enlarged view of sheared blank edge.

The location of the burnished side and of the burr side of the blank is very important for performing secondary operations such as shaving, burnishing, and the like. In addition, the burr-side position can influence the functioning or the appearance of the finished stamping.

In blanking, the burnished band goes completely around the blank and the breakoff taper extends completely around the blank on the opposite side. This is not the case for blanks produced in cut-off or progressive dies. In such dies, the burnished side may alternate from side to side in a number of positions. Careful study is needed to ensure that no burr will interfere with the function or appearance of the stamping.

Shearing of material occurs in a continuous action. However, to understand the process, it will be necessary to "stop" the action in its various stages and to examine what occurs.

3.3.2 Clearance

Clearance generally is expressed as a free space between two mating parts. In closed contours, clearance is measured on one side.

a) Insufficient Clearance

The inset at **A** in Figure 3.4a shows the four effects of insufficient clearance:

- Radius **R** is smaller than when correct clearance is applied.
- A double burnished band **D** is formed on the blank edge.
- The breakoff angle **B** is smaller than when correct clearance is applied
- Greater pressure is required for producing the blank.

Referring to Figure 3.4a:

1. This figure shows cutting edges of a punch and die in partial penetration. It is obvious that cracks have appeared at the punch.

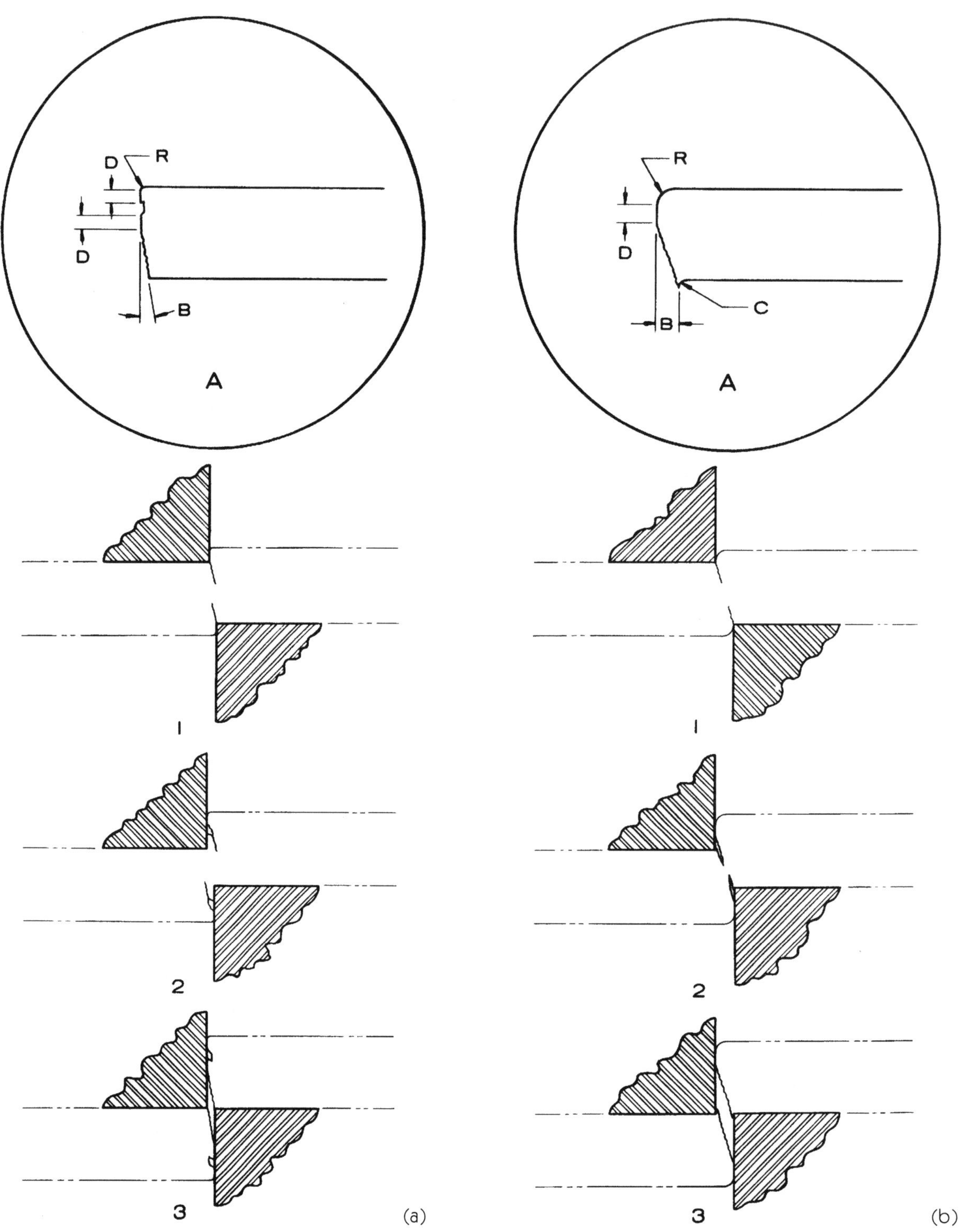

Figure 3.4 Enlarged views of blank edge sheared: a) with insufficient clearance **A** and material undergoing shear with insufficient clearance (1 to 3); and b) with excessive clearance **A** and material undergoing shear with excessive clearance (1 to 3).

Die sides will not meet when extended because the clearance is insufficient.

2. Continued downward descent of the punch causes elongation of the cracks. The uncut area between them will be broken in a secondary fracture.
3. At the bottom of the stroke, the secondary fracture has occurred. A second burnished band has been produced on the blank edge and on the strip edge. The characteristic contour shown in inset **A** has been formed.

b) Excessive Clearance

The inset at **A** in Figure 3.4b shows the four effects of excessive clearance:

- Radius **R** is considerably larger than when correct clearance has been applied.
- Burnished band **D** is narrower.
- The break-off angle **B** is excessive.
- A burr **C** is left on the blank.

Referring to Figure 3.4b:

1. This shows the cutting edges of a punch and die in partial penetration. Cracks have begun to form at opposite sides
2. Continued downward descent of the punch causes elongation and widening of the cracks. Their alignment is fairly good
3. At the bottom of the stroke, separation has occurred, leaving the characteristic blank edges shown in the inset at **A**.

When a die is provided with excessive clearance, less pressure is required to effect cutting of the material. For this reason, more clearance is often specified for blanking the heavy gages of stock to reduce pressure on the press.

3.4 DETERMINING STRIP

The first step in the actual production of stampings is to order standard-size sheets of the proper size and thickness from the mill. These are then sheared into strips, as described above. The widths of the strips into which the sheets are to be cut is specified by the die design department. Therefore, let us go over the steps taken in determining strip width so that your understanding will be complete.

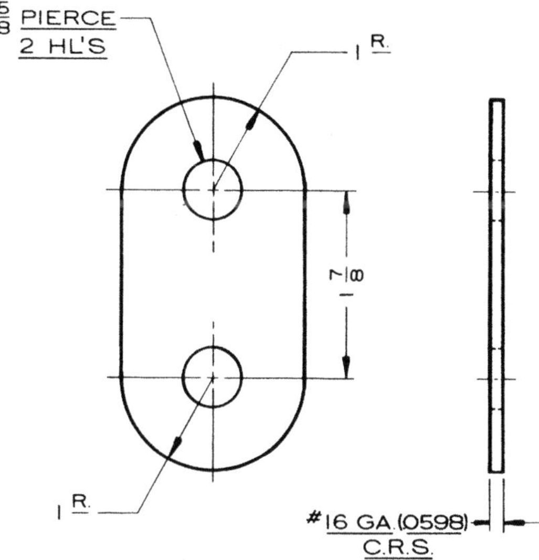

Figure 3.5 Typical part drawing.

Figure 3.5 shows a part drawing of a typical representative stamping to be produced in a pierce and blank die.

3.4.1 Blank Layout

Figure 3.6 shows the two possible ways of running the strip of a typical representative

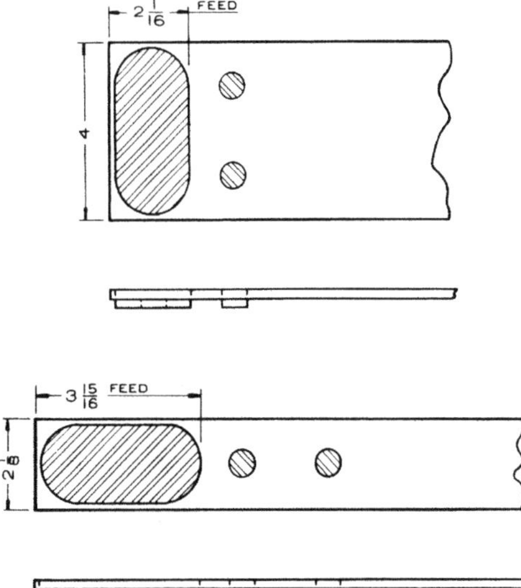

Figure 3.6 Blank layouts necessitating either wide or narrow strips.

stamping (Figure 3.5) through the die. The blanks may be positioned the wide way, necessitating a wide strip, or they may be run the narrow way, permitting the use of a narrower strip. These are called "blank layouts" and it is important that you understand exactly what is meant by the term. A blank layout shows the way in which the designer *proposes to produce the blank.* For both the wide-run and narrow-run layouts, two holes are to be pierced at the first station and the part is to be blanked out at the second station. It is customary to show small piercing punches in solid black. Section lines are applied through larger piercing punches and through blanking punches, as shown. The strip width and the feed are given directly on the blank layout.

Now let us go through the first steps taken in the production of blanks in cut-off dies. Two of the sides of such blanks are originally sides of the material strip, and no scrap bridge is produced as in blanking dies. Figure 3.7 shows a representative stamping having the parallel sides typical of blanks suitable for production in cut-off dies.

Figure 3.8 shows two blank layouts for producing the stamping (Figure 3.7) in a cut-off die. At view **A,** the part is positioned the wide way in the strip. The edges of the strip are notched at the first station and a rectangular hole is punched. The blank is cut off from the strip at the second station. At view **B,** the part is positioned the narrow way in the strip. Observe how notching punches are sectioned. The heels **C,** which prevent deflection of the punches, are shown, but not sectioned. At **D,** short 45 degree

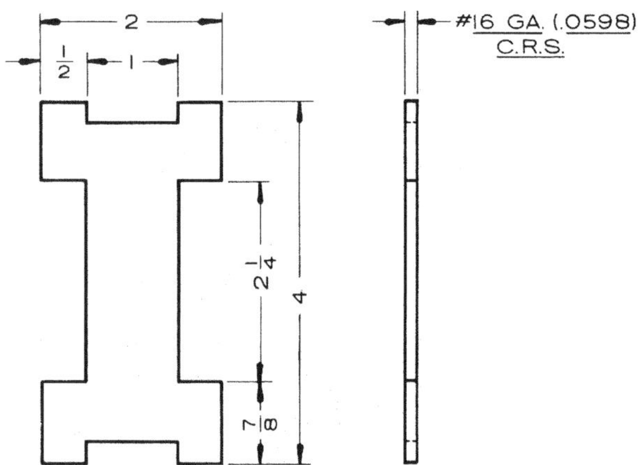

Figure 3.7 Typical part for production in cut-off die.

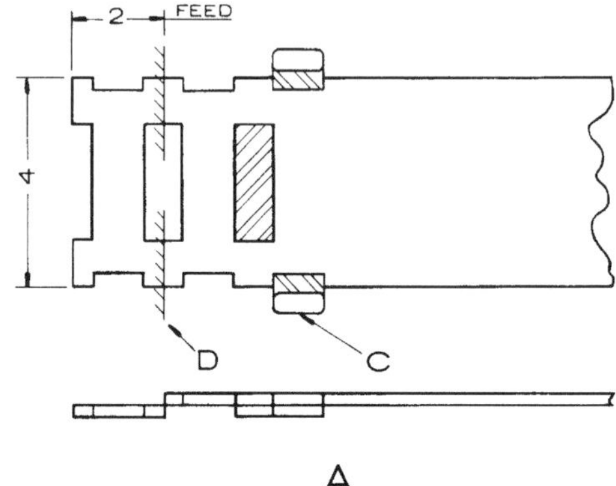

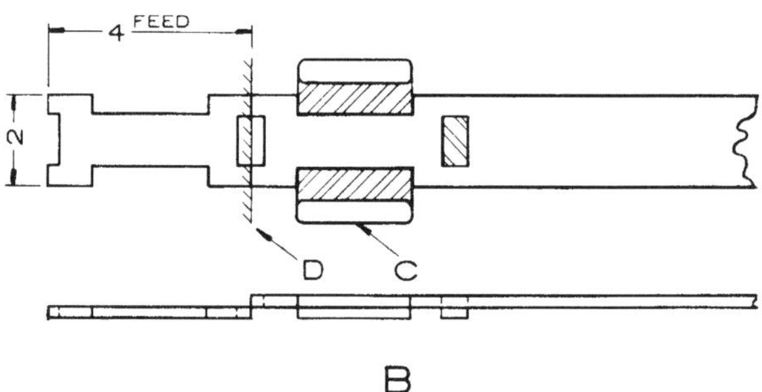

Figure 3.8 Blank layout for part shown in Figure 3.7 run, either the wide (A) or narrow (B) way.

lines and a long vertical line represent the "cut off" line.

Blank layouts are drawn to explain the proposed operation of a die to others. When die designers are given a part print of a stamping for which a die is to be designed, they proceed to lay out a suitable scrap strip. Then they section significant punches and add cut-off lines to make the proposal layout clearer. This is the blank layout, and it must be approved by the group leader or chief engineer before design of the die is begun. When an outside engineering office is doing the work for a manufacturing company, the blank layout is submitted to the customer for approval.

3.4.2 Stripper Sheet

Sizes of sheets as they are manufactured by the mill are given in steel catalogs. Here is a representative list for #18 Gage (0.0478 in. or 1.2 mm) cold-rolled steel:

30 in. × 96 in. (762 mm × 2438 mm)
30 in. × 120 in. (762 mm × 3048 mm)

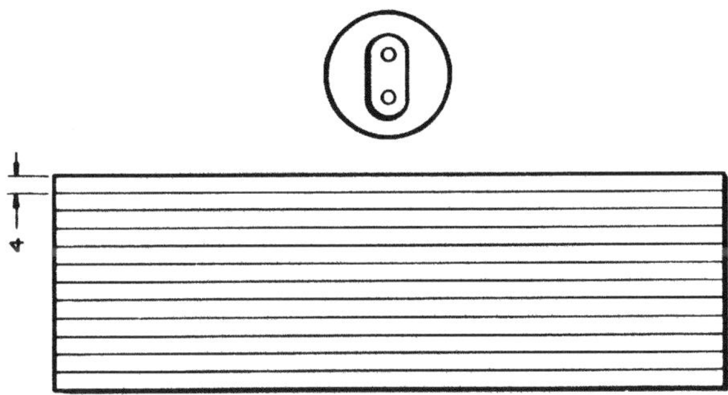

Figure 3.9 Number of strips obtainable with wide strip blank layout.

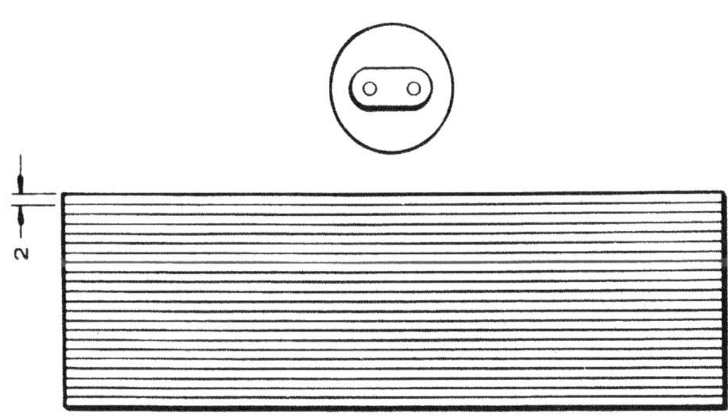

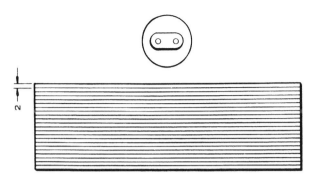

Figure 3.11 Number of strips obtainable with narrow strip blank layout.

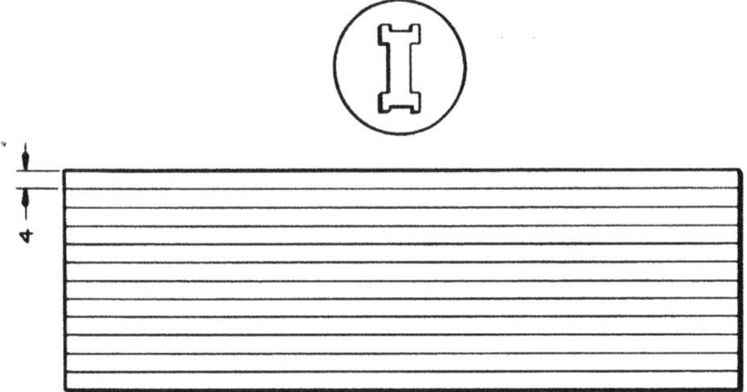

Figure 3.10 Strips per sheet with wide strip blank layout for production in a cut-off die.

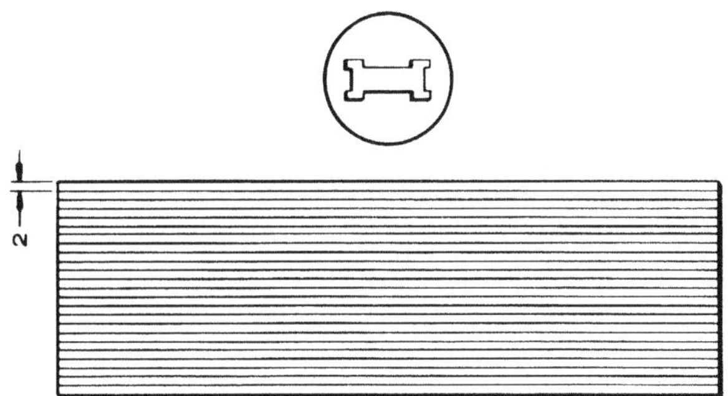

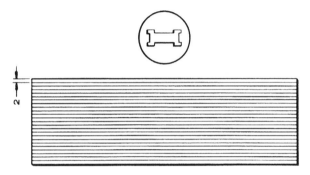

Figure 3.12 Strips per sheet with narrow strip blank layout for production in a cut-off die.

36 in. × 96 in. (914 mm × 2438 mm)
36 in. × 120 in. (914 mm × 3048 mm)
48 in. × 96 in. (1219 mm × 2438 mm)
48 in. × 120 in. (1219 mm × 3048 mm).

The next step is to select the sheet that will be most economical, that is, the sheet from which a maximum number of strips can be cut, leaving a minimum amount of waste.

a) Wide Run

Strip width is taken from the blank layout. Divide the value given into the values for "width of sheet" in the steel catalog, and compare to determine which sheet leaves the smallest remainder. Figure 3.9 shows a sheet 48 by 120 inches (1219 mm × 3048 mm) divided into strips when the typical representative blank is run the wide way.

Figure 3.10 shows the sheet divided into strips for producing parts in a cut-off die when the blank is run the wide way.

b) Narrow Run

Next, we must know how many blanks are produced per sheet with the blanks positioned the narrow way in the strip. With blanks arranged the narrow way, more strips are cut from the sheet, but fewer blanks are contained in each strip.

Figure 3.11 shows the same sheet divided into strips when the typical representative blanks are to be run the narrow way. More strips are produced from the same size of sheet.

Figure 3.12 shows the sheet divided into strips for producing a part in a cut-off die when the blank is run the narrow way.

3.4.3 Strip Layout

After it has been decided how the blanks are to be run (wide or narrow way), a stock layout is prepared complete with the following dimensions:

- Strip width. This dimension is used in selecting the proper width of sheet from which strips are to be cut.

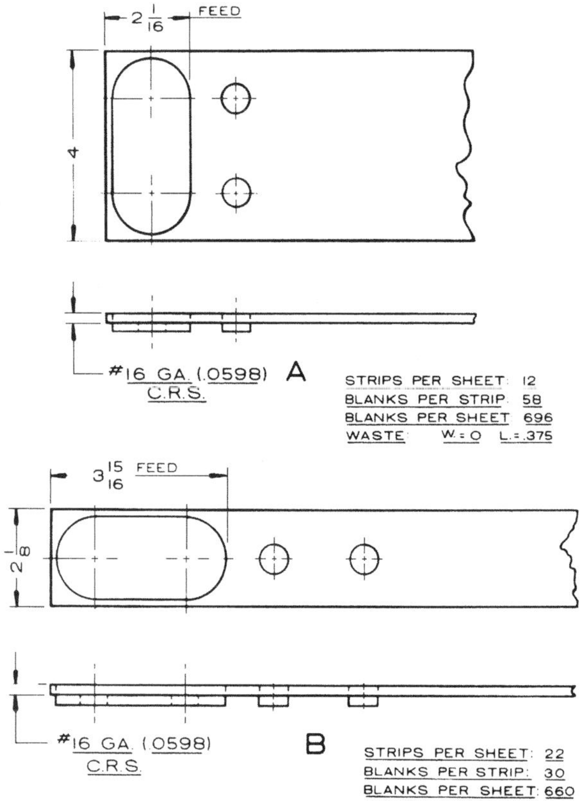

Figure 3.13 Complete strip layouts for blanks run either the wide (A) or narrow (B) way.

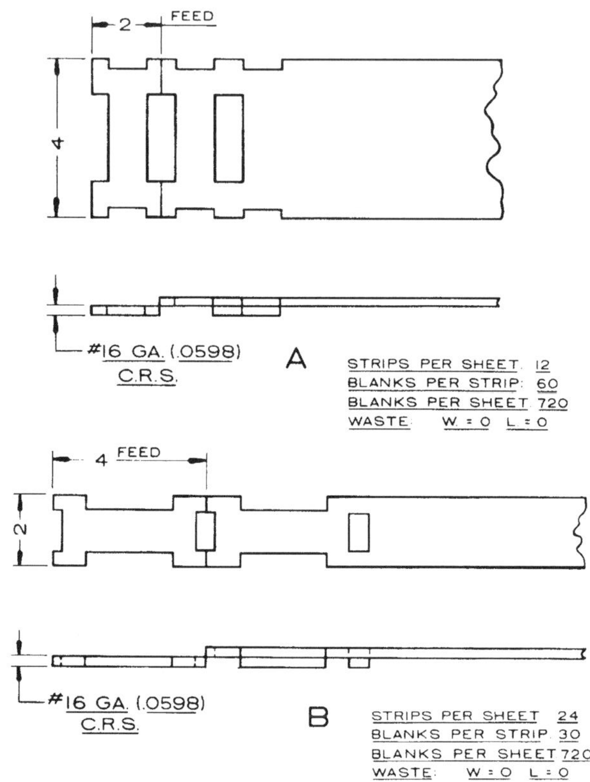

Figure 3.14 Strip layouts for blanks run either the wide (A) or narrow (B) way.

- Feed. This is the amount of travel of the strip between stations. This dimension is used in selecting the proper length of sheet.

Figure 3.13 shows complete strip layouts for the typical representative blanks run either the wide (view **A**) or narrow (view **B**) way. Two views are applied, ordinarily. These are exactly the views of the strip that will be drawn on the die drawing except that an end view of the strip is added to the die drawing. The die is then actually designed around these views.

View **A** illustrates a strip layout for a blanking die in which the blank is run the wide way. View **B** shows a layout in which the blank is run the narrow way. For this particular job, more blanks per sheet are produced when the blanks are positioned the wide way and there is less waste. Therefore, all else being equal, this method of positioning the blanks would be selected.

Figure 3.14 shows strip layout for production of parts in cut-off die. The strip layout is prepared and copies are sent to the purchasing department and to the shear department. From the layout, sheets are ordered and, upon delivery, they are sheared to the strip width given on the layout. View **A** shows a representative strip layout for a blank for a cut-off die positioned the wide way, and view **B** shows a layout for a blank positioned the narrow way.

For this job we find that exactly the same number of blanks are produced with blanks positioned the narrow way as for wide-run positioning; there is no waste in either method. When blanks can be run either way, select the wide run method for three reasons:

- Fewer cuts will be necessary for producing the strips.
- The feed is shorter when running strips through the die, thus reducing the time required.

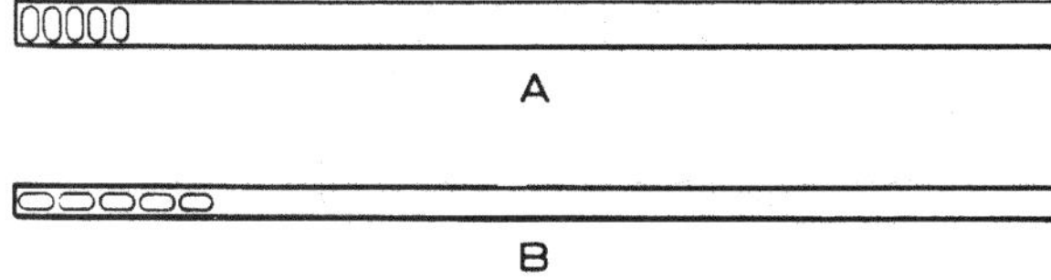

Figure 3.15 Strips ready for feeding either the wide (A) or narrow (B) way.

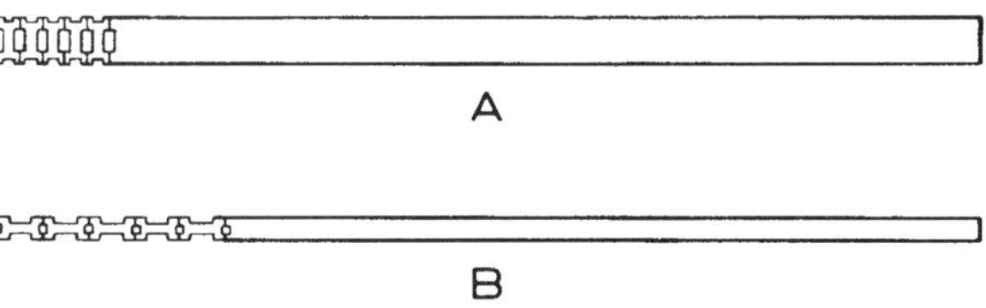

Figure 3.16 Strips ready for feeding either the wide (A) or narrow (B) way in a cut-off die.

- More blanks are produced per strip and fewer strips have to be handled.

3.4.4 Strips

Figure 3.15 shows a strip ready for feeding either the wide or narrow run. View **A** shows one of the strips ready to be fed through the die with blanks to be removed from it positioned the wide way. At view **B,** the blanks are positioned the narrow way in the strip. Five of the parts have been blanked out of each strip. After the strip is run completely through the die, only a narrow scrap bridge is left.

Figure 3.16 shows a strip ready for feeding either the wide or narrow run in a cut-off die. View **A** shows one of the strips ready to be fed through a die, with blanks to be removed from it positioned the wide way. At view **B,** the blanks are positioned the narrow way. Five blanks are shown in each strip. Because they are run in cut-off dies, no scrap bridge is produced.

3.5 METHODS FOR PRODUCING STRIPS

3.5.1 Shearing

The oldest and simplest method of producing metal strips is by shearing. In the steel mill, metal is formed into large sheets by rolling and trimming. A sheet that is to be cut into strips is introduced under the blade of a shear. Gages register

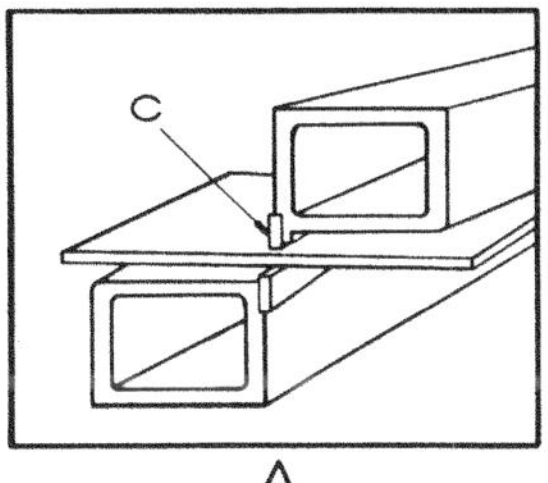

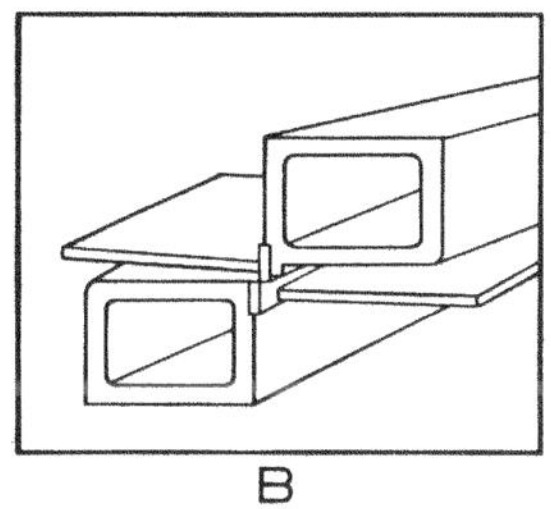

Figure 3.17 Producing metal strips from sheet by shearing.

the edges of the sheet for cutting correct widths of strips. Descent of the shear blade causes each strip to be parted from the sheet. Advancing the sheet against the gages brings it into position for cutting the next strip and this process is repeated until the sheet has been cut entirely into strips. Figure 3.17 at **A** shows a sheet in position under the shear blade **C** ready to be cut. At **B,** the blade has descended and the strip has been cut from the sheet.

The power shear can cut material in any direction—lengthwise of the sheet, across the sheet, or at any angle.

3.5.2 Slitting

Slitting machines (Figure 3.18) are also used for producing material strips. In slitting operations, the sheet is fed though rotating cutting rolls, and all strips are cut simultaneously. In the illustration, cutting rolls **A** are mounted the proper distances apart on arbors **B**. The cutting edges of the rolls are separated by the required amount of clearance to effect cutting of the material as shown in inset **C**. Turning the rolls under power causes the sheet to advance, and it is cut into strips. As many as 20 or more strands can be cut at one time. In other types of slitters, the

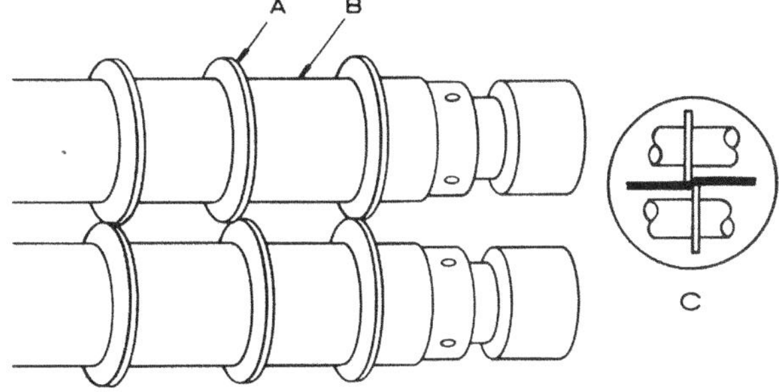

Figure 3.18 Cutting rolls for slitting strips from sheet.

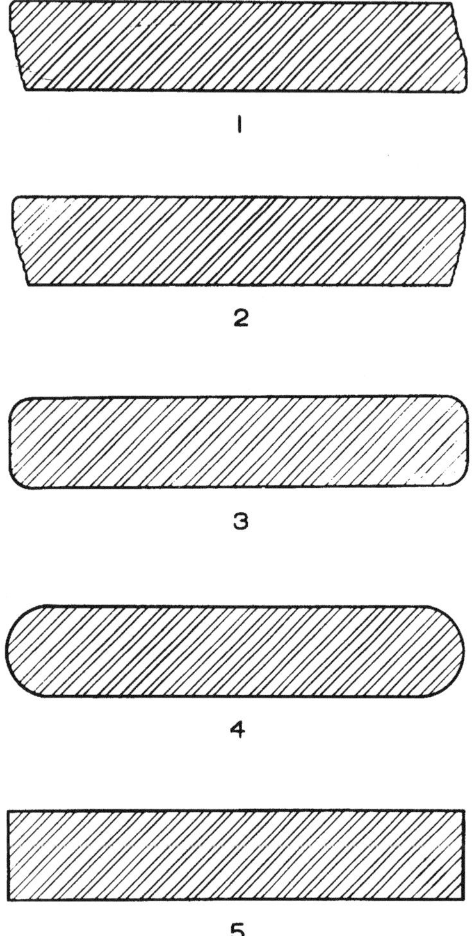

Figure 3.19 The various edge contours shown are the result of different production processes.

sheets are pulled through the rolls instead, and the rolls are free to turn.

Slit strips are very accurate in width, flatness, and parallelism of sides because accuracy is built into the machine instead of depending upon the operator. Unlike the shear, which can cut strips only as long as the blade, the slitter will cut continuously to any length, without limit.

3.5.3 Edge Contour (Contour of the Edge of a Strip)

The contour of the edge of a strip (Figure 3.19) depends upon the process by which the strip is produced. Five contours may be recognized:

1. Strips produced in a shear have the burnished bands along the edges on *opposite* sides of the strip. If burrs are produced because of dull cutting edges, they will also occur on opposite sides of the strip. In addition, sheared strips often become spiraled or curved because the upper blade of the shear is at an angle to the lower blade. This makes the strips difficult to feed through the die unless they are first straightened.
2. Strips produced in the slitter have the burnished bands on the same side of the strip. Blanks produced from these strips in cut-off dies have a better appearance and they are fed more easily because they are straighter. Sheared strips or slit strips may be produced in the shear department of the plant, or they may be ordered directly from the mill.
3. Mill-edge strips have a radius at each corner. They are produced by rolling sheared or slit strips at the mill. Mill-edge strips are used for long stampings, such as for handles, shelf brackets, and other parts where sharp edges would be objectionable.
4. Rolled-edge strips have a full radius at each side, rolled at the mill. They are used for parts where appearance is a deciding factor, such as in ornamental grills, gratings, and the like.
5. Square-edge strips are ordered from the mill when the sides of the strips must be square and smooth. The widths of these strips are held very accurately. Square-edge strips are also specified when blanks are to be bent or formed edgewise. The square edges prevent cracking or splitting in the bending or forming operation.

THE BLANK

4.1 DEFINITION AND TYPES OF BLANKS

A blank is a piece of flat steel or other material cut to any outside contour. The thickness of a blank may range between 0.001 and 0.500 inch (0.025 mm and 12.7 mm) or more depending on its function. However, most stampings are between 0.025 inch and 0.125 inch (0.6 and 3.2 mm) in thickness.

Some blanks have simple round, square, or rectangular contours. Others may be very irregular in shape. Many blanks are subsequently bent, formed, or drawn. It is important to realize, however, that when we refer to a blank, what is meant is the flat part before any deformation has been applied.

There are only two basic types of blanks (see Figure 4.1):

1. Blanks having straight, parallel sides, two of which are originally sides of the material strip (see view 1). Small blanks of this type are produced in cut-off dies. Large blanks are produced by square-shearing and trimming.
2. Blanks having irregular contours cut entirely out of the material strip (see view 2). When they are required in quantity, such blanks are produced in blanking dies. When only a few blanks are required, they may be shaped by contour sawing, nibbling, routing, or other machining operations.

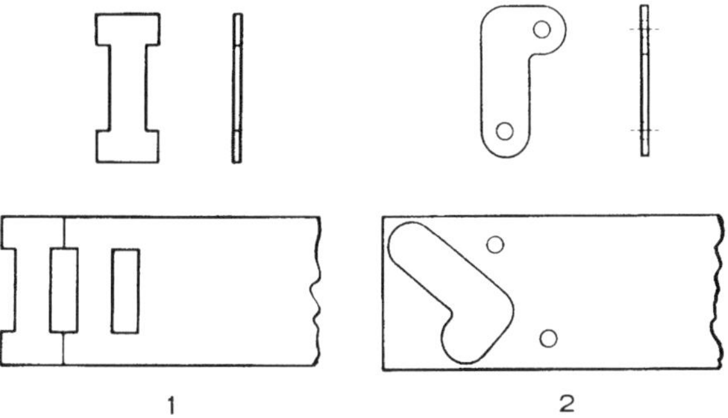

Figure 4.1 Two basic types of blanks.

4.2 PRODUCING BLANKS

To select the best method of producing a particular blank, consider five factors:

1. Contour

If the blank has two parallel sides, determine if it can be produced in a cut-off operation. The width between the parallel sides would then become the width of the strip. Four advantages are realized when cut-off dies are used:

- There is a minimum waste of material.
- Cut-off dies cost less to build.
- Faster press speeds are possible.
- There is no scrap strip to handle.

After you have determined that a blank can be produced in a cut-off operation, consider three additional factors before making a final decision:

Accuracy in strip width. Sheared strips cannot be held to closer accuracy than ±0.010 inch (0.25 mm). If the width dimension between parallel sides of the blank must be held to closer limits, discard the idea of using a cut-off die.

Accuracy of the blank. If the blank must be held to close limits, it should be produced in a blanking die, regardless of the number of straight sides that it may have.

Flatness. If the part print contains the note "MUST BE FLAT," you should plan to design a blanking die because it will produce considerably flatter parts. Cut-off dies produce blanks by a series of piercing, trimming, and cut-off operations. Uncut portions can become distorted, especially for heavier gages of strip. In blanking dies, the entire periphery is cut in one operation and distortion cannot occur.

Blanking dies produce flat, accurate parts. Whatever accuracy has been built into the die is duplicated in the blanks; each is identical to every other blank that the die produces. This is true because the entire blank contour is cut and none of the edges of the strip form any edge of the blank. Blanking is the most widely used method of producing blanks from sheet materials.

If the stamping is intricate and is to be produced complete in a progressive die, the contour of the blank may be formed by trimming away portions of the strip at one or more of the stations.

2. Size

Consider the size of the blank in relation to the number of parts required. This is especially important for large blanks because large dies are very costly to build. Determine if shearing and trimming would do the job, especially if production requirements are low.

3. Accuracy

Study the part print carefully to determine the degree of accuracy required in the blank. Very accurate blanks have to be produced in compound dies in which all operations are performed simultaneously at one station. Blanks requiring a lesser degree of accuracy may be produced in more economical two-station dies.

4. Number required

This information is taken from the design order and it often determines the type of die to be designed, as well as the class of die.

5. Burr side

The burr side must be known when blanks are to be shaved or burnished in a subsequent operation. The same applies for blanks that are to be assembled into other components and those that are to have components assembled into them. The presence of burrs at engaging edges can slow down assembly operations considerably.

4.2.1 Methods of Producing Blanks

Let us now gain an understanding of the various methods of producing blanks. We will begin by considering ways in which blanks may be shaped without the use of dies. These low-cost, but relatively slow methods are employed when only a few blanks are required and it would be uneconomical to design and build special dies for producing them.

a) Circle Shearing

Large, round blanks may be circle-sheared when quantities required are moderate. Square blanks are clamped in the center of a circle shearing machine. Two disk-shaped cutters are adjusted to the required radius. They apply rotation to the blank and, at the same time, cut it to a circular shape.

For larger quantities, it is less expensive to order round blanks precut to the required diameter. Steel companies stock various sizes of round blanks, or they can supply them cut to special sizes.

b) Contour Sawing

When only a few blanks are required, their contours may be laid out directly on sheet material. After lines have been scribed, the blanks are sawed out in a metal-cutting bandsaw. For contour sawing a number of blanks requiring greater accuracy, a short stack of square or rectangular blanks are clamped in a vise, then tack-welded together at several places around the edges. The outline of the blank is laid out on the upper sheet and the blanks are sawed directly to this outline. Thus, all the blanks are identical in contour.

c) Nibbling

The nibbling machine operates by reciprocating a punch up and down at about five strokes per second. The punch is provided with a pilot long enough so it is not raised above the material being cut. As the sheet is moved, the punch cuts a series of partial holes that overlap each other. A jagged edge is left around the edges of the blank and the sharp corners left by the punch must be die-filed after the nibbling operation. The nibbling process is used to produce blanks when only one, two, or a few are required.

d) Routing

A routing machine is provided with a long radial arm that can travel over a large area. Mounted at the outer end of the arm is a rapidly revolving cutting tool, similar to an end-mill cutter, that can cut its way through a stack of blanks. The router bit, as the cutting tool is called, rotates at about 15,000 revolutions per minute and is guided by a template to produce blanks identical to the template. Routing large aluminum blanks is common practice in the aircraft and missile industries.

e) Flame Cutting

Flame cutting or torch cutting means the cutting of thick blanks by the use of an acetylene torch. In operation, the torch heats the metal under its flame tip until it melts. Compressed air then blows the molten metal out, forming a narrow channel called the "kerf." The width of the kerf ranges from 5/64 inch to 1/8 inch (2 mm to 3.2 mm) depending upon stock thickness and the speed of the torch. For producing thick blanks in quantity, a template guides the torch by means of a pantograph. Flame cutting is employed for cutting blanks ranging from 1/4 to 1 inch (6.35 to 25.4 mm) or more in thickness.

Holes in blanks can also be torch cut by the same method. Flame cutting leaves the edges somewhat rough and ridged. However, such edges are satisfactory for some parts for trucks, tanks, ships, and other similar applications.

f) Water-Jet Cutting

Water-jet cutting is a process used to cut materials using a jet of pressurized water. There are two main steps involved in the water-jet cutting process. First, the ultra-high pressure pump or intensifier pressurizes water to pressure levels above 60,000 PSI (400 MPa) to produce the energy required for cutting. Second, water is then focused through a small, precious stone orifice to form an intense cutting stream. The nozzle diameters used to achieve these pressures range from 0.002 inch (0.05 mm) to 0.04 inch (1 mm). The stream moves at a velocity of up to 2.5 times the speed of sound, depending on how the water pressure is exerted.

As in flame cutting, kerf is an important term in water-jet machining. The kerf is the width of the actual water-jet cutting beam. Depending on the nozzle, the kerf width for an abrasive jet ranges from 0.020 inch to 0.060 inch (0.5 mm to 1.5 mm). Plain water jets with no abrasives have a narrow kerf ranging from 0.005 inch to 0.014 inch (0.13 mm to 0.35 mm).

A water jet can cut both hard and soft materials. Soft materials are cut with water only, whereas hard materials require a stream of water mixed with fine grains of abrasive garnet. This method is used in cutting processes of materials including titanium, stainless steel, aluminum, exotic alloys, composites, stone, marble, floor tile, glass, automotive door panels, gaskets, foam, rubber, insulation, textiles, and many others.

Cutting speed is determined by several variable factors, including the edge quality desired. Variables such as amount of abrasive used, cutting pressure, size of orifice and focus tube, and pump horsepower can be adjusted to produce the desired results, whether your priority is speed or the finest cut.

Speed and accuracy also depend on material texture, material thickness, and the cut quality desired. In case of rubber and gasket cutting, water-jet motion capabilities would allow traversing at 0.1 to 200 inch/min (0.0025 to 5 m/min).

g) LASER Cutting

The acronym LASER stands for Light Amplification by Stimulated Emission of Radiation. How does laser cutting work? Laser cutting can be compared to cutting with a computer-controlled miniature torch. Industrial laser cutting is designed to concentrate high amounts of energy into a small, well-defined spot. Typically the laser cutting beam is approximately 0.003 to 0.006 inch (0.07 to 0.15 mm) in diameter in short wavelength lasers. The distance between the nozzle and the material is approximately 0.2 inch (5 mm). The material thickness at which cutting or processing is economical is up to 0.4 inch (10 mm). The resulting heat energy created by the laser melts or vaporizes materials in this small-defined area, and a gas (or a mixture of gases), such as oxygen, CO_2, nitrogen, or helium, is used to blow the vaporized material out of the kerf. The beam's energy is applied directly where it is needed, minimizing the heat's effect on the zone surrounding the area being cut.

There is almost no limit to the cutting path of a laser. The point can move in any direction. Small diameter holes that cannot be made with other machining processes can be done easily and quickly with a laser. The process is forceless. The part keeps its original shape from start to finish.

This method is ideal when production quantities or prototypes do not justify producing tooling for stamping or die cutting.

Lasers can cut at very high speeds. The speed at which materials can be processed is limited only by the power available from the laser. Laser cutting is a very cost-effective process with low operating and maintenance costs and maximum flexibility.

4.2.2 Square Shearing

Large, straight-sided blanks are produced in the shear by cutting sheets into strips, then cutting the strips to required lengths or widths. Blanks larger than 8 by 10 inches (203 by 254 mm) and composed primarily of straight sides are ordinarily produced by shearing because of the high cost of large dies.

Blanks cut in a modern shear can be held to an accuracy of 0.005 inch (0.125 mm). Four factors govern shearing accuracy:

- The shear must have sufficient rigidity to withstand the cutting load without deflection or spring.
- Knife clearance must be set correctly and proper rake selected to reduce twist, camber, or bow. *Rake* is the angle of the upper knife in relation to the horizontal lower knife of the shear. *Twist* is spiraling of the strip; it is more severe in soft, narrow, or thick strips than it is in hard, wide, or thin strips. *Camber* is curvature along the edge in the plane of the strip whereas *bow* is curvature perpendicular to the surface of the strip.
- Good gaging practice must be followed.
- The sheet must be held down securely while shearing occurs.

For producing square and rectangular blanks shown in the upper illustration of Figure 4.2, the sheet is first cut into strips to length **A** of the blanks. The strips are then run through the shear again and cut into blanks having width **B**. Here is the method of listing operations on the route sheet:

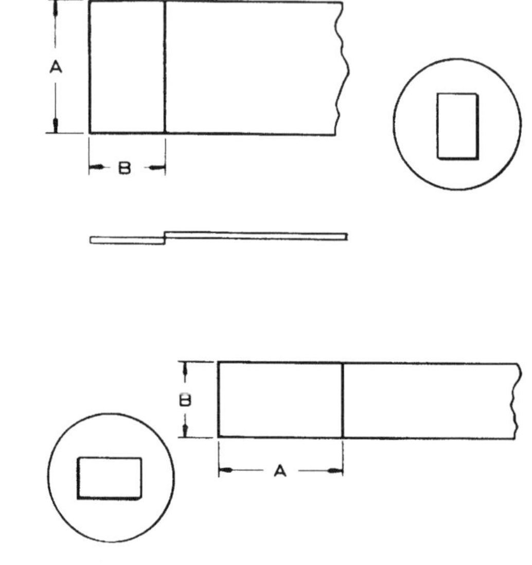

Figure 4.2 Rectangular blanks sheared from wide and narrow strips.

- Operation No. 1. Shear to length (**A**)
- Operation No. 2. Shear to width (**B**)

When the grain of the material must run lengthwise of the blank for extra stiffness, the sheet is cut into strips to width **B** of the blanks (Figure 4.2, lower illustration). The strips are then run through the shear again and cut into blanks having length **A**. On the route sheet, operations are listed as follows:

- Operation No. 1. Shear to width (**B**)
- Operation No. 2. Shear to length (**A**)

4.2.3 Triangular Blanks

Triangular blanks (Figure 4.3) are produced by shearing square or rectangular blanks, then splitting them to produce two blanks. In the upper illustration:

- Operation No. 1. Shear to length (**A**)
- Operation No. 2. Shear to double width (**B**)
- Operation No. 3. Split (**C**)

Although operation No. 2 states "Shear to double width," this does not mean that the strip is to be cut twice the width of a single blank. It simply means that the square or rectangular

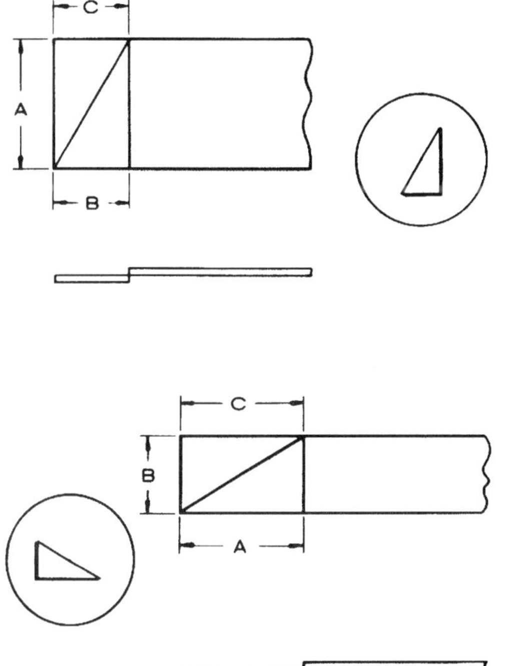

Figure 4.3 Triangular blanks made by shearing rectangular blanks sheared from wide and narrow strips.

blanks are to be made wide enough so splitting will produce two full blanks.

For running strips the narrow way (Figure 4.3, lower illustration), operations are listed as follows:

- Operation No. 1. Shear to width (**B**)
- Operation No. 2. Shear to double length (**A**)
- Operation No. 3. Split (**C**)

4.2.4 Angular Edge Blanks

Wider blanks having one angular edge (Figure 4.4) can be produced by the same method employed for triangular blanks. For the upper illustration:

- Operation No. 1. Shear to length (**A**)
- Operation No. 2. Shear to double width (**B**)
- Operation No. 3. Split (**C**)

The function of some blanks renders it necessary to run them the narrow way (Figure 4.4, lower illustration). Operations are then as follows:

- Operation No. 1. Shear to width (**B**)
- Operation No. 2. Shear to double length (**A**)
- Operation No. 3. Split (**C**)

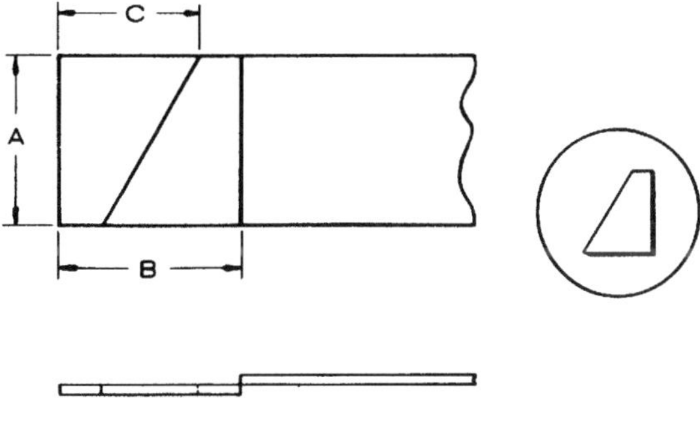

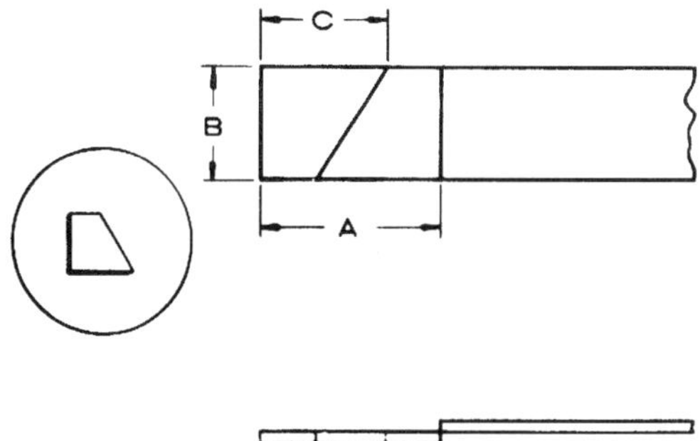

Figure 4.4 Angular edge blanks made by shearing rectangular blanks sheared from wide and narrow strips.

Added cuts

One or more extra cuts may be required to complete the blanks (Figure 4.5).
Here is the order of operations for the blank in the upper inset:

- Operation No. 1. Shear to length (**A**)
- Operation No. 2. Shear to double width (**B**)
- Operation No. 3. Split (**C**)
- Operation No. 4. Trim (**D**)

In the lower illustration:

- Operation No. 1. Shear to width (**B**)
- Operation No. 2. Shear to double length (**A**)
- Operation No. 3. Split (**C**)
- Operation No. 4. Trim (**D**)

4.2.5 Parallelogram Blanks

Blanks in the shape of an angular parallelogram (Figure 4.6) are produced by shearing with

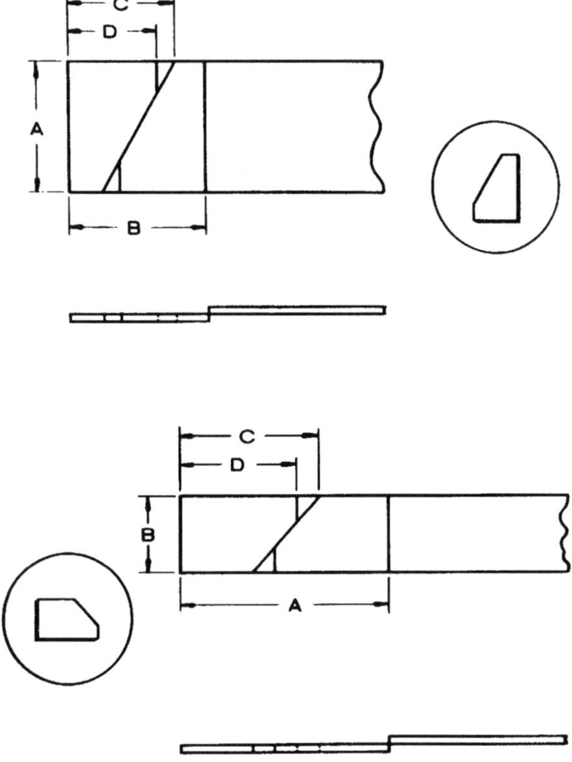

Figure 4.5 Blanks produced by added cutting of angular edge pieces.

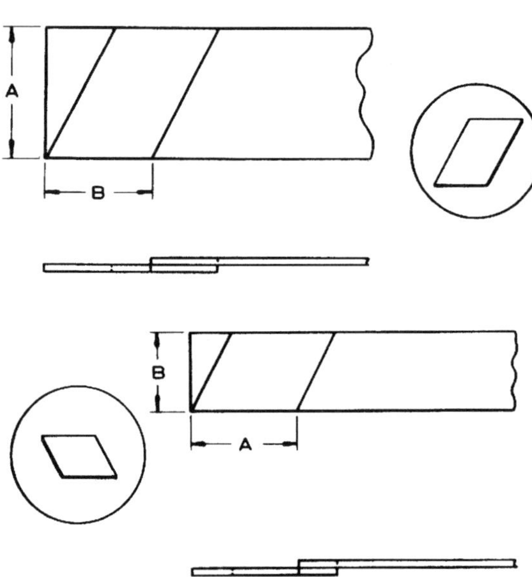

Figure 4.6 Parallelogram-shaped blanks sheared from wide and narrow strips.

the strip positioned at an angle to the shear blade. For the wide strips, upper illustration:

- Operation No. 1. Shear to length (**A**)
- Operation No. 2. Shear to width (**B**)

For narrow strips, lower illustration:

- Operation No. 1. Shear to width (**B**)
- Operation No. 2. Shear to length (**A**)

4.2.6 Triangular Blanks

Triangular blanks with an acute angle at each of the three apexes (see Figure 4.7) are produced by splitting a parallelogram.

For wide strips (Figure 4.7, upper illustration):

- Operation No. 1. Shear to length (**A**)
- Operation No. 2. Shear to double width (**B**)
- Operation No. 3. Split (**C**)

For narrow strips (Figure 4.7, lower illustration):

- Operation No. 1. Shear to width (**B**)
- Operation No. 2. Shear to double length (**A**)
- Operation No. 3. Split (**C**)

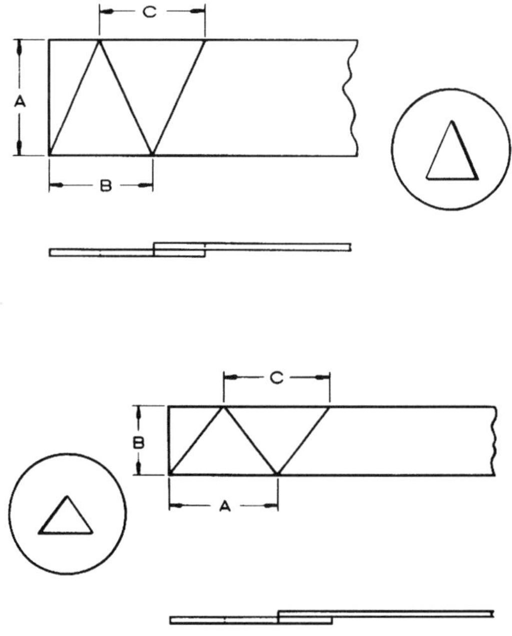

Figure 4.7 Acute-angle triangular blanks made by shearing parallelogram-shaped blanks sheared from wide and narrow strips.

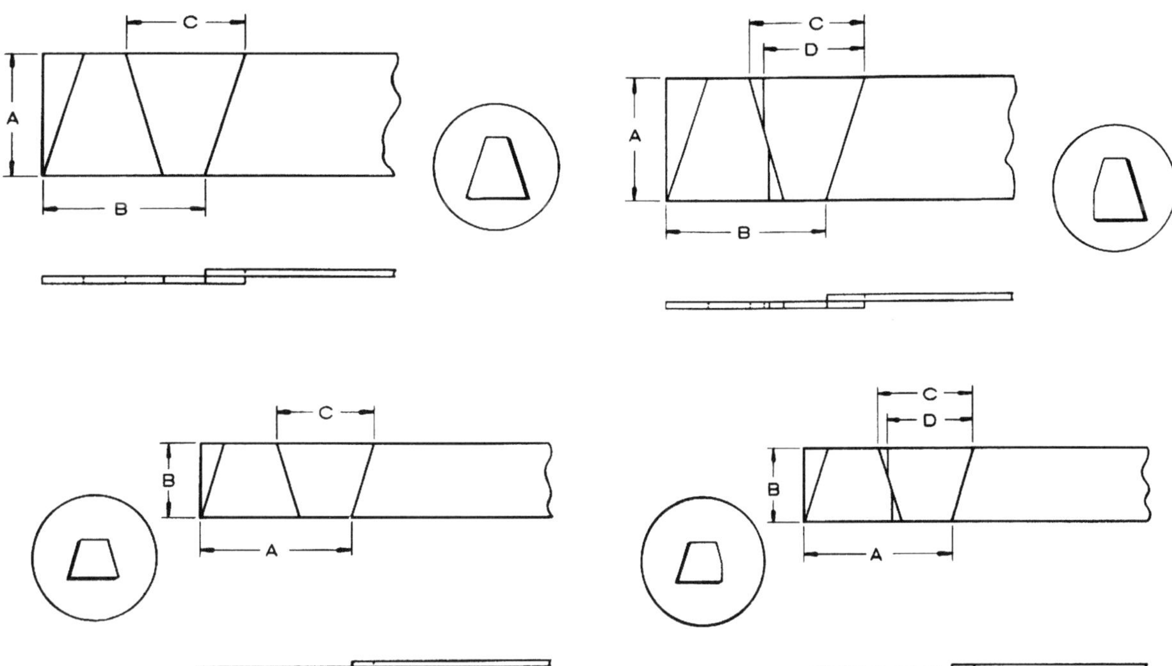

Figure 4.8 Trapezoidal blanks are sheared in a manner similar to acute angle triangular blanks.

Figure 4.9 Blanks produced by added cutting of trapezoidal blanks.

4.2.7 Trapezoidal Blanks

Large trapezoidal blanks (Figure 4.8) are produced by employing the same method as triangular blanks.

For wide strips (Figure 4.8, upper illustration):

- Operation No. 1. Shear to length (**A**)
- Operation No. 2. Shear to double width (**B**)
- Operation No. 3. Split (**C**)

For narrow strips (Figure. 4.8, lower illustration):

- Operation No. 1. Shear to width (**B**)
- Operation No. 2. Shear to double length (**A**)
- Operation No. 3. Split (**C**)

Added cuts

One or more extra cuts (Figure 4.9) may be required to complete the blanks.

For wide strips (Figure 4.9, upper illustration):

- Operation No. 1. Shear to length (**A**)
- Operation No. 2. Shear to double width (**B**)
- Operation No. 3. Split (**C**)
- Operation No. 4. Trim (**D**)

For narrow strips (Figure 4.9, lower illustration):

- Operation No. 1. Shear to width (**B**)
- Operation No. 2. Shear to double length (**A**)
- Operation No. 3. Split (**C**)
- Operation No. 4. Trim (**D**)

4.2.8 Notching

Irregular notches or cuts are applied in trimming dies after the straight sides of the blanks have been sheared (Figure 4.10). It is less expensive to design and build a die for trimming small portions of a blank than it would be to design and build a die for producing the entire blank. This is because the length of cutting edges is shorter for trimming than for blanking and less fitting is required. Also, less material is required in the strip when blanks are sheared because there is no scrap bridge.

In this connection, you should know exactly what is meant by the words *notching* and *trimming*. Notching is the operation of cutting small portions from the edges of a strip or blank. The area of such portions is no greater than that of average pierced holes. Trimming means the removal of larger portions of metal to alter the form in the area of the trimmed contour. The difference

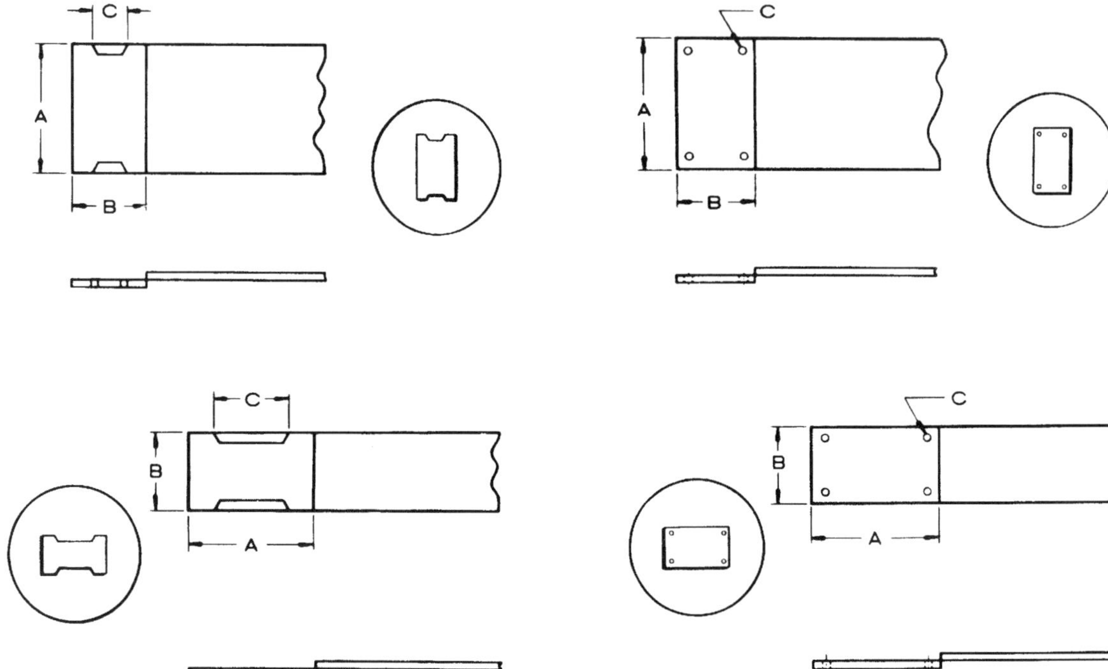

Figure 4.10 Blanks sheared and trimmed from wide and narrow strips.

Figure 4.11 Blanks sheared and pierced from wide and narrow strips.

is one of degree. Trimming thus means the reshaping of blanks on a larger scale than is accomplished by simple notching.

Operations for wide strips (Figure 4.10, upper illustration) are:

- Operation No. 1. Shear to length (**A**)
- Operation No. 2. Shear to width (**B**)
- Operation No. 3. Trim (**C**)

Operations for narrow strips (Figure 4.10 lower illustration):

- Operation No. 1. Shear to width (**B**)
- Operation No. 2. Shear to length (**A**)
- Operation No. 3. Trim (**C**)

4.2.9 Piercing

Holes may be pierced in the blanks in a turret punch when required quantities are low. For greater production, a special piercing die is designed and built for the job. After they have been sheared, the blanks are hand-fed into the piercing die (Figure 4.11).

Operations for wide strips (Figure 4.11 upper illustration) are:

- Operation No. 1. Shear to length (**A**)
- Operation No. 2. Shear to width (**B**)
- Operation No. 3. Pierce (**C**)

Operations for narrow strips (Figure 4.11 lower illustration) are:

- Operation No. 1. Shear to width (**B**)
- Operation No. 2. Shear to length (**A**)
- Operation No. 3. Pierce (**C**)

4.3 BLANKING FORCE

In cutting any material, a force acting on the area to be sheared (called the *shear stress*) is applied to the material. The material offers resistance to separation. Its molecular structure resists the shear stress applied to it and the amount of resistance is called the *shear strength*. To effect cutting, the shear stress applied to the material

Table 4.1 Shear Strength for Various Materials

MATERIAL		SHEAR STRENGTH PSI	MPa
STEEL: 10% Carbon content	Hot-rolled	35,000	241
	Cold-rolled	45,000	310
STEEL: 20% Carbon content	Hot-rolled	45,000	310
	Cold-rolled	55,000	379
STEEL: 30% Carbon content	Hot-rolled	55,000	379
	Cold-rolled	65,000	448
STAINLESS STEEL		60,000	413
SILICON STEEL		65,000	448
TIN		5,000	34
ALUMINUM		10,000	68
COPPER		25,000	172
BRASS		30,000	206
BRONZE		35,000	241

must be greater than the shear strength. The molecular structure will then fail and fracture; then separation will occur.

Shear strengths of various materials have been found by experiment, and they are listed in the table. The shear strength is equivalent to the force required to cut a bar 1-inch square in two. Table 4.1 gives the value of shear strength for various materials. Expressed differently, the shear strength values in the table are a measure of the force required to cut an area 1-inch square. Given this information, it is a simple matter to determine the shear strength for any area to be cut.

4.3.1 Area to Be Cut

The first step in establishing cutting force or blanking force is to determine the area to be cut. For straight cuts as performed in shearing and in some cut-off die operations, the area to be cut is found by multiplying the length by the thickness. Figure 4.12 illustrates the area subjected to shear and is given by this formula:

$$A = L \times T \tag{4.1}$$

where:

A = area to be cut
L = length of material
T = thickness of material.

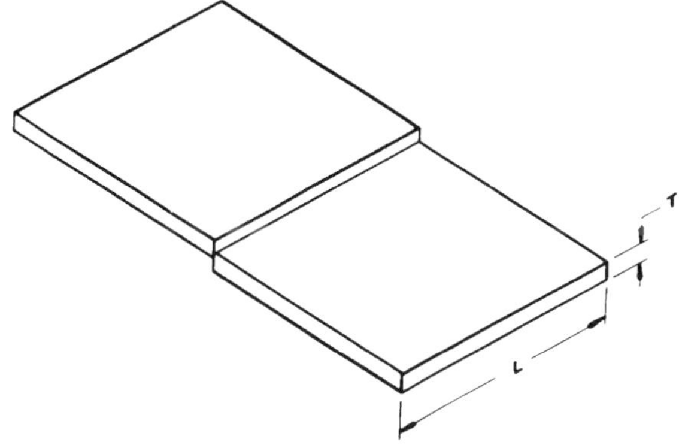

Figure 4.12 Drawing that illustrates area subjected to shear.

In blanking, an area is removed from within the strip. The cut, therefore, is around an enclosed contour. The area to be cut is found by multiplying the perimeter of the blank by the thickness (Figure 4.13).

For round cuts, the area is calculated as:

$$A = 3.1416 \times D \times T \tag{4.1a}$$

where:

D = diameter of blank
T = thickness of material.

For rectangular cuts, the area is calculated as:

$$A = (2L + 2W) \times T \tag{4.1b}$$

where:

L = length of rectangle
W = width of rectangle.

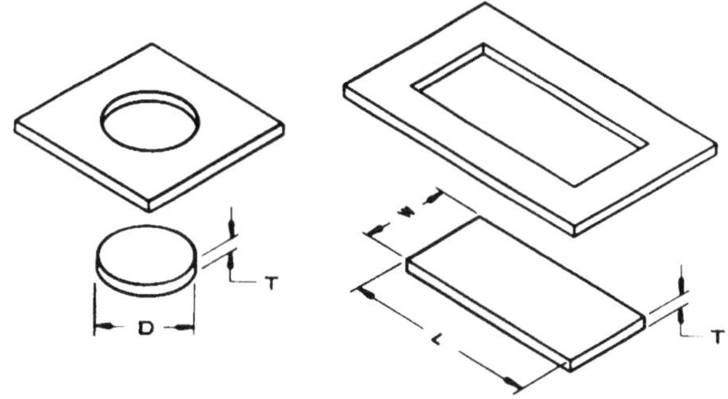

Figure 4.13 Drawing that illustrates areas subjected to shear in blanking.

We are now prepared to use the formula for determining the blanking force:

$$F = S \times A = S \times P \times T \qquad (4.2)$$

where:

S = shear strength as taken from the Table 4.1
P = perimeter (this is the distance around all cut edges)
T = thickness of material to be cut.

Such variables as unequal thickness of the material, friction between the pinch and the workpiece, or poorly sharpened edges can increase the necessary force 15 percent to 30 percent. Therefore, these variables must be considered in selecting the power requirements of the press. That is, the force requirements of the press F_p are:

$$F_p = (1.15 \text{ to } 1.30) \times F \qquad (4.3)$$

Example:
A blank 2 inches by 4 inches (50.8 mm by 101.6 mm) is to be cut in a strip of No.16 gage (0.0598 inch or 1.5 mm) thick, 0.20 percent carbon, cold-rolled steel. In addition, two 1/2 inch (12.7 mm) holes are to be pierced in a previous station; piercing of the holes is to occur simultaneously with blanking of the part.

Solution:
Perimeter of blank:

$$P_b = (2 + 2 + 4 + 4)$$
$$= 12 \text{ inches } (304.8 \text{ mm}).$$

Perimeter of holes:

$$P_h = (2 \times 0.500 \times 3.1416)$$
$$= 3.1416 \text{ inches } (79.8 \text{ mm}).$$

Total perimeter:

$$P = P_b + P_h = 12 \text{ in.} + 3.1416 \text{ inch}$$
$$= 15.1416 \text{ inches } (384.6 \text{ mm}).$$

Blanking force:

$$F = S \times P \times T$$

$$F = 55{,}000 \times 15.1416 \times 0.0598$$
$$= 49{,}800.722 \text{ lb.}$$

Force required of the press is:

$$F_p = 1.2 \times F = 1.2 \times 49{,}800.722$$
$$= 59{,}760.866 \text{ lb.}$$

Because press capacities are rated in tons, this number is divided by 2,000 and the force of the press is:

$$F_p = \frac{59{,}760.722}{2000} = 29.880 \text{ tons.}$$

A 30-ton press would be selected.

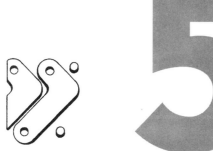

5 FOURTEEN STEPS TO DESIGN A DIE

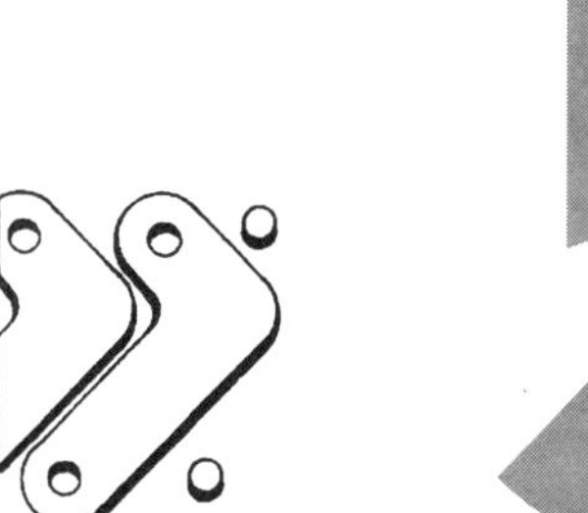

5.1 INTRODUCTION

Having completed the introductory portion of our work, we must now come to an understanding not only of die design itself, but also of the procedures followed by die designers in organizing components of different shapes, sizes, and composition into the unified concept called a *die*.

First, it should be understood that a definite order of steps must be taken in originating any die design. Haphazard design methods waste time and they often result in inefficient press tools. Conversely, systematic procedures will provide:

1. Consistently good designs.
2. Speedy, effortless work.
3. Fewer erasures.
4. Improved appearance of drawings.
5. Stronger punch and die components.

This section illustrates the 14 steps required for designing a die to produce the sample part shown in Figure 1.2. Study this order of steps carefully; by following it closely, you can begin at once to design a die yourself. Then when you have completed your design, the results you have achieved will surprise and please you.

By appropriate substitutions, the same steps can be taken in the design of any die. For example, in designing a multiple-station progressive die, all die blocks would be laid out at Step 2—The Die Block. If a spring stripper must be employed because of the nature of the operation, it simply replaces the solid stripper at Step 10—The Stripper Plate.

In the top view, outlines of the strip as well as all holes or openings are drawn with thin, red phantom lines. In the front and side view, the strip thickness is filled in solid red. The blank and slugs are shown pushed out of the strip and they are also drawn in solid red. Locate their top surfaces in line, or flush, with the bottom surface of the strip. A pictorial drawing of the strip with the blank and pierced slugs pushed out is shown between views. This drawing shows how the strip would be visualized before drawing the actual views. To improve your own faculty of visualization, make a freehand pictorial sketch of the strip on a separate sheet. In Chapter 6, rules for properly laying out scrap strips are given, as are as methods for design verification.

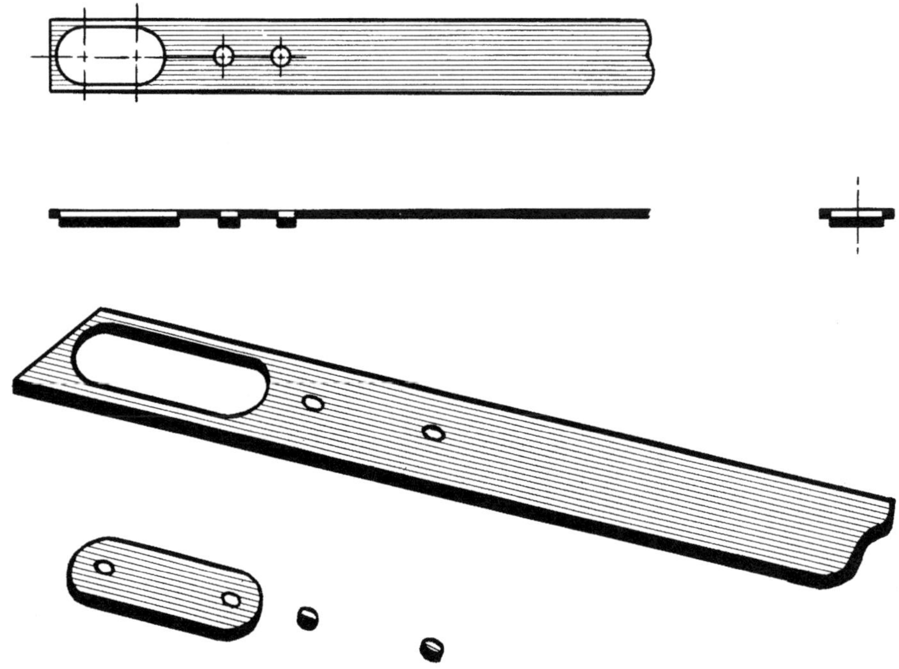

Figure 5.1 Material strip as it appears at the bottom of the press stroke.

In Chapters 6 through 19, each step will be explained in far greater detail. As you study each section, keep referring back to the illustrated list in this section to fix the position of each step firmly in your mind.

5.2 THE SCRAP STRIP (STEP 1)

The first step in designing any die is to lay out the material strip exactly as it will appear at the bottom of the press stroke. Three views are shown (Figure 5.1), and the distance between views must be carefully estimated to prevent views from running into each other as the die grows. Always use red pencil when drawing the material strip so it will show clearly through the maze of black lines that represent the punch and die members.

5.3 THE DIE BLOCK (STEP 2)

Draw the three views of the die block (Figure 5.2, upper illustrations). In the top view, the die block is usually rectangular in shape. The blank opening and holes for punches are represented by black lines. Simply draw with black pencil directly over the red lines. Dotted lines represent edges at the bottoms of blank and slug openings. These are larger because of the tapered walls that are provided for blank and slug relief.

Leave sufficient room for the screws and dowels that will fasten the block to the die set. Because both the front and side views will be section views, the lines representing internal openings are solid lines. The pictorial drawing now shows the die block in position under the strip (Figure 5.2, lower illustration).

5.4 THE BLANKING PUNCH (STEP 3)

The blanking punch is now drawn in position above the die block (Figure 5.3). Its plan view is applied in the upper right corner in an inverted position. That is, the punch is drawn as if removed from above the die block and turned over so cutting edges are viewed directly. When

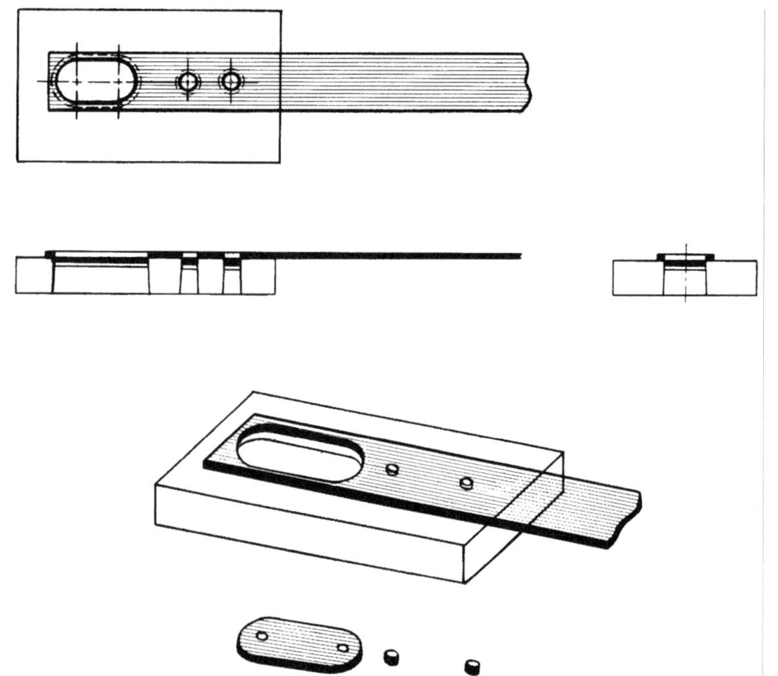

Figure 5.2 Various views of the die block and material strip.

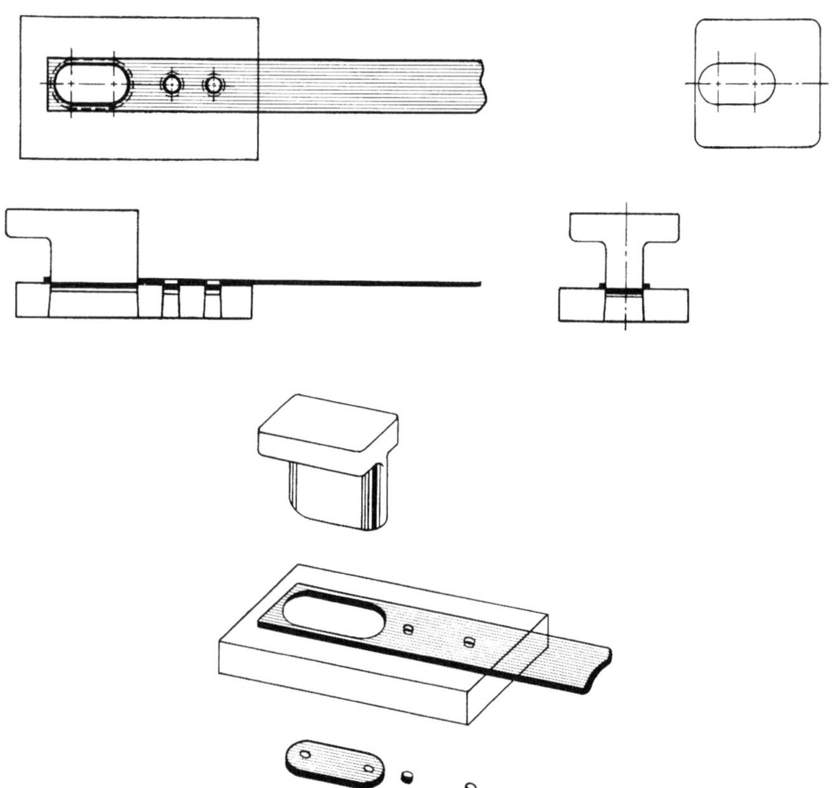

Figure 5.3 Various views of the die assembly with the blanking punch added.

outlining the flange width and thickness, take into consideration the screws and dowels that will fasten the blank to the punch holder of the die set. In the lower section views, the cutting face of the blanking punch is drawn flush with the top of the die block with a single line. To improve your ability to visualize the die in three dimensions, turn to the pictorial drawing and sketch the blanking punch above the blank opening in the die block.

5.5 PIERCING PUNCHES (STEP 4)

Piercing punches are now drawn in their proper positions (Figure 5.4). Include their plan views, represented by concentric circles, in the upper right corner, remembering that this view of the piercing punches and blanking punch is opposite from the plan view of the die block. The

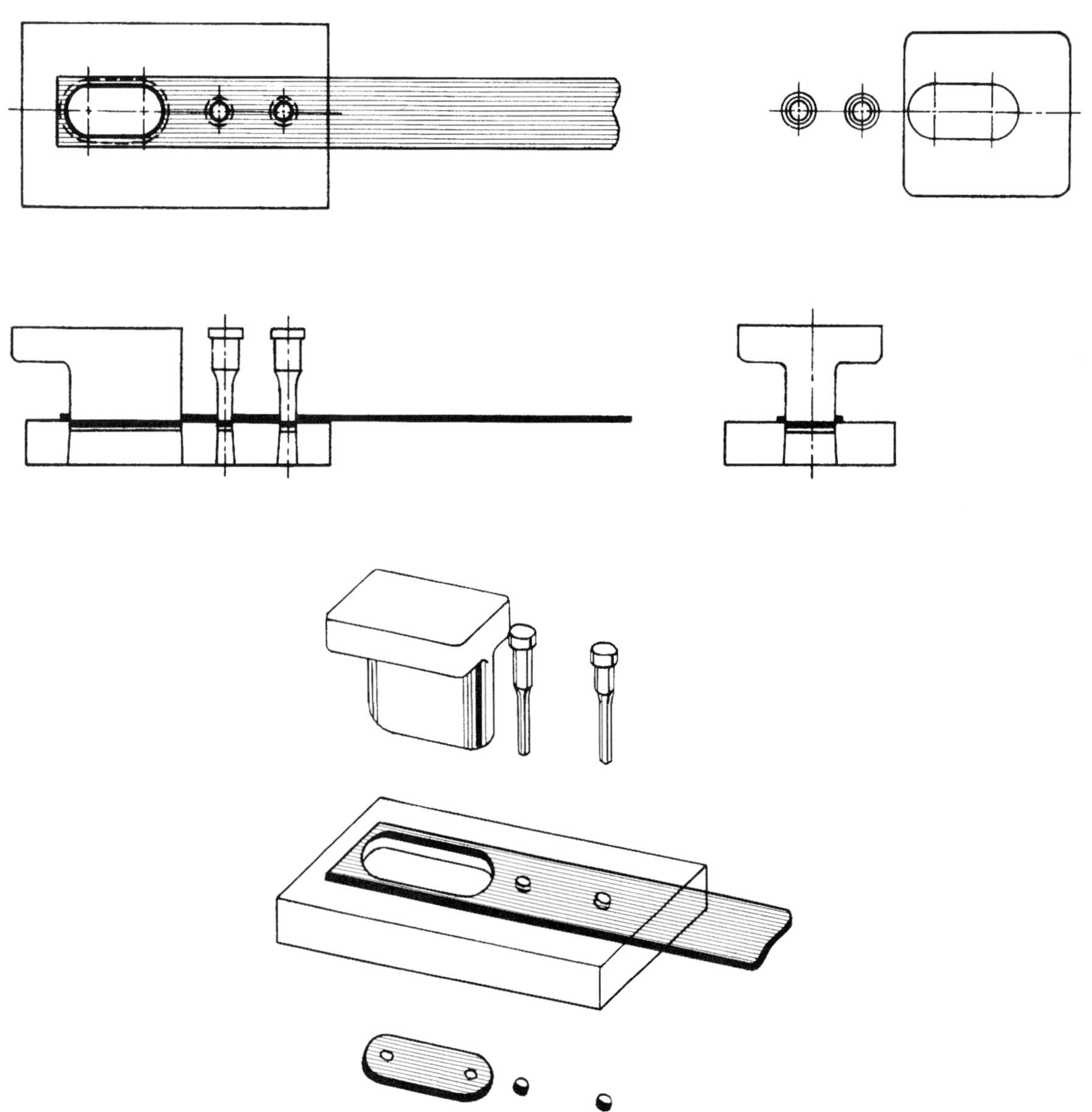

Figure 5.4 Various views of the die assembly with the piercing punches added.

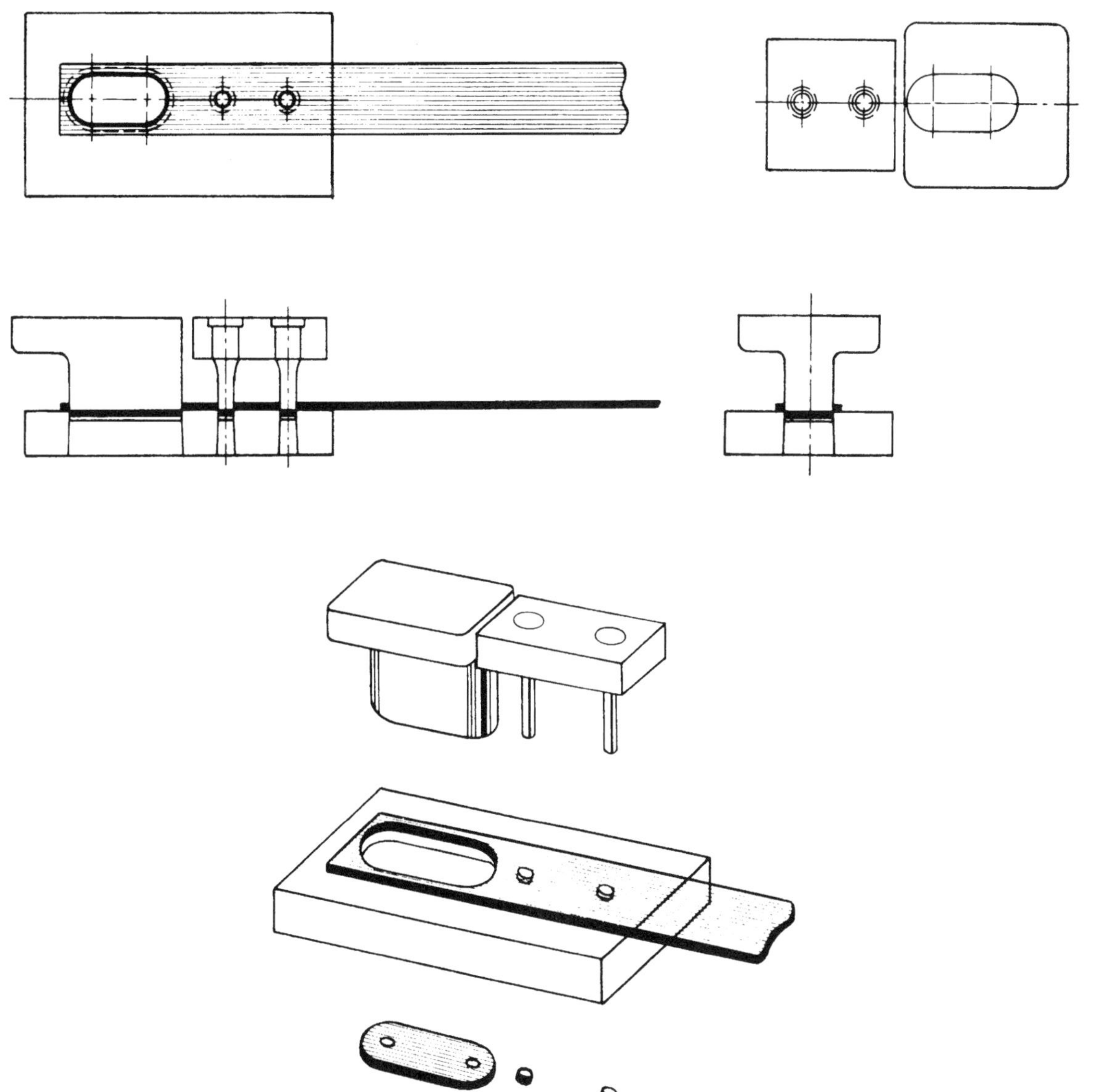

Figure 5.5 Various views of the die assembly with the punch plate added.

finished die will appear as if opened, much in the same manner as a book is opened. Sketch piercing punches on the pictorial drawing.

5.6 THE PUNCH PLATE (STEP 5)

In this step, the punch plate, which retains the piercing punches, is drawn in the front and upper right views (Figure 5.5). This plate is usually made of a good grade of machine steel. Here again, room must be provided for the screws and dowels that will secure it to the punch holder of the die set.

5.7 PILOTS (STEP 6)

Draw the pilots and the nuts that hold them in the blanking punch (Figure 5.6). As you know, the pilots accurately locate the strip by engaging the holes pierced in the first station. Only outlines of the parts are drawn; section lines are not applied until the entire design is completed. Be

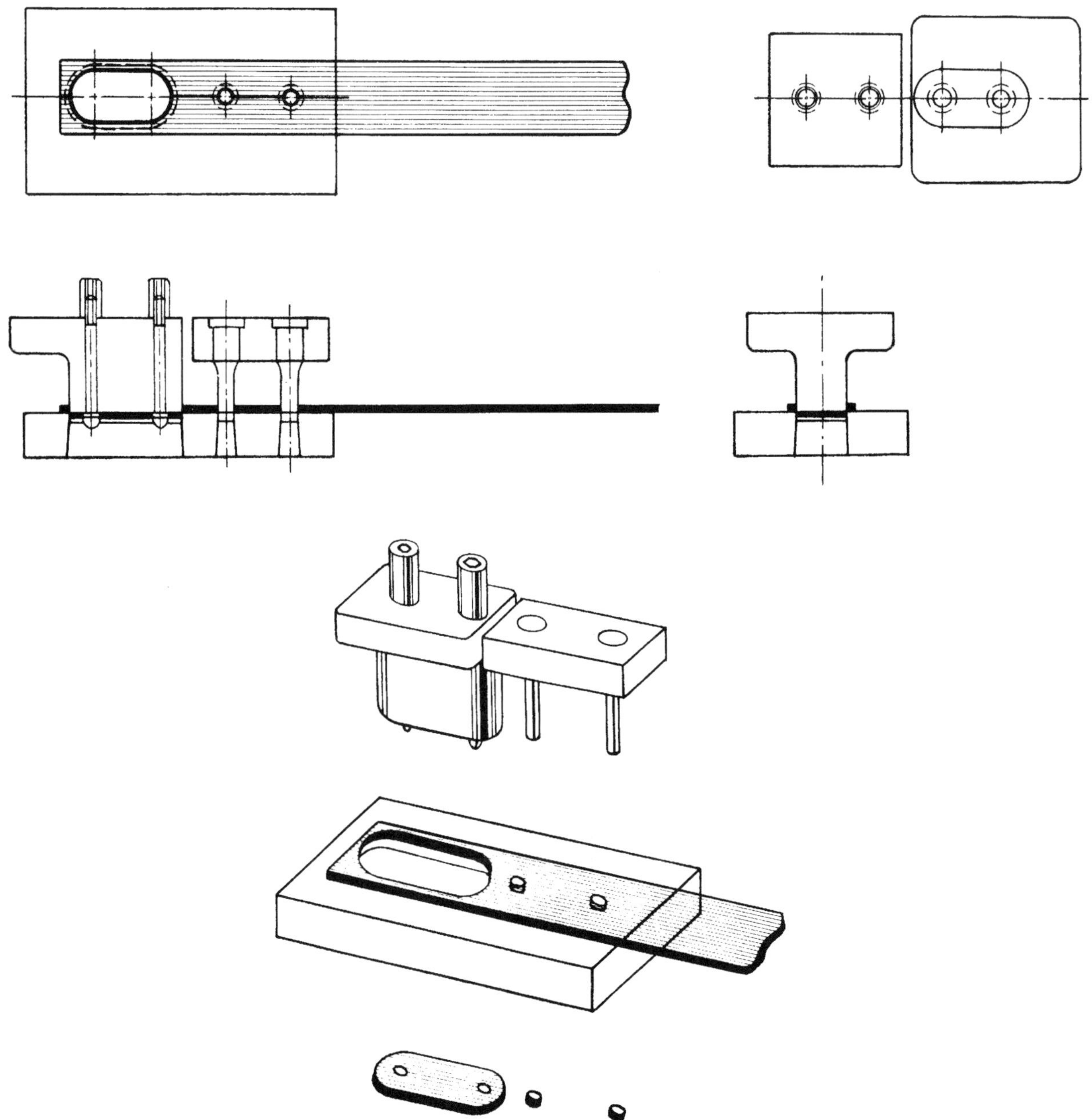

Figure 5.6 Various views of the die assembly with the pilots added.

sure to draw the concentric circles to represent pilots and pilot nuts in the upper right view, in addition to showing them in the front view.

5.8 GAGES (STEP 7)

The strip has not yet been located; it simply lies on top of the die block. We must now draw the back gage and front spacer that guide the strip by its sides (Figure 5.7). Strip support A, if one is used, can be drawn at this time. In the front and side views, draw horizontal lines to represent the top and bottom surfaces of the punch and die holders of the die set. Thicknesses are taken from a die set catalog. For a small die such as this one, a punch holder 1 1/4 inch thick

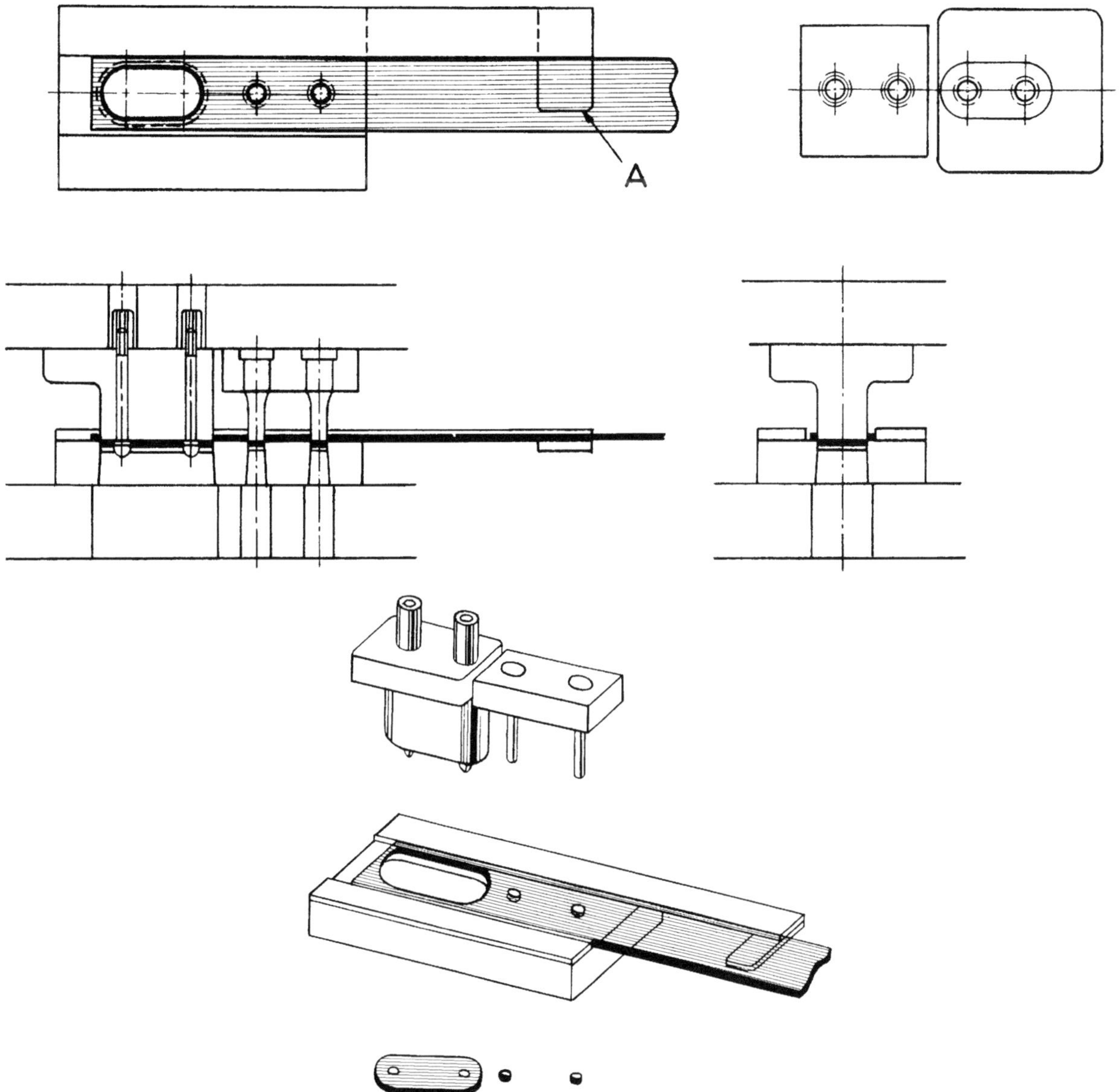

Figure 5.7 Various views of the die assembly with the back gage and front spacer added.

(32 mm) and a die holder 1 1/2 inch thick (38 mm) are sufficient.

5.9 THE FINGER STOP (STEP 8)

Although the strip is now supported sideways between the back gage and front spacer, no provision has been made for endwise location. Draw a finger stop to position the strip for piercing the two holes in the first blank (Figure 5.8). In operation, the finger stop is pushed in and the end of the strip engages it. After the press has been tripped to pierce the holes, the strip is then advanced until its end contacts the automatic stop (see next step).

Also at this time, the punch shank is drawn in the upper right hand view, being represented by a dotted circle. It should be so placed that any clearance hole will appear entirely within, or completely outside of it.

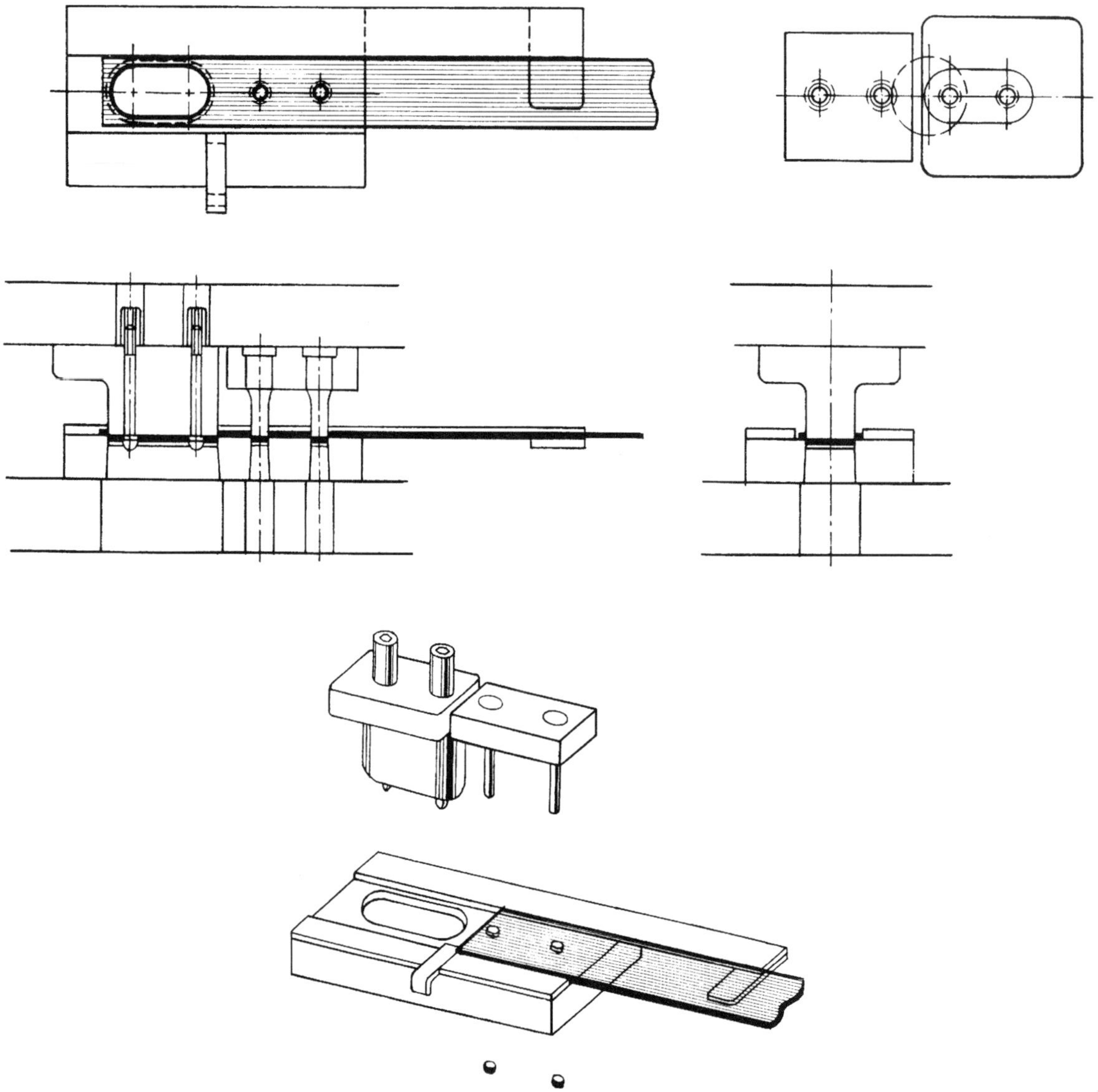

Figure 5.8 Various views of the die assembly with the finger stop added.

5.10 THE AUTOMATIC STOP (STEP 9)

An automatic stop, which locates the strip at every punch stroke, is now drawn (Figure 5.9). There are numerous types of automatic stops, but the one illustrated is perhaps the most widely used. In operation, the end of the strip contacts the toe of the stop, locating it in a set position. When the press ram descends, a square-head set screw, retained in the punch holder, contacts the opposite end of the stop, causing the toe to be raised above the strip. A spring then triggers the stop, causing it to rotate a few degrees so that the toe of the stop comes to a position above the scrap bridge. When the ram goes up, the stop toe falls on top of the scrap bridge, allowing the strip to slide under it until the toe of the stop falls into the blanked opening. Travel continues until the right edge of the blanked opening contacts the toe, moving it and resetting the stop. These motions take place at high speeds, much faster than the time it takes to describe them.

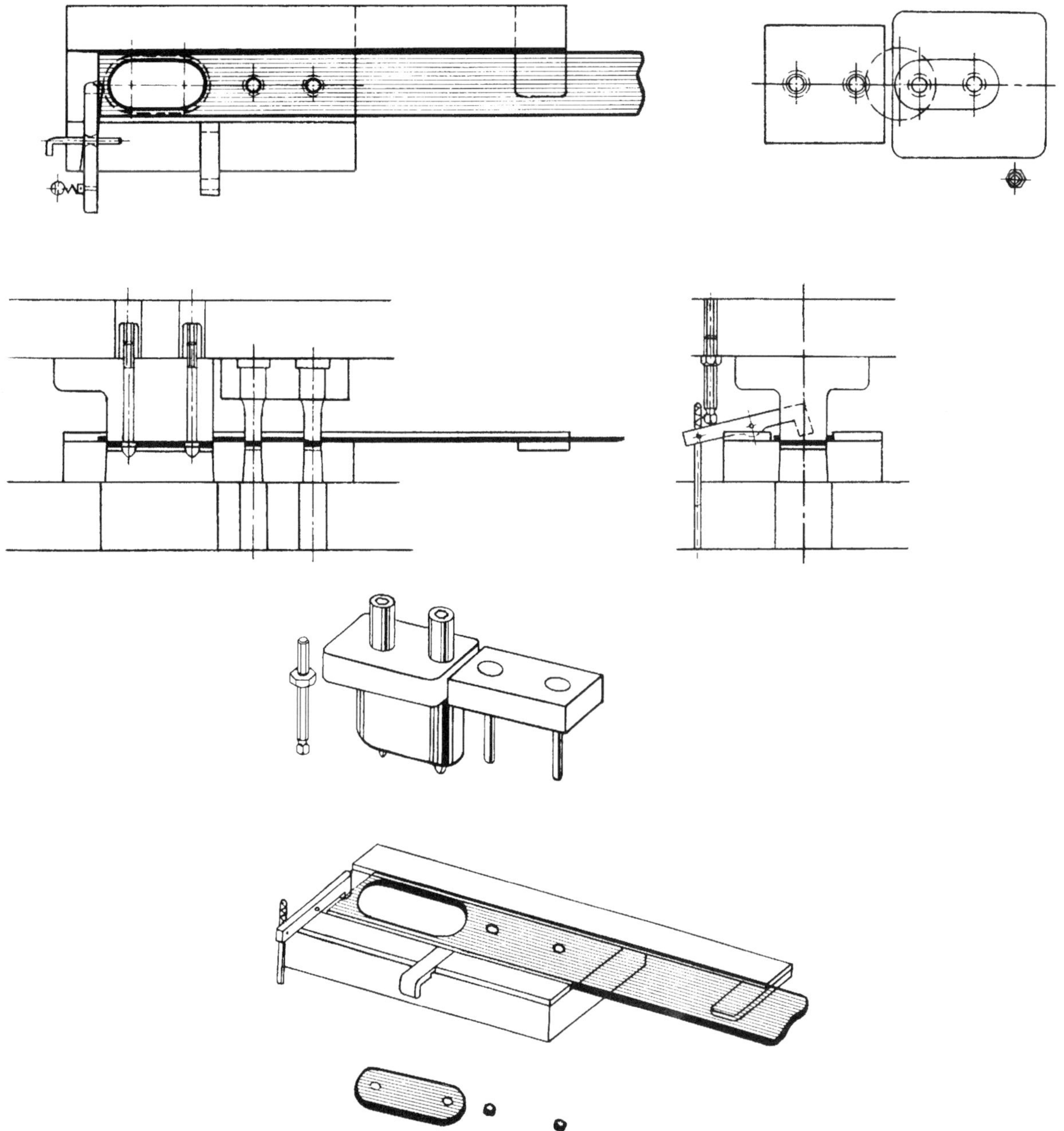

Figure 5.9 Various views of the die assembly with the automatic stop added.

To this point, all lines have been drawn very lightly for easy erasure when sizes of components are altered.

5.11 THE STRIPPER (STEP 10)

The stripper is now applied to all three views (Figure 5.10). Strippers remove the strip from around blanking and piercing punches. For some types of dies, a knockout would be designed at this step. Knockouts are internal strippers and they remove blanks and formed parts from inside cavities of punches and dies. In the upper left plan view of the die block, all lines under the solid stripper now become dotted lines since they now represent hidden edges of surfaces.

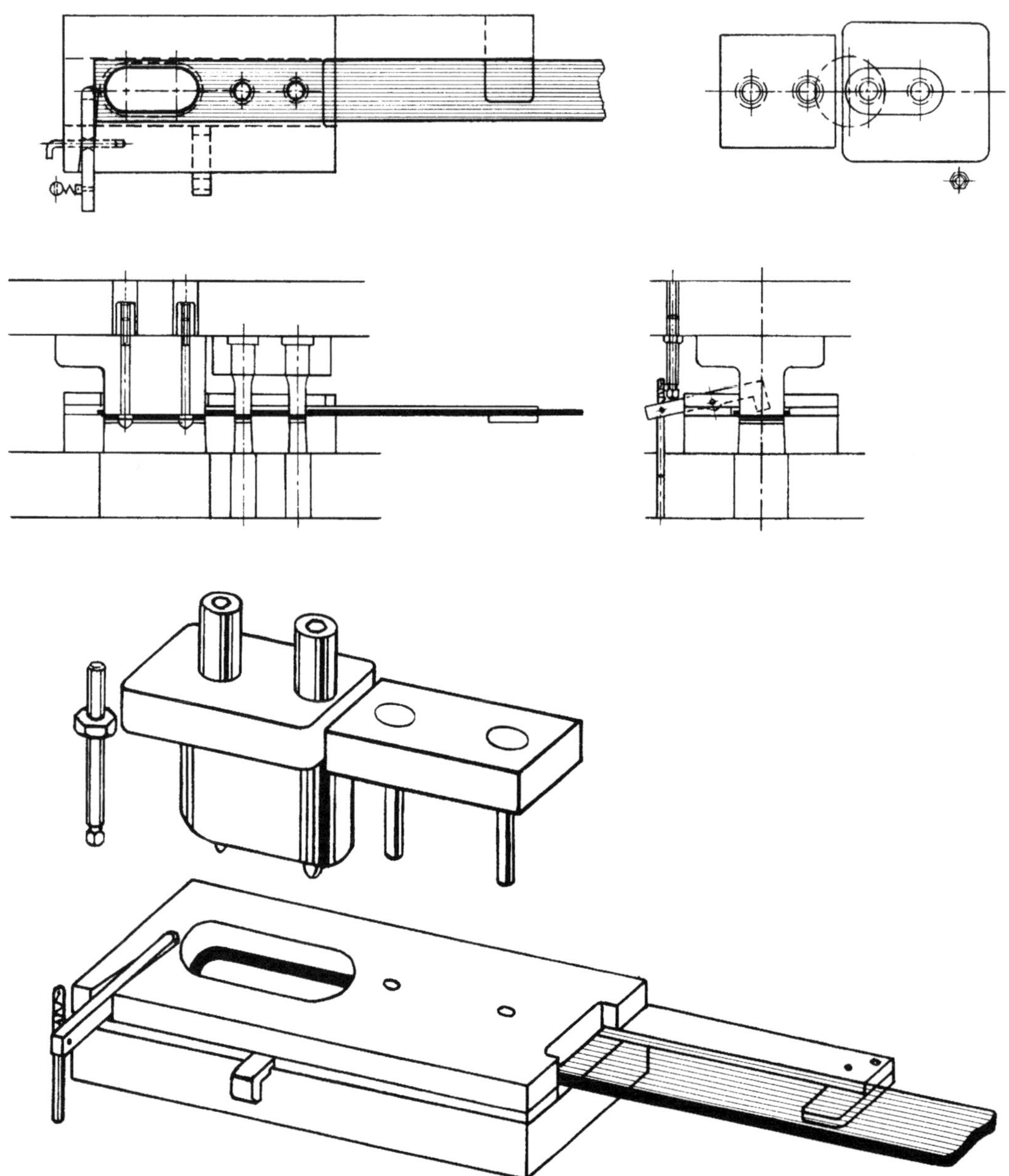

Figure 5.10 Various views of the die assembly with the stripper added.

A stripper is required because the material strip always clings tightly around any punch that penetrates it. Provision must be made to remove it from the punch.

5.12 FASTENERS (STEP 11)

In this step, screws, dowels, and other fasteners are drawn (Figure 5.11). For press tools, all dowels are the same diameter as the screws for any particular member. Note that button-head socket screws are employed to fasten the solid stripper to the die block. Because of their low head height, strength, and clean lines, these are excellent fasteners for this application.

Considerable designing ability is required for applying fasteners properly and there are a number of rules to be learned. This subject will

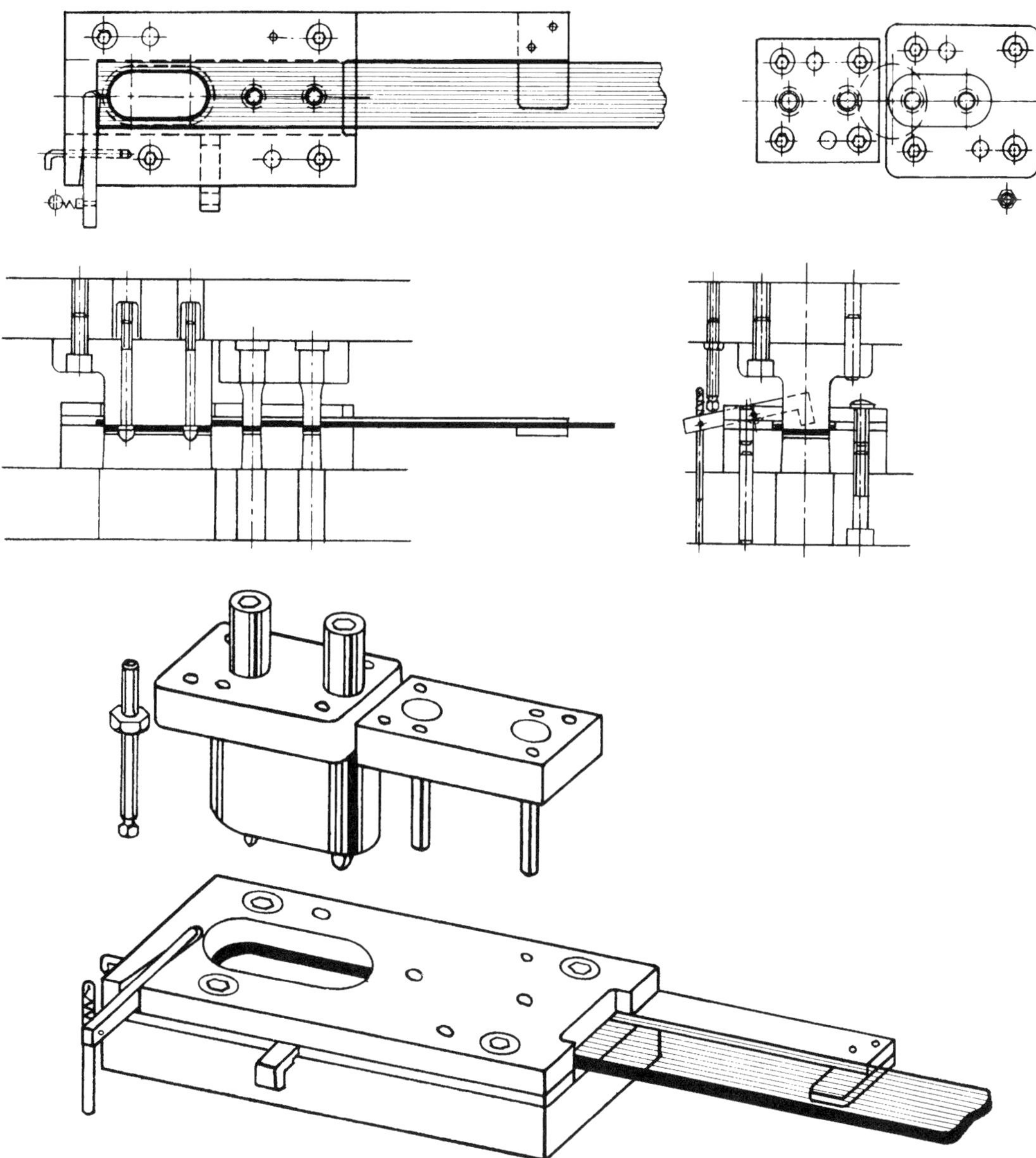

Figure 5.11 Various views of the die assembly with the fasteners added.

be taken up in detail in Chapter 16—How to Apply Fasteners.

5.13 THE DIE SET (STEP 12)

The die set, which was started with horizontal lines in the lower views of Step 7, and with a dotted circle in the upper right view of Step 8, is now completed (Figure 5.12). To draw in the die set before this step will very frequently result in considerable erasing when die sections are found to be too small and their sizes are increased.

Die sets are available in a bewildering variety of shapes and sizes; judgment must be exercised when selecting a suitable one under given conditions. In addition, a definite method of procedure must be followed in applying the views to the drawing. This subject will be explained thoroughly in Chapter 17—How to Select a Die Set.

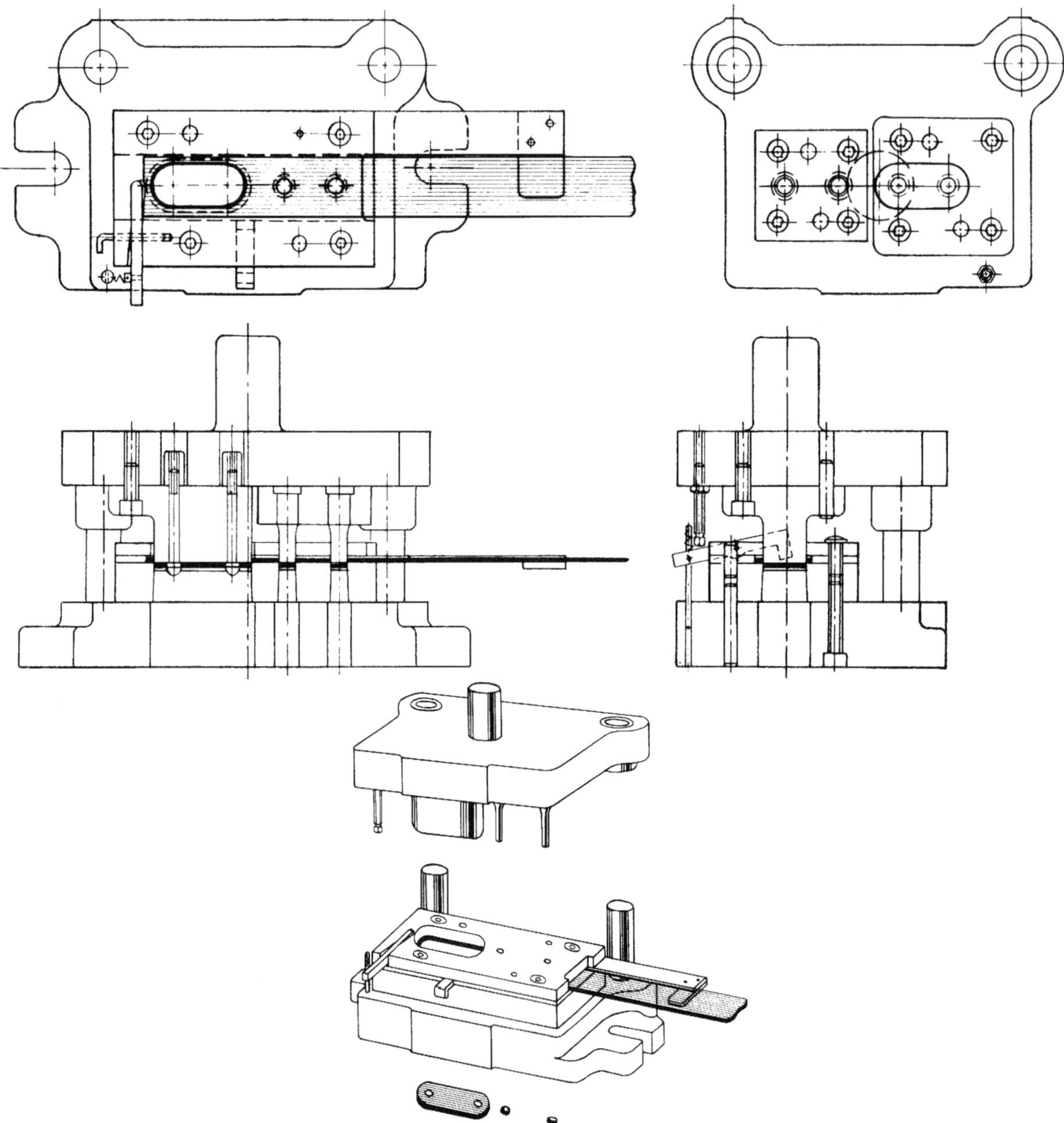

Figure 5.12 Various views of the die assembly including the die set.

5.14 DIMENSIONS AND NOTES (STEP 13)

All dimensions and notes are now lettered in place (Figure 5.13). Die block dimensions are applied as for a jig borer layout. All information that the die maker might require to complete the die should now be on the drawing, unless components are detailed separately, as is done for more complex dies.

5.15 BILL OF MATERIAL (STEP 14)

The final step in designing this die is to fill in the bill of material (Figure 5.14). This lists all

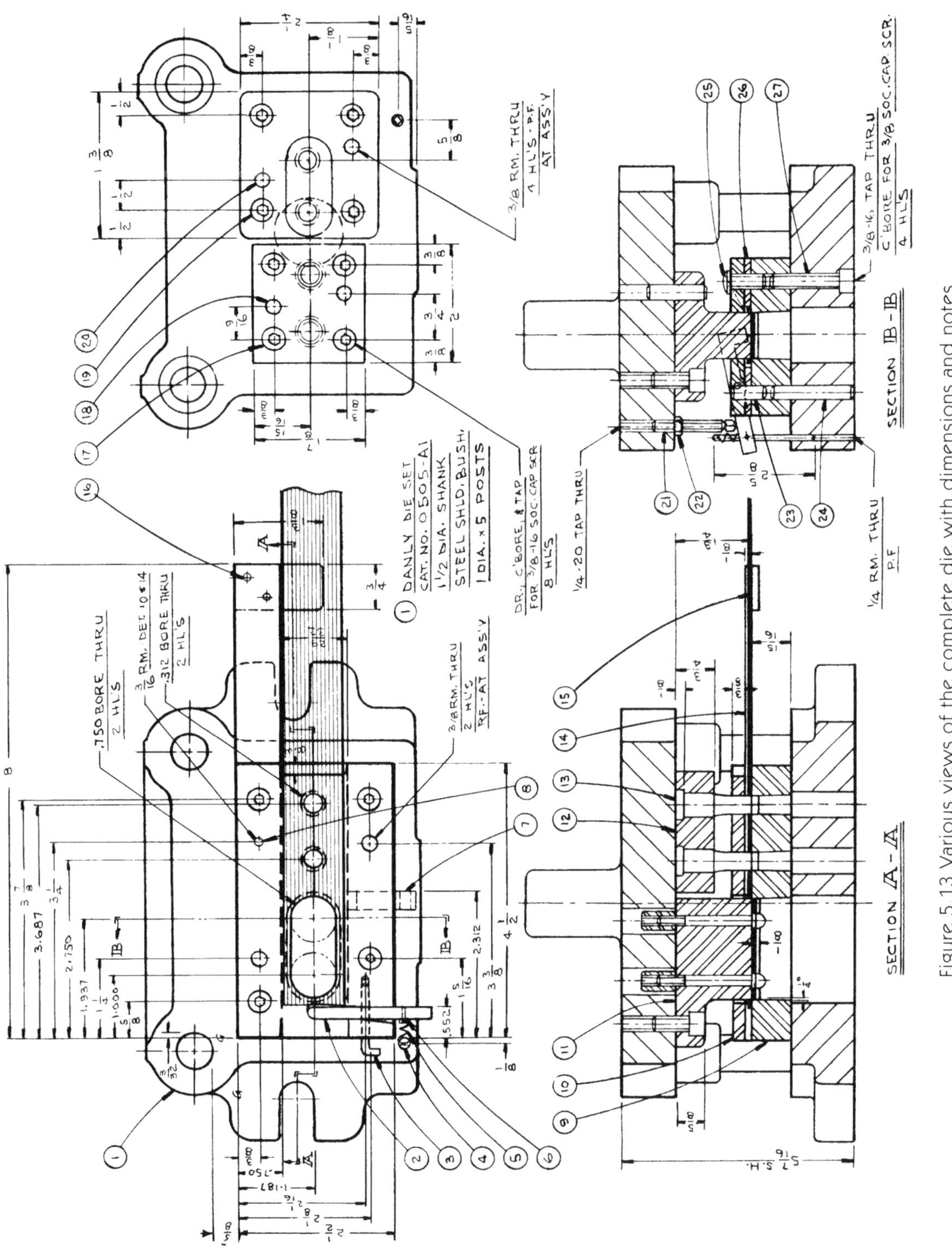

Figure 5.13 Various views of the complete die with dimensions and notes.

components that will enter into building the die. Unless printed sheets are used, the border is applied, the paper is trimmed, and the die design is complete. Over the next several chapters, each of the 14 steps will be taken up in detail.

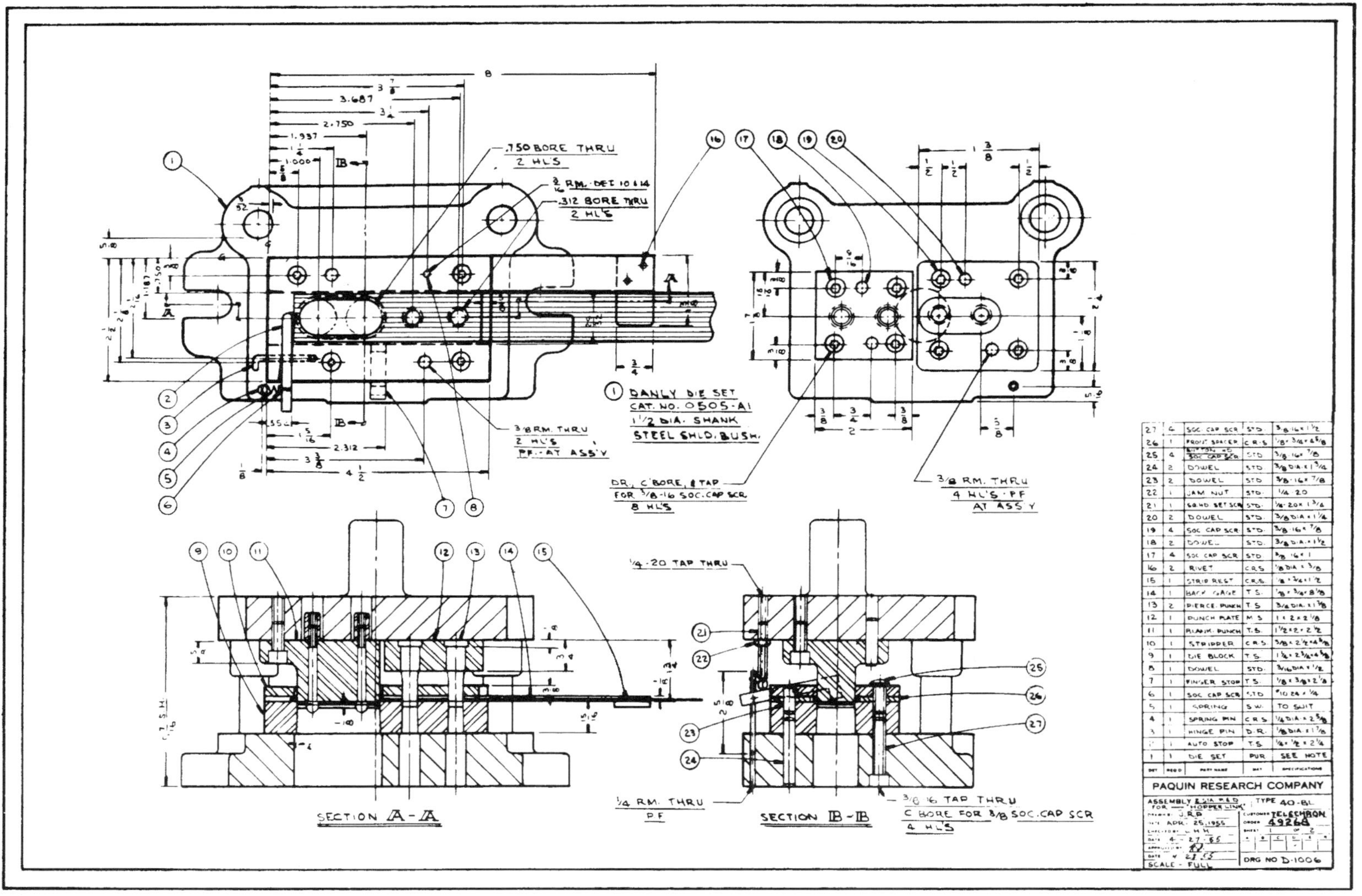

Figure 5.14 The complete die drawing, including the bill of material.

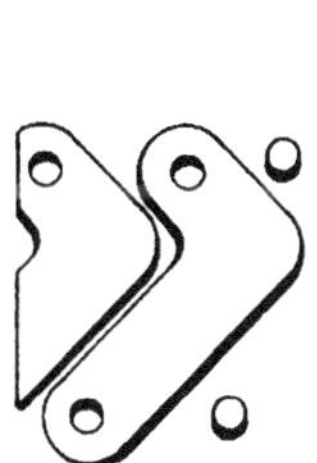

HOW TO LAY OUT A SCRAP STRIP

6.1 Introduction
6.2 Strip Material Economy
6.3 Scrap Strips
6.4 Scrap Strip Allowances
6.5 Strip Area per Blank Calculations

6.1 INTRODUCTION

The first step in designing a die is to lay out the material strip exactly as it will appear after all operations have been performed. It is then called a *scrap strip*. To be successful, scrap strip designing must follow a definite procedure that will ensure nothing has been omitted or left to chance.

In illustrations to follow, we will consider the steps taken in designing a scrap strip for a pierced link, from the first tentative tries to the finished layout. These same steps, applied to the design of a scrap strip for any similar part, should assure successful results.

This section of the book further explains Step 1 in Chapter 5—Fourteen Steps to Design a Die.

6.2 STRIP MATERIAL ECONOMY

The major portion of the cost of stamping is material. Therefore, material economy is of the utmost importance from the standpoint of cost. Fifty to seventy percent of the cost of a stamping goes for material. The method employed for laying out the scrap strip directly influences the financial success or failure of any press operation. The blank must be positioned so that a maximum area of the strip is utilized in production of the stamping. A blank layout is drawn before any work is done on the die design itself. In fact, the scrap strip layout will govern the shapes and sizes of many of the die members.

6.2.1 Blank Positioning

Blanks can be positioned many different ways in the strip. Choice of the correct method depends upon part shape, production requirements, and any bends that must be applied. Figure 6.1 shows both the single-row, one-pass (*a, b, c*) method and the single-row, two-pass (*d, e, f*) method. In the first case, the blanks are arranged in a single row; the strip is then passed through the die once to cut all blanks from it. At A, the parts are located in a vertical position in the strip. This is the preferred method because the maximum number of blanks can be cut from one strip and fewer strips must be handled.

When severe bends are required in subsequent operations, as in flat springs made from strip, layout *b* must be used. This procedure involves handling more strips to produce the same number of blanks.

The shape of the part will often lend itself to angular positioning, as in layout *c*. For some

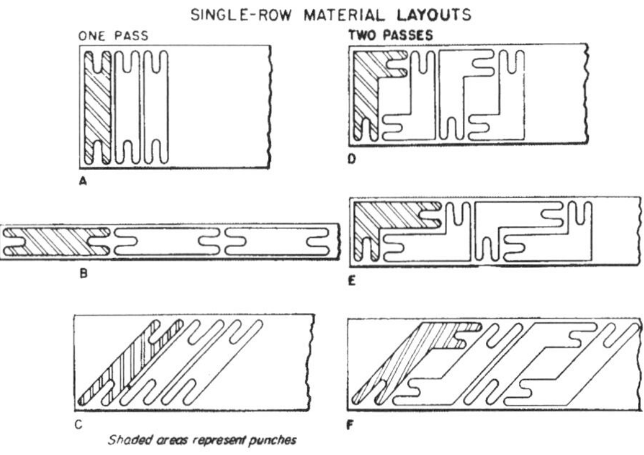

Figure 6.1 Various ways in which blanks can be positioned.

contours, this positioning is economical in terms of material. It has the further advantage of allowing bends to be made without possible fracture.

Material can be saved by the single-run, two-pass method (D, E, and F) when it is used for certain part shapes. In such cases, the blanks are positioned in the strip in a single row. Alternate blanks are turned upside down as shown; the strip must be passed through the die twice to remove all blanks from it. As the strip goes through the first time, the blanks in the upper row are cut. The strip is then turned over and run through the die again, removing the rest of the blanks.

A 10 to 15 percent higher labor cost will occur in double-run layouts. The operator must pass the strip through the die twice and use greater care in gaging it. The extra labor cost is offset by the saving in material when blanks are large and waste is considered.

Vertical positioning is shown at **D,** horizontal positioning at **E,** and angular positioning at **F.** The contour and bends found in many parts will dictate use of angular positioning.

6.2.2 Blank Layout

Further economy in material can be achieved by use of double-row, two-pass layouts (Figure 6.2 at *a* and *b*). The strip is run through the die twice, as in previous examples, but blanking centers are closer together, giving greater operating speed.

The same positioning method may be used for double-row gang dies (Figure 6.2 at *c* and *d*). An extra punch and die opening is applied to the die, cutting two blanks with each stroke, and the strip is run through only once. Gang dies are high speed tools; the method should be selected only when the added expense of the extra punch and die hole is warranted.

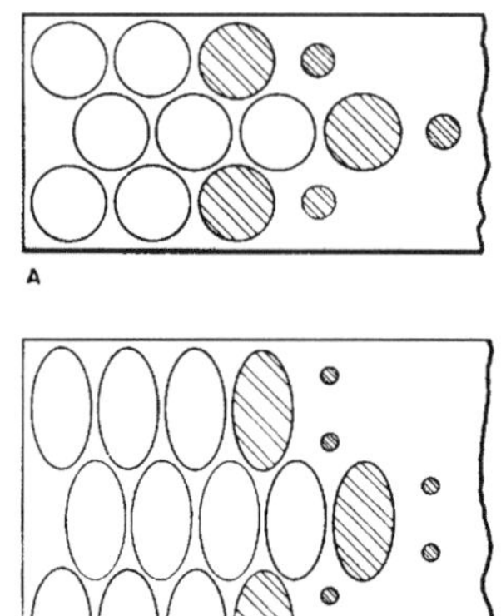

Figure 6.3 Triple-row blank layouts.

Figure 6.3 shows two triple-row layouts. Part *a* shows the layout for a die used to produce washers at high speed, whereas Part B shows a die to produce elliptical blanks. Such dies may have more than three rows; the number is limited only by press size and production requirements.

6.3 SCRAP STRIPS

A scrap strip is to be designed, run, tested, evaluated, and corrected for producing small, elbow-shaped parts called *offset links* (Figure 6.4).

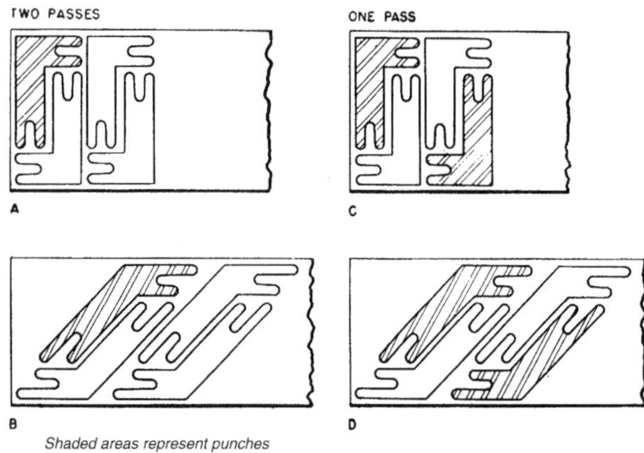

Figure 6.2 Double-row blank layouts.

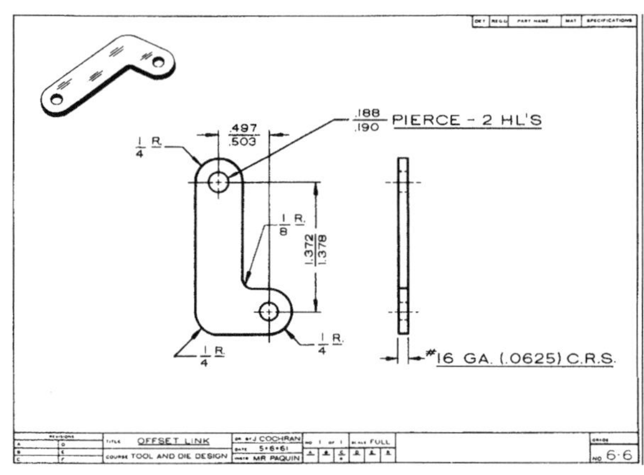

Figure 6.4 Typical part print.

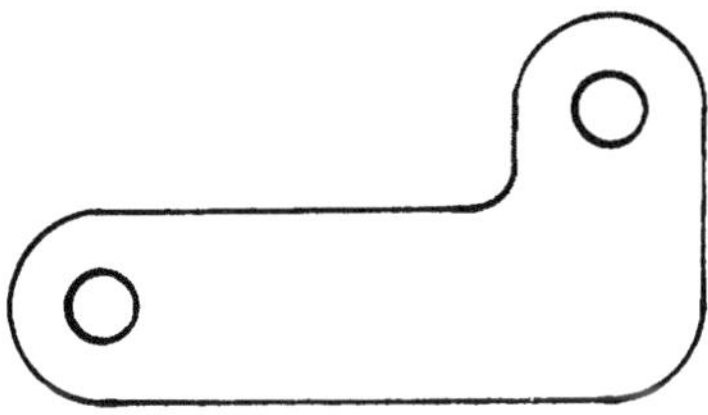

Figure 6.5 Part layout on paper.

6.3.1 Designing

Study of the part print shows that dimensional allowances are liberal enough so that a two-station progressive die can be used, instead of a compound die. In such dies, the holes are pierced at one station and the part is blanked out at the succeeding station.

a) Part Layout

The first step is to lay out the part accurately on paper (Figure 6.5). Check dimensions carefully at this time as any mistake will result in a considerable amount of useless work.

b) Part Positioning

Wide-run positioning. Usually, the first thought that comes to mind is to lay out the blanks side by side, as shown in Figure 6.6 at *a*. Blanks arranged in this manner can be run through the die in any of four different ways, shown further at *b, c,* and *d*. Any of these layouts is wasteful of material.

Narrow-run positioning. If a bend were to be applied to the long arm of this part (Figure 6.7), the first thought would be to lay it out the narrow way, as shown, because bends should be made across the grain, preferably. However, this layout is also wasteful of material. As with wide-run positioning, it is possible to run the strip in four ways.

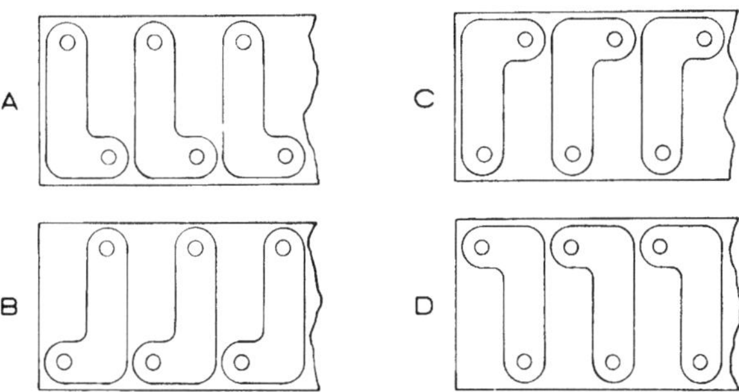

Figure 6.6 Part positioned for wide run.

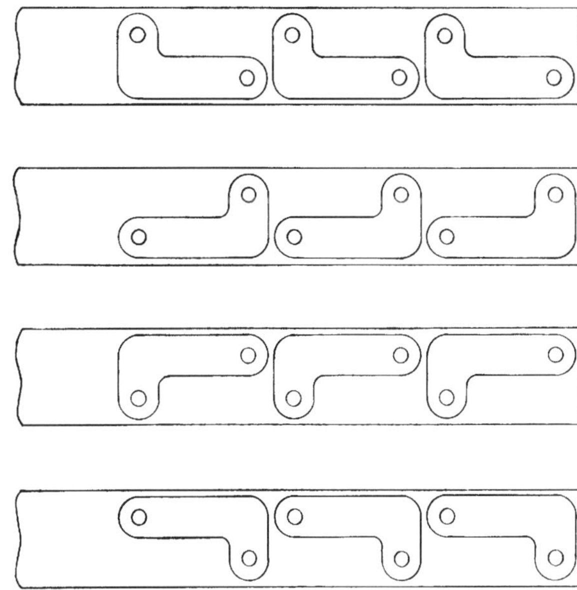

Figure 6.7 Part positioned for narrow run.

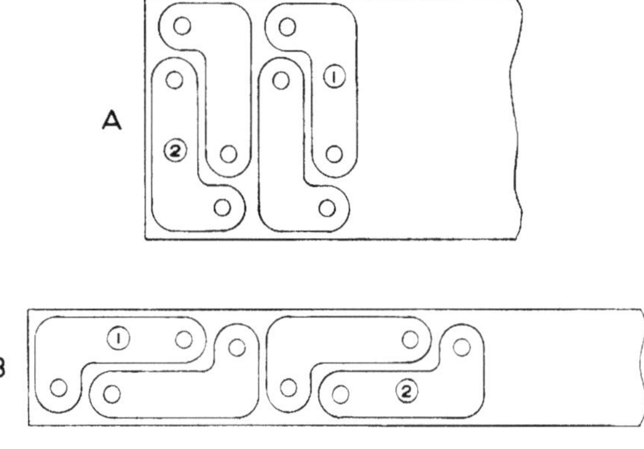

Figure 6.8 Double-row layouts.

Double-row layouts. Double-row layouts (Figure 6.8) are economical of material, but they may lure the designer into trouble. Part A shows a wide-run, double-row layout. The strip would be run through the die, blanking all holes in the upper portion of the strip (layout **1**). The strip would then be reversed and run through the die a second time, blanking all holes (layout **2**). In running the strip for the first row, however, it becomes curved and distorted, causing the strip to stick and bind when running the second time. In fact, there is a 10 to 15 per cent increase in labor cost for double-row dies.

Double die layouts, in which both layouts **1** and **2** are blanking at the same time, present fewer difficulties. But few parts have the high-production requirements that would justify the expense of providing two blanking punches and die openings, as well as an extra set of piercing punches. A double-row, narrow-run layout is shown in part *b*. This can be either a two-pass layout, in which the strip is run through the die twice to extract all blanks, or it can be a double die layout to extract blanks from both layouts **1** and **2** and the same time.

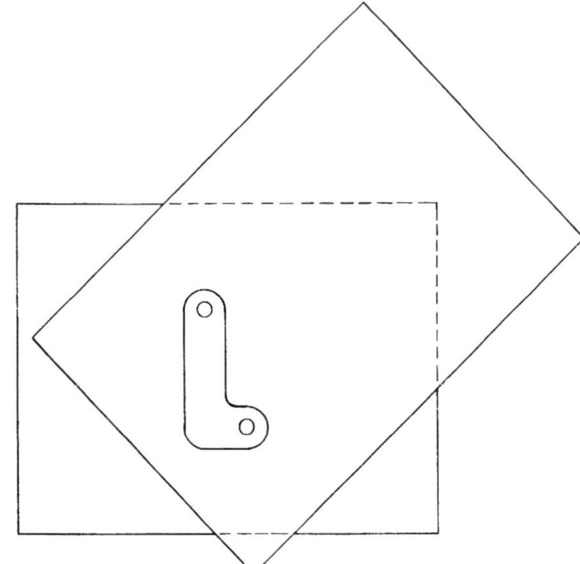

Figure 6.9 Tracing another part layout from Figure 6.5.

c) Nesting

To summarize: Side-by-side layouts, except for some rectangular or round parts, waste material. Double-run layouts can be a source of trouble and are seldom used for small parts. Only for very large blanks are the savings in material great enough to justify their use. Double-die layouts are quite expensive except for extremely large quantities.

Most blanks with irregular contours will nest (fit snugly one against another) when placed side by side in the correct position. The best nesting position for the particular part must be found before proceeding with the strip layout. Here is the traditional method:

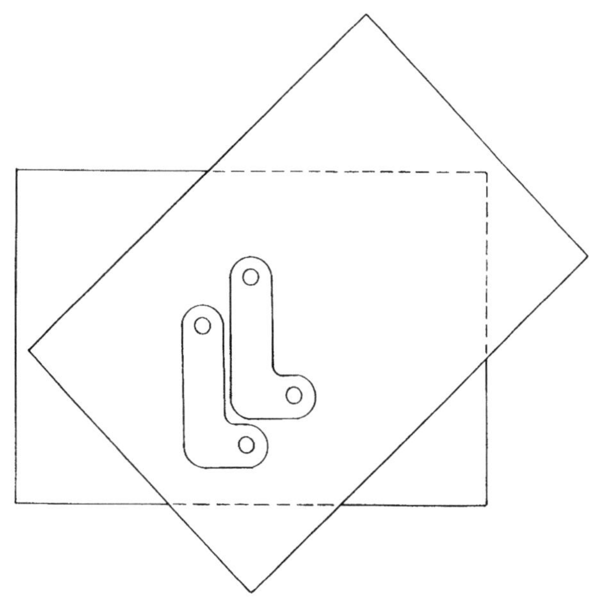

Figure 6.10 Moving the tracing to the best nesting position.

1. *Placing a sheet of thin paper.* Over the blank layout shown in Figure 6.5, place a sheet of thin, translucent paper and trace the part outline (Figure 6.9).
2. *Moving the tracing.* Move the paper over the blank layout until the best nesting position is found. Figure 6.10 shows the best nesting position for this part. Correct scrap bridge allowances must be allowed between blanks, as will be explained later.
3. *Correcting the layout.* Fasten this layout to the drafting table, straighten inaccurate lines, and check dimensions. Draw guide lines **A** and, leaving the correct scrap bridge allowance, part line **B** of the next blank to help position the paper for tracing it (Figure 6.11).
4. *Drawing the third blank.* Place the sheet back over the part layout shown in Figure 6.5. Then draw the third blank as accurately as possible (Figure 6.12).
5. *Completing the layout.* The blank layout is now complete and accurate (Figure 6.13).

It might appear that our task is finished, but much work remains before we can be sure that we have the best possible layout. We must know that it will perform satisfactorily.

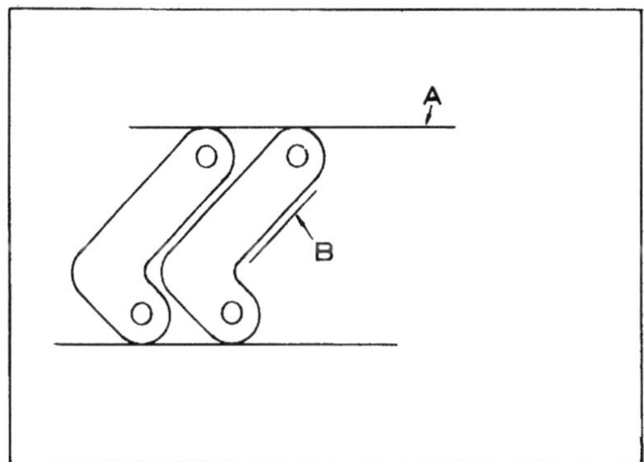

Figure 6.11 Corrected layout.

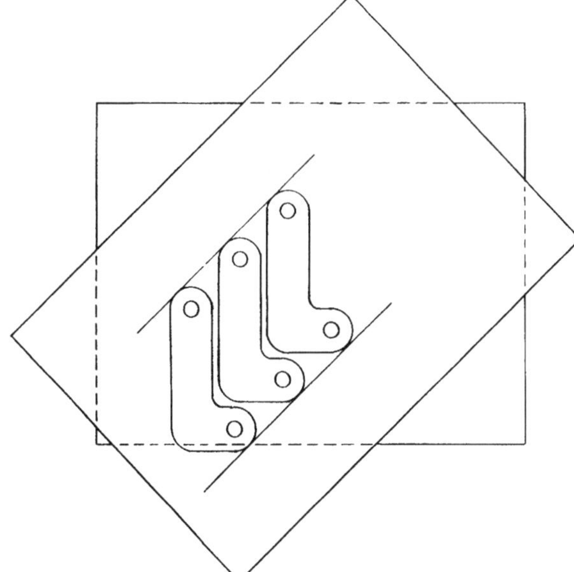

Figure 6.12 Drawing the third blank.

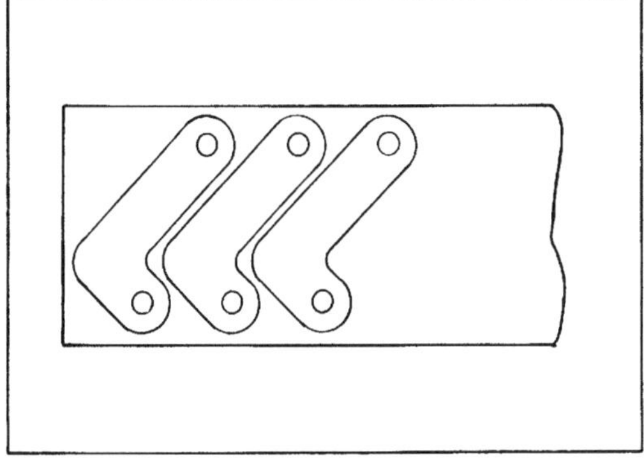

Figure 6.13 Completed layout.

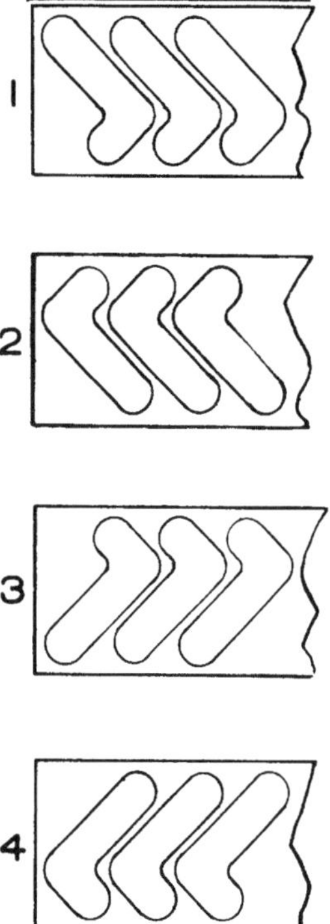

Figure 6.14 Four ways in which the strip can be run through the die.

Finding the nest position for part layout is much easier if you use a CAD system for die design. Just use command "Copying with Move" or "Rotating and Moving," if it is necessary.

6.3.2 Running the Strip

There are four ways of running the strip through the die (Figure 6.14) using the layout just completed. Numbered 1, 2, 3, and 4, one of these will be the best way for a given part, depending on its contour. From the layout in Figure 6.14, it is possible to trace directly all four ways of running the strip.

Figure 6.15 shows how to rotate the strip in order to obtain the various running methods. Strip 2 is arrived at by turning strip number 1 around as shown, and tracing it in position. Strip 3 is arrived at by turning strip number 1 upside down as shown, then tracing it in this newer position. Strip 4 is arrived at by turning strip number **3** around as shown, then tracing it in its new position.

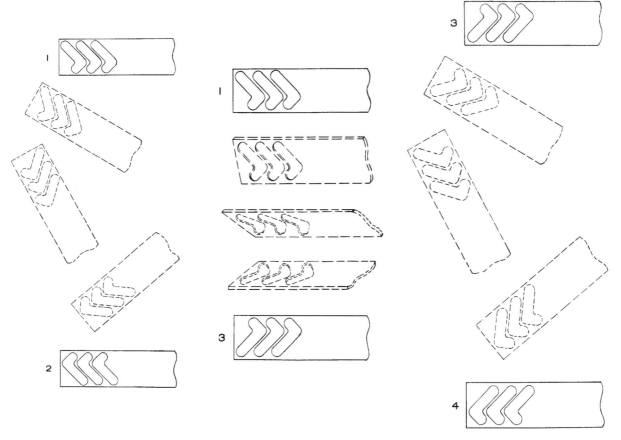

Figure 6.15 Ways of rotating the strip to obtain the various running methods.

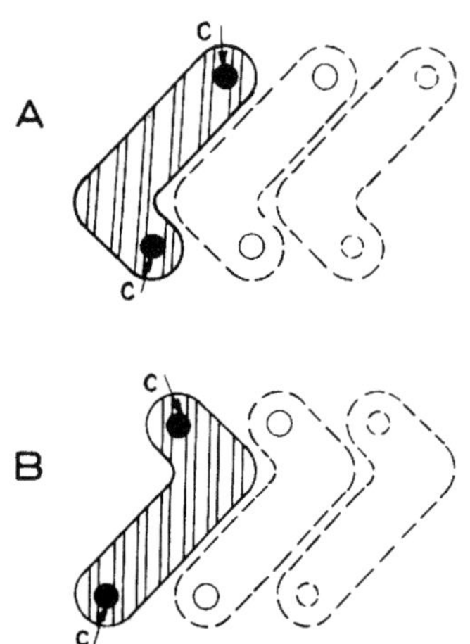

Figure 6.16 Use of the layout for making the piercing and blanking punch layout.

a) Pierce and Blank Layout

The strip laid out in Figure 6.13 is used in making piercing punch and blanking punch layouts. Figure 6.16 at a is a layout that corresponds to strip 4 in Figure 6.14. The layout at B in Figure 6.16 corresponds to strip 3 in Figure 6.14. Piercing punches are usually drawn solid black. (The white used here is for convenience.) Circles ***c*** are pilots in the blanking punch, used to locate the strip prior to blanking. These are drawn in solid red. The importance of drawing three blanks in the layout in Figure 6.14 should now be clear.

b) Tracing Pierce and Blank Layouts

In Figure 6.17, four piercing punch and blanking punch layouts are traced from the layout in Figure 6.13 to correspond to the four methods of running the strip shown in Figure 6.14. The line

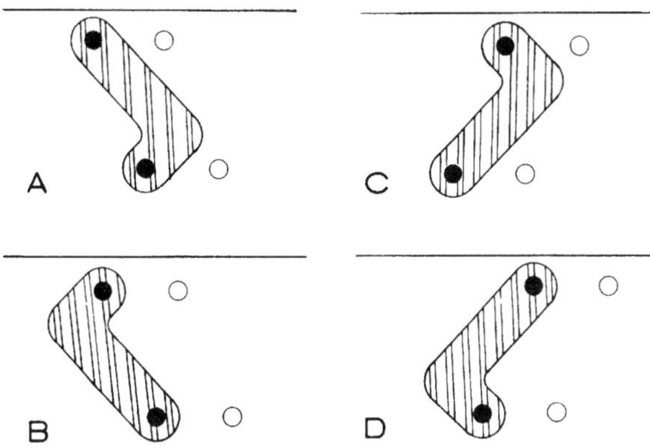

Figure 6.17 Four piercing and blanking punch layouts traced from Figure 6.13.

above each of these layouts represents the front edge of the back gage against which the strip is positioned in its travel through the die.

The trial scrap strip is drawn in Figure 6.18. On another sheet of thin paper, draw the scrap strip, as shown at *a*. For experienced designers, the two lines shown at *b* are sufficient for this tryout work.

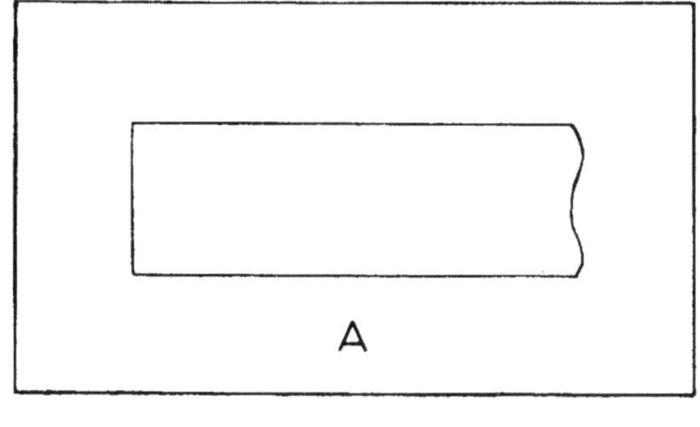

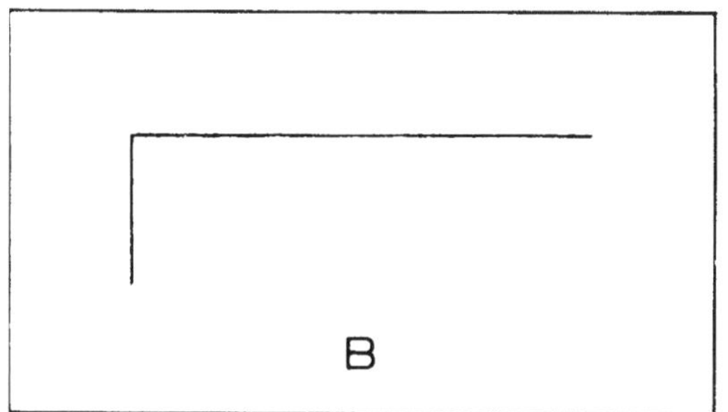

Figure 6.18 Drawing the trial scrap strip.

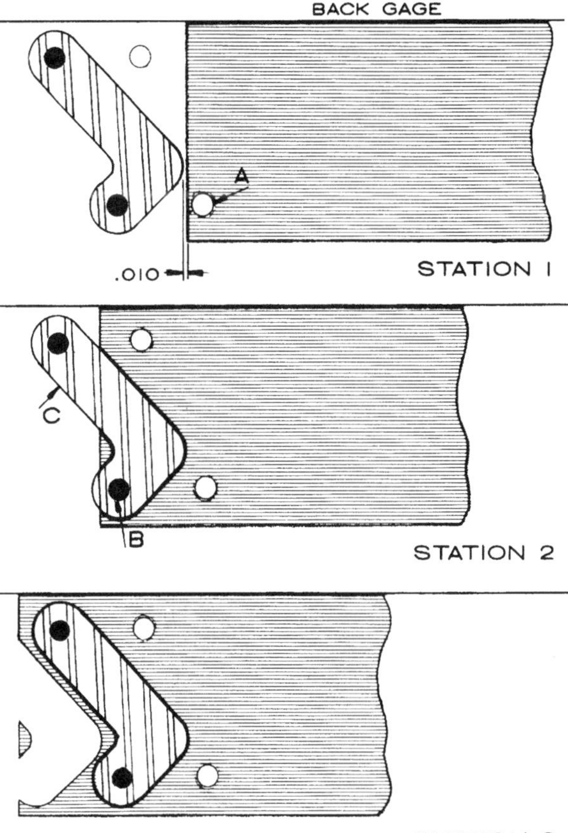

Figure 6.19 Testing the scrap strip for the first layout from Figure 6.14.

6.3.3 Testing and Evaluating the Strip

The four layouts from Figure 6.14 can now be tested and evaluated to determine which one will be the best.

a) The First Layout

Testing. Figure 6.19 shows the strip being run through the piercing and blanking punch layout **A** of Figure 6.17. At station 1 the strip is advanced as close to the blanking punch as possible. In dimensioning the die, the finger stop should locate the strip 0.010 inch (0.25 mm) away from the edge of the blanking punch. If the press were tripped, piercing punch **A** would pierce the first hole. In this paper layout, this hole is drawn on the strip as a circle.

At station 2, the strip is advanced until the pierced hole is in line with pilot **B** of the blanking punch. If the press were now tripped, blanking punch **C** would produce a partial blank while the two piercing punches would pierce holes in the strip. In our paper strip, we duplicate these operations by tracing around the blanking punch to show this partial removal of a blank, and around the piercing punches, just as the press would do it on an actual strip. This can be done freehand for the time being.

At station 3 the strip is advanced again until the holes are in line with the pilots. Tripping the press would produce a complete blank and two more pierced holes at station 1. Draw the outline of the blanking punch and the two holes on the paper strip. With practice these operations can be performed very quickly.

Evaluating results. Our scrap strip layout now looks like the one in Figure 6.20 at A, whereas the parts removed are shown at B. By analyzing this scrap strip, we see a serious fault. Tab **C**, cut at station 2, has remained on top of the die block where it can get under either the cutting edge of the blanking punch to chip it, or under the piercing punches to break them. The tab would be produced only at the beginning, and possibly at the end of each strip, but many hundreds of strips will be run during the life of the die.

Rule Number 1

Arrange the first partial blank so that an unattached piece of metal never remains on top of the die block.

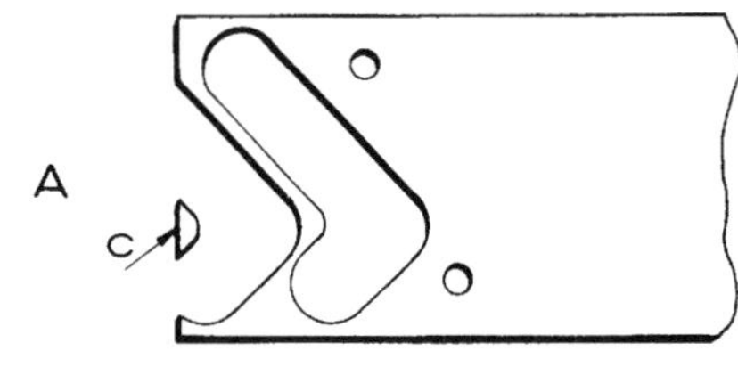

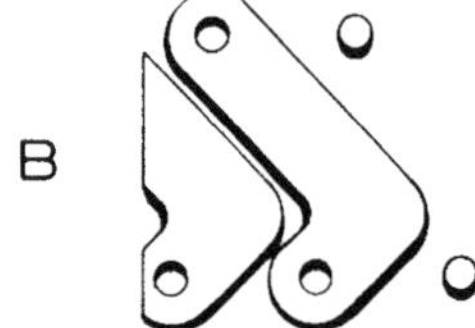

Figure 6.20 The resulting scrap strip (A) and parts removed (B) in testing the first layout.

b) The Second Layout

Testing. Figure 6.21 shows the strip being run through layout **B** of Figure 6.17. At station 1, we are unable to bring the strip very close to the blanking punch; to do so would mean a partial cut by piercing punch **A**, which would break it. Instead, use a finger stop to locate the strip close to the piercing punch. Draw the punch **B** in position. At station 2, pilot **C** locates the strip and blanking punch **D** makes a partial cut. Draw the blanking punch outline and the two piercing punch circles. At station 3, the blanking punch produces a full blank. Draw the blanking punch and the two piercing punch circles.

Evaluating results. Figure 6.22 at *a* shows the scrap strip produced by the die, whereas *b* illustrates the partial blank, full blank, and pierced

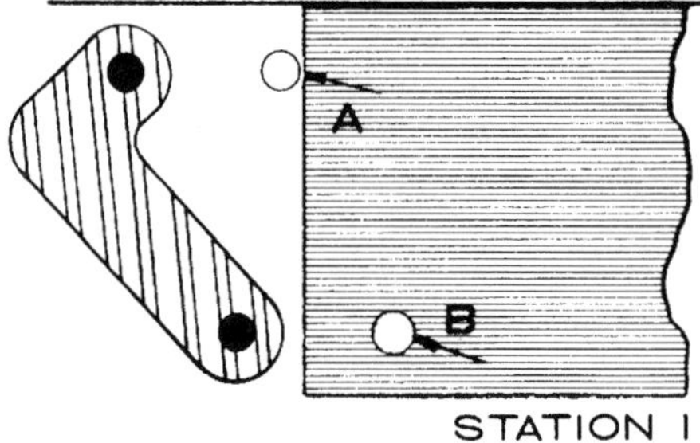

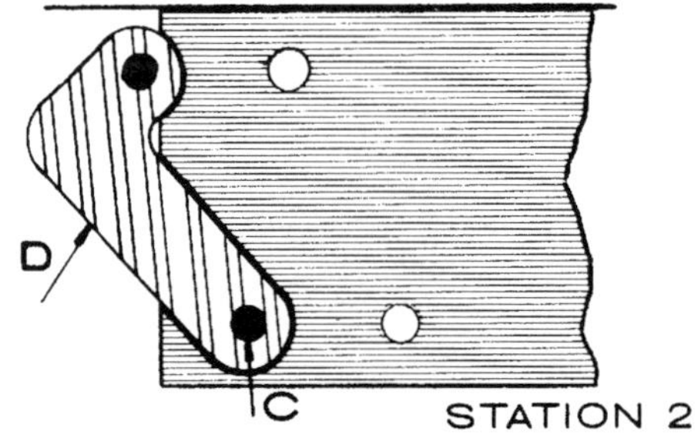

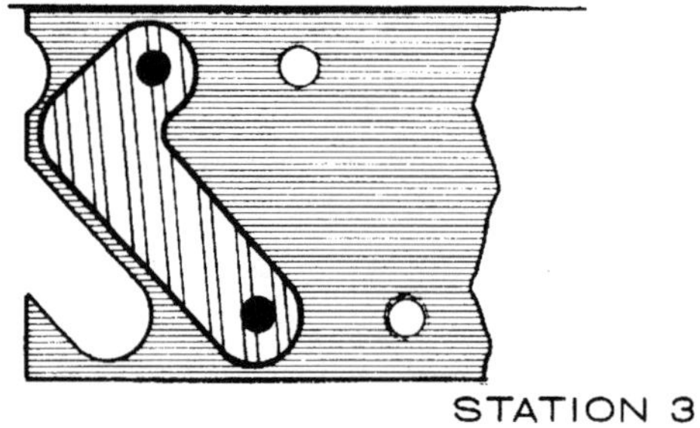

Figure 6.21 Testing the scrap strip for the second layout from Figure 6.14.

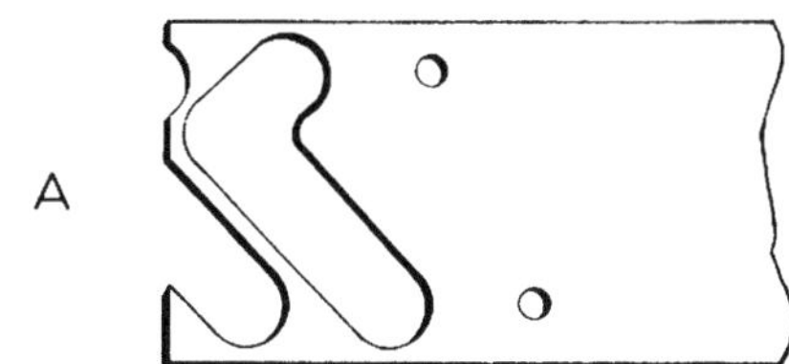

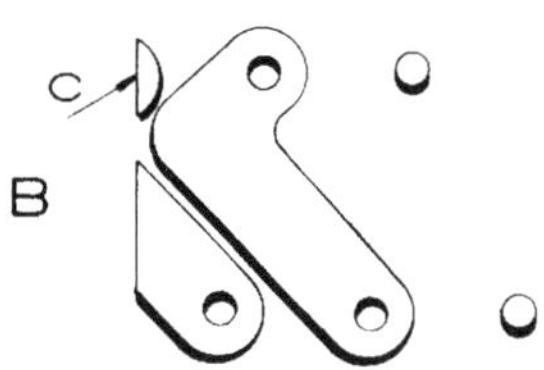

Figure 6.22 The resulting scrap strip (A) and parts removed (B) in testing the second layout.

slugs removed from it. No slug is left on top of the die block as in the previous example. However, it is not good practice to leave a partial slug as at **C**. Because the slug is not gripped securely in the die opening, it can cling to the face of the blanking punch when the press ram goes up and then drops on top of the die block.

Another bad feature of this layout is the long, *unsupported cut* made by the blanking punch on one side. An unsupported cut, or *unbalanced cut,* is one in which cutting occurs along one edge only. The cut is not balanced or supported by a cut at the opposite side. As the punch penetrates the material, side thrust is developed, tending to cause the punch to back away. Consequently, wear of the guide posts and bushings of the die set are caused by the increased side pressure. In time, as wear of posts and bushings increase, the blanking punch will be deflected sufficiently to strike the opposite edge of the die block, resulting in nicked cutting edges and shortened die life.

The effect of side deflection may be observed when we try to cut material with scissors that have become somewhat dull. As pressure is applied to cut the material, the blades have a tendency to spread apart or deflect.

Rule Number 2

Avoid cutting long, one-sided portions of a blank unsupported by a balancing cut on the other side.

Rule Number 3

Avoid cutting small portions of the strip not confined in the die block.

c) The Third Layout

Testing. The layout in Figure 6.23 is opposite to the layout for Figure 6.19, but the same rules and recommendations apply.

Evaluating results. This scrap strip, shown in Figure 6.24, is produced by the die in Figure 6.23. Although it does not seem to apply for this particular part, it would be the one to use for some parts with a different contour.

d) The Fourth Layout

Testing. The layout in Figure 6.25 is flipped top to bottom from the layout for Figure 6.21. The same rules and recommendations apply.

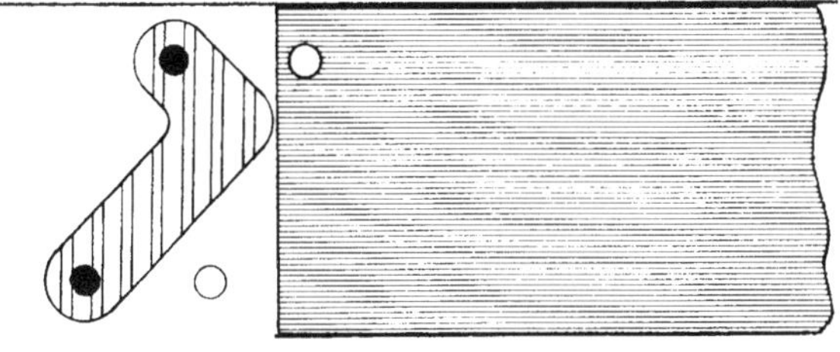

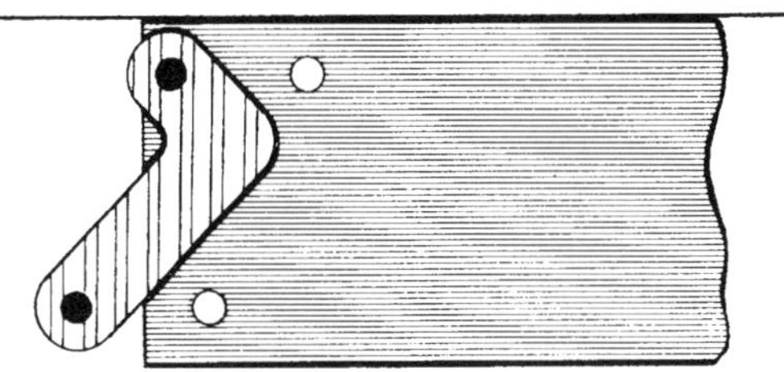

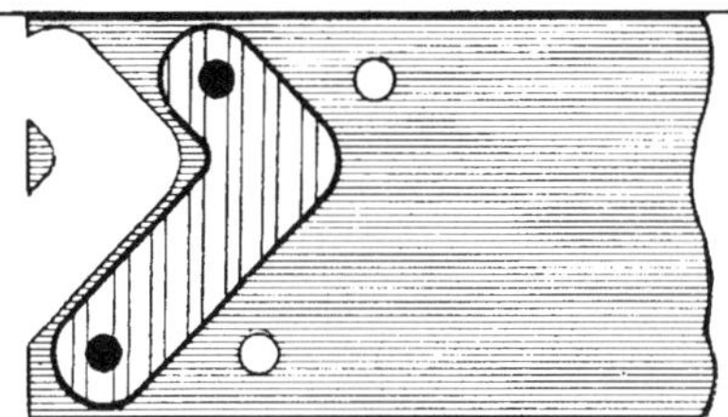

Figure 6.23 Testing the scrap strip for the third layout from Figure 6.14.

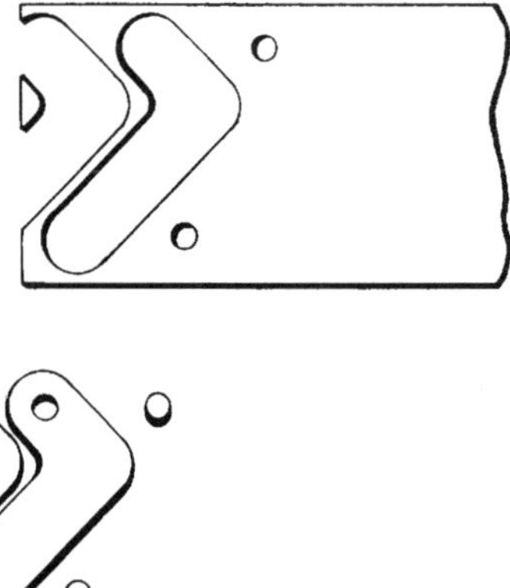

Figure 6.24 The resulting scrap strip and parts removed in testing the third layout.

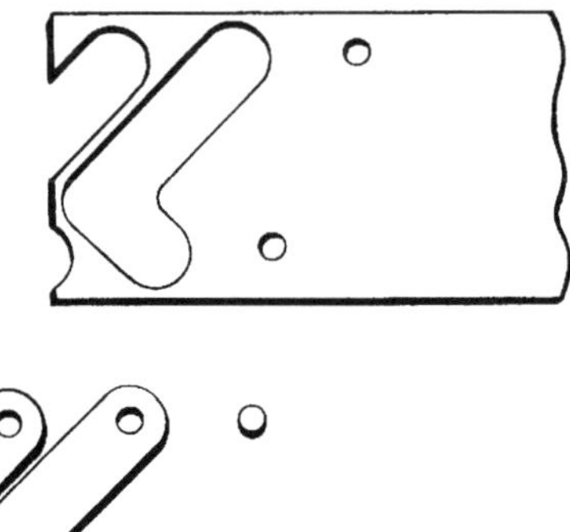

Figure 6.26 The resulting scrap strip and parts removed in testing the fourth layout.

Evaluating results. The scrap strip in Figure 6.26 is produced by the die in Figure 6.25. None of the four methods produces a trouble-free die. However, these layouts do provide a clue to development of an excellent scrap strip. The layout in Figure 6.19 appears promising. Inclination of the part is good for gaging purposes. The automatic stop should push the strip against the back gage, rather than pull it forward. But we must find a way to keep tab **C** attached to the strip, possibly by changing the angle of the blank.

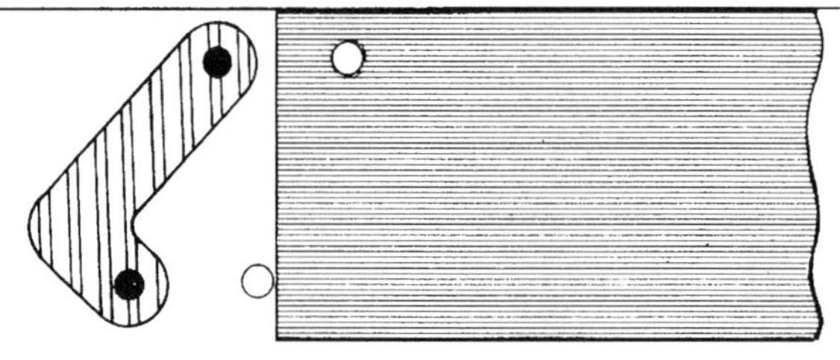

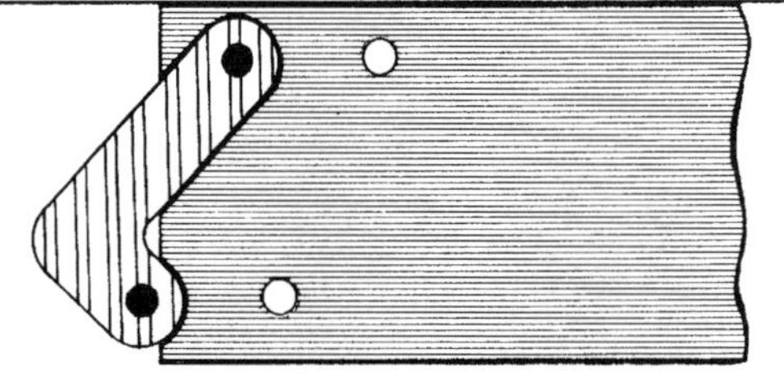

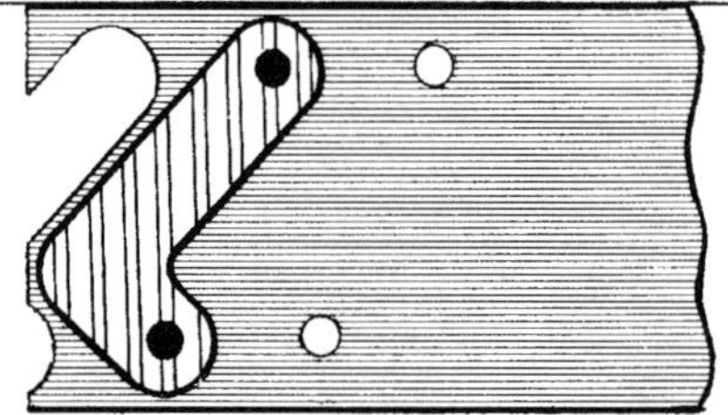

Figure 6.25 Testing the scrap strip for the fourth layout from Figure 6.14.

6.3.4 Altering the Layout

Place another sheet of translucent paper over the layout in Figure 6.5, turned over. Next, draw two blanks, leaving a greater distance **A** between them (Figure 6.27). Using a protractor to go over the outlines, straighten them accurately (Figure 6.28).

a) New Layout

After tracing the third blank, we have a new layout (Figure 6.29). The part is shown with a smaller angle to the horizontal than we had in Figure 6.13. From this tracing, we make another piercing and blanking punch layout and run the strip through it.

Testing the New Layout. At station 1 of Figure 6.30, the strip is brought close to the blanking punch to get the maximum "bite," and the first circle is drawn for piercing punch **A**. The altered angle creates a scrap bridge to hold the previously loose tab to the strip. Vary this angle until scrap bridge **B** is the same as the scrap bridge **C** between blanks. Cutting forces are fairly well balanced. There is little tendency for the blanking punch to

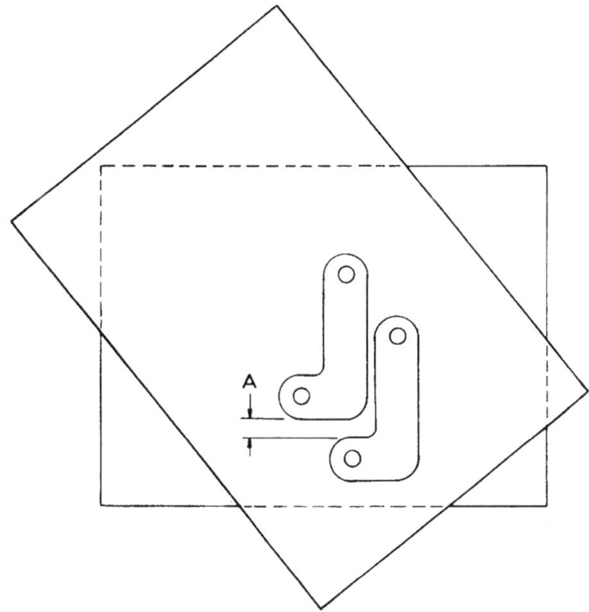

Figure 6.27 Tracing a new layout that allows for a greater space at **A**.

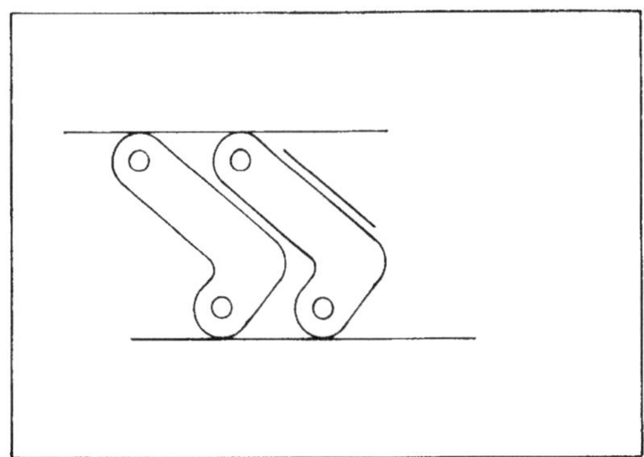

Figure 6.28 Corrected outlines.

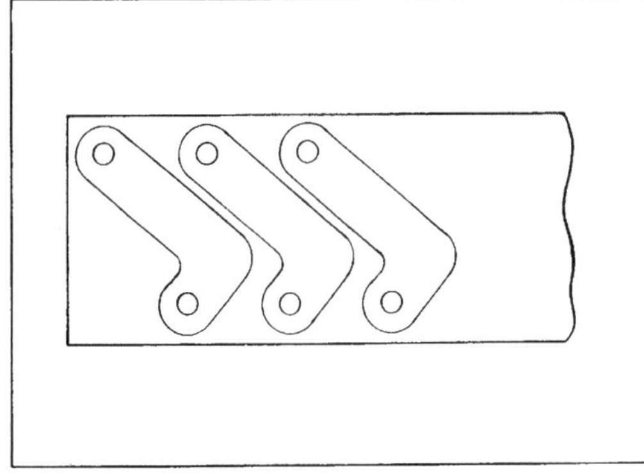

Figure 6.29 Completed new layout.

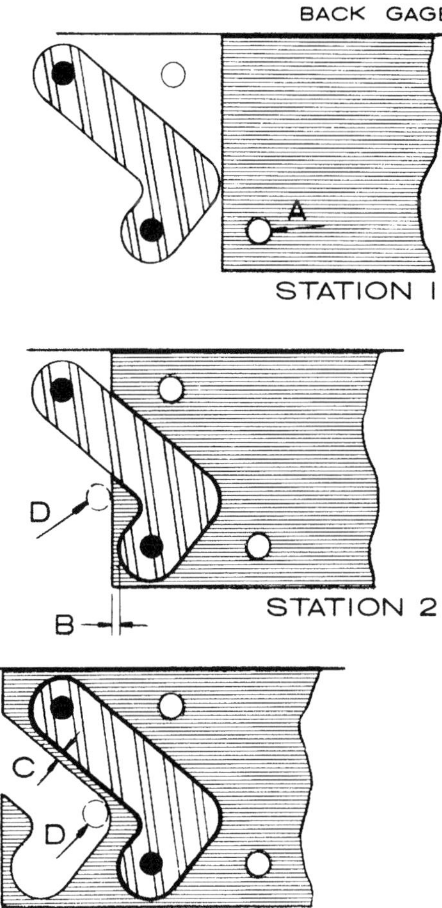

Figure 6.30 Testing the new scrap strip layout.

deflect with this arrangement. At station 3, a full blank is cut from the strip. Note that the toe of the automatic stop **D** is placed to stop the strip for the first blanking cut at station 2, and it also stops the strip for all succeeding cuts.

Completed scrap strip layout. Figure 6.31 shows the scrap strip resulting from testing the new layout, including our completed scrap strip, along with the partial blank, full blank, and pierced slugs which were removed from it. Die design for this part can now be safety started with full knowledge that a good tool will result.

6.4 SCRAP STRIP ALLOWANCES

It is important that correct bridge allowances be applied not only between blanks but also

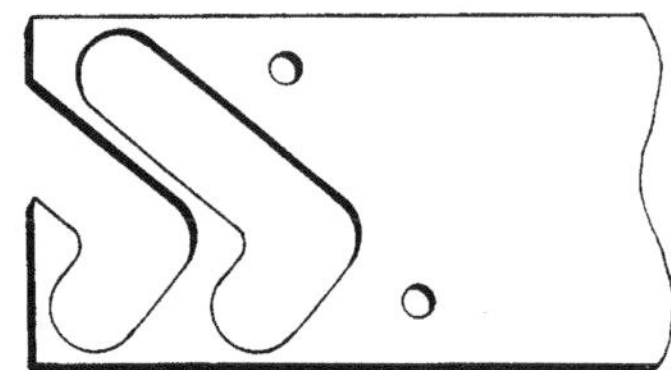

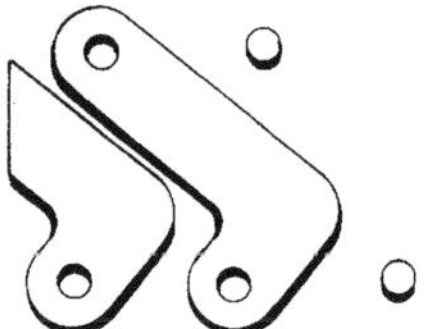

Figure 6.31 The resulting scrap strip and parts removed in testing the new layout.

between blanks and edges of the strip. Excessive allowance is wasteful of material. Insufficient allowance results in a weak scrap strip subject to possible breakage, with consequent slowdowns on the press line. In addition, a weak scrap area around the blank can cause *dishing* of the part.

6.4.1 Scrap Strip Allowance-One Pass Layout

Figure 6.32 shows peripheries of blanks classified under four distinct outline shapes. In the illustration:

1. *Curved outlines.* For these, dimensions **A** are given a minimum allowance of 70 percent of the strip thickness **T**.

2. *Straight edges.* Dimensions of **B** and **B′** depend upon the length of the bridge, dimensions **L** and **L′**, respectively:

Where **L** or **L′** is less than $2^1/_2$ inches (63.5 mm), **B** or **B′** = 1**T**, respectively.

Where **L** or **L′** is $2^1/_2$ to 8 inches (63.5 to 203 mm), **B** or **B′** = $1^1/_4$**T**, respectively.

Where **L** or **L′** is greater than 8 inches (203 mm), **B** or **B′** = $1^1/_2$**T**, respectively.

3. *Parallel curves.* For work with parallel curves, the same rules apply as for straight edges:

Where **L** is less than $2^1/_2$ inch (63.5 mm), **C** = 1**T**.

Where **L** is $2^1/_2$ to 8 inches (63.5 to 203 mm), **C** = $1^1/_4$**T**.

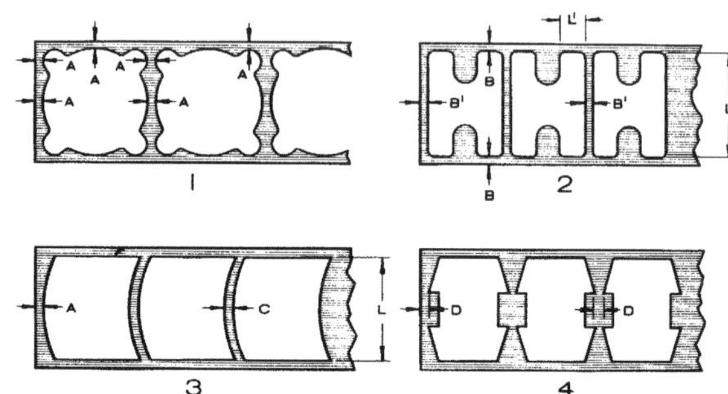

Figure 6.32 Four classifications of blank peripheries aid in determining scrap strip allowances for one-pass layouts.

Where **L** is greater than 8 inches (203 mm), **C** = $1^1/_2$**T**.

4. *Adjacent sharp corners.* These form a focal point for fractures. Minimum allowance is $1^1/_4$ **T**, dimension **D** on the drawing.

6.4.2 Scrap Strip Allowance-Two-Pass Layout

When the strip is to be run through the die twice in order to remove all the blanks from it, more allowance must be provided than is required for one-pass layouts. Figure 6.33 shows minimum scrap bridge allowances **A**, which should be applied under given conditions. Note that single-row layouts at *a* are given more allowance than is required for double-row layouts at *b* and *c*. This is because the wider double-row strips do not distort as much in cutting the first row and less allowance is required. Because of this minimum strip distortion, double-row layouts are preferred.

6.4.3 Minimum Allowances

Figure 6.34 shows minimum scrap strip allowances for both one-pass and two-pass layouts. These values apply for thin gages (less than 3/64 inch or 1.2 mm) of stock where use of the previous rules would give such small allowances as to be impractical. Select the value for space **A** opposite the appropriate strip width.

6.4.4 Application Allowances

Figure 6.35 shows how allowances are applied for two representative parts. The upper blank is laid out for single-row, one-pass through the die.

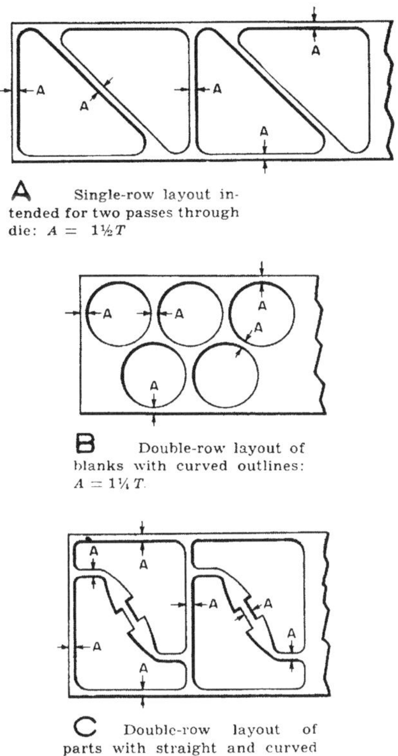

A Single-row layout intended for two passes through die: $A = 1\frac{1}{2}T$

B Double-row layout of blanks with curved outlines: $A = 1\frac{1}{4}T$

C Double-row layout of parts with straight and curved outlines: $A = 1\frac{1}{4}T$

Figure 6.33 Scrap strip allowances for two-pass layouts.

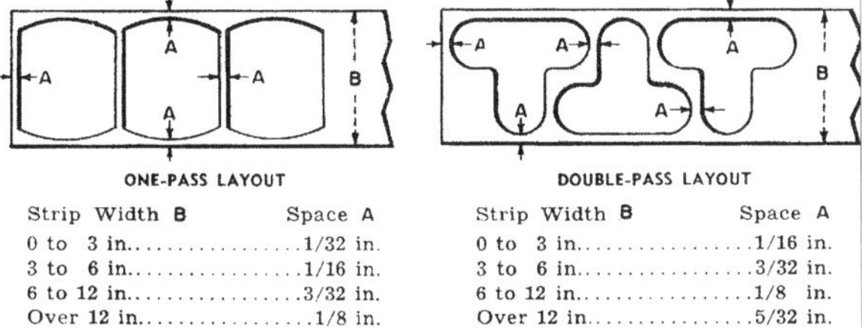

ONE-PASS LAYOUT

Strip Width B	Space A
0 to 3 in.	1/32 in.
3 to 6 in.	1/16 in.
6 to 12 in.	3/32 in.
Over 12 in.	1/8 in.

DOUBLE-PASS LAYOUT

Strip Width B	Space A
0 to 3 in.	1/16 in.
3 to 6 in.	3/32 in.
6 to 12 in.	1/8 in.
Over 12 in.	5/32 in.

Figure 6.34 Minimum scrap-strip allowances.

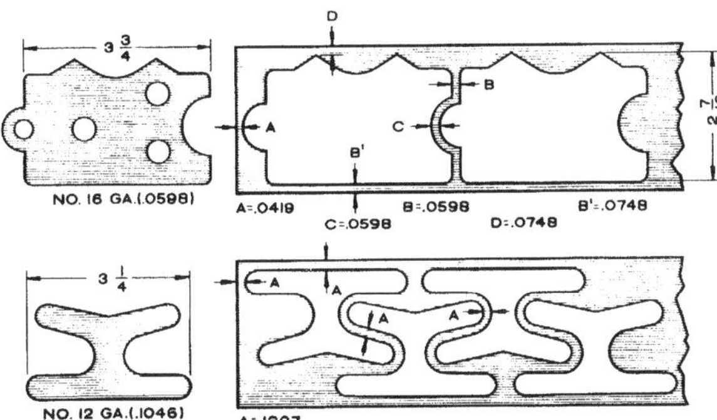

Figure 6.35 Application of allowances for two representative parts.

The lower blank is laid out for double-row, two-passes through the die. Calculated values are listed under each view. Realize that these are minimum allowances, the next larger fractional dimension would actually be used.

6.5 STRIP AREA PER BLANK CALCULATIONS

When it is possible to position blanks in two or more ways in the strip (Figure 6.36), the strip area per blank for each method is found. The one most economical of material is selected for the die. The top part of Figure 6.36 shows a blank laid out for single-row, one-pass positioning, the most frequently used method.

The area of the strip which is used for one blank is found by the following formula:

$$A_s = A \times B \tag{6.1}$$

where

A_s = area of the strip per one blank
A = width of strip
B = length of one piece

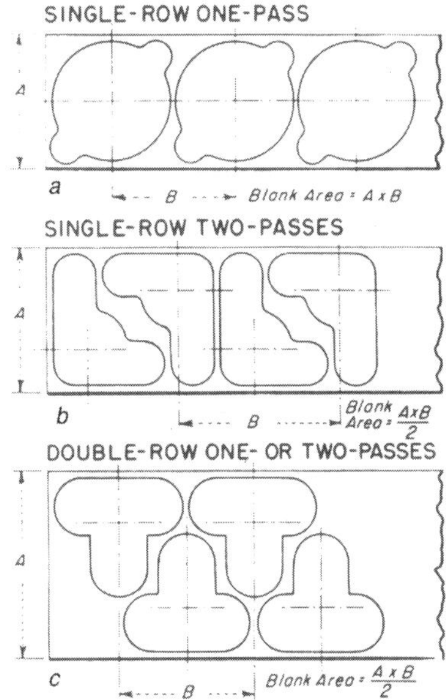

Figure 6.36 Representative illustrations for calculating blank areas.

Single-row, blanks, which must pass through the die twice, have their strip area per piece determined by the formula:

$$A_s = \frac{A \times B}{2} \tag{6.2}$$

This formula applies to the blank layout *b* in the middle of Figure 6.36 and also for double-row blanks shown at the bottom of the figure, layout *c*.

6.5.1 Number of Blanks per Strip

Representative illustrations for determining the number of blanks in a strip is shown in Figure 6.37. For large blanks, it is often necessary to determine the number of blanks in each strip to establish the extent of the waste end **D**. This has an influence on the blank layout because too great a waste end is uneconomical.

The number of blanks per strip for a single-pass layout (Figure 6.37 upper illustration) is found by the formula:

$$N = \frac{S-(X+Y+2E)}{B} + 1 \tag{6.3}$$

The waste at the end of the single-pass strip (Figure 6.37 upper illustration) is:

$$D = S - [B(N-1) + X + Y + 2E] \tag{6.4}$$

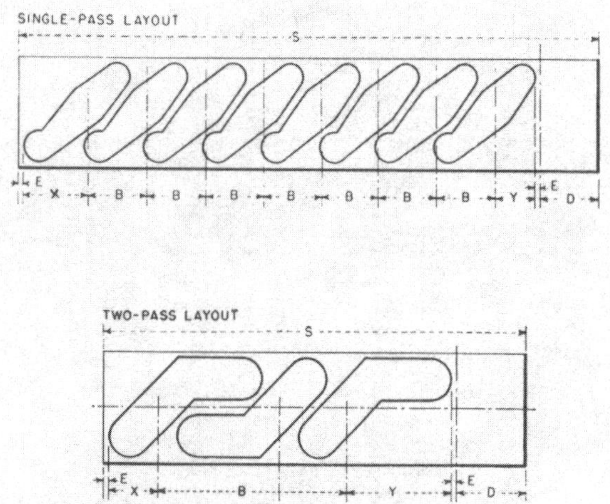

Figure 6.37 Representative illustrations for determining the number of blanks in a strip.

When strips must make two passes through the die (Figure 6.37 lower illustration), the following formulas determine the number of blanks per strip and the extent of the waste end:

$$N = \frac{1}{2} \times \frac{S-(X+Y+2E)}{B} + 1 \tag{6.5}$$

For the waste end:

$$D = S - \left[\frac{1}{2}B(N-1) + X + Y + 2E\right] \tag{6.6}$$

Example:

Figure 6.38 illustrates a flat blank A and a bent blank B for which scrap strips are to be laid out. The flat blank could be positioned vertically, horizontally, or at an angle in the strip. The bent blank should be positioned angularly to prevent possible fracture in the subsequent bending

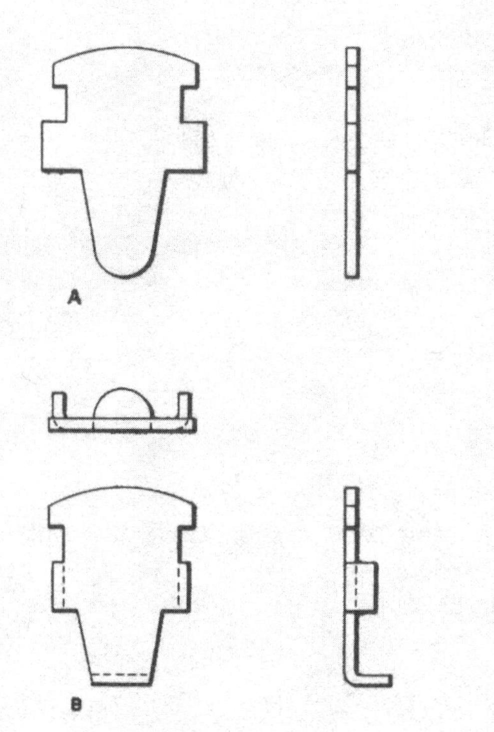

Figure 6.38 Flat A and bent B blanks.

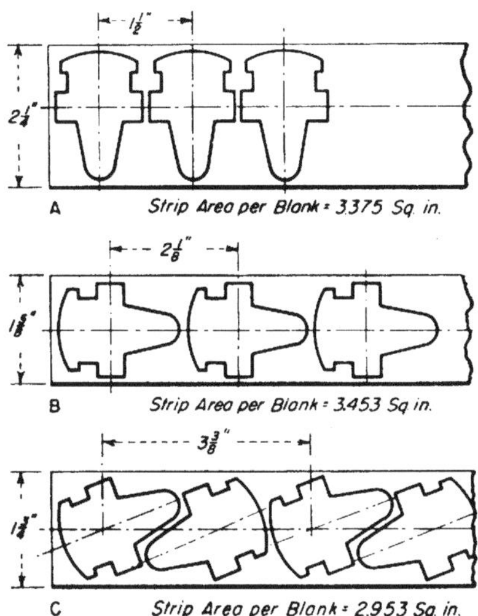

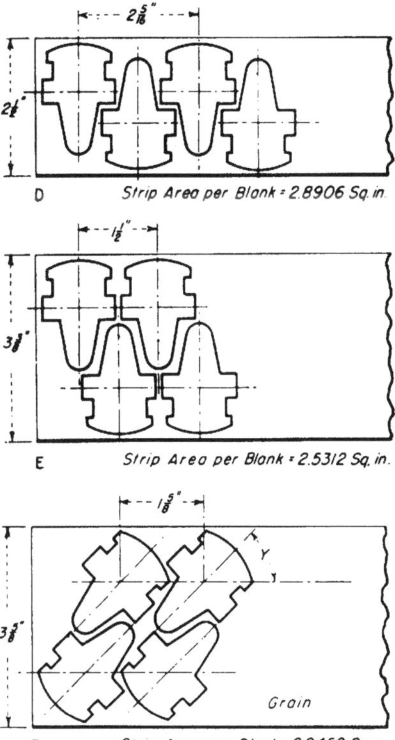

Figure 6.39 Six ways of positioning blanks.

operation. Calculate the area of the strip per one blank and six different positions of blank.

Solution:

Figure 6.39 shows six ways in which the parts illustrated in Figure 6.38 can be positioned for blanking. The strip width for each position, as well as the feed or distance from station to station, are noted. Observe the considerable difference in strip area per blank for the different methods of positioning. The most economical position is layout *e*.

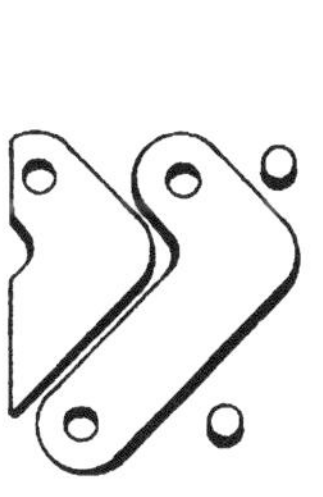

HOW TO DESIGN DIE BLOCKS

7.1 INTRODUCTION

Four factors influence the design of a die block for any particular die. They are:

1. Part size
2. Part thickness
3. Intricacy of part contour
4. Type of die

Small dies, such as those for producing business-machine parts, usually have a solid die block. Only for intricate part contours would the die block be sectioned to facilitate machining. Large die blocks are made in sections for easy machining, hardening, and grinding. The illustrations that follow show more methods of applying die blocks to small, medium, and large cutting dies.

Figure 7.1 show a photograph of a die block for a large piercing die. Large die blocks such as this one are composed of sections for easier machining, hardening, and grinding. Observe that each section is provided with working holes, that is, holes which engage punches to perform cutting operations on the material and with screw and dowel holes that fasten each section to the die holder.

This section of the book further explains Step 2 in Chapter 5—Fourteen Steps to Design a Die.

7.2 APPLYING THE DIE BLOCK

Figure 7.2 shows a method of applying a die block to a die. Machined into the die block (lower left illustration) are blanking opening **C** and piercing holes **D**. Holes **E**, located at each corner, are tapped completely through, and two dowel holes **F** are reamed completely through the block.

Section views A-A and B-B show the fastening method. Four socket cap screws **G** securely hold the die block to the die holder of the die set. Two dowels **H**, pressed into the die holder and partly into the die block, prevent any possible shifting in operation. Four button-head socket screws **I** fasten the stripper plate and gages to the die block, while two dowels **J** maintain accurate positioning. Distance **K**, usually 1/4 inch (6.35 mm), is the grinding allowance, used up when the screws and dowels are lowered in repeated sharpening of the die face.

Courtesy of Bethlehem Steel Co.

Figure 7.1 Die block for a large piercing die.

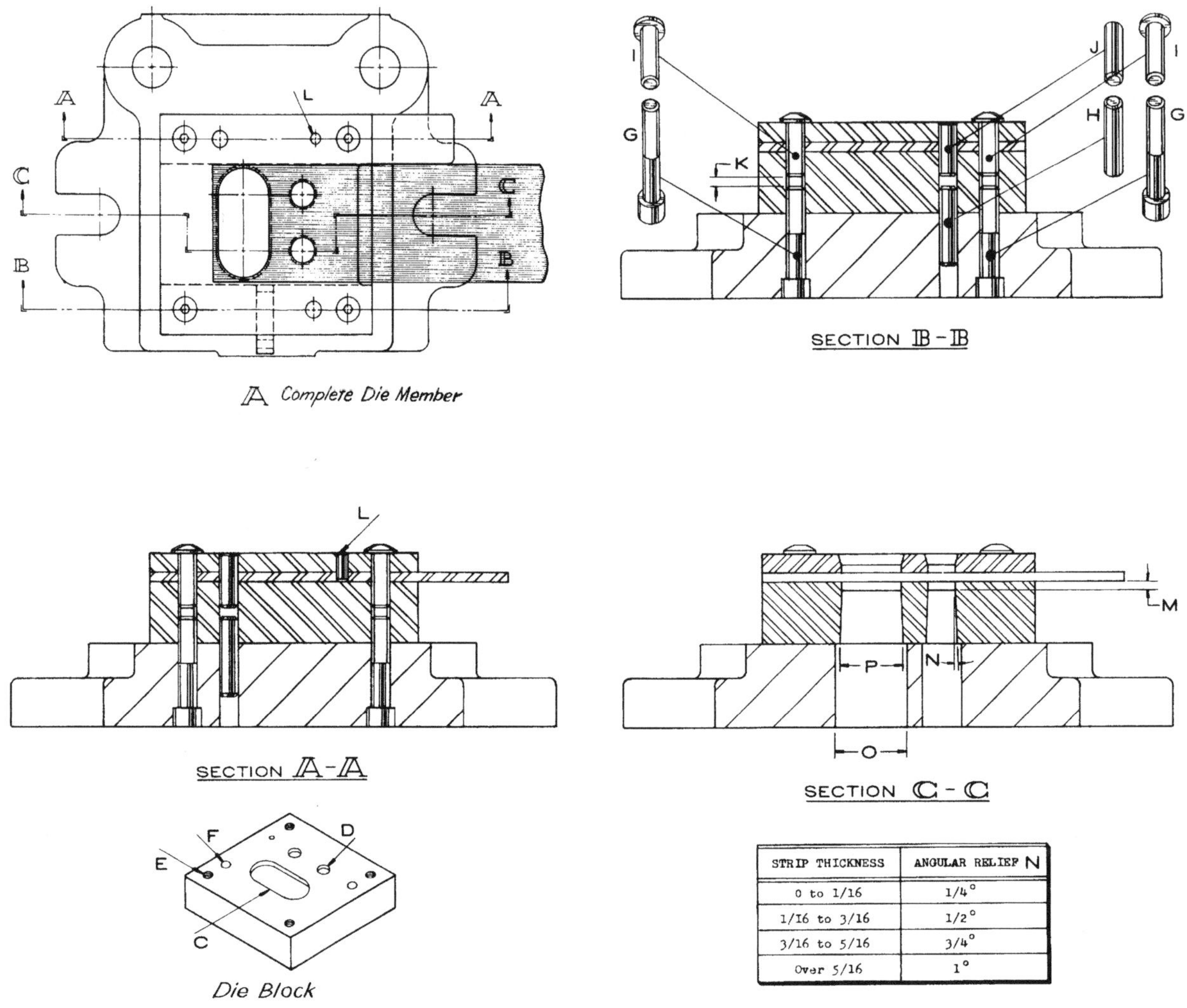

STRIP THICKNESS	ANGULAR RELIEF N
0 to 1/16	1/4°
1/16 to 3/16	1/2°
3/16 to 5/16	3/4°
Over 5/16	1°

Figure 7.2 Method of applying a die block to a die.

A small dowel **L** locates the right end of the back gage to the stripper plate. This dowel is made 3/16 inch (4.8 mm) in diameter. The other screws and dowels are never less than 5/16 inch (8.0 mm) diameter for any die. Dowels **J** are made a press fit in the stripper and gages, but a sliding fit in the die block. With this construction the stripper, back gage, and front spacer can be removed quickly for sharpening the die face without removing the die block from the die holder.

Section C-C shows the method of machining the die hole and the piercing punch holes. Straight land **M** is 1/8 inch (3.2 mm). Angular relief **N** is made according to strip thickness values given in the table in Figure 7.2. Hole **O** in the die set is 1/8 inch (3.2 mm) larger than opening **P** in the die block to provide 1/16 inch (1.6 mm) clearance per side.

7.2.1 Alternate Method

Because it conserves tool steel, the method of applying a die block in Figure 7.3 is recommended when material shortages exist. Tool steel die block **A**, made one-half the normal thickness, has a machine steel spacer **B** under it. Two long socket cap screws **C** fasten the die block to the die set, while two dowels **D** accurately locate it. Socket button-head screws **E** fasten the stripper plate to the die block, while dowels **F** and **G** are applied as in Figure 7.2. Section view C-C shows

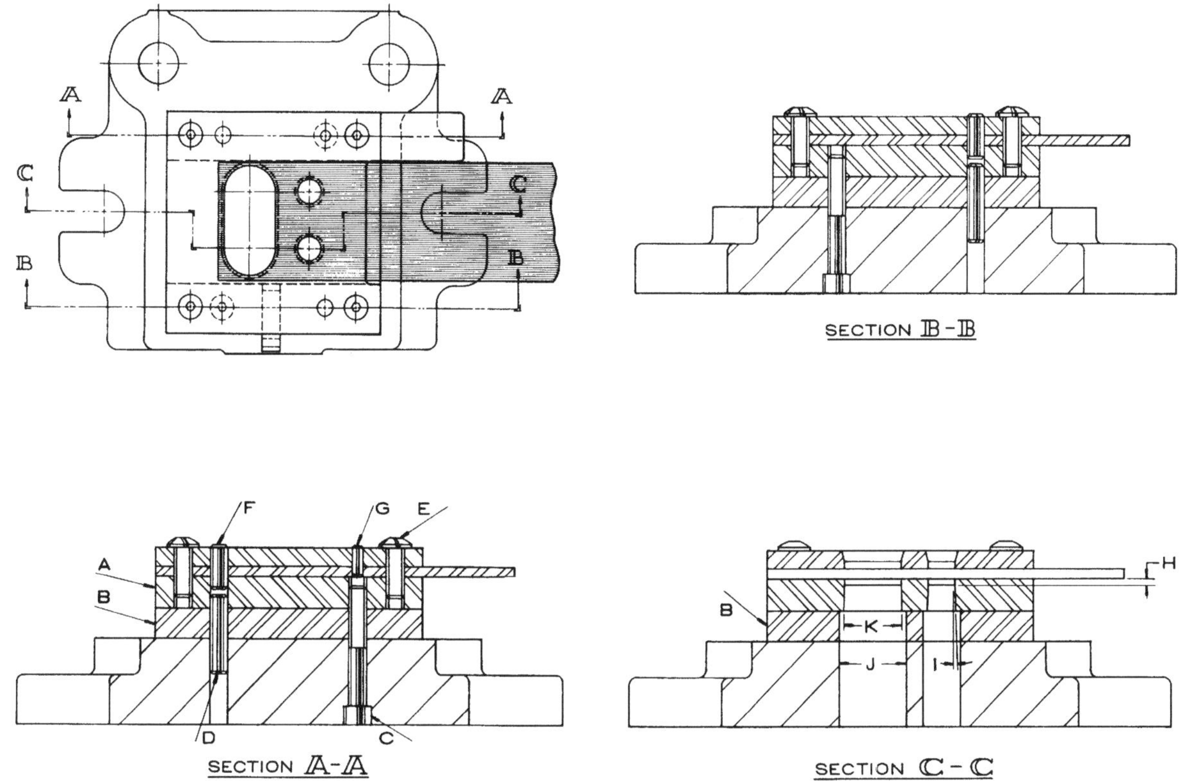

Figure 7.3 Alternate method of applying a die block to a die.

the method of machining the die hole and piercing punch hole. Straight land **H** is made 1/8 inch (3.2 mm). Angular relief **I** follows the values given in the table in Figure 7.2. Slug clearance hole **J**, 1/8 inch (3.2 mm) larger than **K**, is carried through both the die holder and spacer **B**.

7.3 RECOMMENDATIONS FOR DESIGNER

7.3.1 Proportion of Die Block

Figure 7.4 shows recommended minimum **C** distances from the die hole to the outside edge of the die block. These minimum distances depend almost entirely on the contour of the die hole.

Secondary considerations include:

Extent of Straight Lands

When straight lands are long, the die block is subjected to considerable side pressure during the cutting process. It is often advisable to use a minimum distance **C** in the thicker ranges, regardless of die hole contour.

Size of the Die Hole Opening

When openings are large and the material strip is thick, it is wise to go to the thicker **C** ranges, regardless of die hole contour.

For die holes with smooth or rounded contours, view Figure 7.4, part 1, where minimum distance **C** is generally 1 1/8 times the thickness **B** of the die block. When the die hole contains inside corners, view part 2; here, minimum distance **C** is generally 1 1/2 times die block thickness **B**. Part 3 of Figure 7.4 shows severe inside corners in the die hole contour. Minimum distance **C** is then made twice the thickness of the die block. These minimum distances are tabulated for recommended die block heights which, in turn, are governed by strip thickness **A**.

The table in Figure 7.4 also gives recommended die block heights **B** for various strip thicknesses.

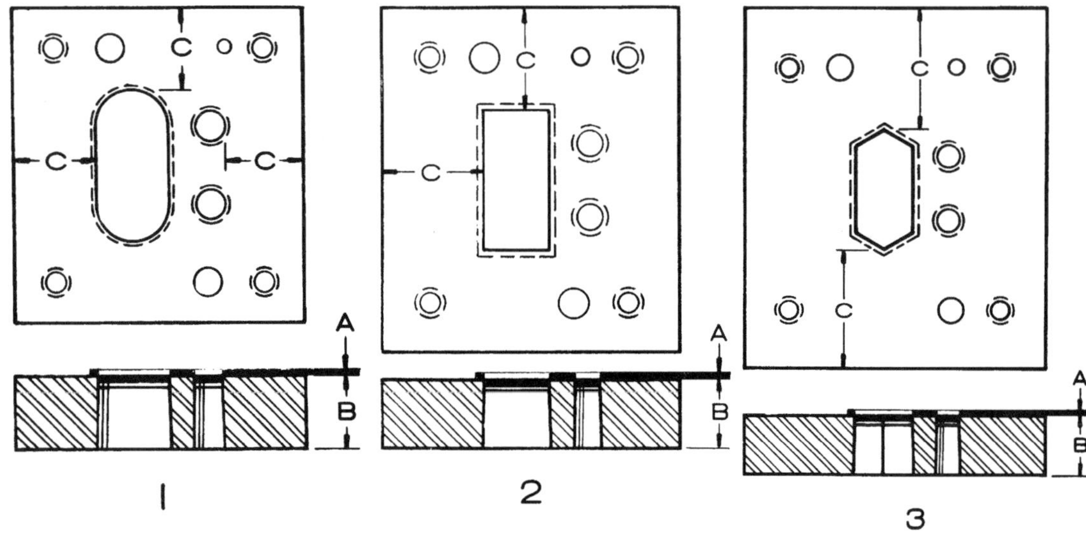

A	B	C MINIMUM DISTANCE - DIE HOLE TO OUTSIDE EDGE		
STRIP THICKNESS	DIE BLOCK HEIGHT	1 SMOOTH DIE HOLE CONTOUR (1 1/8 B)	2 INSIDE CORNERS (1 1/2 B)	3 SHARP INSIDE CORNERS (2 B)
0 to 1/16	15/16	1.0547	1.4062	1.875
1/16 to 1/8	1 1/8	1.2656	1.6875	2.250
1/8 to 3/16	1 3/8	1.5469	2.0625	2.750
3/16 to 1/4	1 5/8	1.8281	2.4375	3.250
over 1/4	1 7/8	2.1094	2.8125	3.750

Figure 7.4 Recommended minimum **C** distances for various die hole contours and die block heights **B**.

7.3.2 Foolproofing

Dies with symmetrical openings can be assembled incorrectly after repair. This occurrence can result in nicked cutting edges because of slight mismatching of punch and die members. Prevent this by foolproofing the die block. When dimensioning, place one of the dowels a different distance than the other from its nearest screw hole, as shown in Figure 7.5.

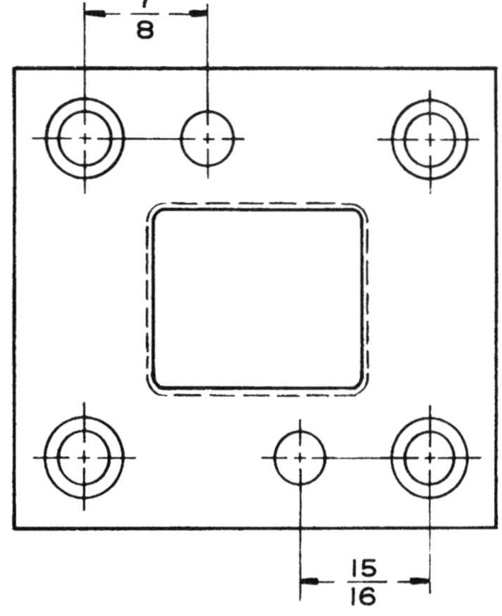

Figure 7.5 Mismatching of fastener holes called "foolproofing."

7.3.3 Blank Enlargement

When close part tolerances are present, you should know how much the blank or hole size will increase in die sharpening *past the straight land*. The table in Figure 7.6 lists the amount of growth for each 0.005 inch (0.127 mm) removed from the face of the die block when 1/4-degree angular relief is applied. For 1/2-degree relief, multiply the values given by 2. For 3/4-degree relief, multiply by 3, and for 1-degree relief, multiply by 4. Of course, there will be no increase in

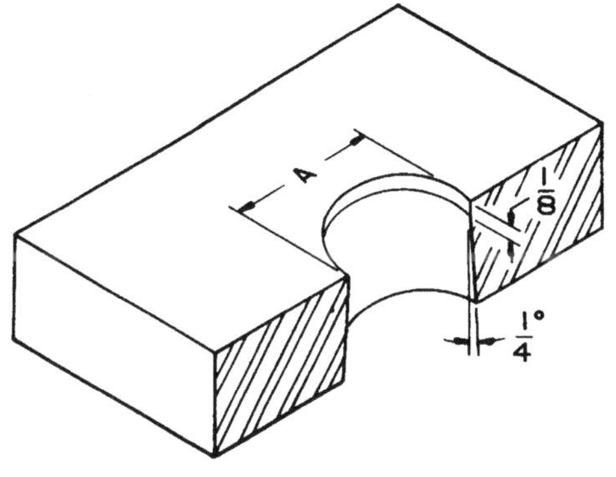

Removed from face of die in sharpening	Blank opening A will increase by this amount
0.005	0.000043
0.010	0.000087
0.015	0.000130
0.020	0.000174
0.025	0.000218
0.030	0.000261
0.035	0.000305
0.040	0.000348
0.045	0.000392
0.050	0.000436
0.055	0.000479
0.060	0.000523
0.065	0.000566
0.070	0.000610
0.075	0.000654
0.080	0.000697
0.085	0.000741
0.090	0.000784
0.095	0.000828
0.100	0.000872
0.105	0.000915
0.110	0.000959
0.115	0.001002
0.120	0.001046
0.125	0.001090

Figure 7.6 Table for determining hole size when a die is sharpened past the straight land.

part size for the first 1/8 inch (3.2 mm) removed from the face of the die block because this is a straight land.

7.3.4 Standardizing

Where large quantities of dies are built, time and expense can be saved by standardizing on die block sizes. These standard blocks can be machined during slow periods, or by apprentices, then stocked ready for finish-machining when required. An inexpensive drill jig will assure interchangeability of screw holes.

The die block can then be specified on the drawing by part number, reducing time required for dimensioning. The table in Figure 7.7 shows the most commonly used sizes for small dies.

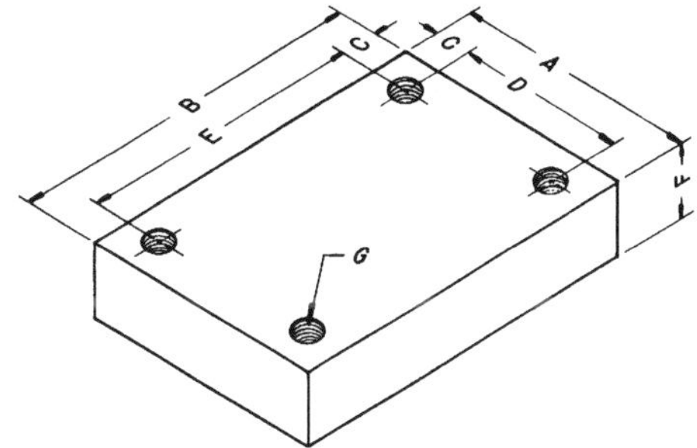

A	B	C	D	E	F	G
3	3 1/2	5/8	1 3/4	2 1/4	15/16	#I (.272) Drill, 5/16-24 Tap Thru
3	5	5/8	1 3/4	3 3/4	15/16	#I (.272) Drill, 5/16-24 Tap Thru
4	4	5/8	2 3/4	2 3/4	15/16	#I (.272) Drill, 5/16-24 Tap Thru
4	5	5/8	2 3/4	3 3/4	15/16	#I (.272) Drill, 5/16-24 Tap Thru
4	6	5/8	2 3/4	4 3/4	15/16	#Q (.332) Drill, 3/8-24 Tap Thru
5	5	3/4	3 1/2	3 1/2	15/16	#Q (.332) Drill, 3/8-24 Tap Thru
5	6	3/4	3 1/2	4 1/2	15/16	#Q (.332) Drill, 3/8-24 Tap Thru

Figure 7.7 Tabulation of suggested standard die block sizes.

Tabulated data in Figure 7.8 are the most commonly used sizes of standard die blocks for medium size die. Tapped holes are usually National Fine Thread because they resist loosening under vibration better than the Coarse Thread Series.

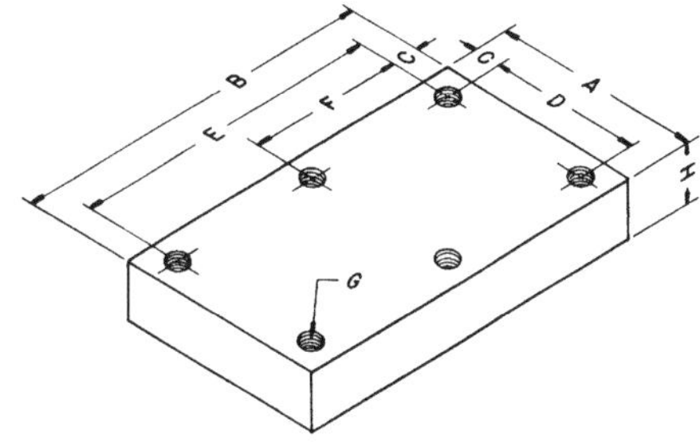

A	B	C	D	E	F	G	H
4	7	3/4	2 1/2	5 1/2	2 3/4	#Q (.332) Drill, 3/8-24 Tap Thru	1 1/8
4	8	3/4	2 1/2	6 1/2	3 1/4	#Q (.332) Drill, 3/8-24 Tap Thru	1 1/8
5	8	3/4	3 1/2	6 1/2	3 1/4	#Q (.332) Drill, 3/8-24 Tap Thru	1 1/8
5	10	3/4	3 1/2	8 1/2	4 1/4	#Q (.332) Drill, 3/8-24 Tap Thru	1 1/8
6	8	3/4	4 1/2	6 1/2	3 1/4	#Q (.332) Drill, 3/8-24 Tap Thru	1 1/8
6	10	3/4	4 1/2	8 1/2	4 1/4	#Q (.332) Drill, 3/8-24 Tap Thru	1 1/8
7	11	3/4	5 1/2	9 1/2	4 3/4	#Q (.332) Drill, 3/8-24 Tap Thru	1 1/8

Figure 7.8 Tabulation of suggested standard medium-size die block sizes.

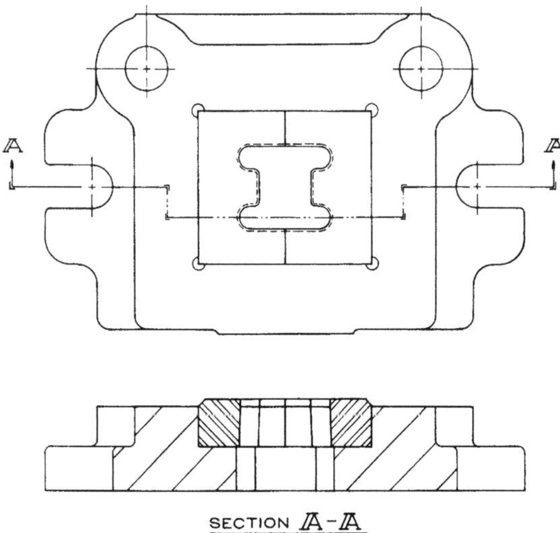

Figure 7.9 This die block is split for easier machining.

7.4 TYPES OF DIE BLOCK

7.4.1 Split Die Blocks

When the die block is split for easier machining of the die opening, it is set into the die holder. Figure 7.9 shows a split block, pressed into a recess machined in the die holder.

Wedge Lock

A tapered wedge held down by socket cap screws is often used to clamp split die blocks set into the die holder (Figure 7.10). Knockout pins can be inserted through the holes **A** for pressing out the wedge.

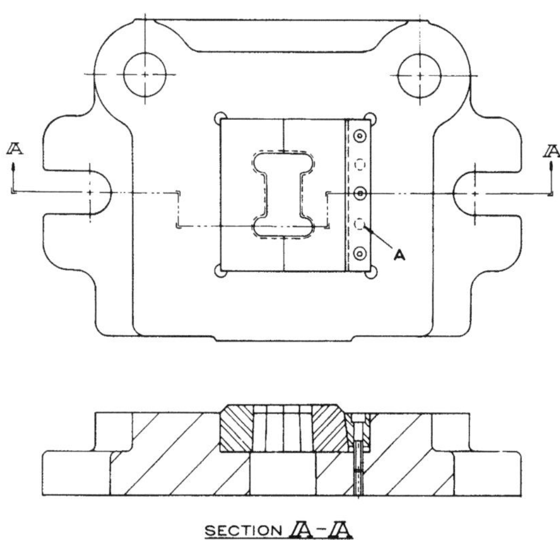

Figure 7.10 The split die block is held in place by a tapered wedge.

7.4.2 Inserted Die Bushings

Hardened die bushings, held in machined steel retainers, are commonly used in large dies where piercing occurs. They conserve tool steel and are easily replaced should edges become chipped or worn. Figure 7.11 shows several versions of die bushings used for piercing. The straight press-fit (**P.F.**) bushing at ***a*** is the type most commonly used. The bushing is made a slip fit (**S.F.**) for approximately 1/8 inch (3.2 mm) at its lower end. This assures correct alignment when pressing in, an important feature for accurate center-to-center distance between holes. The inside of the bushing hole is ground straight for 1/8 inch (3.2 mm). The remainder is given angular relief for proper slug disposal.

With spring strippers, the top of the bushing is made flush with the face of its retainer, as shown at ***b***. For heavy material strip, the flanged type of bushing shown at ***c*** is used to provide greater area against the die set. It is also used when upward pressure would tend to work the bushing upward. Shown at ***d***, a bushing with an irregular hole is kept from turning by a flat ground in the shoulder. This bears against one edge of a slot machined in the retainer. Where space is limited, die bushings can be kept from turning by a key bearing against a flat ground in the shoulder, as shown at ***e***.

7.4.3 Die Block with Bevel-Cut Edges

a) Angular Shear

For large blanks, shear is applied to the face of the die to reduce shock on the press, force required, and blanking noise. Properly applied, shear reduces the blanking force by 25 percent for metals thicker than 1/4 inch (6.35 mm). When thinner stock is blanked, the reduction in force is as large as 33 percent. The shear shown in Figure 7.12 is the most common method of providing for shear on the face of a large die block. Shear depth **A** is made 2/3 the thickness of the material strip. Radius **B** removes the sharp edge to avoid a focal point for stock fracture.

b) Alternate Method

Figure 7.13 shows an alternate die block for employing shear in a piercing die. This is a better method of providing angular shear. The material,

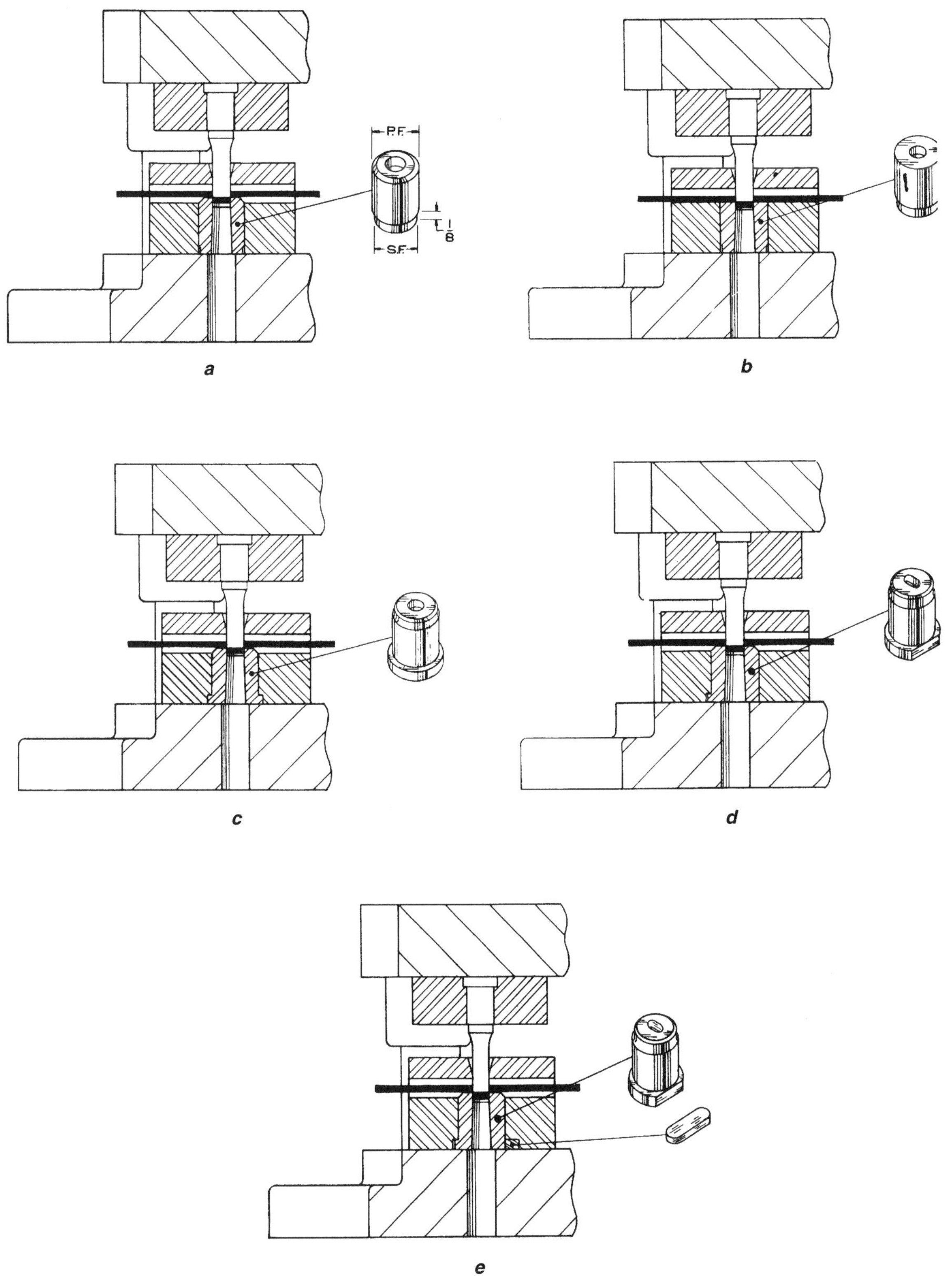

Figure 7.11 Several versions of hardened die bushings used for piercing.

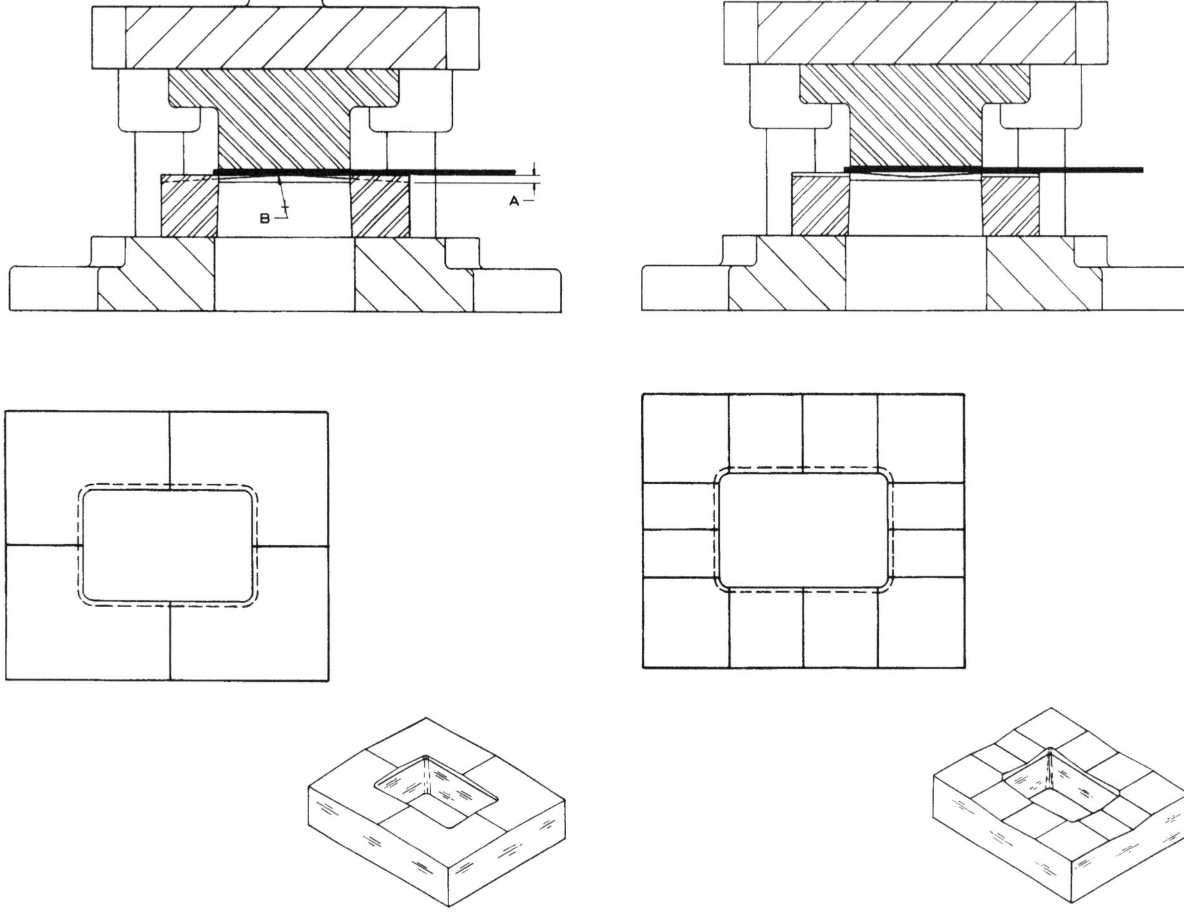

Figure 7.12 Die block configuration for employing shear in producing large blanks.

Figure 7.13 Alternate die block configuration for employing shear in producing large blanks.

securely gripped at the corners between the punch and die members, is gradually cut towards the center. Shear is usually applied when blanks are larger than 5 inches by 5 inches (127 by 127 mm), and over 1/16 inch (1.6 mm) thick.

c) Shear in Round Die Block

Round die blocks may be given a shearing action by scalloping the face in a series of waves around the periphery (Figure 7.14). Here again, shear depth should not exceed 2/3 the thickness of the stock.

7.4.4 Sectioning Die Blocks

In Figures 7.9 and 7.10, the subject of sectioning die blocks for ease in hardening and grinding was introduced; it may be well to enter into a more precise understanding of it at this time. For some die hole shapes, sectioning the die block provides the following distinct advantages:

- Difficult machining is avoided.
- Less distortion occurs in hardening.
- Any distortion can be corrected by grinding.
- Inserts can be replaced more quickly in case of breakage.

a) Fastening Die Segments

Figure 7.15 illustrates two ways of fastening tool steel bars for blanking parts of square or rectangular shape.

In part **A**, the bars are set on top of the die set. Clamping members **C** hold the longer side components **D** from spreading under the cutting thrust. The short members are always used as clamps

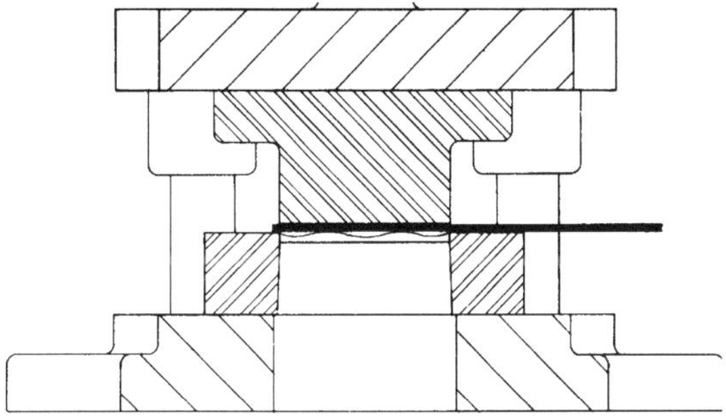

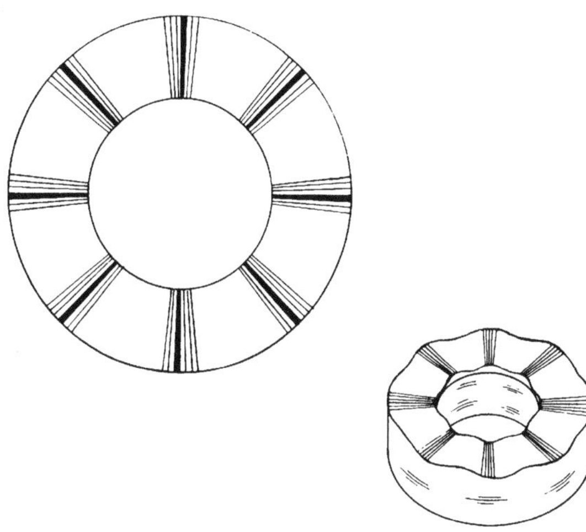

Figure 7.14 Scalloped round die block for employing shear.

because the greatest amount of side pressure occurs against the longer sides.

Part **B** shows a rectangular sectional die for long runs. It is composed of four bars **E** arranged, as shown. This provides for setting the bars to the exact opening required and it obviates the necessity of accurate machining of the members. The assembly is set into the die holder of the die set, as in Figure 7.9.

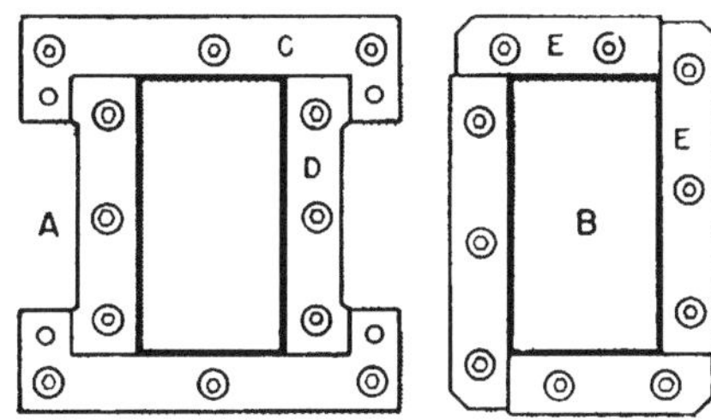

Figure 7.15 Methods of fastening die steels.

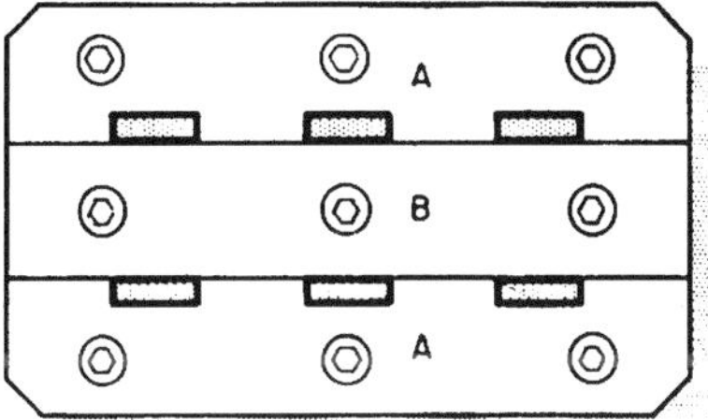

Figure 7.16 Die block sectioned for piercing two rows of rectangular slots.

b) Sections for Slots

Die blocks for piercing rows of square or rectangular slots are often sectioned (Figure 7.16). In this example, slots are machined in parts **A**, and part **B** is a spacer of proper width. The entire assembly is set into a recess in the die set to prevent deflection under cutting pressure.

c) Center Inserts

Figure 7.17 shows various groupings of square and rectangular holes that can be punched with

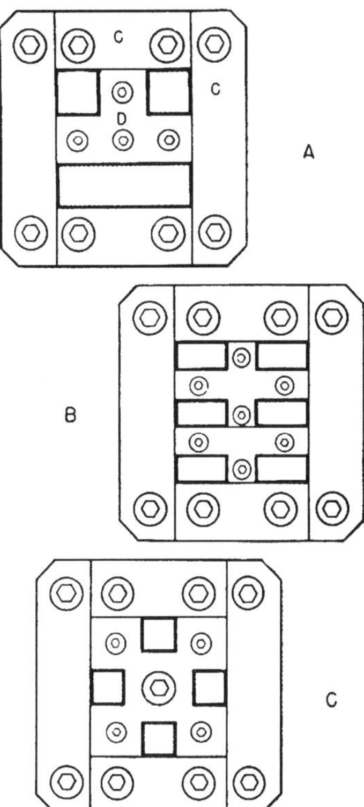

Figure 7.17 Various sectioned die blocks with center inserts for punching different arrays of square and rectangular holes.

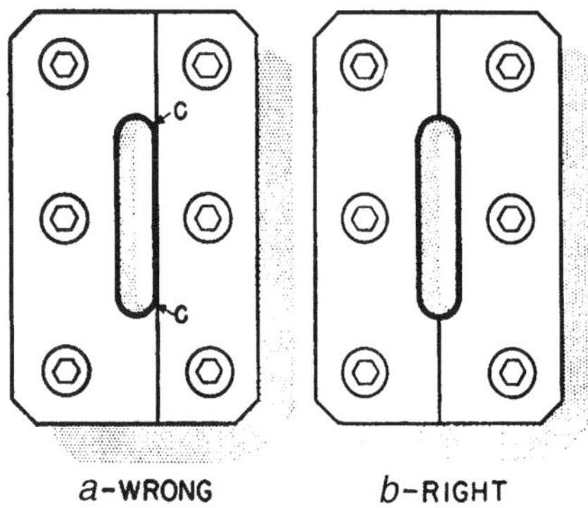

Figure 7.18 Correct and incorrect ways of applying break lines in sectioned die blocks.

square sectional dies. The die block shown at illustration *a* consists of four tool steel parts **C** and the center insert **D**, both hardened and ground. Construction of the die blocks shown in illustrations *b* and *c* is similar. Each would be retained in a pocket milled in the die holder of the die set.

d) Applying the Break Line

The first rule in sectioning die blocks having openings with curved outlines is never to apply the break line tangent to an arc, as shown in Figure 7.18 at illustration *a*. The reason is sharp projections **C** would be weak and subject to breakage. Break lines should cross the centers of radii, as at illustration *b*.

e) Inserts

Frail projections should be designed as inserts for easy machining, heat treating, grinding, and replacement in case of breakage. In the upper illustration of Figure 7.19, the die block is composed of two side parts **A** and an insert **B** with a slender, easily broken projection at one end. In the lower illustration, the die block is composed of four corner parts and a replaceable insert **E**.

f) Die Segments

When a portion of the blank outline is circular in shape, the die block may be bored to size and inserts applied to define the remainder of the shape (Figure 7.20). In the illustration, tool steel block **A** is machined for screws and dowels, and the large center hole is bored. The block is then hardened and subsequently ground to size.

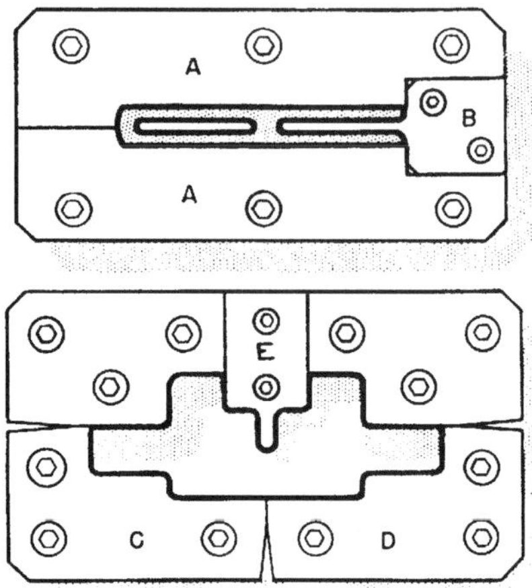

Figure 7.19 Use of inserts in providing for frail projections.

Segments **B** are machined, hardened, ground, and placed in position. The assembly is then fastened to the die holder.

g) Round Sectional Die Blocks

A round sectional die block consists of a hardened and ground tool steel ring employed to retain inserts or segments, preventing their deflection under cutting pressure. Six representative examples are shown in the three illustrations of Figure 7.21, as follows:

Illustration 1: The insert may consist of a single part. At drawing **A**, the center plug **D** of a lamination die is retained in ring **C**. Slot openings and a center hole have been machined in the

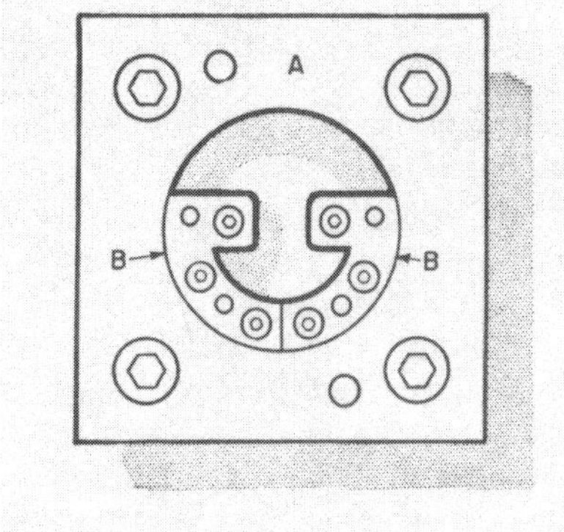

Figure 7.20 Use of inserts in a die block when a portion of the blank is circular in shape.

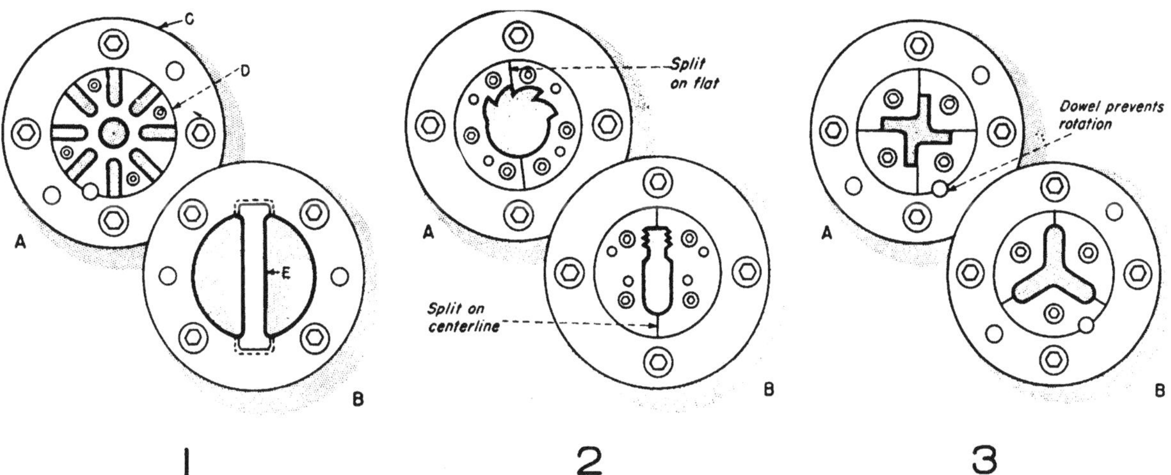

Figure 7.21 Six representative examples showing the use of round sectional die blocks in retaining inserts or segments.

plug **D** prior to hardening, grinding, and assembly within ring **C**. Drawing **B** shows a die block for punching two semicircular openings in a blank. Fitted insert **E** is provided with a shoulder at the bottom to prevent it from pulling out.

Illustration 2: The insert may consist of two parts. At drawing **A** of illustration 2, two ground segments are retained in a ring. The die blanks a small ratchet wheel; sectioning simplifies machining of the center opening. At drawing **B**, a die block for electrical contacts is split into two segments for grinding the opposed halves, which define the die opening. A ring keeps the segments from deflecting under cutting pressure.

Illustration 3: The insert may consist of several parts. The shape of the blank opening may require several parts. At drawing **A**, four identical sections compose the die block. At drawing **B**, the shape of the opening is such that a three-section division is best.

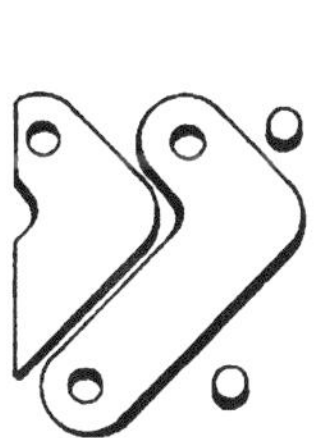

HOW TO DESIGN BLANKING PUNCHES

8.1 INTRODUCTION

Blanking punches range from tiny components for producing watch and instrument parts to large, multi-unit members for blanking such parts as automobile fenders, doors, and tops. The size of the blank to be produced determines the type of punch to use. Design considerations include:

- Stability, to prevent deflection.
- Adequate screws, to overcome stripping load.
- Good doweling practice, for accurate location.
- Sectioning, if required, for proper heat treatment.

Illustrations that follow show numerous methods of applying blanking punches to small, medium, and large cutting dies. Several considerations are covered, such as keying the punch in order to keep it from turning, use of inserts for ease and economy of replacement, use of sectioning to facilitate heat-treating and minimize distortion, use of shedders to prevent clinging of blanks to punch faces, and proper proportions of and construction of blanking punches. These methods further explain Step 3 in Chapter 5—Fourteen Steps to Design a Die.

8.2 APPLYING BLANKING PUNCHES

8.2.1 Small Blanking Punches

Figure 8.1 shows the method of applying a blanking punch for small instrument washers.

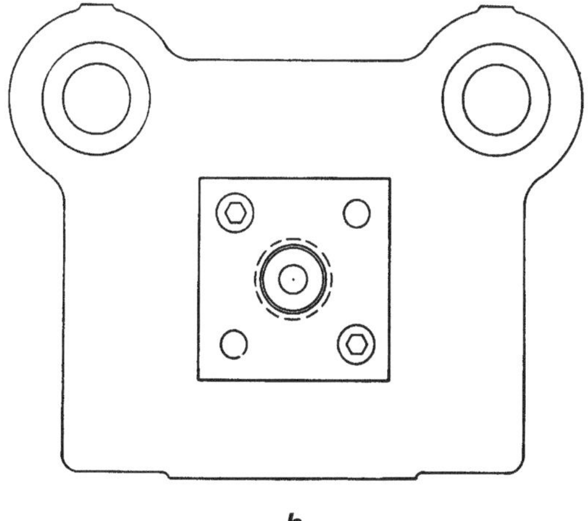

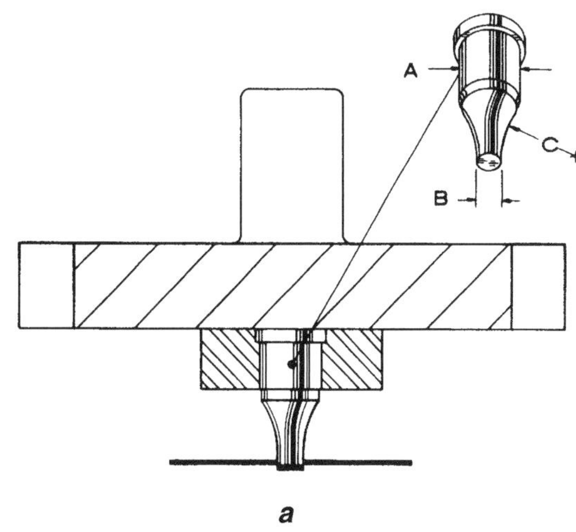

Figure 8.1 A method of applying a blanking punch for small washers.

Body diameter **A** is made considerably larger than blanking diameter **B**; radius **C** provides supporting material to give rigidity and prevent deflection of the punch upon contact with the material strip. Section view *a* shows the punch, punch plate, and punch holder as they would appear in the lower left section view of the complete die drawing. Plan view *b* shows the same components inverted as they would appear in the upper right view of the complete die drawing.

Methods of applying slightly larger blanking punches for producing circular parts are shown in Figure 8.2. These blanking punches are made in much the same manner as piercing punches. Rules for designing piercing punches will be given in the next chapter.

a) Keying the Punch

Small blanking punches having a cutting end of irregular contour can be kept from turning by a round-end key (Figure 8.3). This key is retained in an end-milled slot that is machined in the punch plate; it bears against a flat ground in the punch head. Common applications of this construction are small blanking punches for producing clock, watch, and instrument gears.

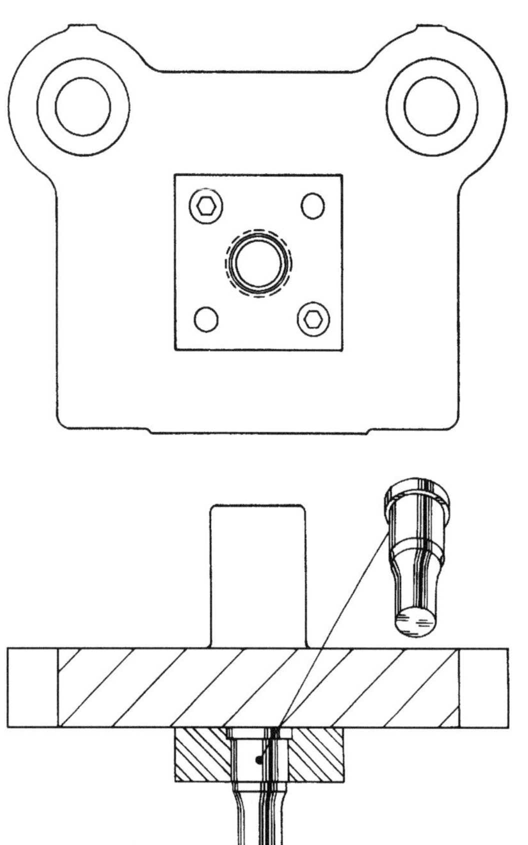

Figure 8.2 A method of applying a slightly larger blanking punch.

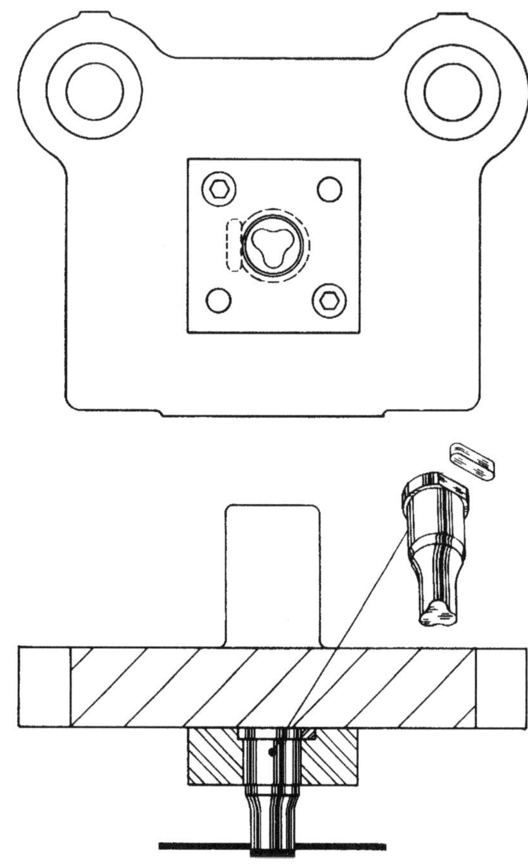

Figure 8.3 One method of keying a blanking punch.

b) An Alternate Method

Figure 8.4 shows another method of keeping small, irregular blanking punches from turning by machining the body to be square or rectangular in shape. The retainer hole is machined to fit the square or rectangular punch body.

c) Inserts

Weak areas of small blanking punches are best applied as inserts (Figure 8.5) for easy and inexpensive replacement of members, in case of breakage. When inserts are used, the blanking punch assembly is backed up by a hardened plate.

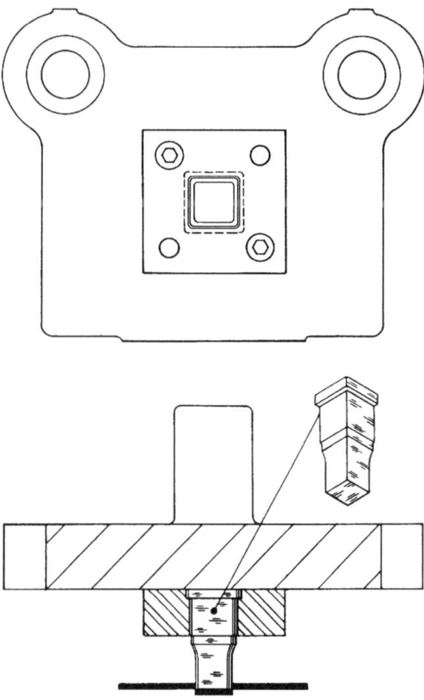

Figure 8.4 A square-machined punch body for keeping punch from turning.

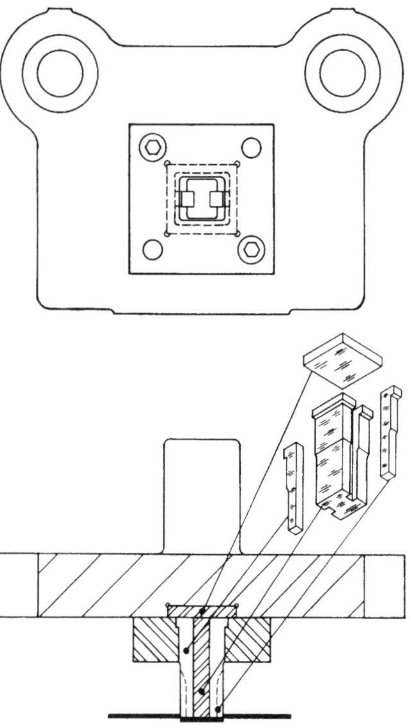

Figure 8.5 Inserts used in the weak areas of small blanking punches.

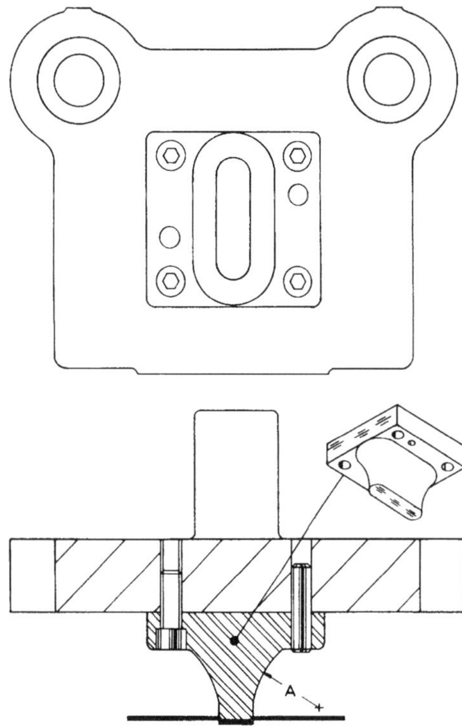

Figure 8.6 Radius A provides extra support in this narrow-and-long punch.

8.2.2 Medium-Size Blanking Punches

The sides of narrow-and-long blanking punches (Figure 8.6) are provided with a radius **A**. By providing extra supporting material for stability, this radius helps prevent deflection of the cutting end upon contact with the material strip.

a) Flanged Punches

The flanged type of blanking punch (Figure 8.7) is the most widely used because it is employed to produce average-size blanks. As in Figure 8.6, flanges are provided for fastening the blanking punch to the punch holder with screws and dowels. Only the cutting end of the blanking punch is hardened; the flanges are soft for accurate fitting of dowels at assembly. Foolproof the punch by offsetting one of the dowels to assure correct reassembly in die maintenance.

Clearing other components. When space is limited, a portion of the flange (Figure 8.8) can be removed to provide room for other die components. For good stability, flange width **A** should not be smaller than punch height **B**.

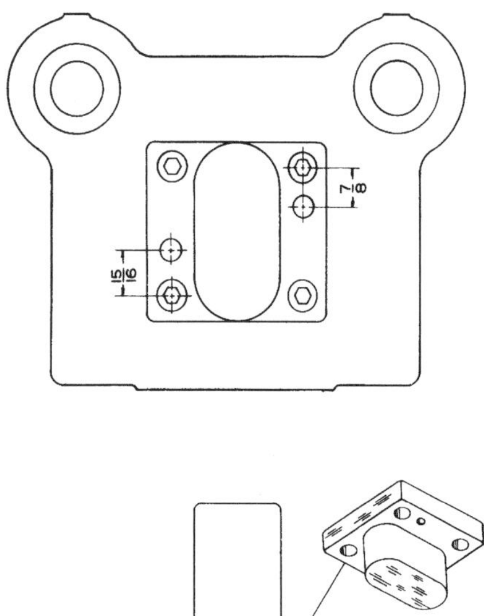

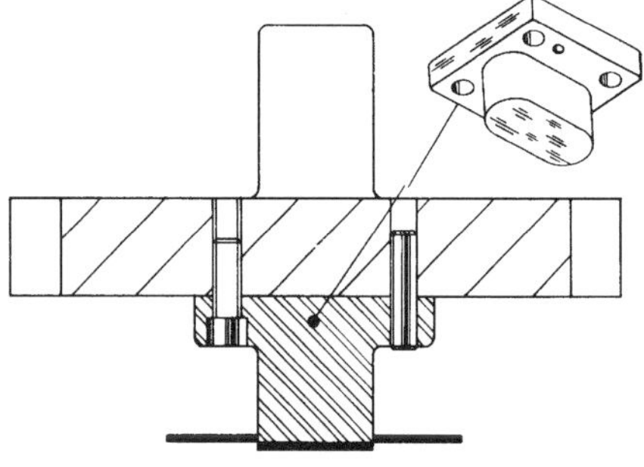

Figure 8.7 A widely used blanking punch for producing average-size blanks.

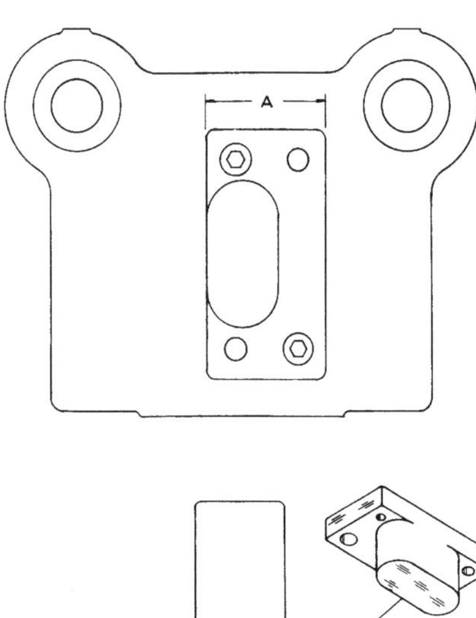

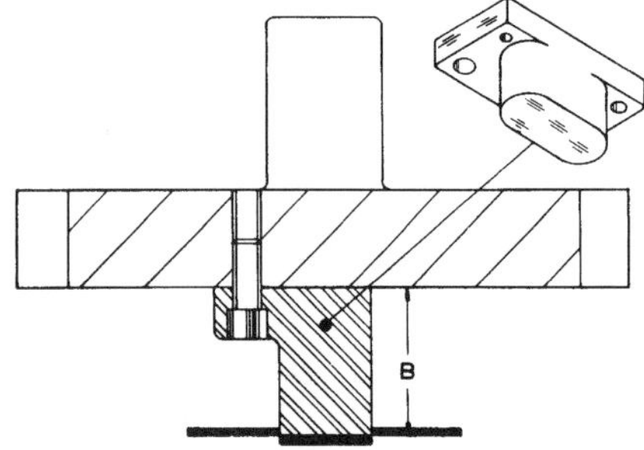

Figure 8.8 Removing a portion of the flange to provide clearance for other die components.

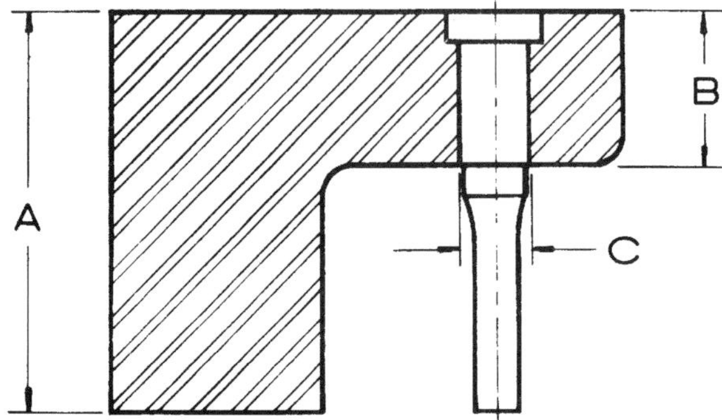

Figure 8.9 Typical proportions of medium-size blanking punches.

b) Proportions

Proportions often used for medium-size blanking punches (Figure 8.9) are as follows: a typical punch length **A** is 1 5/8 inch (41.2 mm); flange height **B**, 5/8 inch (15.9 mm). However, when the flange is employed to retain a small piercing punch, height **B** should not be less than 1 1/2 times the piercing punch diameter **C** for stability.

8.2.3 Large Blanking Punches

Larger blanking punches do not require flanges. They can be fastened to the punch holder with screws and dowels applied from the back (Figure 8.10). Dowel holes are shown all the way through the punch, either full size or as a smaller hole, to enable pressing out the dowels during die maintenance.

a) Sectioning Large Punches

Still larger blanking punches can be made in sections (Figure 8.11) to facilitate heat-treating and minimize distortion. Each section is individually screwed and doweled to the punch holder of the die set.

b) Cutting Ring

Blanking punches for large circular blanks (Figure 8.12) can be made in two parts to conserve tool steel and simplify heat treating. Cutting ring **A** is located by a round boss which is part of spacer **B**. This spacer, in turn, is fitted into a recess machined in the punch holder of the die set. With this construction, dowels are not necessary for positioning the punch.

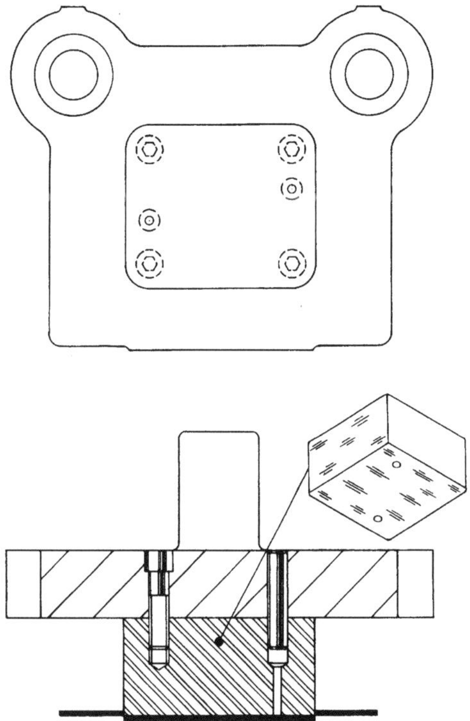

Figure 8.10 Large blanking punches do not require flanges.

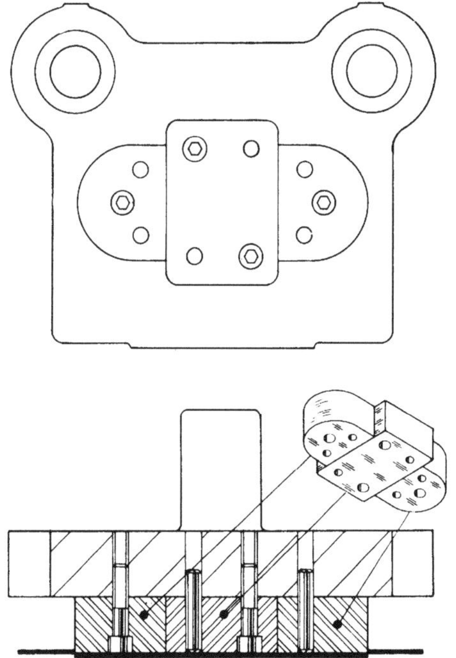

Figure 8.11 Large blanking punches are sectionalized to facilitate heat-treating and minimize distortion.

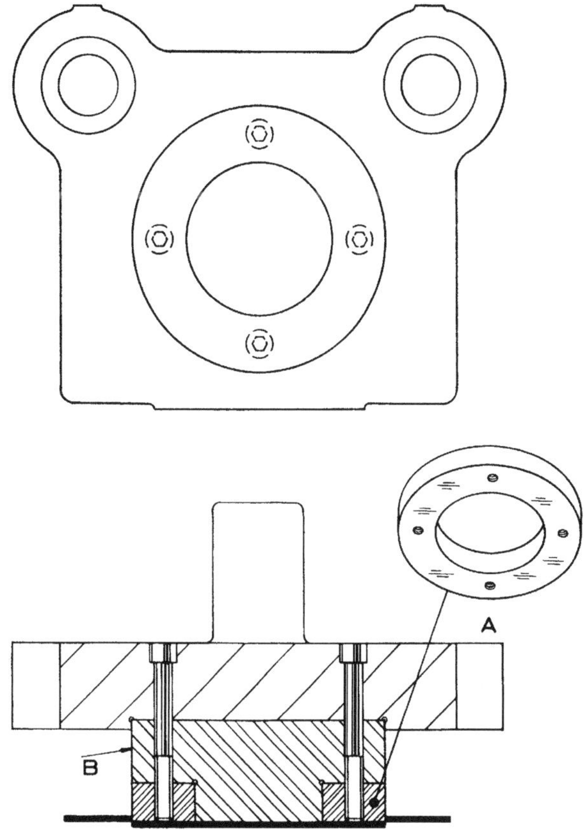

Figure 8.12 This large circular blanking punch is made in two parts to conserve tool steel.

c) Cutting Tool Steel Pads

The construction method illustrated in Figure 8.12 can also be used for blanks with irregular contours (Figure 8.13). Spacer **A** is machined to accommodate hardened tool steel pads **B** and **C**. Long socket screws threaded into the tool steel pads fasten both the pads and the machine steel spacer to the punch holder. Dowels accurately locate the entire assembly.

d) Spacer

Still another method of conserving tool steel in large dies is to make the cutting member **A** in one piece, but relatively thin, then back it up with a machine steel spacer **B** (Figure 8.14). The assembly is fastened to the punch holder, as in Figure 8.13, and dowels locate the blanking punch in much the same manner.

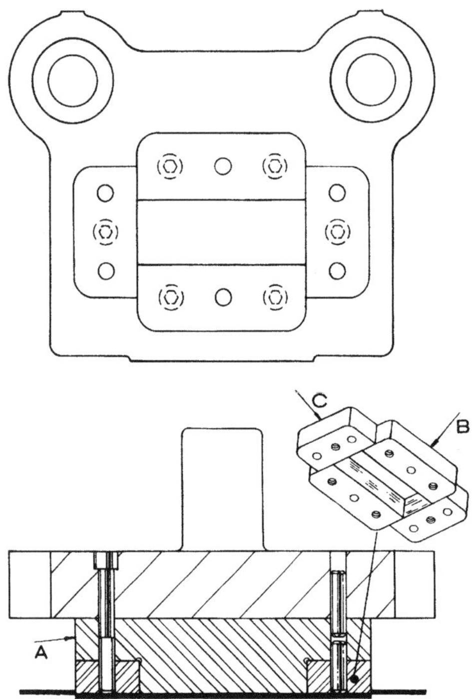

Figure 8.13 Construction method for large blanking punches with irregular contours.

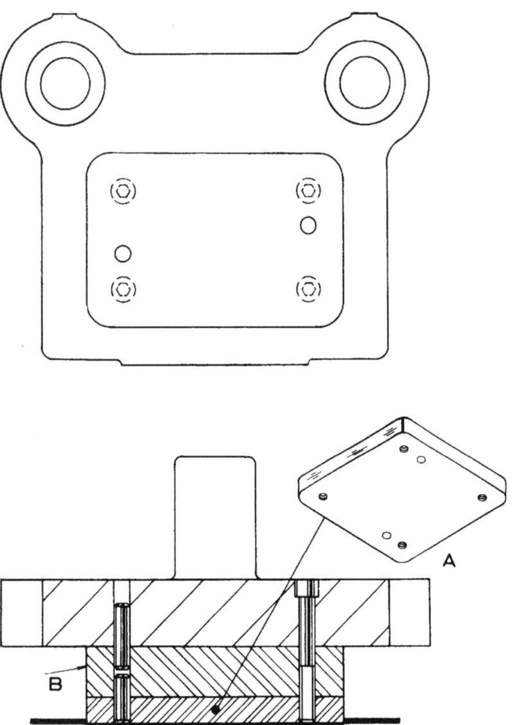

Figure 8.14 Use of a spacer **B** to conserve tool steel.

8.3 BLANKING PUNCHES WITH BEVEL-CUT EDGES

a) Angular Shear

Shear is the machining punch face set so that the cutting edge will be presented at an angle to the surface of the material to be cut. For cutting large openings in blanks, shear is applied to the face of the punch (Figure 8.15) to reduce shock on the press, force required, and blanking noise. Properly applied, shear reduces by one-quarter the force required to blank metals thicker than 1/4 inch (6.35 mm). When thinner stock is blanked, the reduction in force is as large as one-third. Figure 8.15 shows the most common method of shearing with the face of a large punch. Shear depth **A** is made two-thirds the thickness of the

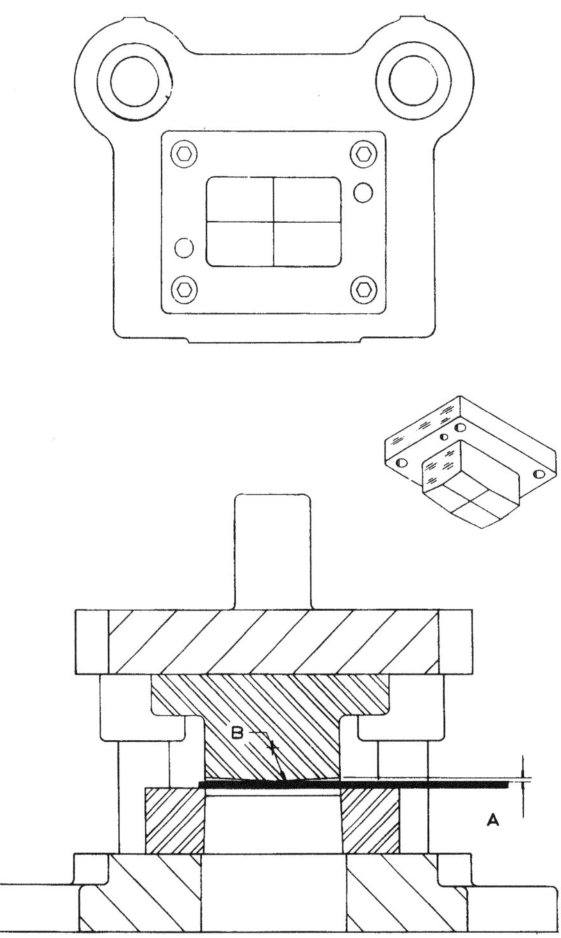

Figure 8.15 A punch configuration for applying shear in cutting large openings.

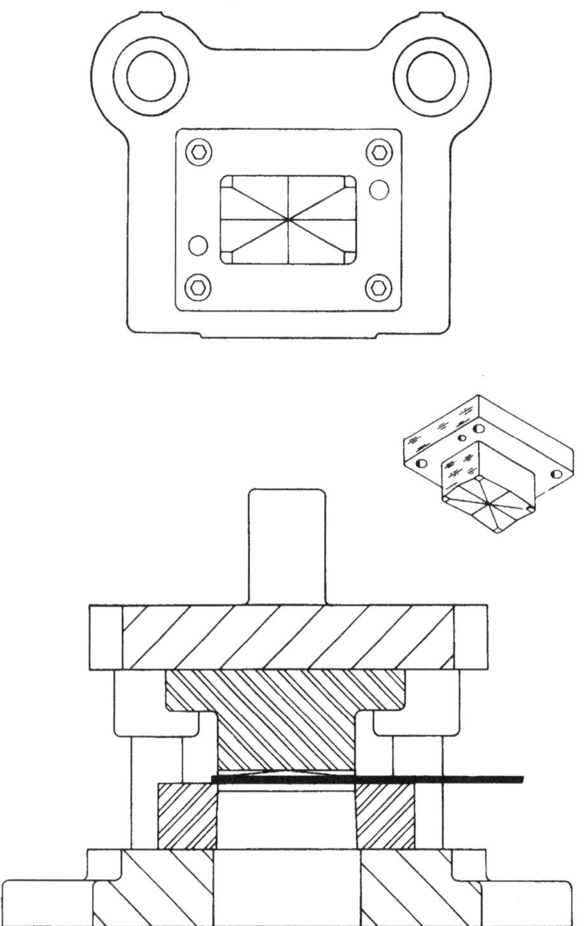

Figure 8.16 An alternate punch configuration for applying shear.

Figure 8.17 Scalloping provides a means of applying shear in cutting round holes.

material. Radius **B** removes the sharp corner to avoid a focal point for stock fracture.

b) Alternate Method

Figure 8.16 shows an alternate punch configuration for applying shear. This is a better method of providing angular shear to a punch. The material is securely gripped at the corners between the punch and die members, and is gradually cut toward the center. Shear is usually applied for openings larger than 5-by-5 inches (127-by-127 mm), and over 1/16 inch (1.6 mm) thick.

c) Round Punch

Round punches can provide a shearing action by scalloping the face of the punch in a series of waves around the periphery (Figure 8.17). A hole is first drilled through the center to provide relief for the cutting tool, which applies the scalloping. Here again, shear depth should not exceed two-thirds the thickness of the stock.

8.4 SHEDDERS

Blanks produced from oiled stock tend to cling to the face of the blanking punch. Spring-loaded shedder pins, applied to one side of the punch, will break the adhesion and free the blank from the punch face to prevent double blanking (Figure 8.18).

Three methods of applying shedder pins are available. At illustration 1, the shedder pin for a small blanking punch is made short, headed, and backed up by a spring and socket set screw. The pin is made from drill rod, hardened, and polished. A clearance hole is provided in the punch holder for removal of the assembly for punch sharpening.

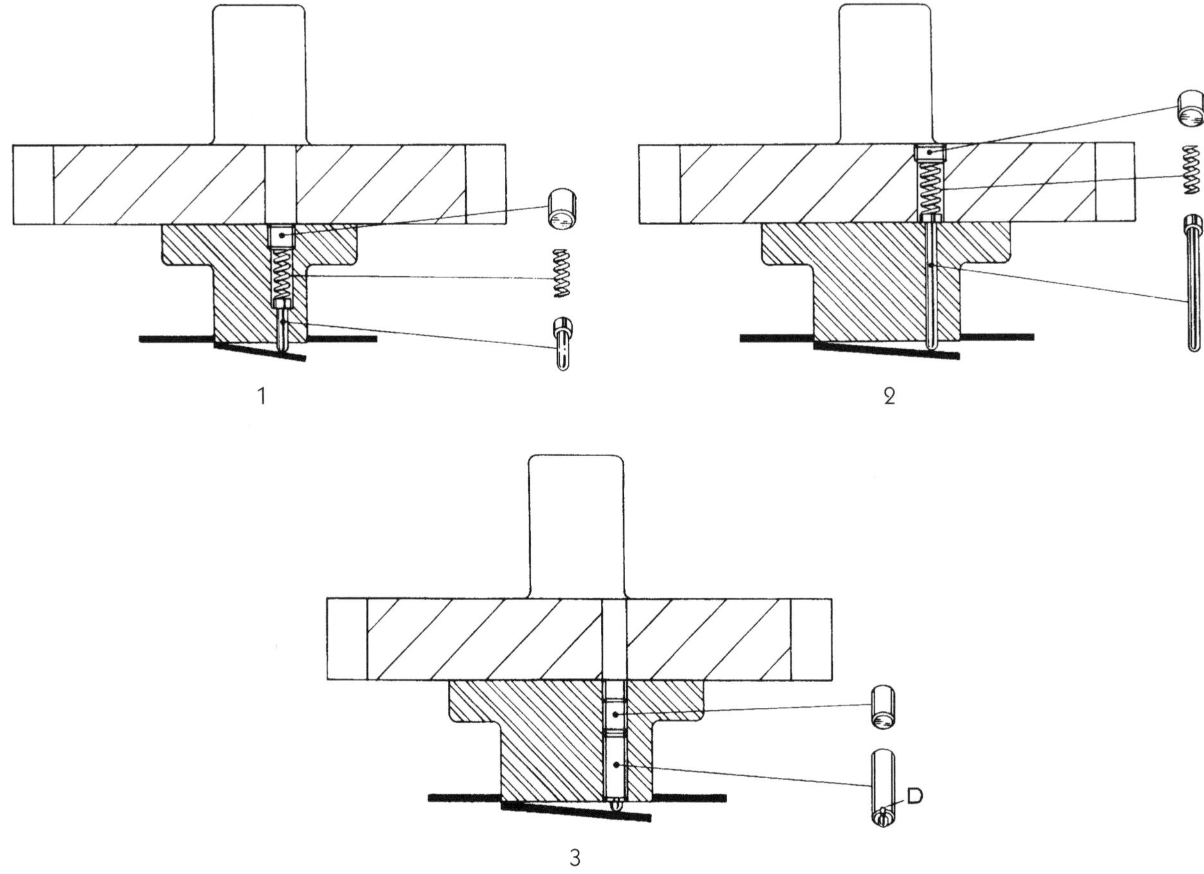

Figure 8.18 Different versions of a shedder assembly, which prevents clinging of the blank to the punch face.

In another method, the shedder can be made long, as at illustration 2. Its head bears against the top of the blanking punch, and it is similarly backed up by a spring and socket set screw.

A Vlier shedder is illustrated at illustration 3. This device is available from Vlier Engineering Co. A self-contained assembly, the shedder consists of a threaded housing containing a shouldered shedder pin and back-up spring. The assembly is inserted in a hole tapped through the blanking punch and is backed up by a socket set screw. Screwdriver slot **D** is used to remove it for punch sharpening.

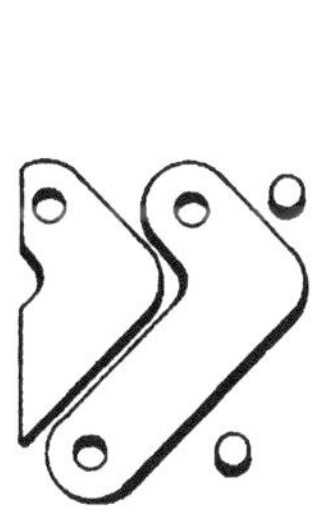

HOW TO DESIGN PIERCING PUNCHES

9.1 INTRODUCTION

Piercing punches are usually the weakest link in any die design. Therefore, the following factors must always be taken into consideration:

1. Make the punches strong enough so that repeated shock in operation will not cause fracture.
2. Slender punches must be sufficiently guided and supported to insure alignment between punch and die members and also to prevent buckling.
3. Make provision for easy removal and replacement of punches in the event of breakage.

The illustrations that follow show numerous methods of designing and applying piercing punches; these figures should help the designer select the best type for the particular job.

9.2 METHODS OF DESIGNING PUNCHES

a) Shoulder Punches

Probably the most commonly used type of punch, shoulder punches are made from a good grade of tool steel, hardened and ground all over (Figure 9.1). They are readily available from a number of suppliers.

Diameter **A** is a press fit in the punch plate. Diameter **B**, which extends at least 1/8 inch (3.2 mm), is a slip fit for good alignment while pressing.

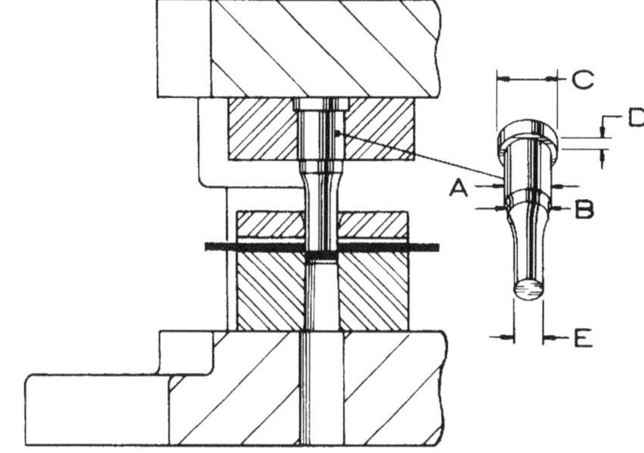

Figure 9.1 Commonly used shoulder punch.

Shoulder **C** is usually made 1/8 inch (3.2 mm) larger in diameter than **A**. Shoulder height **D** is 1/8 to 3/16 inch (3.2 to 6.35 mm), depending on size. Piercing diameter **E** is always on the high side of tolerance. For example: If the hole is dimensioned 0.501/0.500 inch (12.725/12.700 mm) diameter on the part print, the punch diameter would be made 0.501 inch (12.725 mm). The blending radius that connects diameters **B** and **E** should be as large as possible and the surface polished smooth because ridges would present focal points for fracture.

b) Guiding the Punch in the Stripper

In first class dies, small punches under 3/16 inch (4.8 mm) in diameter are usually guided in hardened bushings pressed into the stripper (Figure 9.2). Larger punches are often guided, particularly when cutting heavy stock, if there is a possibility of punch deflection upon contact

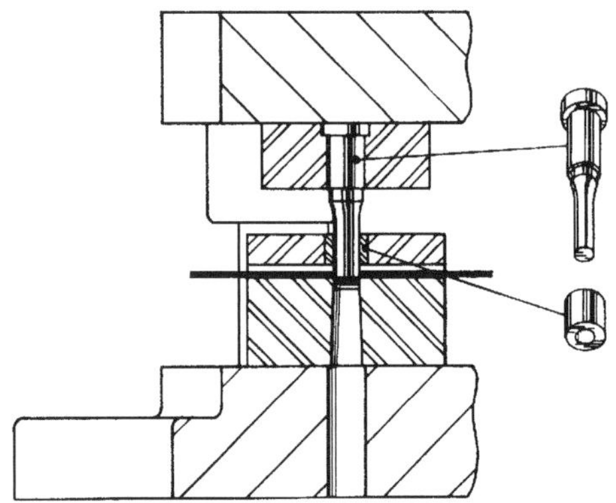

Figure 9.2 Hardened bushings pressed into the stripper are used for guiding small punches.

with the strip. These guide bushings are hardened, then ground on both inside and outside surfaces. Headless drill bushings can often be used for the same application.

c) Backing Plate

When heavy stock is being pierced, the punch head is backed up by a backing plate (Figure 9.3) pressed into a counterbored hole in the punch holder. This plate distributes the thrust over a wider area, thereby preventing the punch head from sinking into the relatively soft material of the die set. Backing plates are made of tool steel, hardened and ground. The small hole through the punch holder serves two purposes:

1. To accommodate the pilot on the counterbore.
2. To allow the backing plate to be pressed out for die repair.

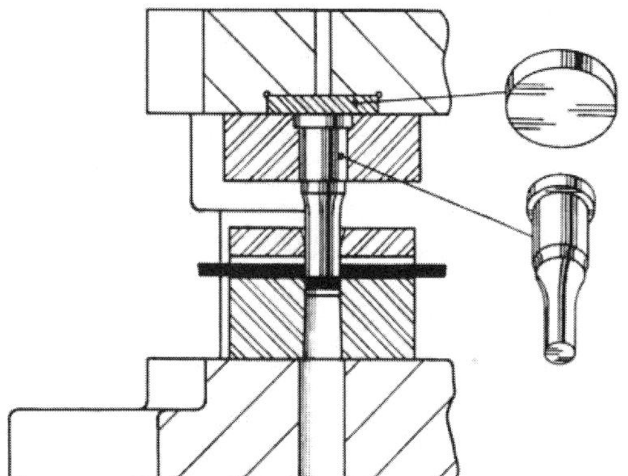

Figure 9.3 The backing plate distributes thrust over a wide area.

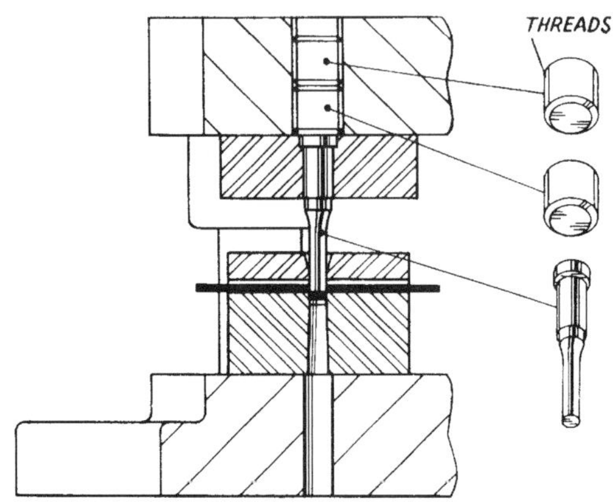

Figure 9.4 Set screws are used for backing up a piercing punch.

d) Set Screw to Back Up the Punch

Two socket set screws back up the piercing punch in this application (Figure 9.4). This method of retaining piercing punches has become widely used in recent years, and it should be selected whenever possible. Its main advantage is the ease with which a broken punch can be removed and a new one inserted, without the necessity of dismantling any part of the die. When dimensioning, make sure that the tap drill for the threaded hole is larger than the punch head.

e) Large Punch

Punches over 1 1/4 inch (31.75 mm) in diameter are relieved at the center of the cutting face for ease of sharpening (Figure 9.5). When cutting heavy stock, the annular cutting ring **A** is often scalloped to relieve the shock on the press.

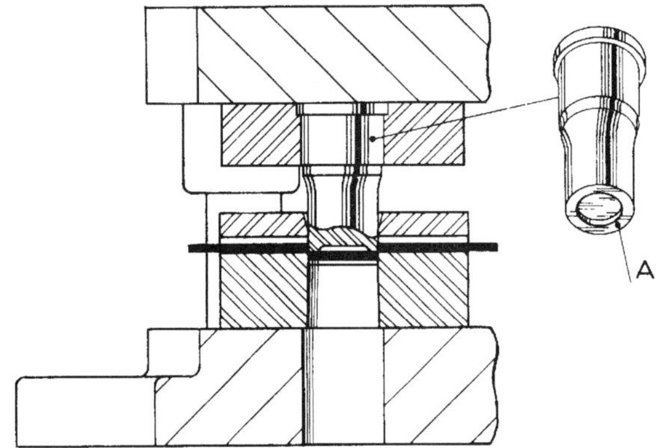

Figure 9.5 Large punches are relieved at center for ease of sharpening.

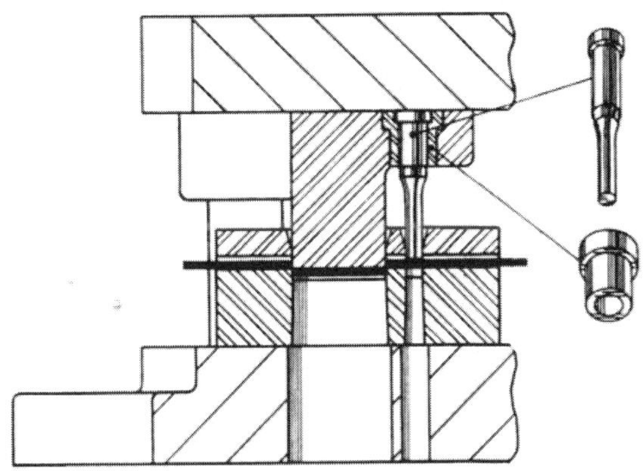

Figure 9.6 A soft-shouldered plug is used to retain a small piercing punch in the flange of a large blanking punch.

f) Soft Plug

When a small piercing punch is retained in the flange of a hardened blanking punch (Figure 9.6), a soft-shouldered plug is pressed into the flange, then bored to retain the piercing punch. Modern practice is to harden only the cutting end of the blanking punch, leaving the flange soft. No bushing is then required.

g) Peening the Head

It is difficult and expensive to machine shoulders on punches of very small diameter. For this reason, they are made of plain drill rod for their entire length. One end is peened to form a head (Figure 9.7) and this is finished to an 82-degree included angle to fit standard countersunk holes. At assembly, the top of the punch plate is ground so that the peened heads come flush with its surface. These punches should invariably be backed up by a hardened plate and guided in the stripper. When properly supported in this manner, these punches will pierce holes with diameters as small as twice the thickness of the stock.

9.3 METHODS OF HOLDING PUNCHES

a) Using a Set Screw

For short-run dies, when low accuracy requirements exist, this inexpensive type of punch may be used (Figure 9.8). It is made of drill rod, machined to size at the cutting end. An angular flat is ground on the punch body for the point of the set screw. However, the set screw has a tendency to throw over the punch by the amount of clearance. Therefore, this construction should never be used in first class dies.

b) Bridging the Gap

When the punch is located a greater distance from the punch plate edge than standard set screw lengths, a drill rod spacer is used between the punch and the set screw (Figure 9.9).

c) Using a Spring-Backed Steel Ball

These commercial standardized punches are held in their retainers and also kept from turning by spring-backed steel balls (Figure 9.10). These bear against spherical seats machined in the

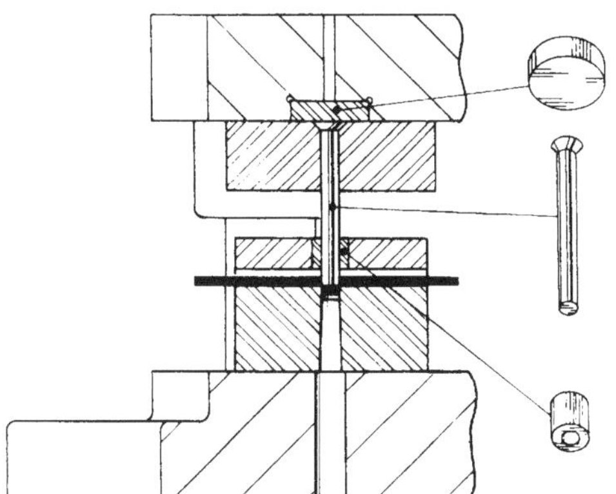

Figure 9.7 A piercing punch is made by peening drill rod head on, then grinding to shape and length.

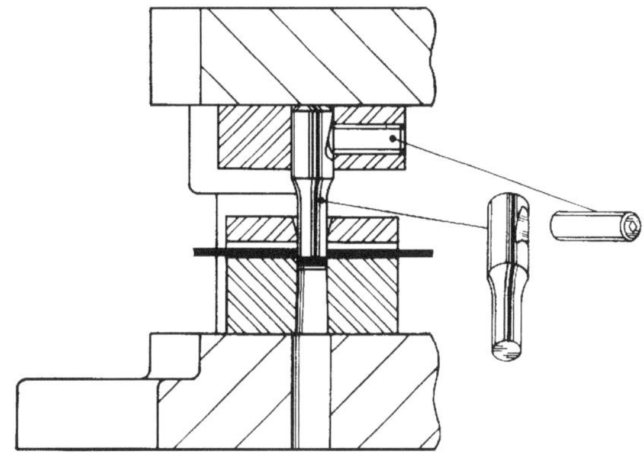

Figure 9.8 A set screw is used to hold the punch in short-run dies.

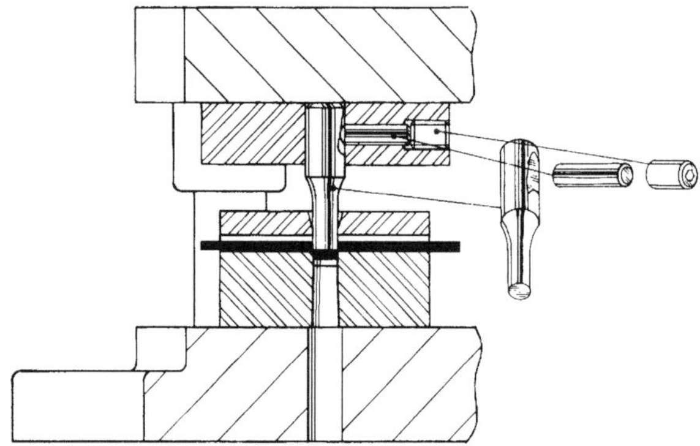

Figure 9.9 A drill rod spacer is used when the punch is a great distance from the punch plate edge.

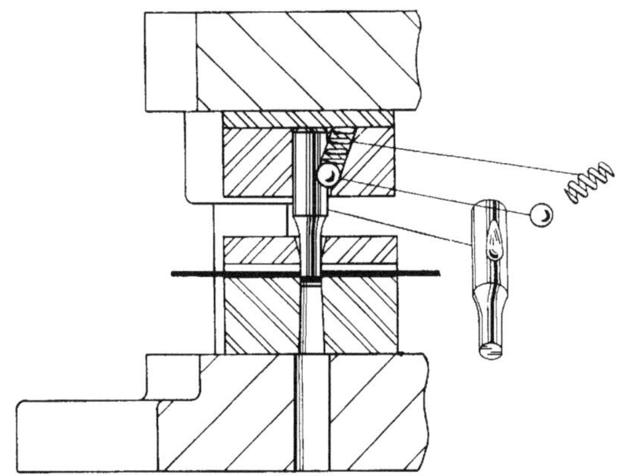

Figure 9.10 The punch is held in place by spring-backed steel balls.

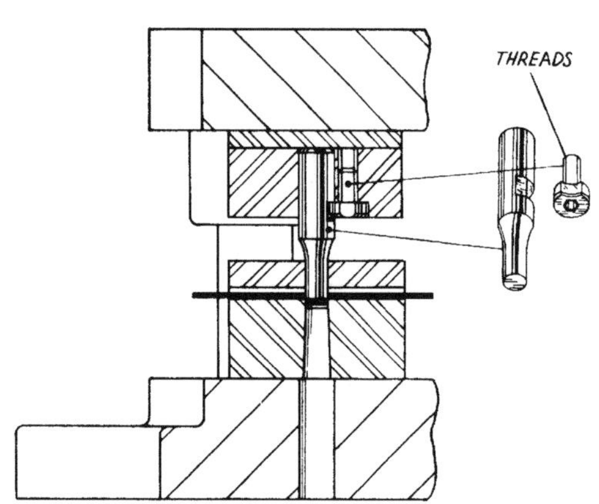

Figure 9.11 The punch is held in place by a special socket head cap screw.

punches. A simple tool pushes up the ball, depressing the spring for removal of the punch for sharpening or replacement.

d) Using a Special Socket Head Screw

Some commercial standardized punches are held in their retainers and also kept from turning by special socket head cap screws (Figure 9.11), the heads of which bear against grooves machined in the punches. The heads of the retaining screws have milled clearance radii so that it is only necessary to give them a quarter turn for removal of the punches.

9.4 METHODS OF SUPPORTING PUNCHES

a) Quill Support

Very small peen-head punches can be held in quills and guided in hardened bushings pressed into the stripper (Figure 9.12). When replacing the punches, simply remove them from their quills and insert new ones. They are backed up by either a hardened plate or socket set screws.

b) Guiding the Quill

Another method of providing lateral support to a small peen-head punch is to guide the quill itself in a hardened bushing in the stripper plate (Figure 9.13). In this way, the punch extends from the quill only a very short distance for maximum stiffness.

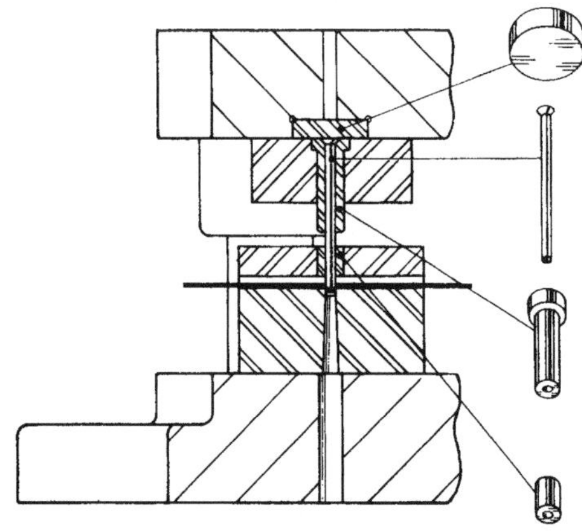

Figure 9.12 Support for small peen-head punches is provided by a quill in the punch plate and a hardened bushing in the stripper.

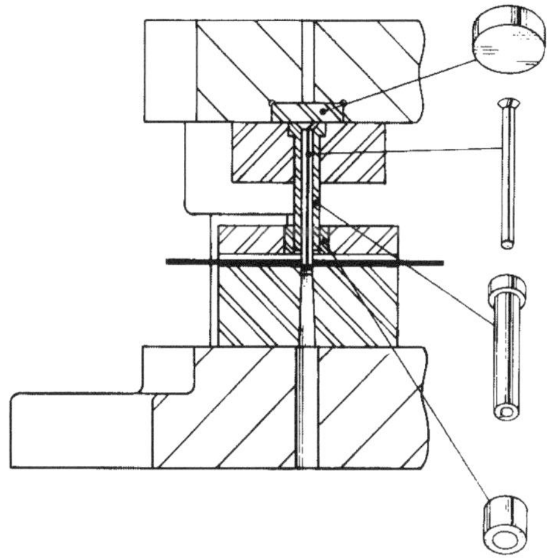

Figure 9.13 Sometimes the quill itself is quilled by a hardened bushing in the stripper.

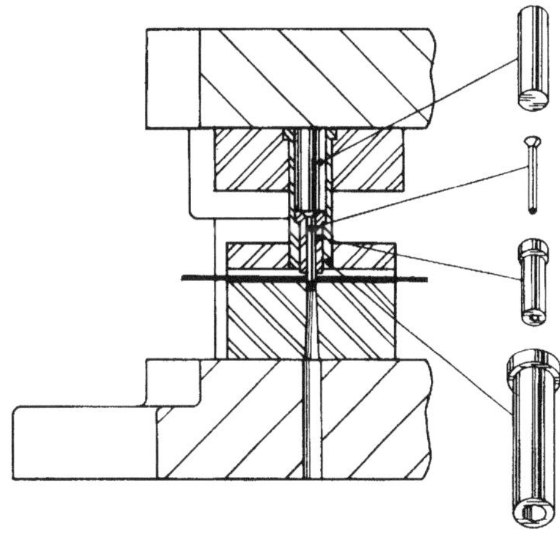

Figure 9.15 A bit punch, enclosed in a small quill and, in turn, enclosed in a larger quill and backed up, may be made of expensive material.

c) One Quill, Two Perforators

Two or more closely spaced peen-head punches can be held in a single quill (Figure 9.14). The quill is kept from turning by a flat machined on one side of its head. This flat bears against one edge of a slot machined in the punch plate. The quill is guided in a hardened bushing in the stripper and the punches are backed up by a backing plate.

d) Bit Punches

Tiny bit punches made of top-grade material are held in a small quill, backed up by a hardened plug. These components are, in turn, held in a larger quill (Figure 9.15). In this way, the punches can be made of expensive material, but the cost compares favorably with larger punches made of lower grade steel.

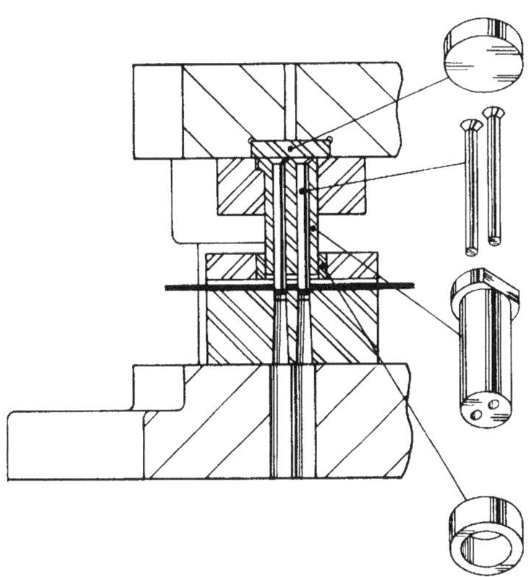

Figure 9.14 Two very closely spaced peen-head punches held in a single quill.

e) Durable Punch and Sleeves

An ingenious method of supporting and guiding slender punches was patented by the Durable Punch and Die Company (Figure 9.16). Headed punch **A** is guided and supported for its entire length in two intermeshing sleeves **B** and **C**. The punch extends out of these supporting sleeves only when actually going through the

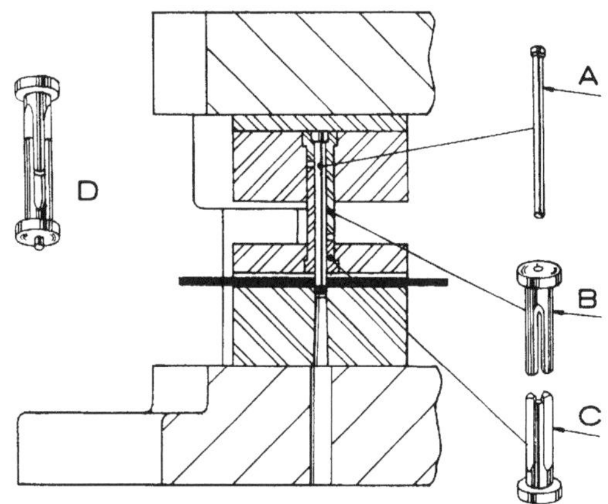

Figure 9.16 Method of supporting punches with intermeshing sleeves.

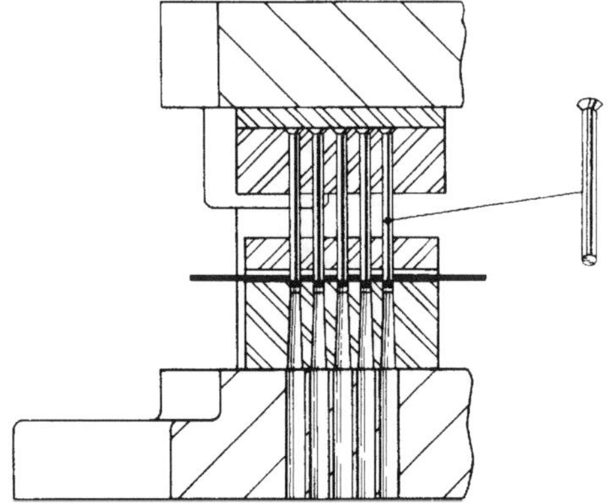

Figure 9.17 An arrangement of tiny peen-head punches for perforating strainers.

material to be cut. As you may know, a slender needle, passed through a cork, can be driven entirely through a coin by a single hammer blow because of the support given by the surrounding cork. The intermeshing action of Durable sleeves supports the punch in much the same manner, allowing the piercing of extremely small holes whose diameters are as little as one-half of the material thickness. A complete Durable assembly is shown at **D**.

f) Guiding Punches in the Stripper

Tiny peen-head punches are often arranged in great numbers at very close center distances, as when perforating coffee strainers and similar parts (Figure 9.17). Such multiple units are backed up by one large hardened plate; the punches are then guided by holes in the stripper.

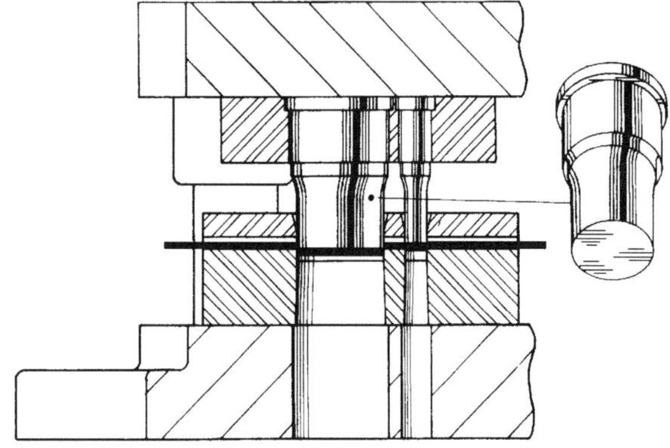

Figure 9.18 A small punch is made shorter to prevent deflection, nicking, and breakage.

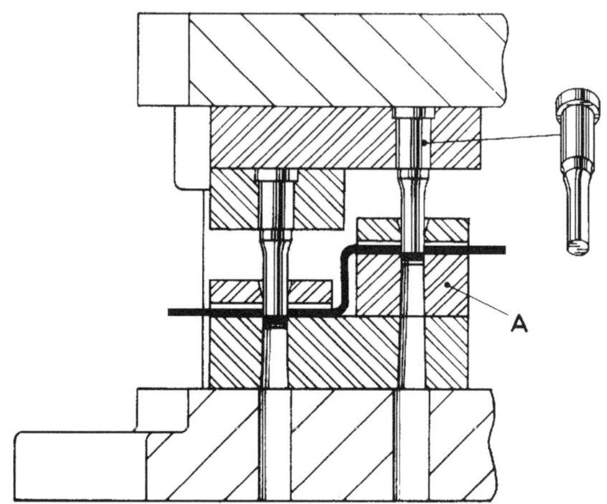

Figure 9.19 Use of auxiliary die block (**A**) avoids use of excessively long piercing punches.

9.5 STEPPING PUNCHES

a) Small Punches

A small punch adjacent to a large one is made shorter by two-thirds of the stock thickness (Figure 9.18). Some material flow occurs as the large punch penetrates the strip. With both punches the same length, the small one would deflect slightly, resulting in a nicked cutting edge and eventual breakage.

b) Split-Level Piercing

To avoid excessively long piercing punches when punching stepped shapes, add an auxiliary die block (Figure 9.19), as at **A**. In this manner, both punches can be made the same length for maximum strength and ease of construction.

c) Stepped Perforators

Closely spaced peen-head punches, when piercing thicker stock, should be staggered (Figure 9.20) so that not all punches enter the strip at once. This practice prevents excessive punch breakage due to crowding of the metal. Two-thirds of the thickness of the stock is generally allowed as an offset on punch length.

Figure 9.21 shows another method of staggering closely spaced punches. The center punch enters the material first. Succeeding punches at either side enter by steps. Again, these are made shorter by two-thirds of the thickness of the stock.

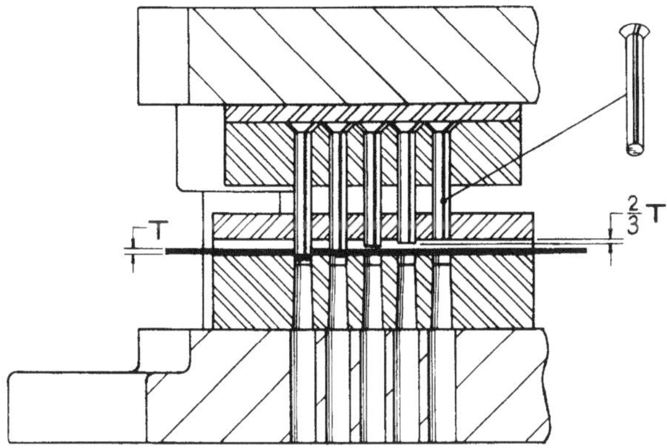

Figure 9.20 Punches are staggered to prevent breakage and crowding of metal.

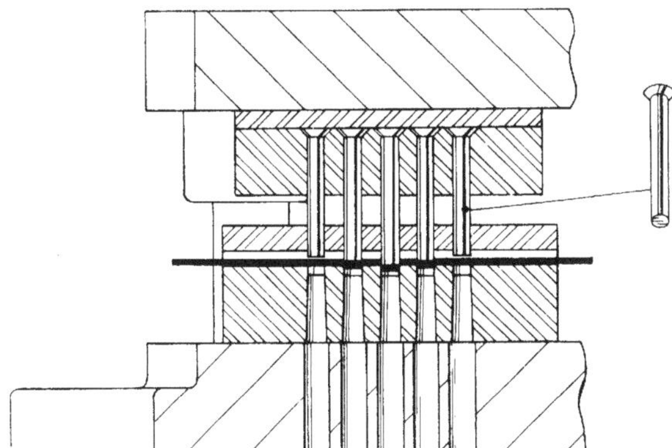

Figure 9.21 Another method of staggering closely spaced punches.

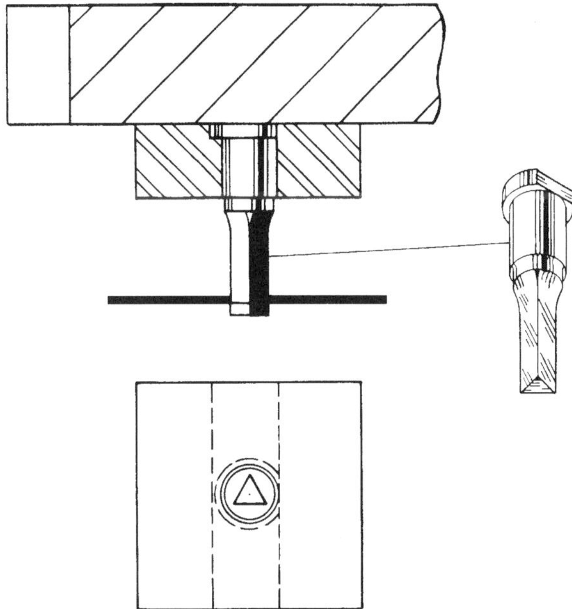

Figure 9.22 The flat on the punch head of an irregular punch keeps it from turning.

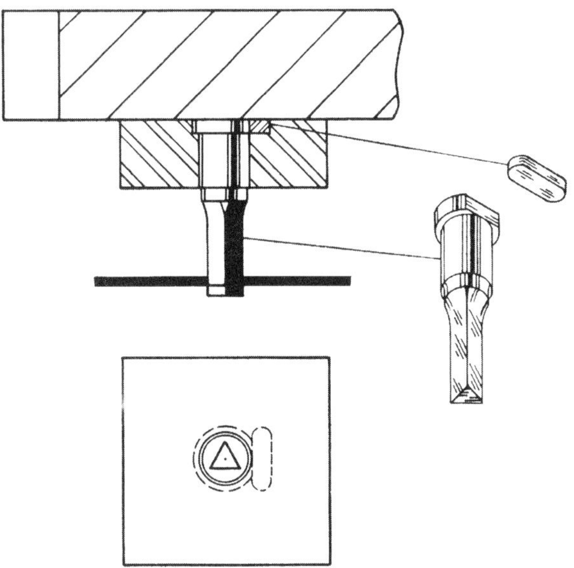

Figure 9.23 A retaining key keeps the punch from turning.

9.6 KEEPING THE PUNCH FROM TURNING

A good method of keeping an irregular punch from turning is to machine a flat in the punch head (Figure 9.22). This flat bears against one edge of a slot machined in the punch plate. A snug fit is required.

a) Retaining Key

Another good method of keeping an irregular punch from turning is to machine a flat in the punch head, as in Figure 9.22, then insert a small round-end key in a slot end-milled in the punch plate (Figure 9.23). This method is particularly useful in multiple station progressive dies where space is limited.

b) Mating Flat

Where irregular punches are close together, an excellent method of keeping them from turning is to machine flats on both punch heads

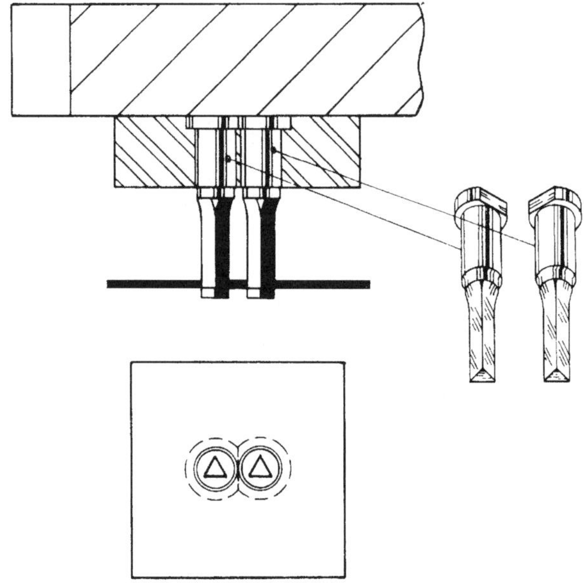

Figure 9.24 Mating flats of irregular punches keep them from turning.

(Figure 9.24). These flats bear against each other with a good fit, keeping the punches in alignment.

c) Doweled Punch Head

If an irregular piercing punch is provided with a large head, then a small dowel, pressed

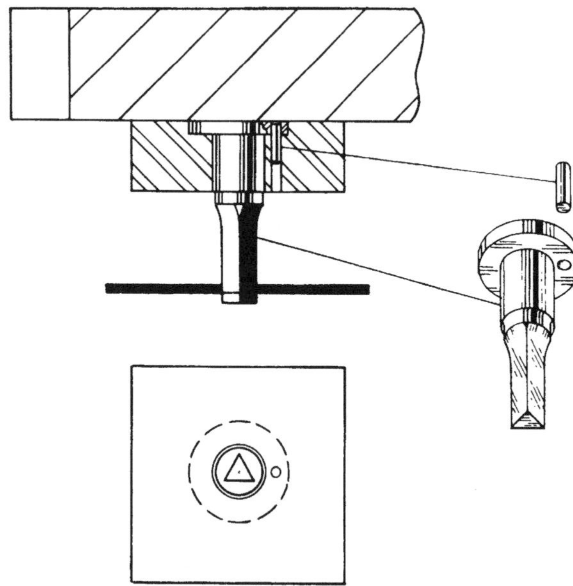

Figure 9.25 A large punch head may be doweled to keep the punch from turning.

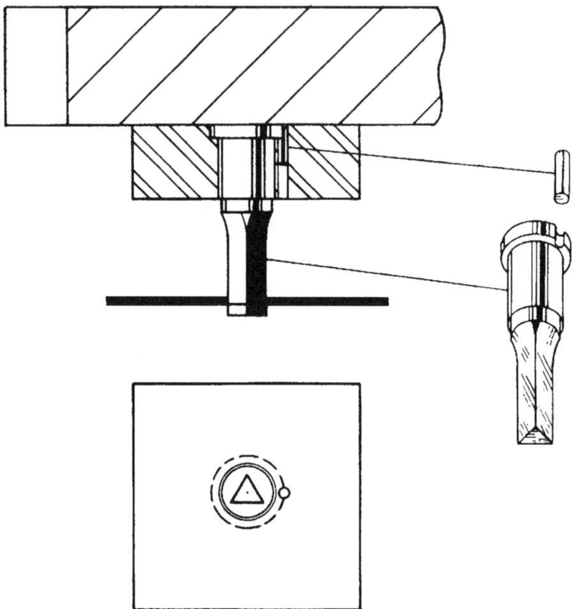

Figure 9.26 A pin (small dowel) may be pressed in place on the periphery of the punch head to keep the punch from turning.

through both the head and the punch plate, will keep the punch in good alignment (Figure 9.25). This is not considered as good a method as those previously described, but it is used frequently for low- and medium-production dies.

d) Pin

A pin, that is, a small dowel pressed half in the punch head and half in the punch plate, will keep the punch from turning (Figure 9.26). As in the previous example, this method should not be used for first class dies. It is more suitable for temporary dies and dies for low production.

9.7 SHEDDERS

a) Shedder Pin

When piercing some thin materials, the slug tends to follow the piercing punch out of the die hole. This can be a serious problem in progressive dies. To overcome the problem, incorporate a shedder pin in the punch (Figure 9.27). This headed pin is backed up by a spring and set screw. Piercing punches provided with shedders are available from a number of suppliers.

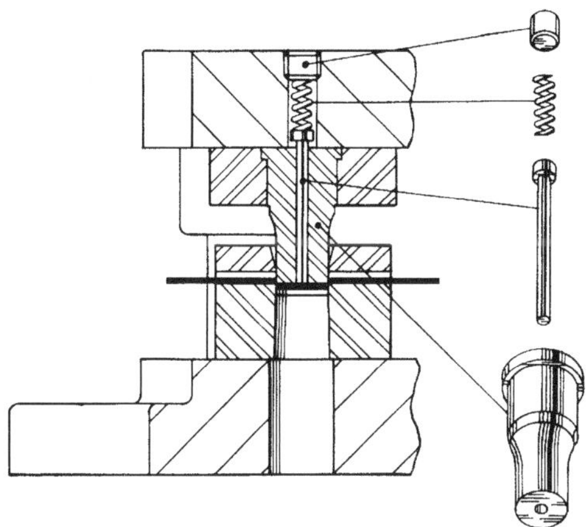

Figure 9.27 A shedder pin is incorporated into the piercing punch to overcome the tendency of a slug to follow a piercing punch out of the die hole.

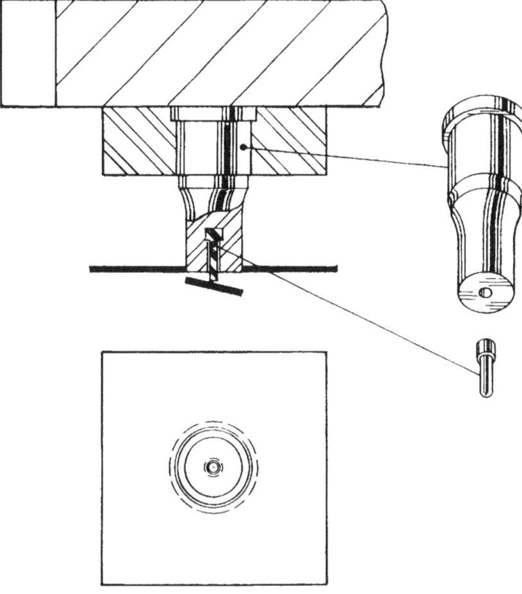

Figure 9.28 A shedder pin made of rubber.

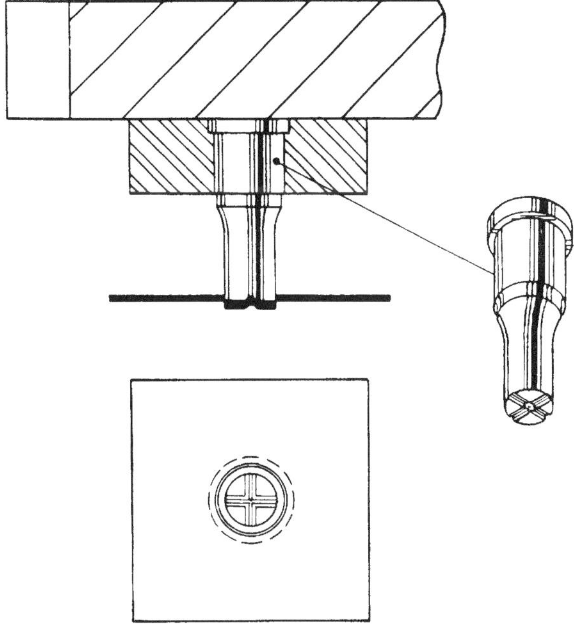

Figure 9.29 Slots ground on the end of a punch temporarily serve the same purpose as shedder pins.

b) Rubber Shedder

Another method of preventing slugs from pulling up out of the die hole is to insert a rubber bumper in a recessed hole in the punch (Figure 9.28). The rubber shedder is made with a head to keep it from falling out in operation.

c) Temporary Solution

If slugs are pulling up in the die tryout stage, grind two slots in the cutting face of the punch, as shown in Figure 9.29. This is a temporary measure and a shedder pin should be applied later.

HOW TO DESIGN PUNCH PLATES

10.1 INTRODUCTION

Punch plates hold and support piercing, notching, and cut-off punches. They are usually made of machine steel, but can also be made of tool steel that has been left soft for high grade dies. Punch plates range from small simple blocks for holding single piercing punches to large, precision-machined plates for holding hundreds of perforators. Important design considerations include:

1. Adequate thickness for proper punch support.
2. Good doweling practice to insure accurate location.
3. Sufficient screws to overcome stripping load.

Numerous methods of designing punch plates and applying them to various types of dies are described in the remainder of this chapter. These methods further explain Step 5 in Chapter 5—Fourteen Steps to Design a Die.

10.2 METHODS OF DESIGNING AND APPLYING PUNCH PLATES

10.2.1 Design Conception

a) Punch Plates for Single Punches

A punch plate for holding a single punch (Figure 10.1) is made square and with sufficient thickness for good punch support. Two socket cap screws, applied at the corners, resist stripping pressure, while dowels at the other two corners provide accurate location. Minimum distance from plate edges to screw centers **A** is 1 1/2 times screw diameter **B**.

Stripping force. When center distances are small, a single punch plate can hold a number of piercing punches. When a large number of punches are retained, the stripping force should be calculated to insure that sufficient screws are used to fasten the punch plate to the die set. Stripping force can be calculated by the formula:

$$F = \frac{\sum_1^n P}{0.00117} \times T \qquad (10.1)$$

where

$\sum_1^n P$ = the sum of the perimeters of all (1 to n) piercing punch faces

T = thickness of material

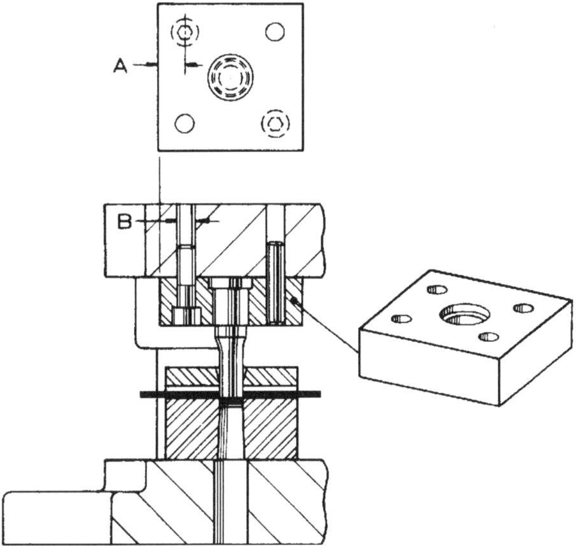

Figure 10.1 Punch plate for a single punch.

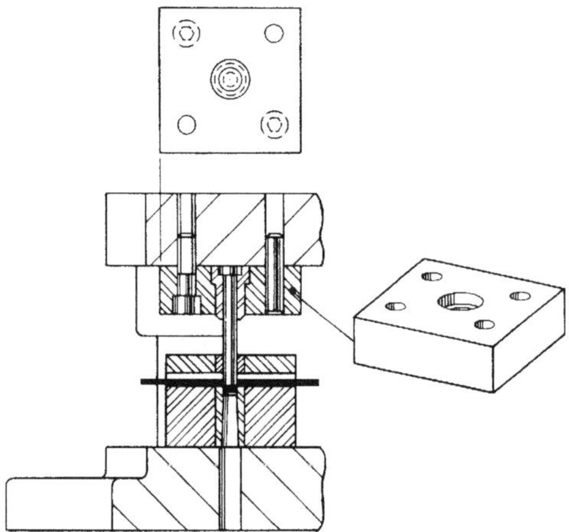

Figure 10.2 Hardened bushing retains slender punch, subject to breakage and replacement.

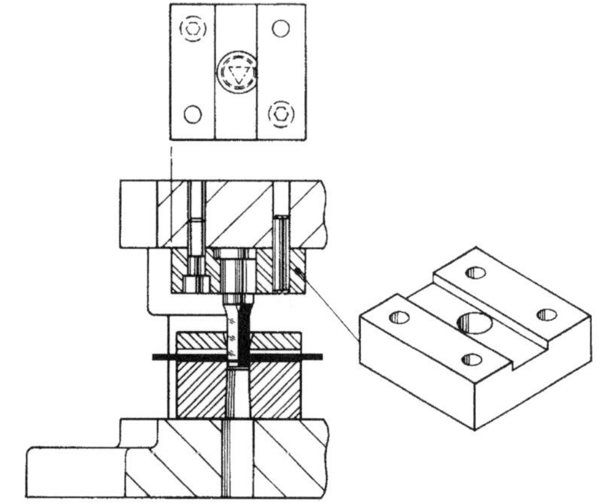

Figure 10.3 The slot in the top of the punch plate provides means of preventing turning of the punch.

b) Punch Plates for Slender Punches

Slender punches, which are subject to breakage and replacement, are retained in hardened bushings pressed into the punch plate (Figure 10.2). The top of the guide bushing is ground flush with the top of the punch plate after pressing in.

c) Punch Plate for Irregular Punches

Punch plates that retain irregular piercing punches should have in the top surface a slot machined to the same depth as the punch head (Figure 10.3). A flat, ground in the punch head, bears against one edge of this slot to prevent turning.

Two or more irregular punches that are in line can be kept from turning by a single slot machined in the top of the punch plate, against which flats on the punch head bear (Figure 10.4). This slot should be made large enough to clear the opposite sides of the punch heads.

Where space is limited, a slot is end-milled in the top of the punch plate to the same depth as the counterbored hole for the punch head (Figure 10.5). A small round-end key is inserted in this slot. The key engages a flat, machined in the punch head, to keep it from turning.

d) Individual Punch Plates

When piercing punches are some distance from each other, they are preferably held in individual punch plates, as shown in Figure 10.6. This condition is encountered in large piercing dies.

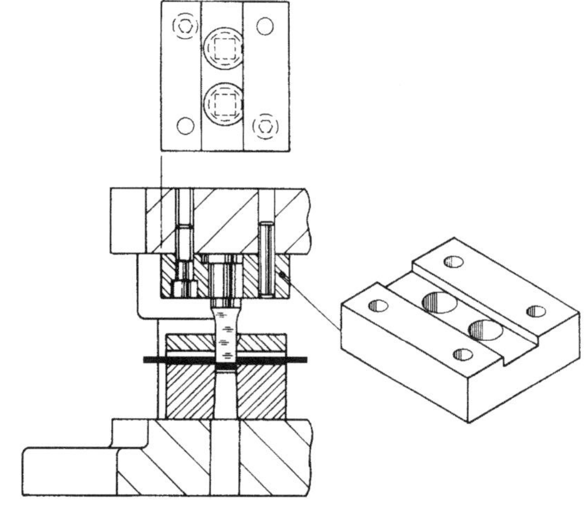

Figure 10.4 The slot may also be used to keep two or more punches from turning.

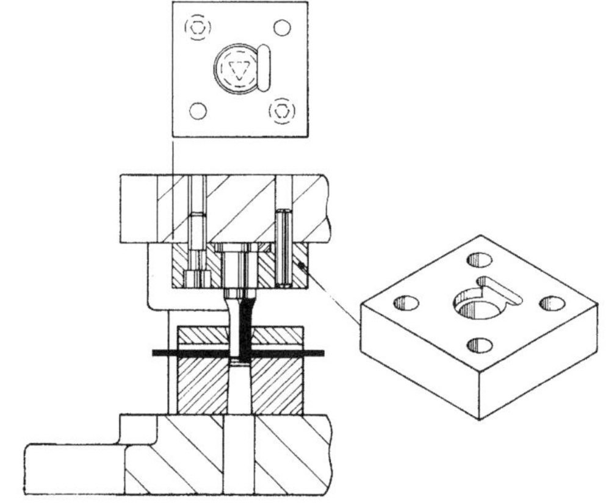

Figure 10.5 A slot milled in the top of the punch plate retains the key for preventing turning.

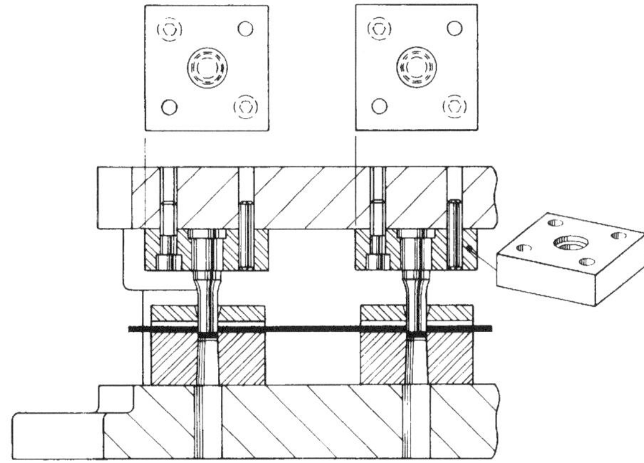

Figure 10.6 Individual punch plates are used when punches are some distance from each other.

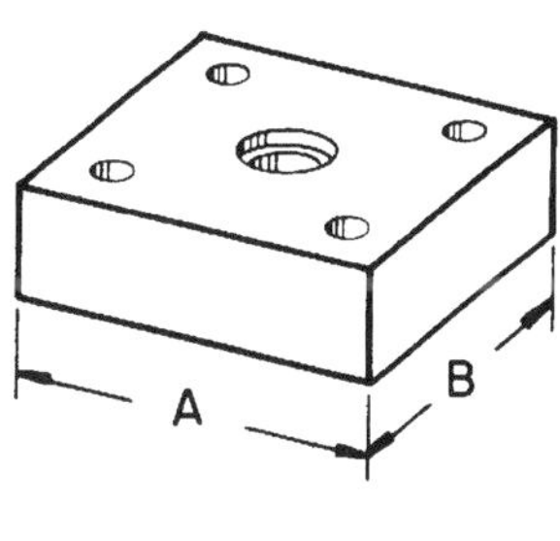

A	B
2	2
2	3
3	3
3	4
3	5
4	4
4	5
4	6
5	5
5	6
5	7
6	6
6	7
6	8
6	10
7	7
7	9
7	11

Figure 10.8 Commonly used standard punch plate sizes.

e) Stepped Punch Plates

When the part to be pierced has two levels (Figure 10.7), double punch plates avoid long piercing punches. Socket cap screws fasten lower to upper plates. The latter is held to the punch holder from the back. Note the double dowels. It is considered poor practice to use extremely long dowels.

f) Standardized Punch Plates

Where large quantities of dies are produced, time and money can be saved by standardizing punch plate sizes and assigning part numbers to them. Figure 10.8 shows, tabulated, the most commonly used sizes of punch plates.

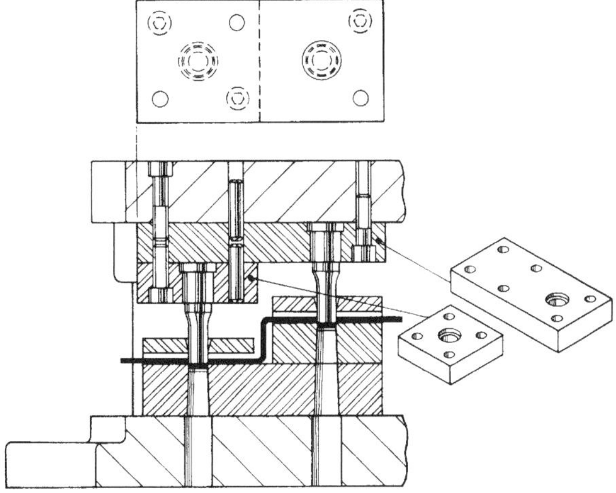

Figure 10.7 Punch plates are stepped to avoid long piercing punches.

Thickness of punch plate. Punch plate thickness **B** should be approximately 1 1/2 times diameter **A** of the piercing punch (Figure 10.9). With the table, the die designer can quickly establish the punch plate's thickness.

g) Punch Plate for Notching Punches

Small notching punches are usually held in their punch plate by means of slots machined in the top surface of the plate (Figure 10.10). Three sides of the punch head bear against the bottom of the machined slot to prevent the punch from pulling out.

10.2.2 Applying Punch Plates

a) Cut-Off Dies

Figure 10.11 illustrates three commonly used methods of retaining small cut-off punches. At *a*, a large slot machined in the top surface of the punch plate provides clearance for the punch head. Where space limitations exist, a

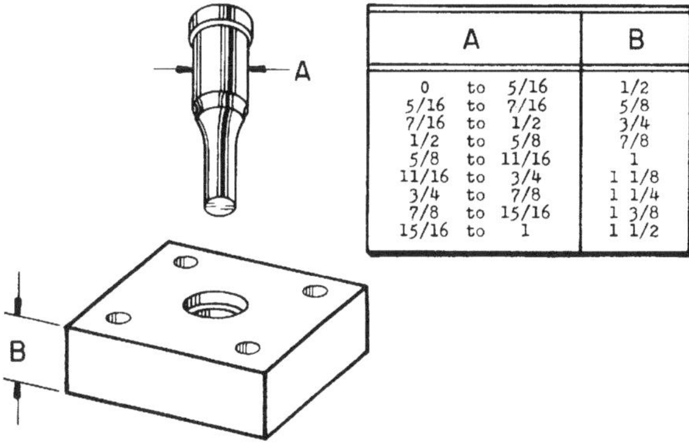

A	B
0 to 5/16	1/2
5/16 to 7/16	5/8
7/16 to 1/2	3/4
1/2 to 5/8	7/8
5/8 to 11/16	1
11/16 to 3/4	1 1/8
3/4 to 7/8	1 1/4
7/8 to 15/16	1 3/8
15/16 to 1	1 1/2

Figure 10.9 Table for determining punch plate thickness.

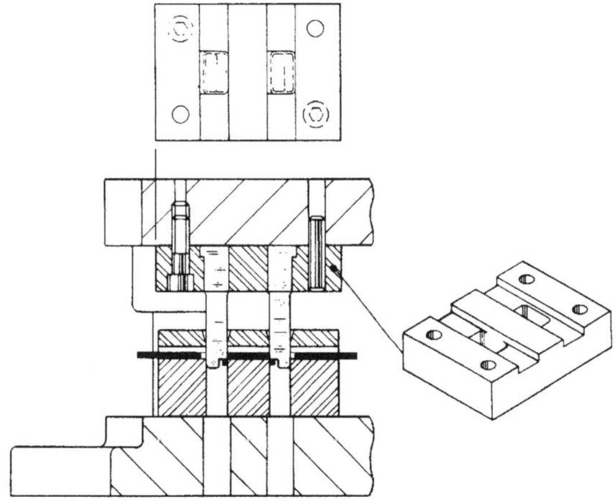

Figure 10.10 Punch plate for notching punches.

narrower slot is machined, as shown at *b*. End flanges (**D**) prevent the cut-off punch from pulling out. Where space is still more limited, a recess, slightly larger than the punch head, is end-milled in the top of the punch plate, as shown at part ***c***.

b) Compound Dies

Punch plates used in compound dies are made large so that they can act as spacer as shown in inverted die block **A** (Figure 10.12). Long socket cap screws applied from the top of the die set pass through clearance holes in the punch plate and are threaded into the die block.

c) Indirect Knockout

When indirect knockouts are used, the punch plate is recessed, as at **A**, to accommodate the knockout plate **B** (Figure 10.13). In this case, the recess is shaped in the form of the letter X.

d) Inverted Dies

In inverted piercing or shaving dies, the punch plate is fastened to the die holder of the die set instead of being placed in its usual position under the upper punch holder. Figure 10.14 illustrates shaving die for two holes. The punches have been inverted to make use of the pressure attachment of the press for more effective stripping.

e) Horn Dies

Piercing punches used in horn dies are retained in a punch plate fastened to the punch holder of the die set (Figure 10.15). This die pierces a hole in the side of a drawn cup.

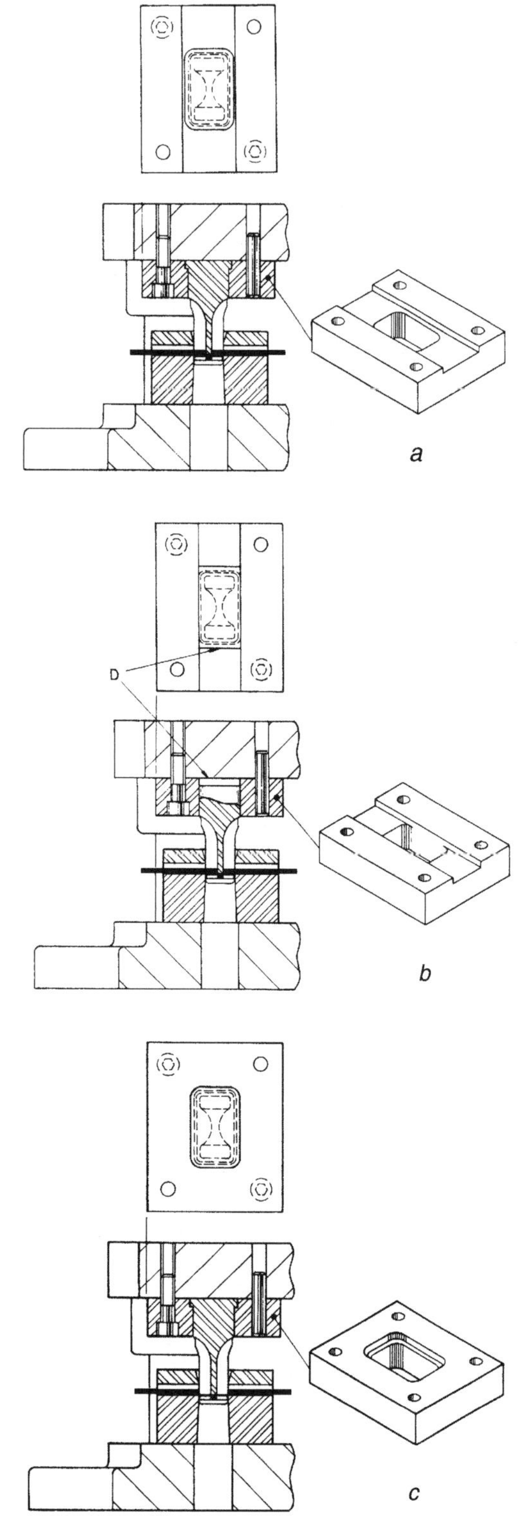

Figure 10.11 Three commonly used methods of retaining small cut-off punches.

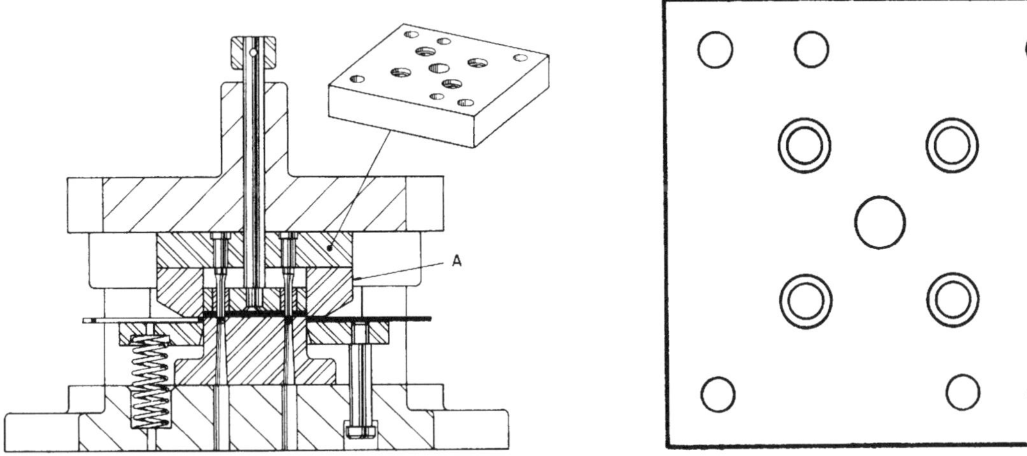

Figure 10.12 Punch plate for compound die.

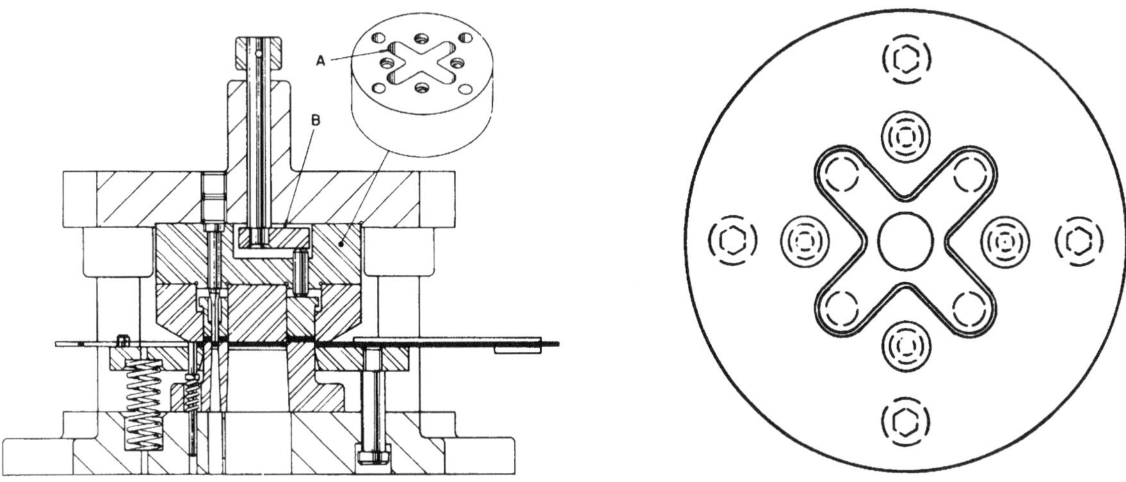

Figure 10.13 Punch plate for indirect knockouts.

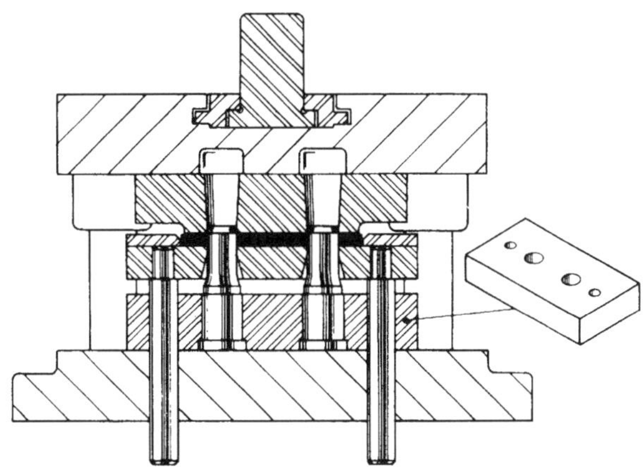

Figure 10.14 Punch plate for inverted die.

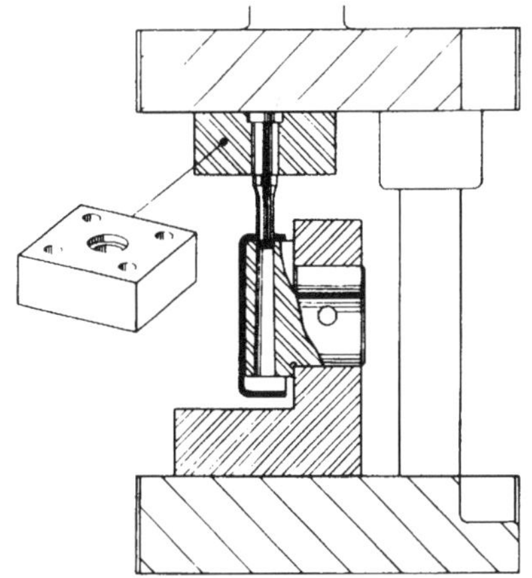

Figure 10.15 Punch plate for horn die.

f) Side Cam Dies

Punch plates for side cam dies usually have a turned boss that engages a counterbored hole in the slide for accurate punch location (Figure 10.16). With this construction, no dowels are required. The die shown at illustration A pierces holes in the sides of shells when the burr is required to be inside the shell. The die shown at illustration B pierces in the sides of shells when the burr is required to come outside the shell. Construction of the punch plate is similar to A.

g) Piercing Angular Holes

For piercing angular holes in shells, the punch plate is made large and guided on the die set guided posts (Figure 10.17). The punch plate retains the spring-backed piercing punch **A**, which is guided in a hardened bushing **B** pressed into the punch plate. Hardened buttons limit downward travel of the punch plate.

10.3 COMMERCIAL PUNCH PLATES

Punch plates can be purchased from commercial sources; standard punch plates can be purchased as components to hold the piercing punches. Three styles are illustrated in Figure 10.18.

Illustration A shows a round punch plate. It is machined to accommodate a hardened, spring-backed ball that engages a recess machined in the

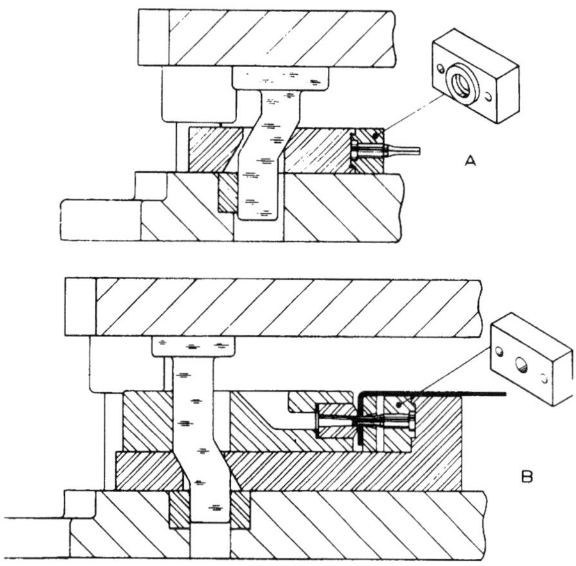

Figure 10.16 Punch plate for side cam die.

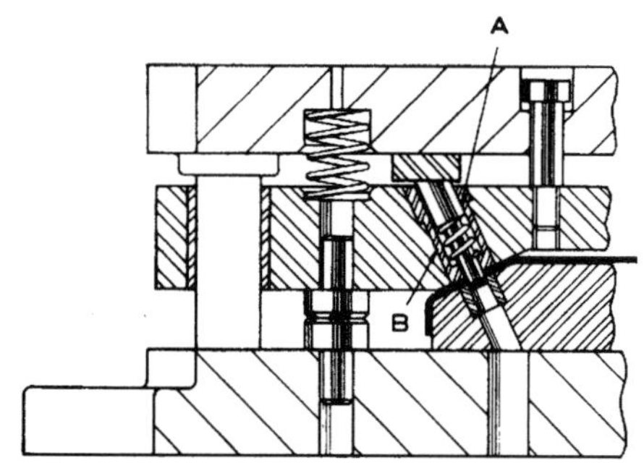

Figure 10.17 Typical large punch plate that is guided on die set guide posts.

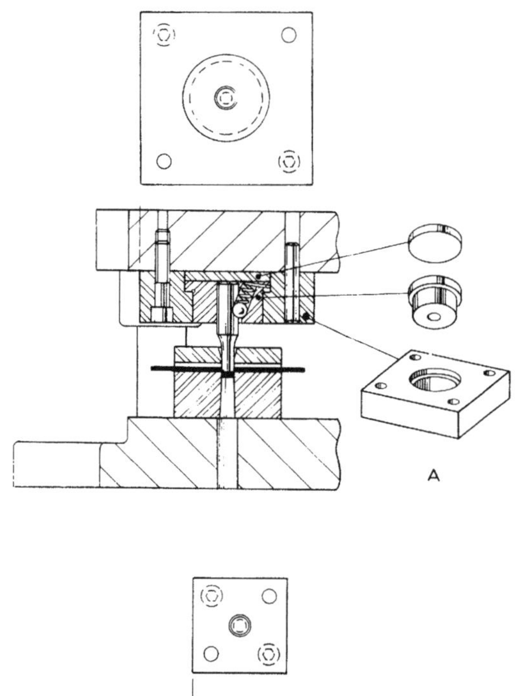

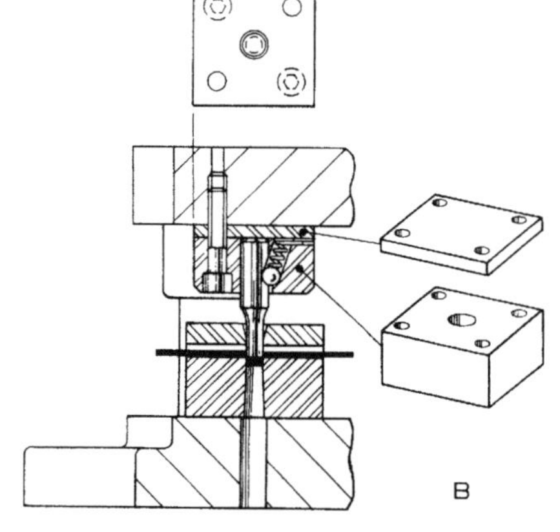

Figure 10.18 Various types of punch plates that are commercially available.

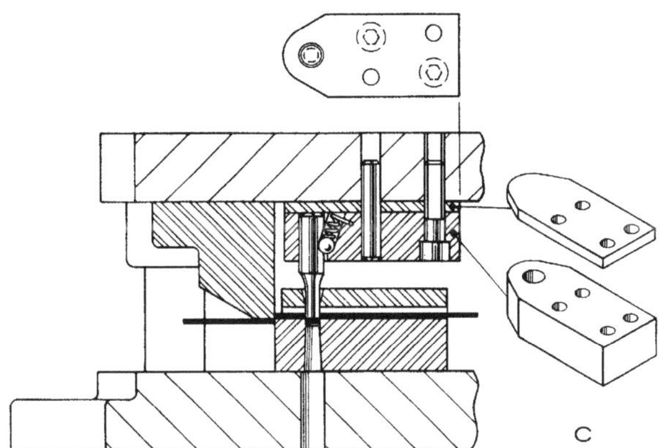

Figure 10.18 (Continued)

piercing punch. A hardened backing plate, made the same diameter as the punch plate flange, prevents the punch from sinking into the soft material of the punch holder of the die set. Illustration B shows the more commonly used square punch plate, backed up by a hardened plate of the same size, while illustration C shows the end-retaining punch plate, used where space is limited.

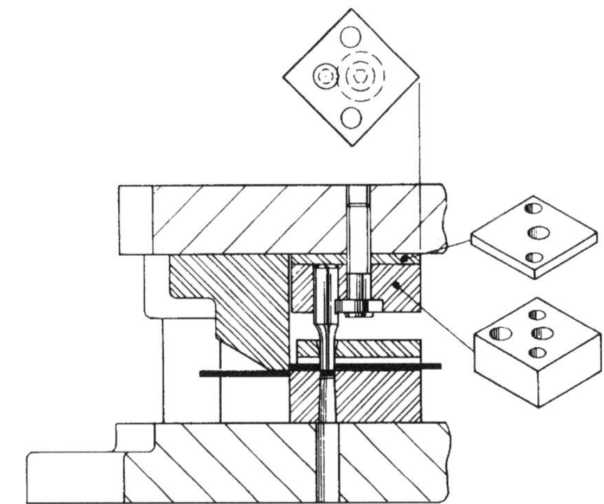

Figure 10.19 Another type of commercially available punch plate.

Another type of commercial punch plate is shown in Figure 10.19. These are used to hold a line of piercing punches. They are retained by special quarter-turn screws that provide quick removal of piercing punches for replacement.

HOW TO DESIGN PILOTS

11.1 INTRODUCTION

Pilots play a vital role in the operation of multiple-station dies. Many press-line troubles can be traced to pilot faulty design. When applying pilots, the following factors should always be considered:

1. Pilots must be strong enough so that repeated shock will not cause fracture. Severe shock is applied to the pilot point more often than is realized. Consider that the pilot moves a heavy material strip almost instantly into register. Pilot breakage increases cost of the stamping because hundreds of inaccurate parts may be produced before failures are discovered. Also, there is the danger of costly jams resulting from a broken pilot falling between the cutting edges or forming members of the die.
2. Slender pilots must be sufficiently guided and supported to prevent bending, which can cause faulty strip-positioning. They should be made of a good grade of tool steel, heat-treated to 57-to-60 Rockwell C for maximum toughness and hardness.
3. Provision should be made for quick and easy removal of the pilots for punch sharpening.

There are two methods of piloting in progressive dies:

- *Direct piloting* consists of piloting in holes pierced in that area of the strip that will become the blank.

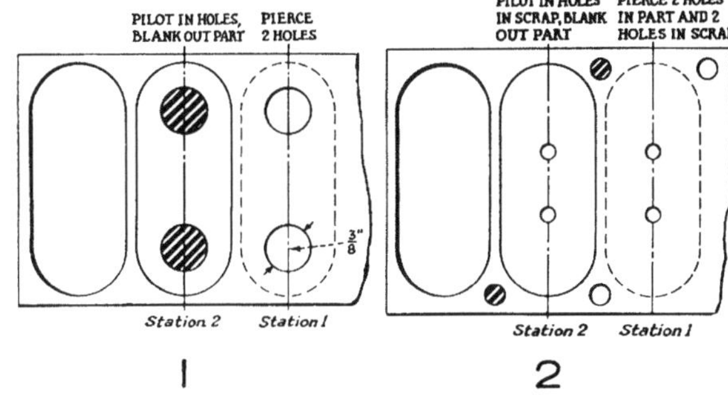

Figure 11.1 Examples of direct piloting (1) and indirect piloting (2).

- *Indirect piloting* consists of piercing holes in the scrap area of the strip, then locating in these holes at subsequent operations.

Direct piloting is the preferred method, but certain blank conditions require indirect piloting. Figure 11.1 shows examples of direct piloting (1) and indirect piloting (2).

This chapter illustrates numerous methods of designing and applying both direct and indirect pilots to help designers select the best type to use for particular jobs.

11.2 DIRECT PILOTING

11.2.1 Shoulder Pilots

The frequently used shoulder pilot **A** in Figure 11.2 is retained in blanking punch **B** by a socket pilot nut **C**. Pilot holes are pierced at the first station. The strip is then located by pilots at the

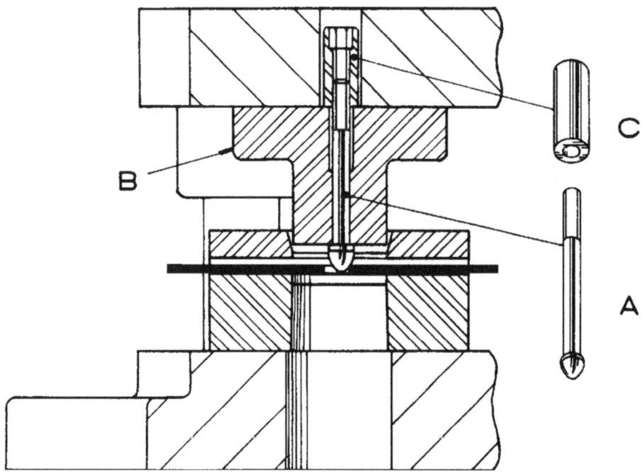

Figure 11.2 Typical shoulder pilot.

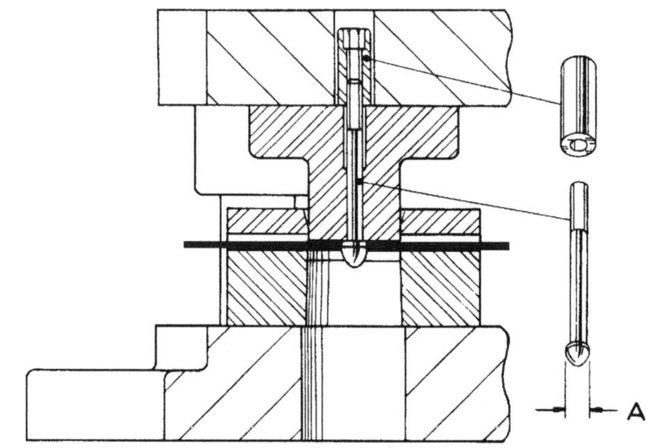

Figure 11.3 A shoulder pilot in fully descended position accurately locates the strip.

second and succeeding stations. The automatic stop is positioned so that the strip is stopped with a previously pierced hole 0.010 inch (0.25 mm) past its final location. The pilot moves the strip back this amount to bring it into correct register. This over-travel prevents possible cramping of the strip between pilot and automatic stop. As shown, the pilot is just contacting one side of the hole in the strip preparatory to bringing it back to true position.

a) Locating the Strip

Descent of the upper die has caused the bullet-shaped nose of the pilot to move the strip back 0.010 inch (0.25 mm). The pilot has entered the strip hole, accurately locating it for the blanking operation (Figure 11.3). The diameter of pilot shoulder **A** is an important dimension. Too small a diameter relative to the hole in the strip will produce inaccurate parts with varying dimensions between hole and part edges. Too large a dimension results in a tight fit in the strip, with a consequent tendency for the blank to be pulled up out of the die hole. This tendency can be a serious problem in progressive dies. A successful formula is: Diameter **A** equals the diameter of the piercing punch, less 3 percent of the strip thickness. The formula is

$$A = D_{pp} - 0.03T \qquad (11.1)$$

where

D_{pp} = diameter of the piercing punch
T = thickness of material.

Example Find the pilot diameter if the piercing punch diameter is 0.500 inch, and the strip thickness is #16 gage (0.0625 inch or 1.6 mm).

Solution: The pilot diameter is:

$$A = D_{pp} - 0.03T = 0.500 - 0.03 \times 0.0625$$
$$= 0.4981 \text{ in. } (12.652 \text{ mm})$$

b) Pilot Proportions

Further descent of the upper die has now caused the blanking punch to penetrate the material strip to produce a blank. The shoulder pilot (Figure 11.4) is used for holes **A** from 1/4 to 3/4 inch (6.35 to 19.05 mm) diameter.

The pilot has the following general proportions: Straight engagement length **B** should be from 1/3 to 2/3 of the stock thickness. For example, for #16 gage (0.0625 inch or 1.6 mm) this length could be made 1/32 inch (0.8 mm). Diameter **C** is made an accurate gage fit. The length of this gage fit—engagement **D**—is made approximately three times pilot diameter **C**. Clearance relief diameter **E** is usually 1/32 inch (0.8 mm) larger than body diameter **C** of the pilot.

$$B = (0.33 \text{ to } 0.66) \times T$$
$$D = 3C \qquad (11.2)$$
$$E = C + 0.032 \text{ inch}$$

where

B = the straight engagement length
D = the length of gage fit engagement

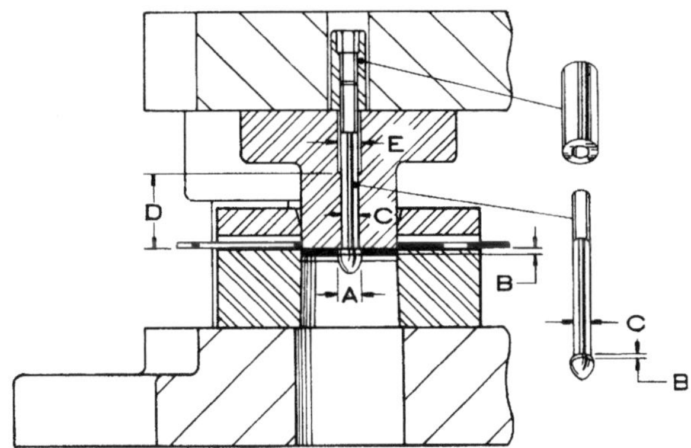

Figure 11.4 With the shoulder pilot accurately locating the strip, a punch descends to a blank part.

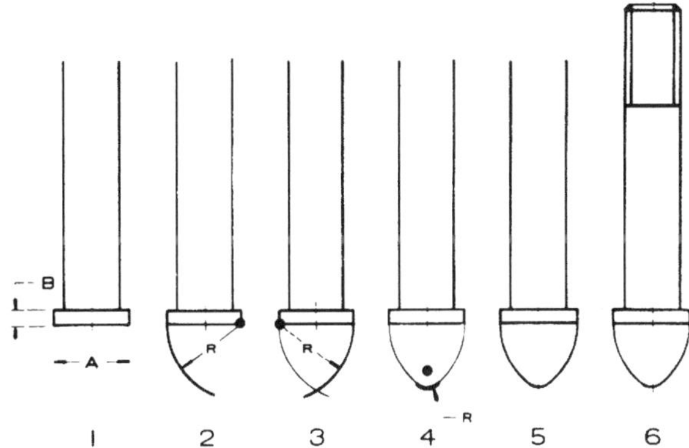

Figure 11.5 Six steps in drawing the head of a shoulder pilot.

C = pilot diameter
E = clearance relief diameter.

c) Drawing the Pilot

Six steps are taken to draw the acorn-shaped head of a shoulder pilot (Figure 11.5):

1. The shank and diameter **A** of the head are drawn with light lines. This is the diameter that actually engages the hole in the strip for register. As previously noted, this diameter is 3 percent of the stock thickness less than the diameter of the punch that pierces the hole in the strip. Thickness **B** is from 1/2 to 1/3 the stock thickness.
2. The point of the compass is applied at the lower right corner and radius **R** is drawn.
3. The point of the compass is applied at the lower left corner and an opposed radius **R** is drawn.
4. A circle template is carefully applied and a smaller blending radius is drawn. This radius should be approximately 1/4 diameter **A**. It should be darkened and the line drawn to finished width at this time.
5. Lines of the two large radii are now blackened and widened to blend with the small radius.
6. The shank is completed and the pilot is now fully drawn.

11.2.2 Press-Fitted Pilot

For many years, press-fitted pilots (Figure 11.6) were almost the only type used in press work, but now they are usually specified only for short-run jobs and temporary dies. They can work loose and fall between cutting edges or forming members of the die with disastrous results. They can also fall through the die, allowing perhaps hundreds of parts to be run through the die and spoiled because of faulty location.

a) Proportions of Press-Fitted Pilot

Although press-fitted pilots are not used as much as in the past, they do have a place in the design of low-production dies for running strip stock. In addition, they are frequently used as locators in secondary operation dies, engaging previously pierced holes for performing other operations. A table of suggested dimensions of press-fitted pilots from 1/8 inch to 1 1/2 inches (3.2 mm to 38.1 mm) in diameter is given in Figure 11.7.

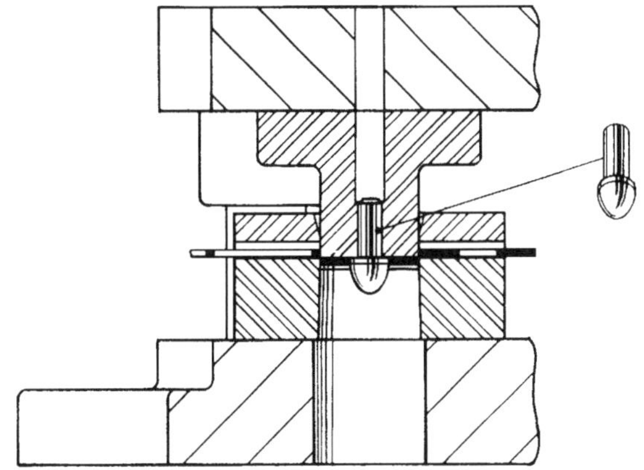

Figure 11.6 Press-fitted pilot for short runs.

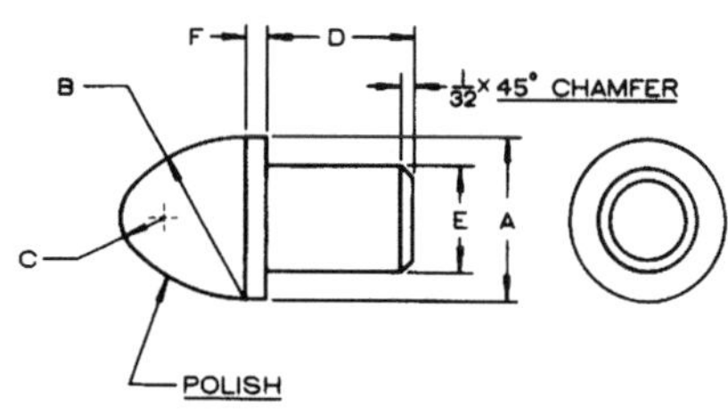

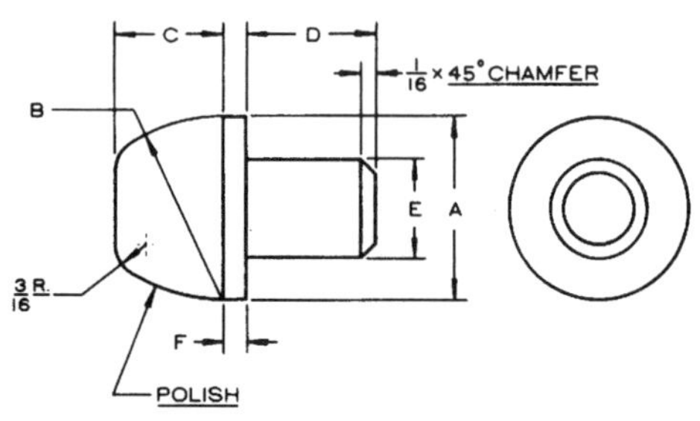

A	B	C	D	E	MAT.
1/8	1/8	1/32	5/32	3/32	D.R.
3/16	3/16	3/64	3/16	1/8	D.R.
1/4	1/4	1/16	9/32	3/16	D.R.
5/16	5/16	5/64	3/8	1/4	D.R.
3/8	3/8	3/32	7/16	9/32	D.R.
7/16	7/16	7/64	1/2	5/16	D.R.
1/2	1/2	1/8	9/16	3/8	D.R.
9/16	9/16	9/64	5/8	7/16	D.R.
5/8	5/8	5/32	11/16	15/32	D.R.
11/16	11/16	11/64	3/4	1/2	D.R.
3/4	3/4	3/16	7/8	9/16	D.R.

A	B	C	D	E	MAT.
13/16	13/16	1/2	15/16	5/8	T.S.
7/8	7/8	17/32	1	11/16	T.S.
15/16	15/16	9/16	1 1/8	3/4	T.S.
1	1	5/8	1 1/4	13/16	T.S.
1 1/16	1 1/16	21/32	1 5/16	7/8	T.S.
1 1/8	1 1/8	11/16	1 7/16	15/16	T.S.
1 3/16	1 3/16	23/32	1 1/2	1	T.S.
1 1/4	1 1/4	3/4	1 5/8	1 1/16	T.S.
1 5/16	1 5/16	13/16	1 11/16	1 1/8	T.S.
1 3/8	1 3/8	27/32	1 3/4	1 3/16	T.S.
1 1/2	1 1/2	15/16	1 7/8	1 1/4	T.S.

Figure 11.7 Table of suggested dimensions of press-fitted pilots.

11.2.3 Irregular Pilot

Piloting is often done in holes that have an irregular shape (Figure 11.8). The pilots must be radially located. In this example, the large, oval-shaped pilot is kept from turning by a pin inserted half in the pilot body and half in the blanking punch. This pin is made a press fit in the pilot and a slide fit in the punch, thereby providing for easy removal of the pilot in order to sharpen the punch face.

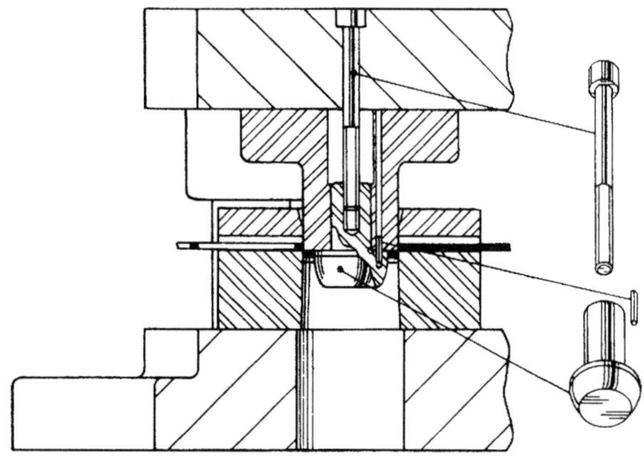

Figure 11.8 Fixed oval-shaped pilot.

11.2.4 Method of Fastening Pilots

a) Pilot Nuts

Pilot nuts (Figure 11.9) represent the fastest and most convenient method of fastening pilots. Their use allows the pilots to be quickly removed for sharpening without the necessity of disassembling other die parts. Standardized and stocked in the tool crib, the pilot nuts effect substantial savings in both design and manufacture of progressive dies, as well as further savings on the press line in making possible quick pilot removal and replacement. The most-used sizes will vary from shop to shop, depending on the general sizes of stampings to be run. The double hexagon hole should be cold-formed before final turning of the outside diameter. The nuts should be made from drill rod, heat-treated; or machine steel, cyanide-hardened.

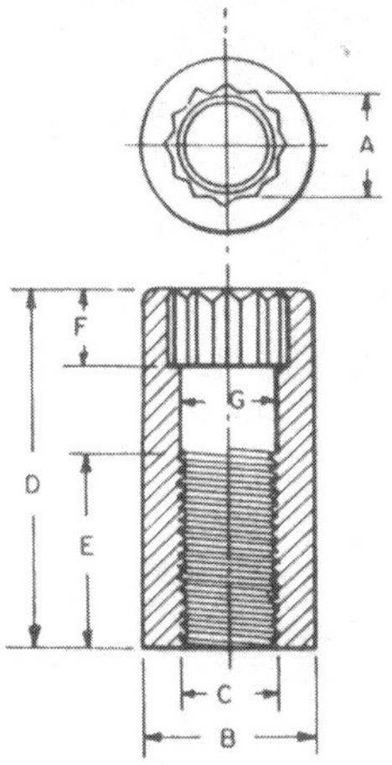

DIMENSIONS OF PILOT NUTS

Nut Size	Socket Width Across Flats A	Body Dia B	Tap Size NC-3 C	Body Lgth. D	Thd Lgth. E	Socket Depth F	Clear Hole G
No. 4	1/8	0.187	#4-48	1	¼	0.076	#32(0.116)
No. 5	1/8	0.187	#5-44	1	¼	0.082	#30(0.1285)
*No. 6	5/32	0.257	#6-40	1	5/16	0.100	#27(0.144)
*No. 8	3/16	0.280	#8-36	1	3/8	0.112	#18(0.1695)
*No. 10	3/16	0.302	#10-32	1	7/16	0.142	#9 (0.196)
*¼	¼	0.403	¼-28	1	½	0.172	#F(0.257)
5/16	5/16	0.498	5/16-24	1	5/8	0.231	#P(0.323)
3/8	3/8	0.590	3/8-24	1	¾	0.250	#W(0.386)

Note: Hexagon "A" must be double, as shown
*Most frequently used sizes

Figure 11.9 Suggested dimensions for standardized pilot nuts.

Slotted pilot nuts. Instead of being driven by internal hexagon wrenches, pilot nuts can be made with a screwdriver slot (Figure 11.10). Although not considered as good as socket nuts, the slotted nuts do provide a quick way of making these components. Dimensions **A** and **B** are grinding allowances, normally 1/4 inch; they are consumed as the blanking punch is sharpened from time to time during the life of the die. Pilots are removed from the punch before sharpening. The gage fit (instead of a press fit) and the pilot nut allow quick dismantling.

b) Set Screws Back the Pilot

Straight, headed pilots (Figure 11.11) are used when diameter **A** is from 3/16 to 1/4 inch (4.8 to 6.35 mm). The head is backed up by two socket set screws. Grinding allowance **B** is 1/4 inch (6.35 mm).

Spacer washers are applied under the pilot head each time, after the blanking punch is sharpened, to maintain the proper relationship between the end of the pilot and the face of the blanking punch. Gage fit length **C** is again made approximately three times the pilot diameter.

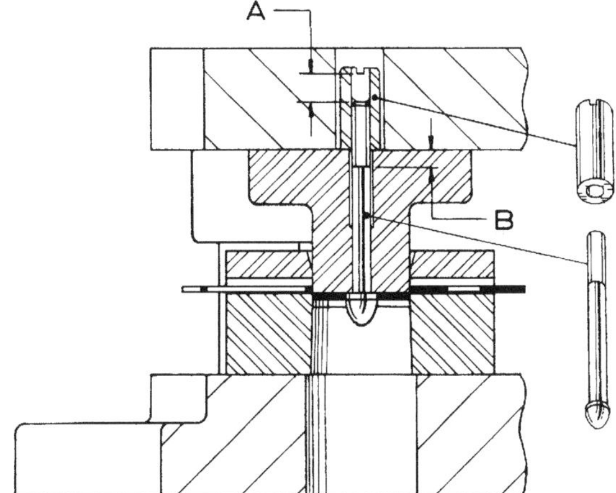

Figure 11.10 A pilot nut with a screwdriver slot may be made quickly.

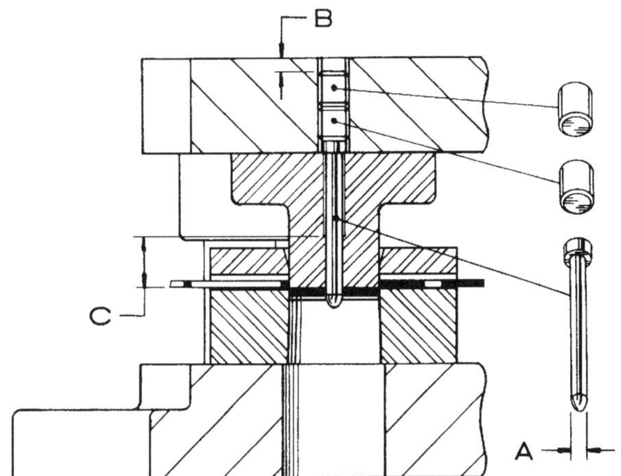

Figure 11.11 Straight, headed pilots are used when the head is backed up by two socket set screws.

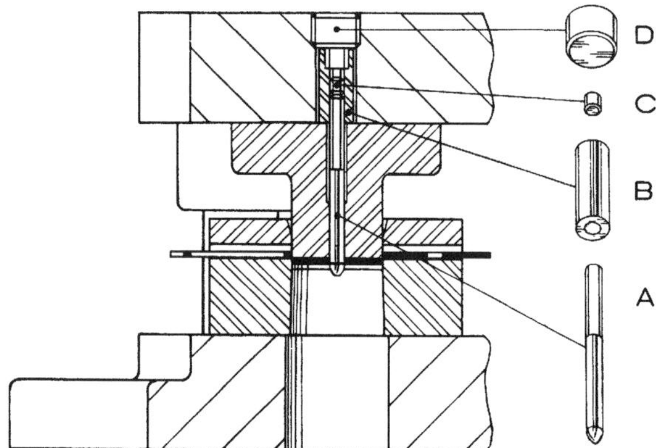

Figure 11.12 Backing up a small pilot with two set screws.

Small pilot. For pilots under 1/8 inch (3.2 mm) in diameter, shank **A** is threaded into pilot nut **B** and is backed up by a small set screw **C** (Figure 11.12). The entire assembly, in turn, is backed up by a large set screw **D**, threaded into the punch holder of the die set. The assembly is removed from the punch for sharpening by simply removing set screw **D**. The pilot is moved up to compensate for the amount removed from the punch face in sharpening by loosening the small set screw, raising the pilot slightly, and again locking it in place.

c) Large Pilot

Pilots over 3/4 inch (19.05 mm) in diameter are held by a long socket cap screw drilled and counterbored into the punch holder of the die set (Figure 11.13). The pilot is threaded an extra distance **A** to provide take-up for punch sharpening.

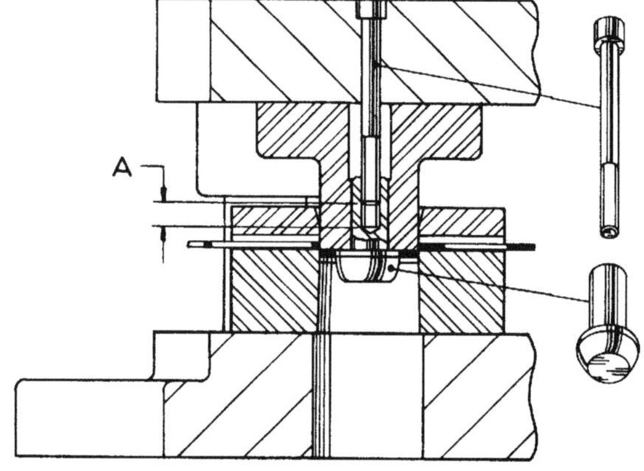

Figure 11.13 Holding a large pilot in place using a long socket cap screw.

11.3 INDIRECT PILOTING

There are seven part conditions that require indirect piloting (Figure 11.14):

1. *Close tolerance on holes.* Pilots can enlarge holes in pulling a heavy strip to position.
2. *Holes too small for sturdy pilots.* Frail pilots can break or deflect in operation.
3. *Holes too close to edges of the blank.* Distortion can occur in the blank because of enlargement of holes.
4. *Holes located in weak portion of piece.* Piloting in projecting tabs is impractical because they may deflect before the strip is pulled to position.
5. *Holes too close for relationship with edge.* Piloting in closely spaced holes does not provide an accurate relationship between holes and outside edges of the blank.
6. *Blanks lacking holes.* Piloting is done in the scrap area whenever the blank does not contain holes.
7. *Pilot to fit opening may potentially bend tongues.* Whenever the hole in the blank contains weak projections that could be bent down by the pilot, indirect piloting should be selected.

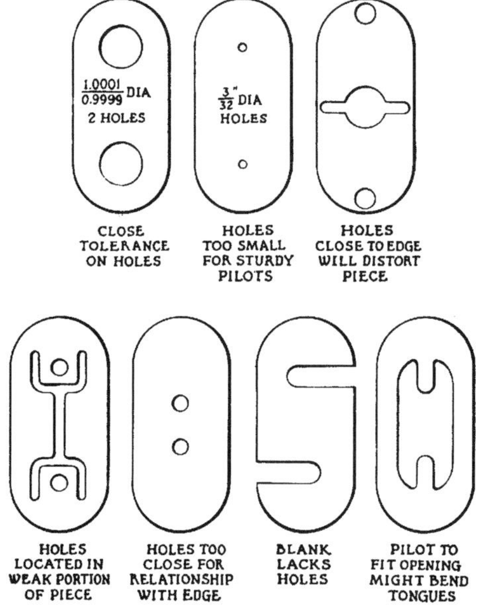

Figure 11.14 Part conditions that require indirect piloting.

11.3.1 Indirect Pilot

Indirect pilots are similar in construction to piercing punches. In Figure 11.15, the pilot is shown contacting the strip before moving it back 0.010 inch (0.25 mm) into correct register. Body diameter **A** is made a light press fit in pilot plate **B**. Diameter **C**, usually 1/8 inch (3.2 mm) long, is a sliding fit in the pilot plate for accurate alignment while pressing in. Shoulder diameter **D** is generally made 1/8 inch (3.2 mm) larger than body **A**. Shoulder height **E** is from 1/8 to 3/16 inch (3.2 to 4.8 mm), depending on pilot size.

a) Positioning the Strip

Descent of the upper die to the lowest position causes the pilot to move the strip back 0.010 inch (0.25 mm) to the correct position (Figure 11.16). The pilot then enters the hole in the die block to position the strip accurately. This die block hole is made the same as for piercing punches, with a 1/4-degree taper per side. Directly underneath, a clearance hole is applied to the die holder of the die set. In cases of misfeed, pilots will punch the strip like a piercing punch, and provision must be made for disposal of slugs. Otherwise, they would pile up and, in time, the pilot would be broken. Pilot diameter **A** should be a good sliding fit in the pilot hole in the die block, with from 0.0002 to 0.0005 inch (0.005 to 0.013 mm) maximum clearance allowed.

b) Guiding in the Stripper

Where considerable accuracy is required, and when moving a large, heavy strip, the pilot is guided and supported in a hardened bushing pressed into the stripper plate (Figure 11.17). The pilot must be a good sliding fit within this bushing, with from 0.0002 to 0.0005 inch (0.005 to 0.013 mm) maximum clearance specified.

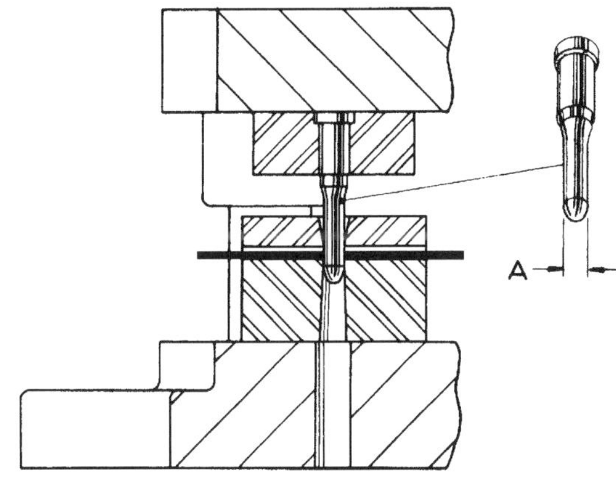

Figure 11.16 An indirect pilot in full descended position accurately locates the strip.

c) Backing Plates

Considerably more pressure is required for an indirect pilot to pierce a strip than for an equivalent piercing punch. When the possibility of repeated misfeeds is present, the pilot should be backed up by a hardened plate to prevent its head from sinking into the punch holder (Figure 11.18).

d) Small Indirect Pilot

Smaller pilots, 3/16 to 1/4 inch (4.8 to 6.35 mm) in diameter, are made with a turned head. They

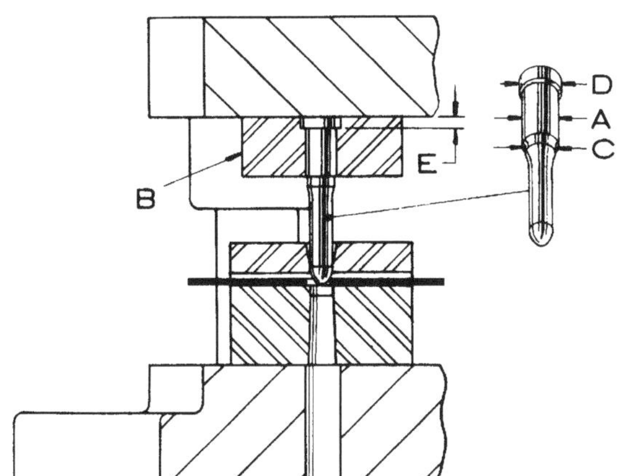

Figure 11.15 An indirect pilot is similar to a piercing punch.

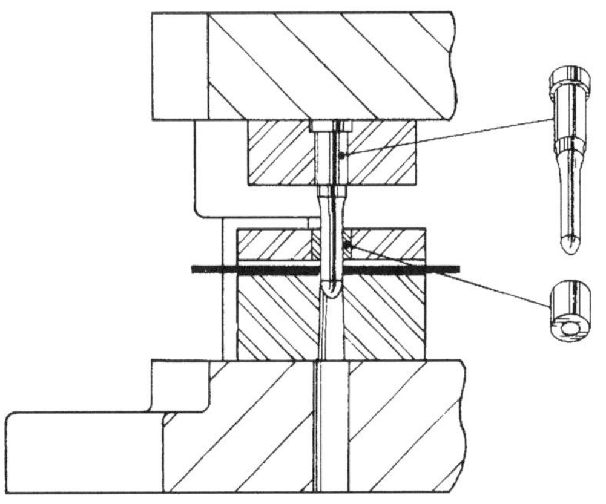

Figure 11.17 A hardened bushing in the stripper provides greater accuracy.

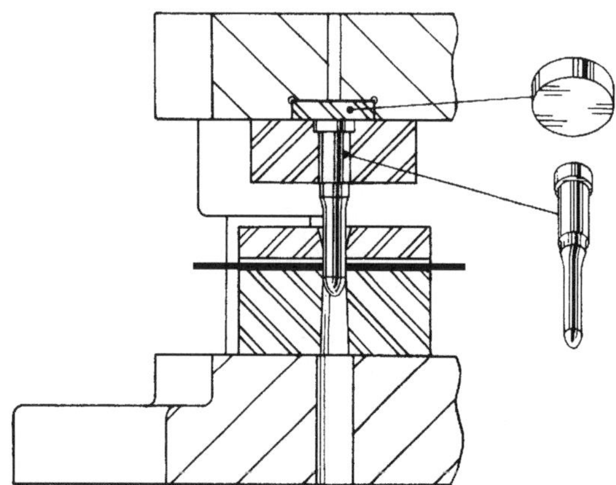

Figure 11.18 A pilot backed up with a hardened plate when the chance of repeated misfeeds is present.

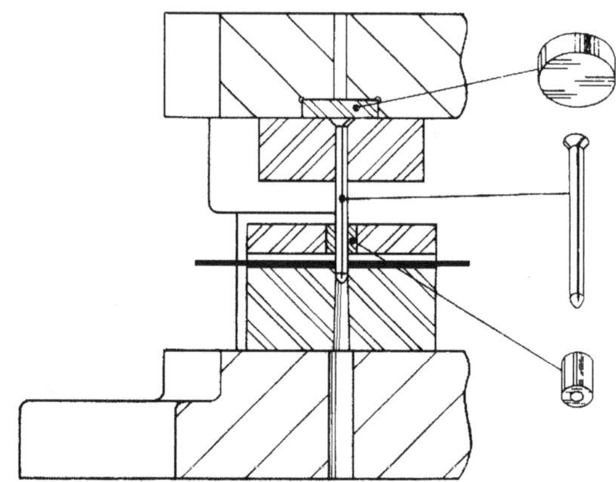

Figure 11.20 Small indirect pilot made of drill rod with a peened head.

are then guided in a hardened bushing in the stripper plate and backed up by two socket set screws (Figure 11.19). In every instance, the designer should pay particular attention to grinding allowances, remembering that, as punches are sharpened, the positions of the pilots will be altered. In this case, grinding allowance **A** is applied to assure that, as the pilots are raised with spacer washers when punches are sharpened, the upper socket set screw will not protrude above the punch holder.

Smaller indirect pilots, 1/8 to 3/16 inch (3.2 to 4.8 mm) in diameter, can be made of drill rod with a peened head (Figure 11.20). Such pilots are guided in hardened bushings in the stripper; they are always backed up, either by a hardened plate or with socket set screws. After peening, the head is machined to an 82-degree included angle to fit standard countersunk holes. At assembly, the top of the pilot plate is ground so that the peened head is flush with its surface.

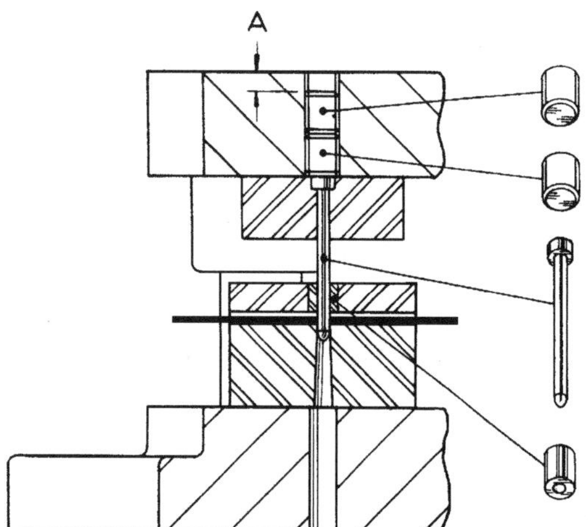

Figure 11.19 A small indirect pilot is guided through the stripper by hardened bushing and is backed up by two set screws.

11.3.2 Method of Fastening Indirect Pilots

a) Set Screw Holds the Pilot

For temporary dies, and where low accuracy and low production requirements exist, pilots can be held by a socket set screw bearing against a tapered flat that has been machined in the pilot body (Figure 11.21). These pilots are made of drill rod, and the bullet-shaped end is machined to suit the hole size. Because the set screw has a

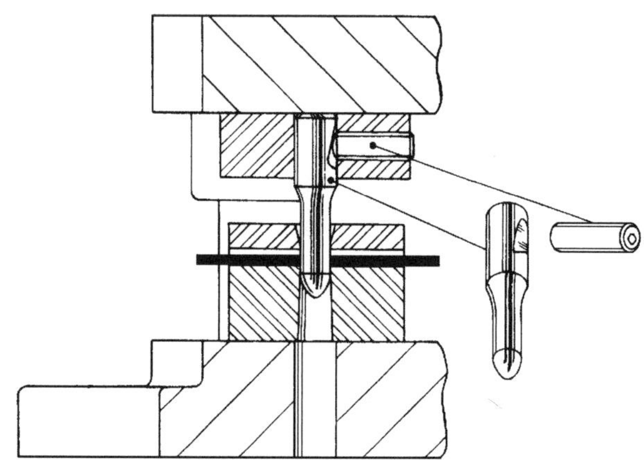

Figure 11.21 Method of holding pilot with set screw in temporary dies.

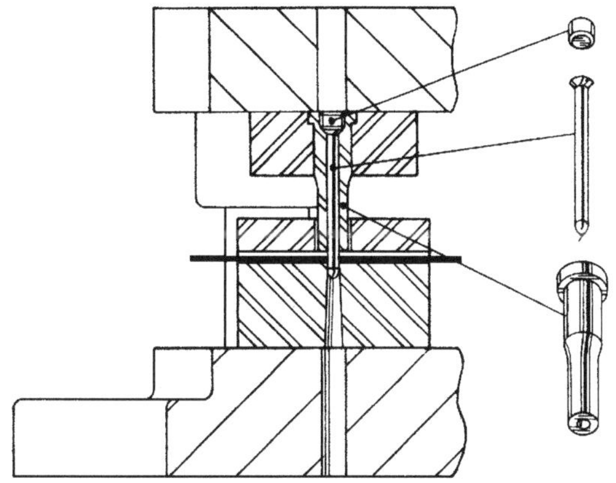

Figure 11.22 A small peened-head type pilot is held in a quill and backed up by a set screw.

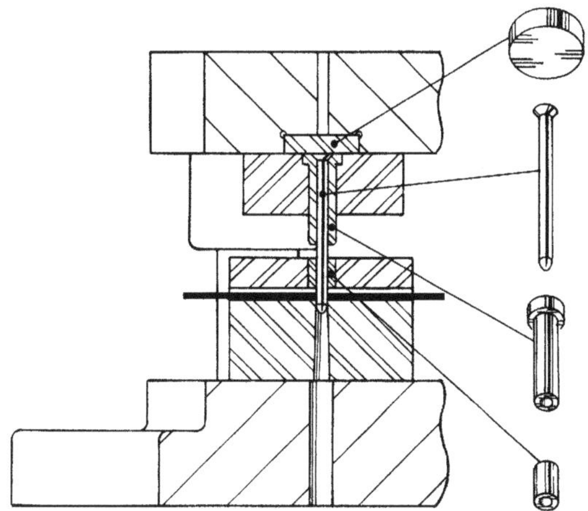

Figure 11.23 Pilot similar to the one shown in Figure 11.22, but with hardened bushing in the stripper plate and a hardened backup plate.

tendency to throw over the pilot by the amount of clearance, this type of pilot should never be used for first class dies.

b) Quill and Pilot

Small pilots, 1/8-inch (3.2 mm) diameter and smaller, are of the peened-head type, held in a quill, and backed up by a socket set screw (Figure 11.22). These quills, made of hardened and ground tool steel, support the pilot close to the piloting end.

c) Quill and Bushing

If the quill is made shorter than shown in Figure 11.22, the pilot is guided and supported in a hardened bushing pressed into the stripper plate (Figure 11.23). The entire assembly is backed up, either by a hardened plate or by two set screws. Whenever a hardened plate is used to back up a pilot, a through hole should be drilled in the punch holder for pressing the plate out. When the punches are sharpened, an equal amount of stock may be removed from the backing plate to maintain the relationship between the ends of the pilots and the cutting faces of the punches.

11.3.3 Methods of Keeping Pilots from Turning

a) Flat on Pilot Head

Some indirect pilots are used at intermediate stations of progressive dies for locating the strip. When these pilots are engaged in holes other than round, they can be kept from turning by grinding a flat on the pilot head (Figure 11.24). This flat bears against one side of a slot machined in the pilot plate.

b) Keyed Pilots

Where space is limited, the pilot can be kept from turning by grinding a flat in the same way as described for Figure 11.24, and locating this flat against a key inserted in the pilot plate (Figure 11.25). Always use a round-end type of key. The fit must be of first class die quality.

c) Pinned Pilot

Another method of keeping an irregular pilot from turning is to provide a large flanged head

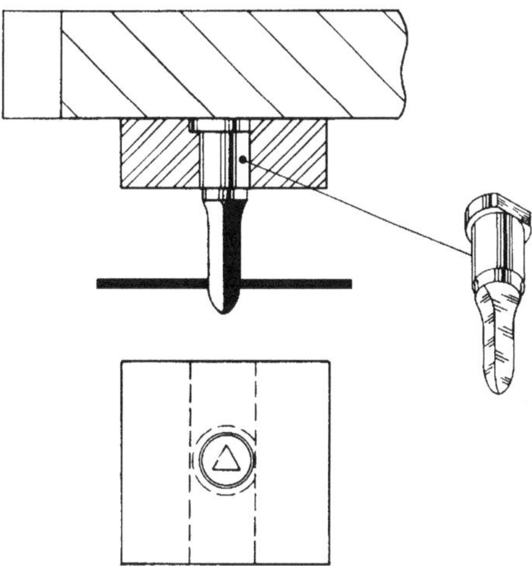

Figure 11.24 Triangular-shaped indirect pilot that is kept from turning by the flat on the pilot head.

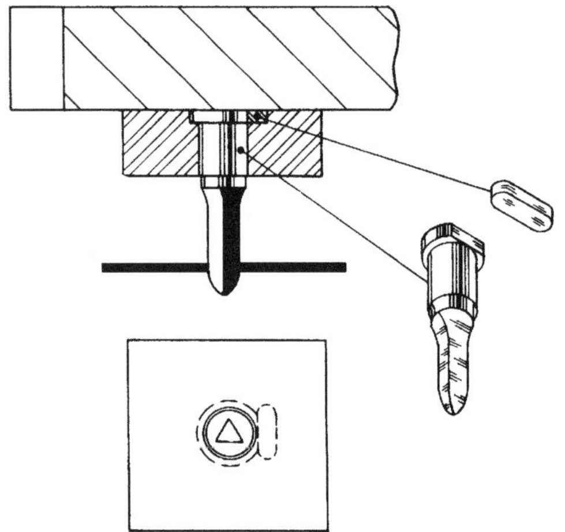

Figure 11.25 The pilot may be keyed with round-end key where space is limited.

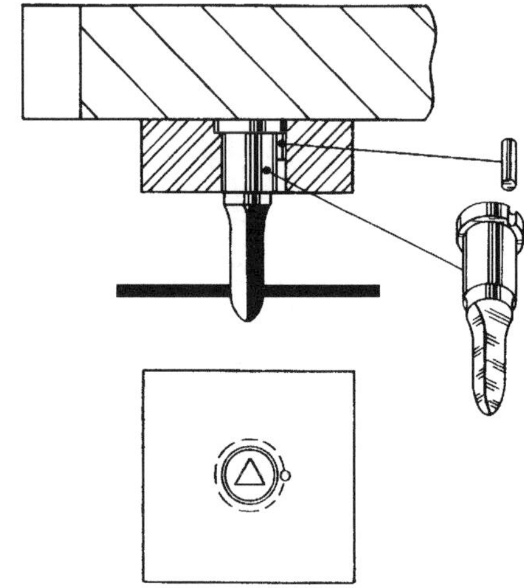

Figure 11.27 Using an anti-rotation pin to keep a pilot from turning in temporary dies.

on the pilot. Dowel this flange to the pilot plate (Figure 11.26).

d) Anti-Rotation Pins

When space is limited, a pin pressed half in the pilot head and half in the pilot plate will keep an irregular pilot from turning (Figure 11.27). This method is not considered as good as those previously described. It should be used only for temporary dies where low production requirements are present.

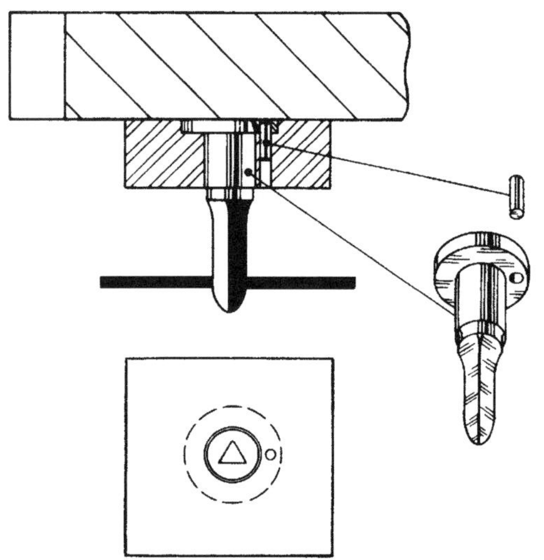

Figure 11.26 Method of pinning an irregular pilot to keep it from turning.

11.4 SPRING-BACKED PILOTS

All of the pilots described thus far have been of the solid variety used for sizes of stock up to and including #16 gage. When piloting in thicker material, it is wise to use the spring-backed type of pilot (Figure 11.28). Pilot **A**, guided in bushings **B** and **C**, is backed up by spring **D**, which in turn is backed up by set screw **E**. The pilot is made a close sliding fit in bushings **B** and **C**, normally with from 0.0002 to 0.0005 inch (0.005 to 0.013 mm) clearance.

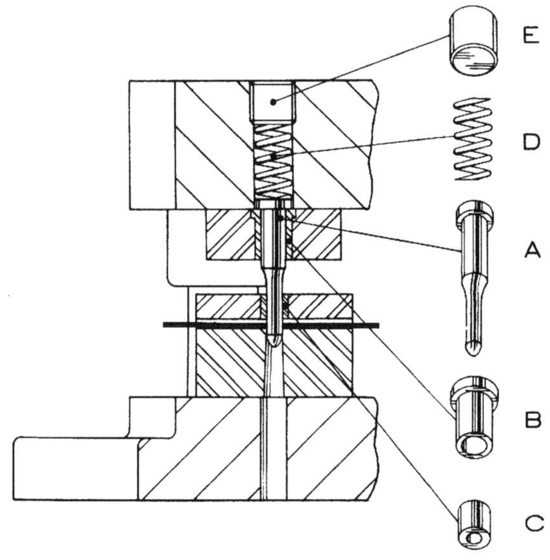

Figure 11.28 Typical spring-backed pilot.

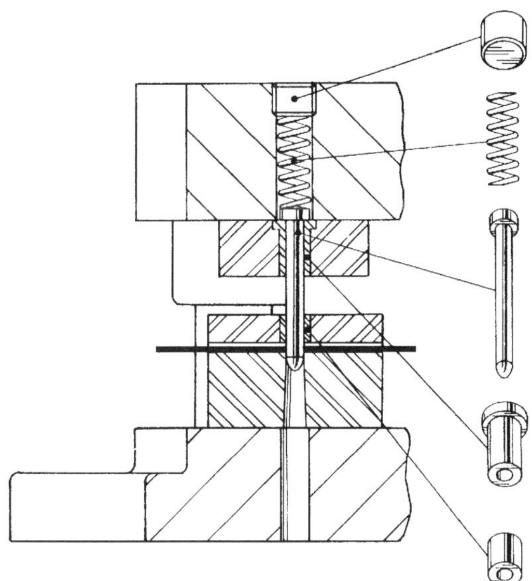

Figure 11.29 This small spring-backed pilot has a shouldered head.

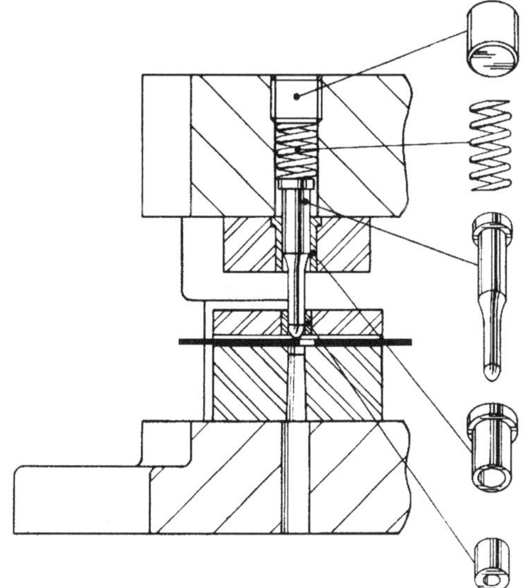

Figure 11.30 Position of pilot during a misfeed.

a) Small Spring-Backed Pilots

Pilots from 3/16 to 1/4 inch (4.8 to 6.35 mm) in diameter are made straight and provided with a shouldered head (Figure 11.29). These also are guided in hardened bushings pressed into the pilot plate and into the stripper plate. A spring and set screw, inserted in the punch holder, back up the pilot.

b) Misfeeds

In the event of a misfeed, the spring-backed pilot retracts harmlessly, as shown in Figure 11.30. Spring pressure is determined by trial and error. Sufficient pressure must be applied to move the strip to correct register at every hit. However, the pressure must not be great enough for the pilot to pierce the strip in the event of a misfeed. Even though a spring-backed pilot is used, the hole in the die block is made as though for a piercing punch. Although the pilot is supposed to retract rather than pierce the strip, the pilot can become frozen in its bushing and pierce instead of retract. For trouble-free operation, the spring should not be compressed for more than 1/3 of its free length, including the initial compression applied when the set screw is inserted.

c) Spring-Backed Quill

Still smaller pilots can be held in quills and backed up by socket-set screws (Figure 11.31).

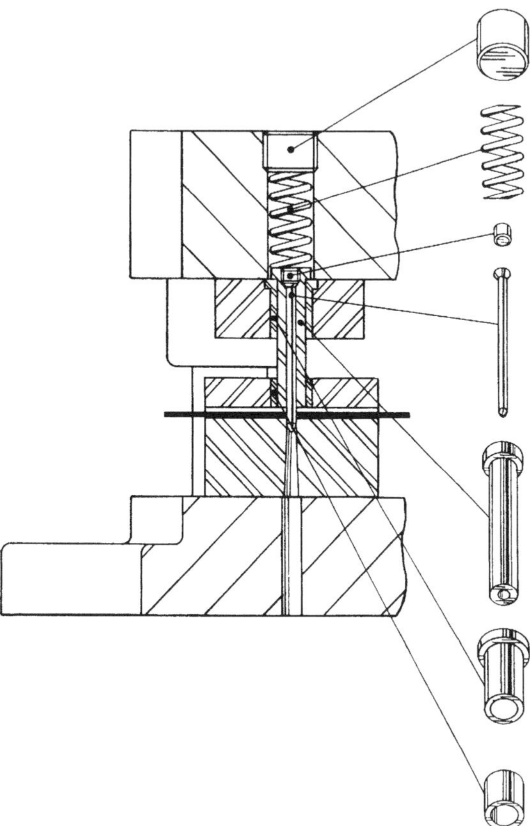

Figure 11.31 Small spring-backed pilot-bearing quill setup with hardened bushings in pilot and stripper plates.

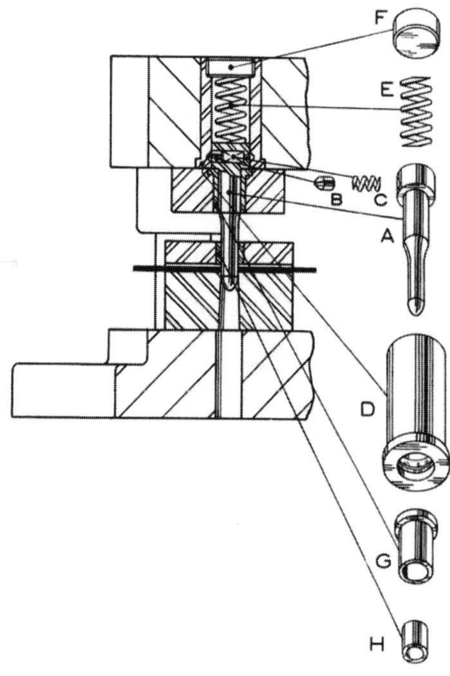

Figure 11.32 Type of pilot used when stock is thick.

The quill is then guided in hardened bushings pressed into the pilot plate and into the stripper plate. A spring backs up the quill and it is, in turn, backed up by a large socket-set screw. This construction will prevent injury to small-pilots. When misfeeds can occur frequently, this design should be used.

d) Piloting in Heavy Strip

This type of pilot is provided when the stock is unusually thick or the strip unusually wide and heavy (Figure 11.32). Pilot **A** is provided with a thick head into which a hole is drilled and reamed. Detent **B**, backed up by spring **C**, is inserted in this hole, and its angular face fits into a groove machined inside of bushing **D**. Spring **E** and socket-set screw **F** back up the assembly. The pilot is guided and supported in bushings **G** and **H**, pressed into the pilot plate and into the stripper plate. The angle on the face of detent **B** is usually made 40 degrees, but it can be varied to release the pilot under any required pressure.

HOW TO DESIGN GAGES

12.1 INTRODUCTION

Gages must be considered in the design of press tools because these components position the strip longitudinally in its travel through the die. In second-operation dies, gages locate the previously blanked or formed part for further processing operations. Design considerations include:

1. *Material choice.* Commercial gage stock or an equivalent finished tool steel is used for gages in first class dies. Cold-rolled steel should be used only when low production requirements exist.
2. *Adequate thickness.* The back gage and front spacer must be thick enough to avoid binding the strip between the stripper plate and the die block because of possible camber in the strip. Camber, or curvature, is more pronounced in coiled stock that has not passed through a straightener.
3. *Good doweling practice.* Because gages locate the strip or part, they should always be doweled in position.
4. *Accuracy of location.* Dimensions from the die hole to the locating surfaces of gages are always given decimally on the drawing.
5. *Accuracy of locating surfaces.* The gaging surfaces that actually bear against the strip (or part) should be ground, and so marked on the die drawing.

This section illustrates numerous methods of applying gages to various types of dies. These methods further explain Step 7 in Chapter 5—Fourteen Steps to Design a Die.

12.2 BACK GAGE AND FRONT SPACER

a) Two-Station Dies

In passing through a two-station pierce and blank die, (Figure 12.1) the strip is positioned against back gage **A** by the operator. Strip support **B** helps to align the bottom of the strip with the top surface of the die block to prevent binding. Dimension **C**, between the back gage and front spacer, is made strip thickness plus 1/32 inch (0.8 mm) when a roll feed is used; strip thickness plus 1/16 to 3/16 inch (1.6 to 4.8 mm) for hand feeding. Thickness **D** of both back gage and front spacer is usually made 1/8 inch (3.2 mm) for strip thicknesses up to #16 gage (0.0625 inch or 1.6 mm). For heavier strip, dimension **D** is the strip thickness plus 1/16 to 1/8 inch (1.6 to 3.2 mm). The three methods of fastening the strip support to the back gage are shown at **E**, **F**, and **G**. At **E** socket button-head screws, passing through the back gage, are threaded into holes tapped in the strip support. At **F** the components are riveted together, whereas at **G** they are welded together.

The stripper is relieved 3/8 inch (9.6 mm), as shown, to help in starting new strips through the die. The relief forms a shelf on which the end of the strip can be dropped, and then advanced over the die block surface.

b) Compound Dies

In compound dies (Figure 12.2), the back end of the spring stripper **A** is extended to provide a pad for fastening back gage **B**. Strip support **C** completes the assembly. A pin **D**, pressed into the spring stripper, helps alignment of the strip.

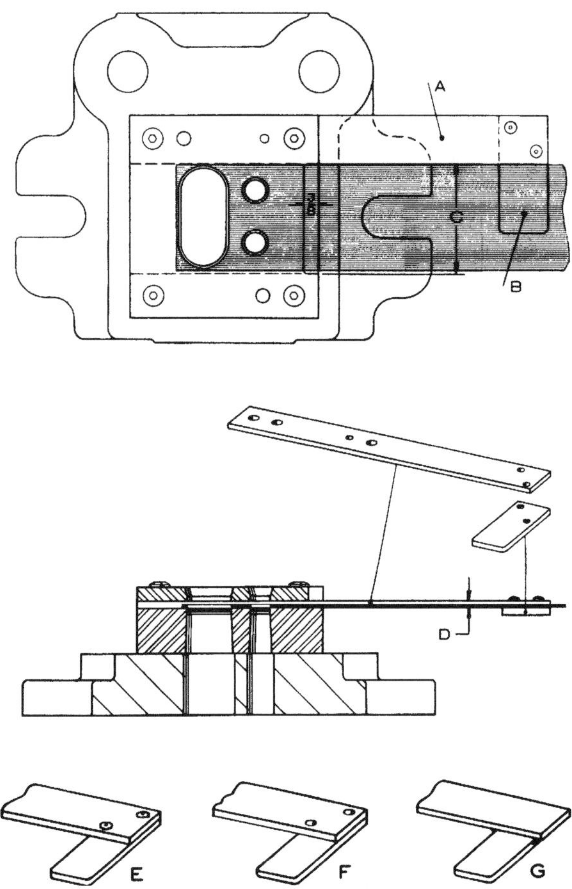

Figure 12.1 Arrangement of the back gage and front spacer in a two-station pierce and blank die.

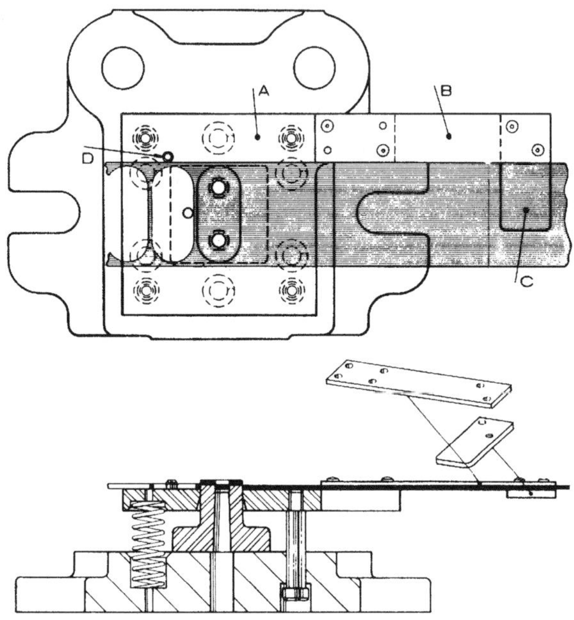

Figure 12.2 Compound die in which spring stripper **A**, back gage **B**, and pin **D** help to provide alignment.

c) Cut-Off Dies

A convenient method of providing a stop for cut-off dies for long parts is to extend the back gage to the left, then the strip stop **A** is fastened to it, with screws and dowels, as shown in Figure 12.3. After they have been cut, the blanks slide to the side of the die by means of the angular relief provided in the die block.

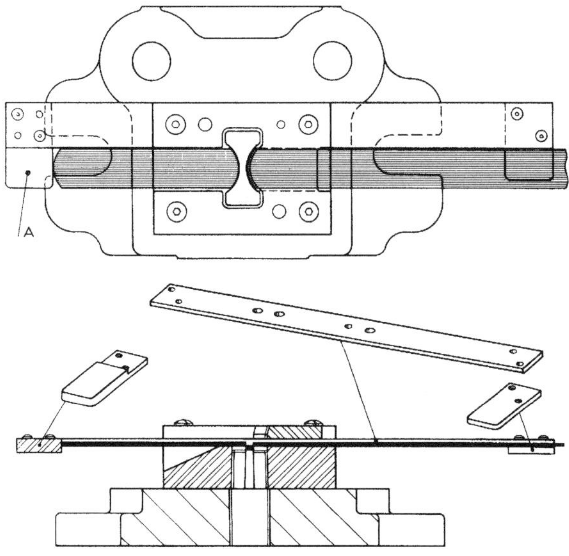

Figure 12.3 The back gage in this cut-off die has been extended to the left to provide mounting for a stop.

12.3 STRIP SUPPORT

a) Feeding Thin Strips

When feeding thin, pliable strips, support **A** is made long and brought up close to the die block (Figure 12.4) to provide better support and guidance. Three socket button-head screws are provided, but no dowels are required. This is a better method of supporting and guiding thin strips and, in fact, may be used for any strip thickness. In Figure 12.5, the front spacer is extended to the right and an ear on strip support **A** is fastened under it with a socket button-head screw to provide rigidity. With this method, it is not necessary to cut a portion of the stripper for starting the strip.

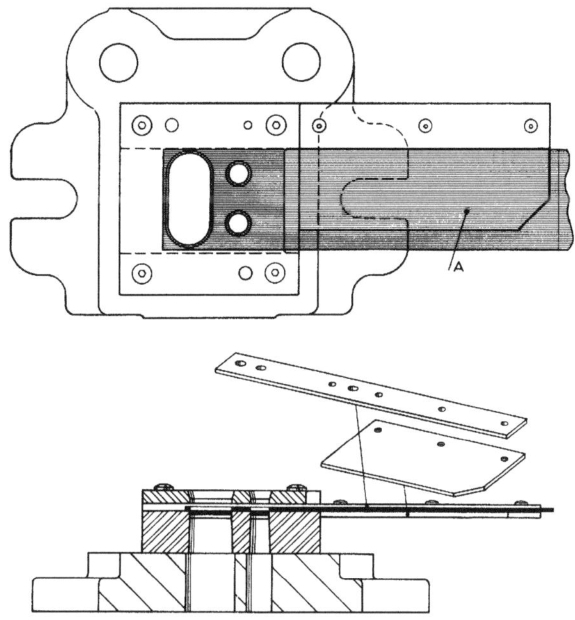

Figure 12.4 Support **A** provides extra support for thin, pliable strips.

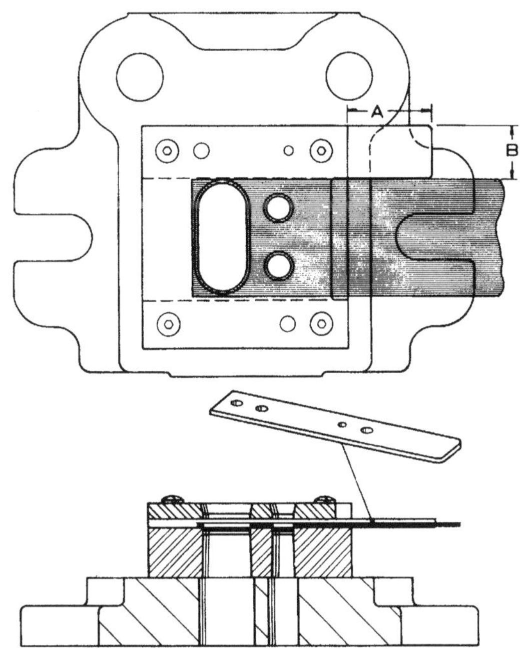

Figure 12.6 Arrangement of the back gage and the front spacer when the roll feed is used.

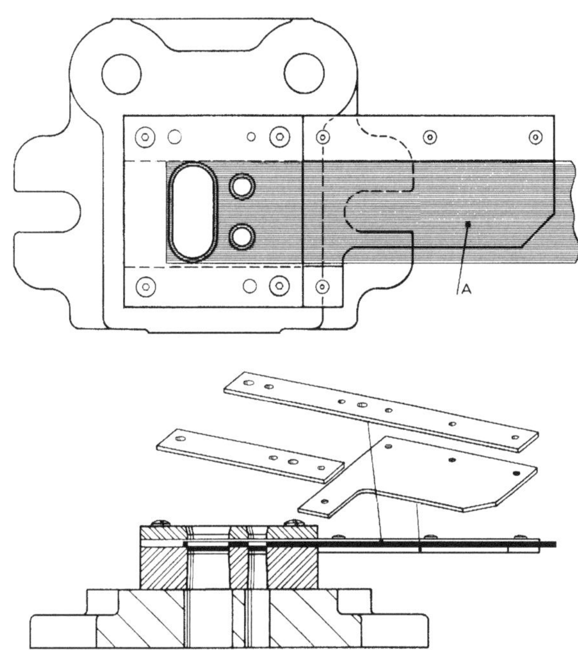

Figure 12.5 Another method of supporting and guiding thin strips.

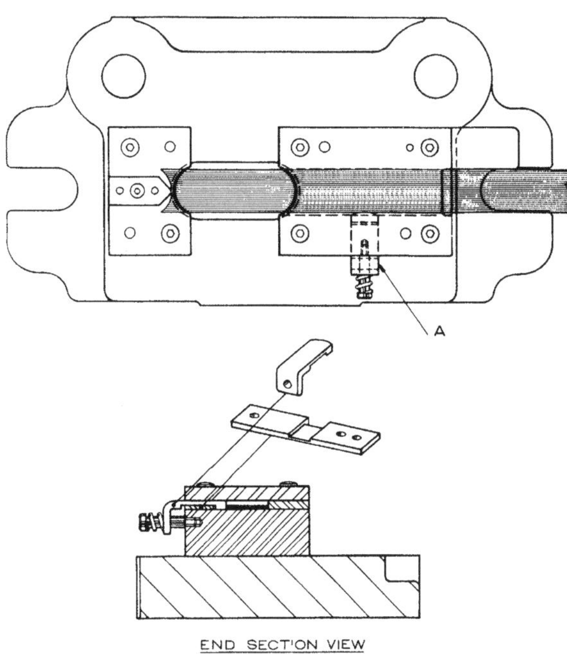

Figure 12.7 Method of keeping the strip firmly against the back gage by employing stock pusher **A**.

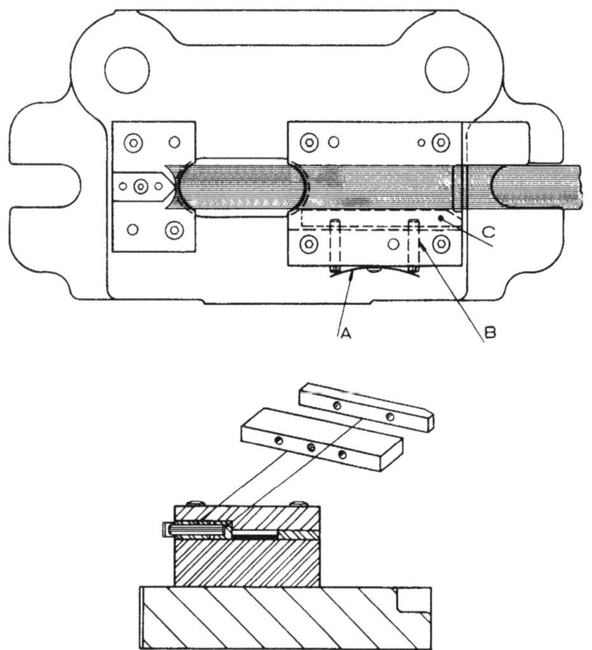

Figure 12.8 Another method of applying a stock pusher to a die.

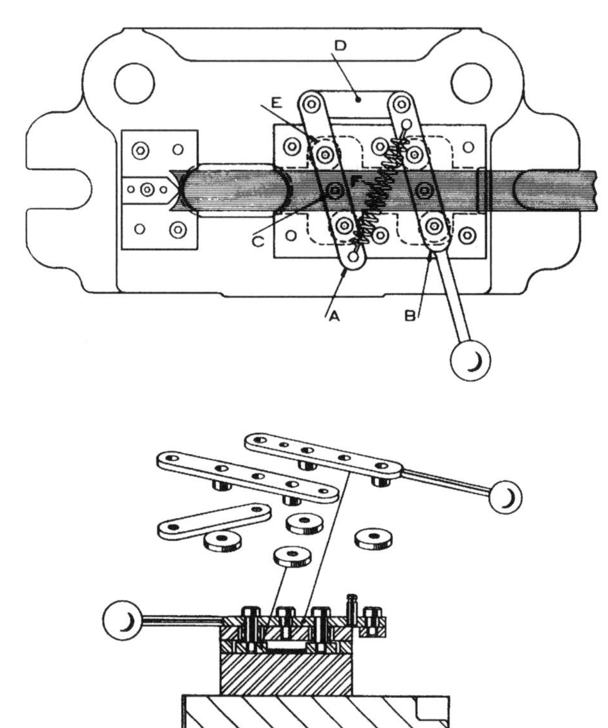

Figure 12.10 Setup employing a strip equalizer instead of a back gage when the strip width varies.

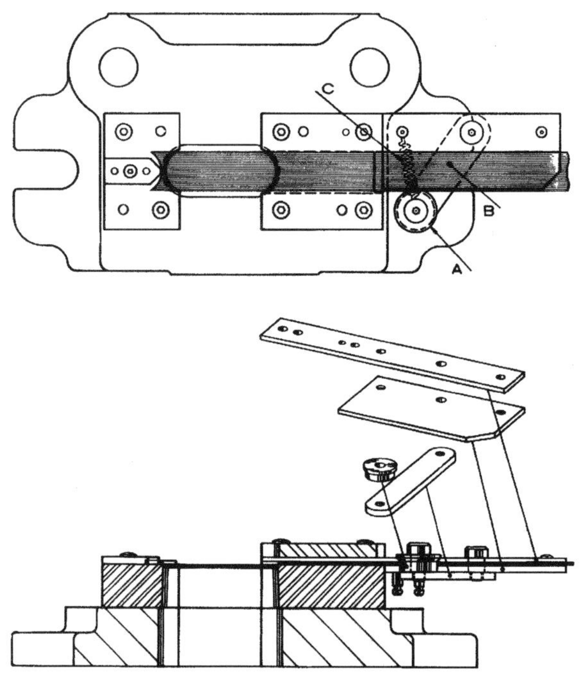

Figure 12.9 Stock pusher that incorporates a roller.

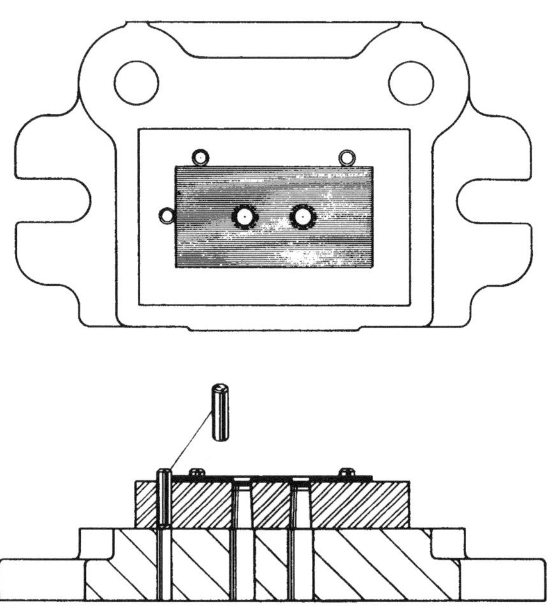

Figure 12.11 The blank location is provided by three dowel pins for low-production secondary requirements.

b) Roll Feeds

Strip supports are not required when a roll feed is used (Figure 12.6). The back gage extends to the right a short distance and the stripper plate is relieved to help start a new strip. Guide length **A** is made approximately 1 1/2 times the back gage width **B**.

12.4 STRIP PUSHERS

Means are often provided to keep the strip firmly against the back gage during its travel through the die, particularly when a roll feed is used. The simplest method is shown in Figure 12.7 to apply a stock pusher **A**, made much like a finger stop. A spring, held by a shoulder screw, applies pressure to register the strip.

a) Bar Pusher

An alternative method of applying a stock pusher to a die (Figure 12.8) is to use a flat spring **A**. This applies pressure through two pins **B** to pusher bar **C** to locate the strip firmly against the back gage.

b) Roller Pusher

The stock pusher in Figure 12.9 incorporates roller **A**, which is mounted on pivoting arm **B** by a shoulder screw. Spring **C** pulls the roller toward the back of the die to position the strip.

12.5 STRIP EQUALIZERS

When the strip width tends to vary, a strip equalizer should be used instead of a back gage (Figure 12.10). This is probably the best type. Arms **A** and **B**, pivoting on shoulder screws **C**, are linked by arm **D**. Four rollers **E** guide the strip, acted upon by spring **F**. Arm **B** is extended to provide a handle for starting the strip through the die.

12.6 POSITIONING A PARTICULAR BLANK

There are many design options for positioning particular blanks. The selection depends on the shape and dimension of the blanks. This section illustrates several methods of locating individual blanks or work parts for secondary-operation dies:

a) Using Three Gage Pins

Positioning methods for secondary-operation dies usually take the form of nests into which the parts are dropped. Figure 12.11 shows a die for piercing two holes in square-sheared blanks. When low-production requirements are present, the blank is simply located against three pins. Always position the two in-line locating pins against the longer side of the blank and the single pin against the short side.

b) Beveled Gage Pins

For higher production requirements, the blank is located in a gage composed of six pins (Figure 12.12). These pins are beveled, leaving a straight land equal to the stock thickness for register. The beveled pins help to slide the part into position quickly.

c) Turned Bevels

Perhaps a better method of applying a bevel to gage pins is to taper turn the ends of the pins, as shown in Figure 12.13. Thus, even if one pin should loosen and turn, blank positioning would not be affected.

d) Disappearing Pins

If the nature of the operation would cause the punch to strike the locating pins, they are made

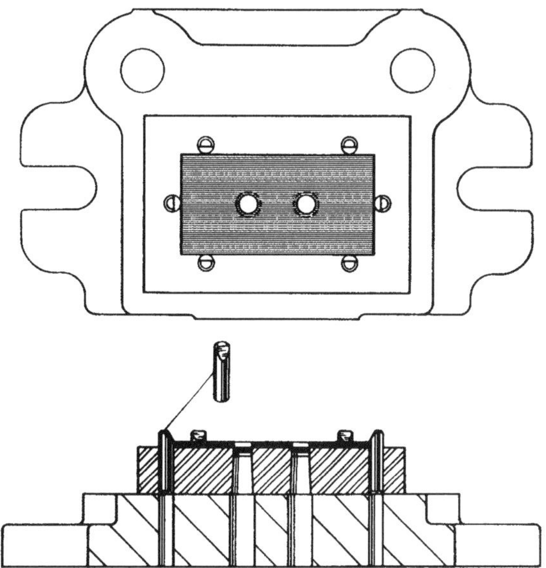

Figure 12.12 Beveled gage pins facilitate quick positioning for higher production requirements.

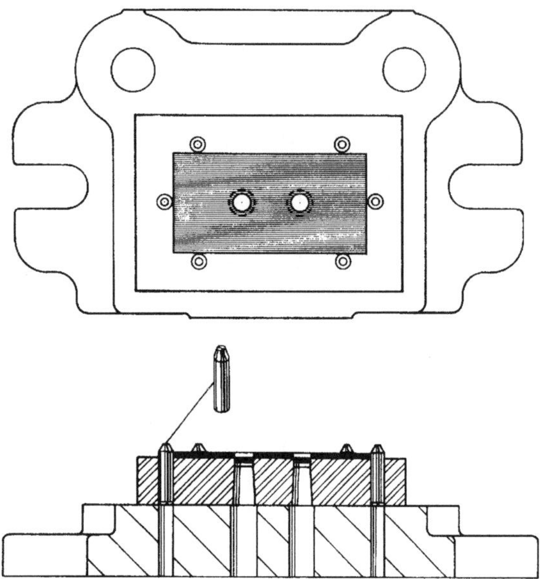

Figure 12.13 Alternate method of applying bevel to pins.

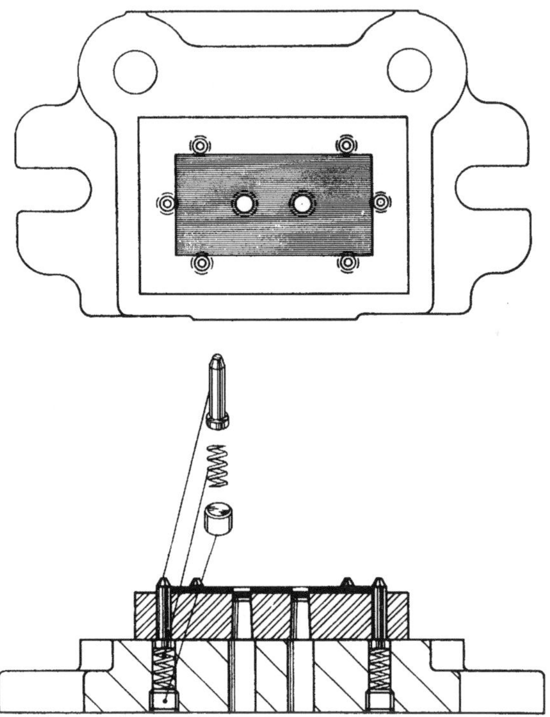

Figure 12.14 Locating pins may be spring-backed if interference is expected.

of the disappearing type (Figure 12.14). Springs, backed up by socket-set screws, act against shoulders machined in the locating pins.

e) Spring Pushers

When dimensions **A** and **B** of the part must be held very accurately, three spring pushers **C**, confined in spring housings **D**, locate the part against gages **E** (Figure 12.15).

f) Locating Pads

Locating pads may be used in lieu of pins for high production and long life (Figure 12.16). The part would be located against locating pads **A**, held in housings **B** instead of against pins, as in Figure 12.15. These pads are hardened and ground and they are retained by socket buttonhead screws.

g) Pilot Positioning

Use pilots for location when the part contains holes (Figure 12.17). Pilots provide quick part positioning and are inexpensive. Frequently used dimensions of position pilots are the same as the press-fit pilots in Chapter 11.

h) Radius Positioning

Many parts contain internal radii somewhere around their peripheries. Pilots, engaging in these

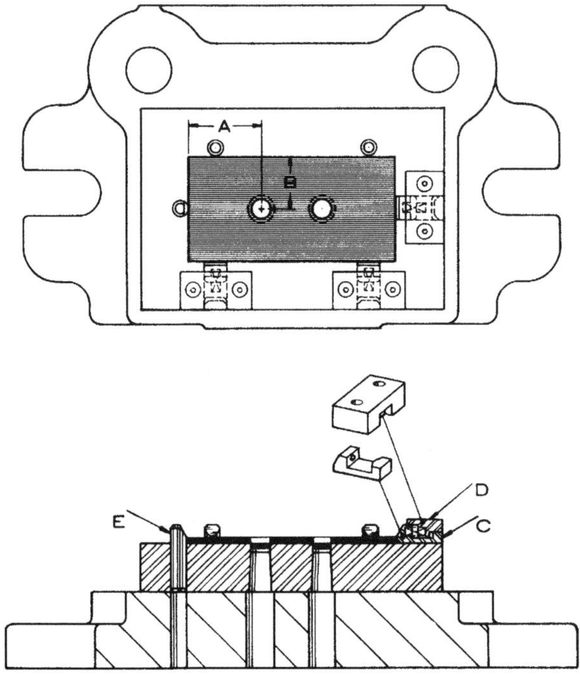

Figure 12.15 Spring pushers **C** provide accurate positioning.

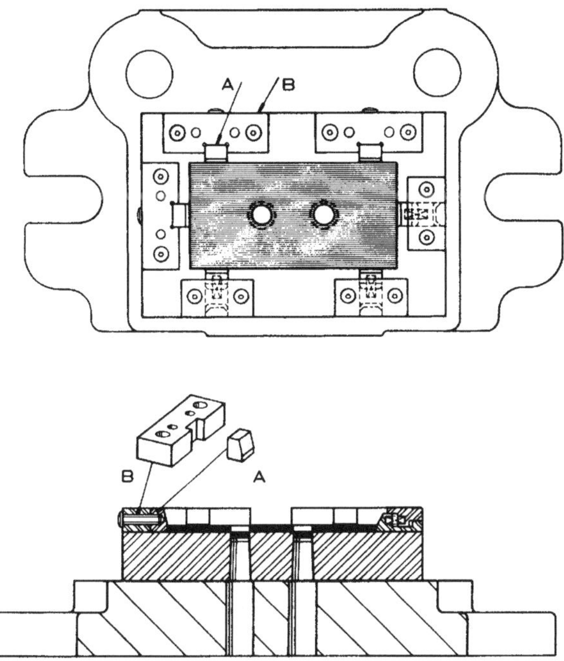

Figure 12.16 Locating pads **A** may be used in lieu of pins for high production and long life.

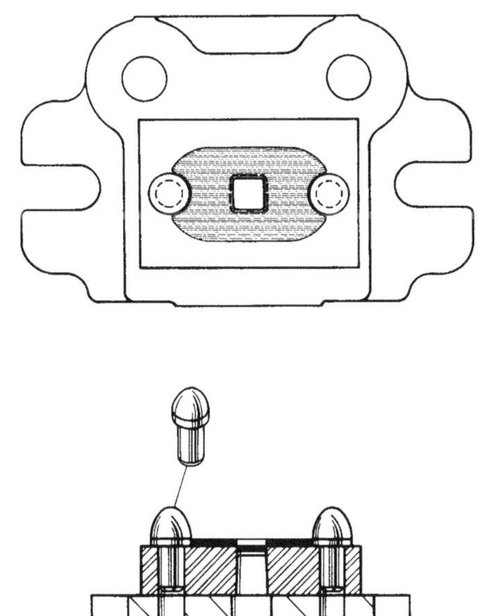

Figure 12.18 Pilots used on outside periphery for positioning part.

radii, effectively locate the blanks for subsequent operations (Figure 12.18).

i) V-Positioning

Parts with opposed outside radii are located in V locators, as shown in Figure 12.19. Engaging

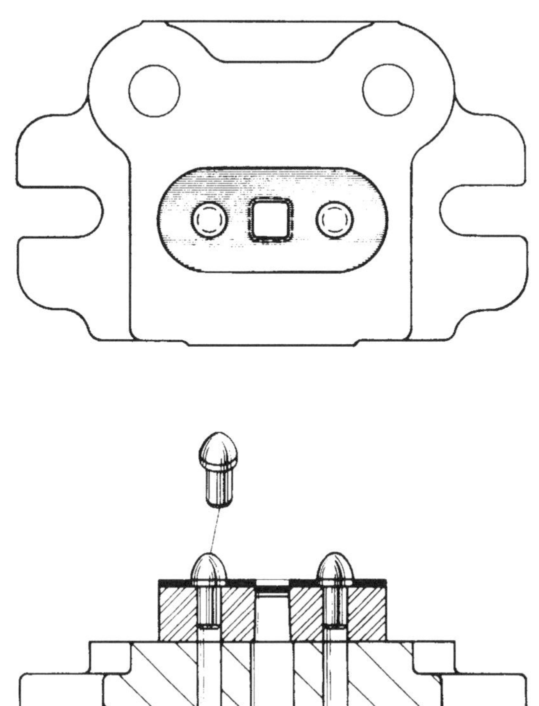

Figure 12.17 Pilots provide quick part positioning.

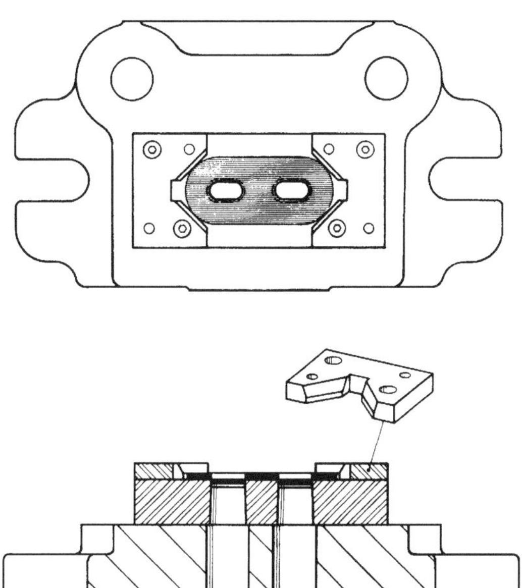

Figure 12.19 Positioning part with V gages.

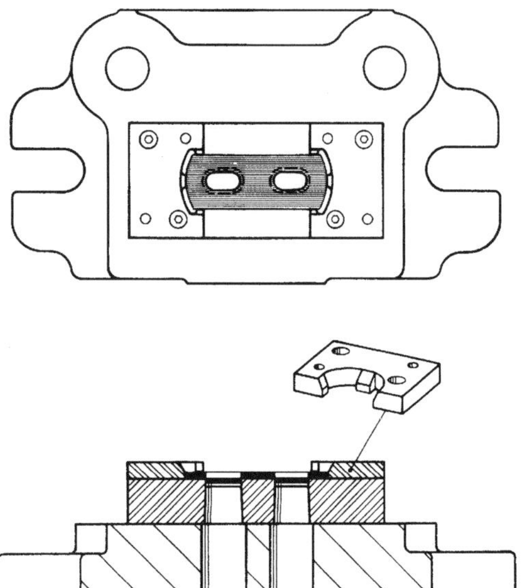

Figure 12.20 Gages machined to the outside contour of the part are used for positioning irregularly shaped pieces.

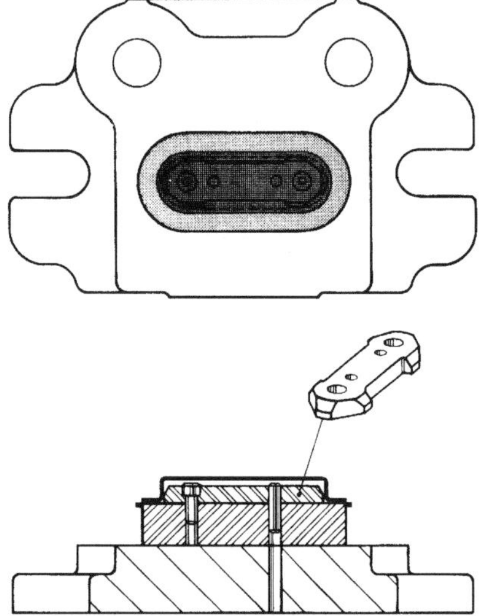

Figure 12.21 Gage machined to fit the inside contour of formed or drawn parts.

faces of each V are tapered for quick part insertion. A straight land, equal in length to the blank thickness, is left for proper positioning of the part.

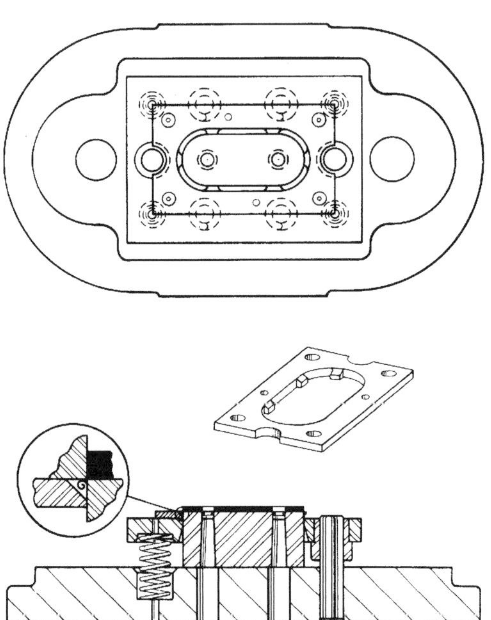

Figure 12.22 Typical gage for shaving die.

j) Relieved Gages

Parts with irregular outside contours are located in gages machined to fit the contour, but relieved so that they bear only at important points (Figure 12.20).

k) Internal Gages

Formed or drawn parts are located for further processing by gages machined to fit the inside contours (Figure 12.21), but relieved so as to bear only at critical points. This die trims the flange of a previously drawn shell.

l) Gages for Shaving Dies

Gages for shaving dies (Figure 12.22) are made to fit the outside contours of the blank, but relieved so they bear only at required points. Bearing areas are beveled to provide room for the curled chip produced by shaving action. Depth of this relief should be two-thirds of the part thickness. For shaving, die sets with floating adapters should be used, as shown.

HOW TO DESIGN FINGER STOPS

13.1 INTRODUCTION

Finger stops, or *primary stops* as they are sometimes called, are used in dies with two or more stations. They register the strip for performing operations prior to strip engagement by the automatic stop or roll feed. The number of finger stops used depends upon the number of stations in the die. For hand feeding, it is always one less than the total number of stations. For automatic feeding, only one finger stop is required. Finger stops are made of cold-rolled steel, cyanide-hardened.

In the early days of press work, numerous vertically acting primary stops were used. The hazard of placing fingers between upper and lower die members caused them to fall into disrepute. Presently, almost the only finger stops used in modern plants are the horizontal types to be described.

This chapter illustrates numerous standardized finger stops; they are fully tabulated to help the designer select the right one for the conditions encountered. Machining data are given for accompanying front spacers. In addition, general rules are outlined to help in the correct selection and application of these die components. This further explains Step 8 in Chapter 5—Fourteen Steps to Design a Die.

13.2 OPERATION OF FINGER STOPS

Taken together, the illustrations in Figure 13.1 and Figure 13.2 explain the operation of a typical finger stop. As shown in this plan view of a two-station die, two holes are pierced at the first station, and the blank is removed from the strip at the second station. In Figure 13.1, finger stop **A** is advanced to the operating position; the strip is pushed against its toe by the operator, and the press is tripped, piercing the two holes. The toe of automatic stop **B** has been pulled to the right by the automatic stop spring.

a) Withdrawing the Stop

The finger stop is now pulled back into position by the operator (Figure 13.2), allowing the strip to be moved toward the left until it contacts the toe of the automatic stop, setting it. Tripping the press produces a full blank and two pierced holes. Subsequently, the operator simply keeps the strip firmly against the automatic stop until all blanks have been removed. Usually, conventional front and side section views of the die are used to show the form of blanking and piercing punches. Thus, on the die drawing, the side view of the finger stop is projected to the right as a partial section in view **A**.

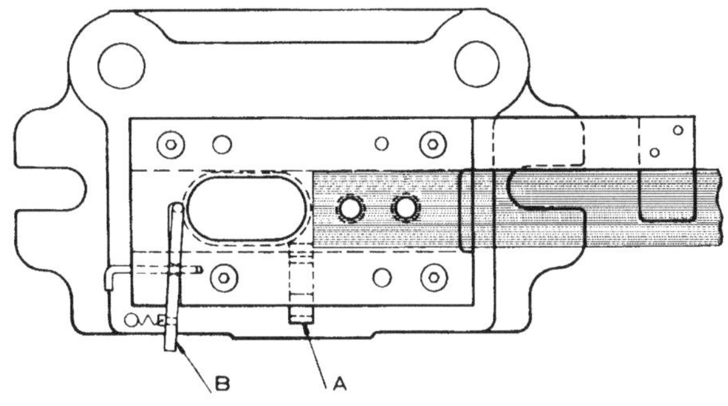

Figure 13.1 Finger stop **A** is in position to stop the strip for the piercing operation.

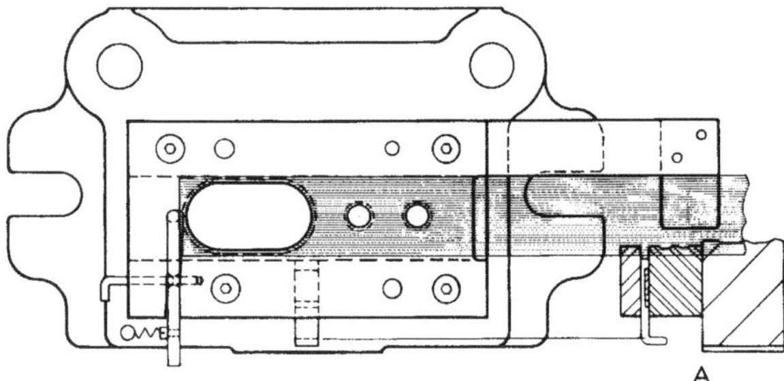

Figure 13.2 The finger stop in pulled back position allows strip to be moved to the automatic stop.

13.3 CONVENTIONAL FINGER STOP

Three common types of machining standard finger stops are illustrated in Figure 13.3. At **A**, the bottom of the stop is milled to provide a slot for limiting stop travel, and to retain the stop in the front spacer. The stop at **B** has an end-milled slot machined along its center. A dowel engages this slot to limit stop travel. Similarly, the stop at **C** has a partial slot cut along its edge to limit travel. Because it does not require a dowel, the stop at **A** is preferred. The designer should always be on the lookout for ways to eliminate unnecessary parts. Also, end-milling of long, narrow slots is slow and therefore expensive.

a) Beveling the Stop

For some part contours, it is necessary to shear the end of the strip at an angle for starting it through the die. In such applications, the end of the finger stop is beveled (Figure 13.4) to the angle applied to the strip end.

Always check carefully to make sure that the corner of the finger stop will not be sheared off by the blanking punch in its downward travel.

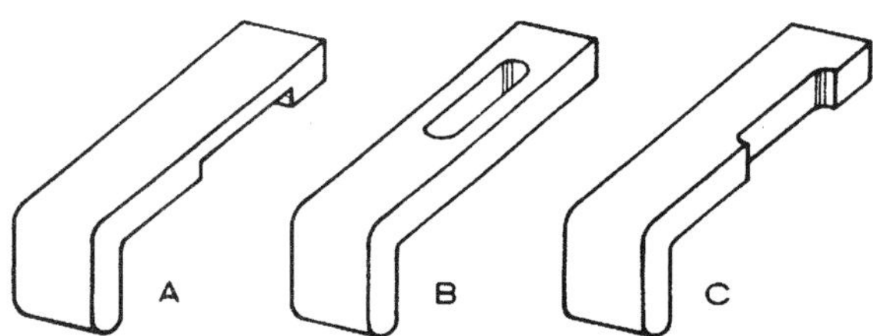

Figure 13.3 Three common types of standard finger stops.

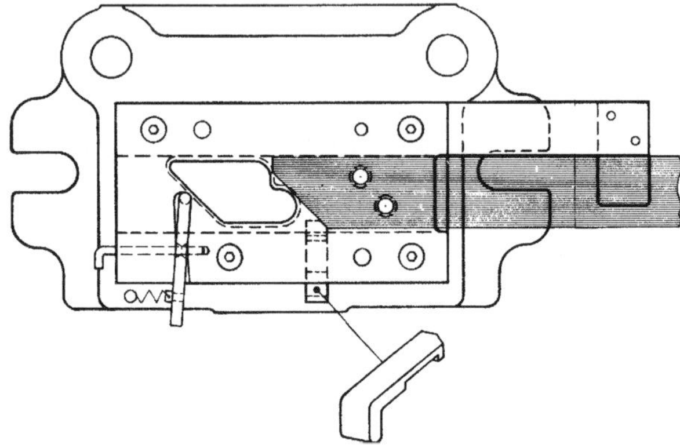

Figure 13.4 Finger stop with bevel to conform to pre-sheared strip.

Where interference occurs, bevel the end of the finger stop (Figure 13.5).

b) Multiple Stops

In multiple station dies designed for hand feeding, finger stops locate the end of the strip for each station except the final automatic stop station (Figure 13.6). In this die, two extruded bosses are applied to the strip at the first station, stop **A** positioning the strip. Two holes are pierced at the second station, stop **B** engaging the end of the strip for locating them. At the third station the part is blanked, automatic stop **C** locating the strip end.

When a roll feed is used, only one finger stop is required, applied at the first station. The roll feed

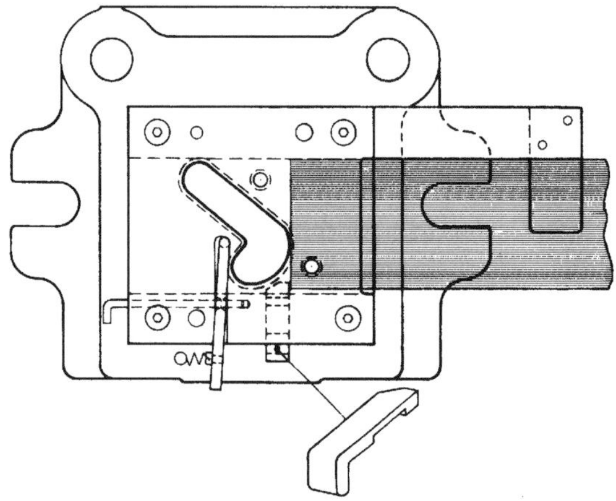

Figure 13.5 End of finger stop is beveled to avoid interference with punch.

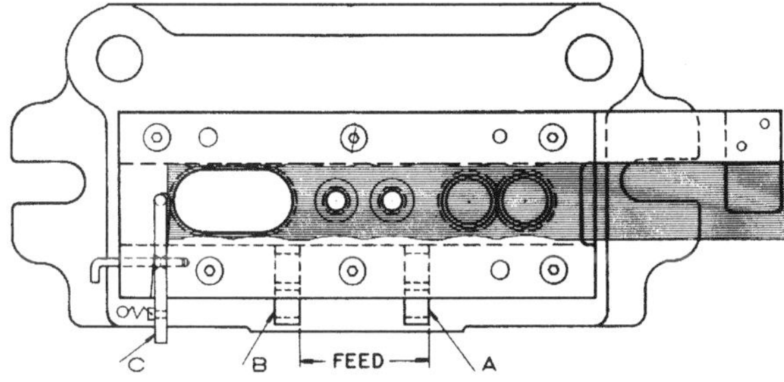

Figure 13.6 Multiple station die that uses two finger stops.

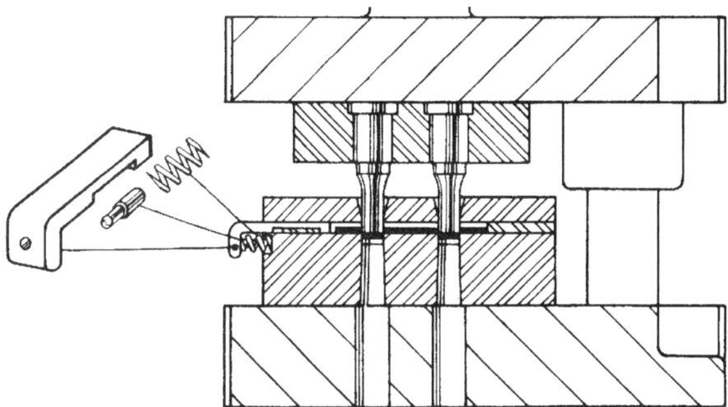

Figure 13.7 Finger stop with return spring for avoiding misfeeds on notched strips.

advances and positions the strip for all subsequent stations. Pilots perform the final accurate positioning.

c) Finger Stop with Return Spring

A return spring is not ordinarily required on standard finger stops. However, when the front edge of the material is notched and a possibility exists that the stop might work inward under vibration and engage the notch to cause a misfeed, a return spring would be used (Figure 13.7). A small guide pin, riveted to the finger stop, holds and guides the spring.

d) Finger Stop with Inverted Slot

When a spring stripper is used, the slot in the finger stop is inverted (Figure 13.8). This provides a convenient means of retaining the stop in the front stock guide.

13.3.1 Proportion of Finger Stops

Finger stop proportions can only be determined after the width of the front spacer has been established. Front spacer width, in turn, is governed by the minimum distance **C** of the die block (defined in Chapter 7, Figure 7.4) that is used for the particular job.

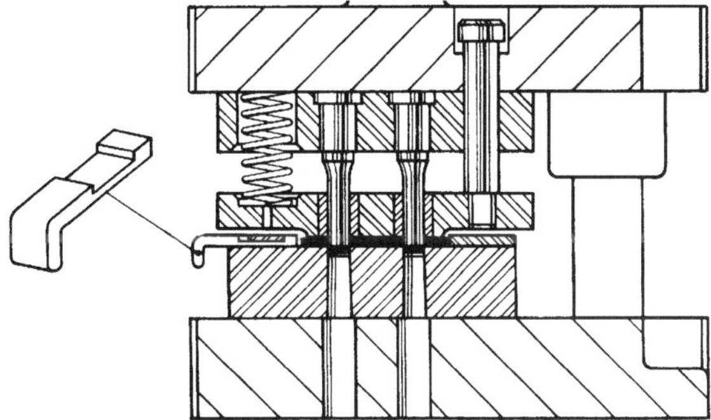

Figure 13.8 The slot in the finger stop is inverted when a spring stripper is used.

Another factor must be known before finger stop proportions can be determined. Distance **A** between the strip edge and the front space (Figure 13.9) must be established because it determines the amount of finger stop travel required to stop the strip in a positive manner. Recommended distances **A** for hand feeding using an automatic stop are tabulated, as well as power feeding using a roll feed or hitch feed. Distances **A** are given for various strip thicknesses.

a) Selection of Finger Stops

The proportions of 15 finger stops given in Figure 13.10 enable ready selection for almost any die design; they assure positive stopping of the strip. In both the forward and return positions, the maximum thicknesses of the stops remain confined in the front spacer for strength.

This table corresponds, for practical purposes, to the one in Figure 7.4. From this table, designers can choose the most suitable finger stop for the conditions confronting them.

Examples

1. If a 3/32-inch (2.4-mm) thick strip is to be run, and the die hole contour is smooth and curved, it would fall in the second line of the table in Figure 7.4 in which strip thickness **A** is 1/16 to 1/8 inch (1.6 to 3.2 mm). The recommended front spacer width (the table in Figure 13.10) for this situation is 1 1/4 inches (31.75 mm); finger stop No. 2 should be used. The 1 1/4-inch (31.75-mm) dimension corresponds roughly to line 2 of

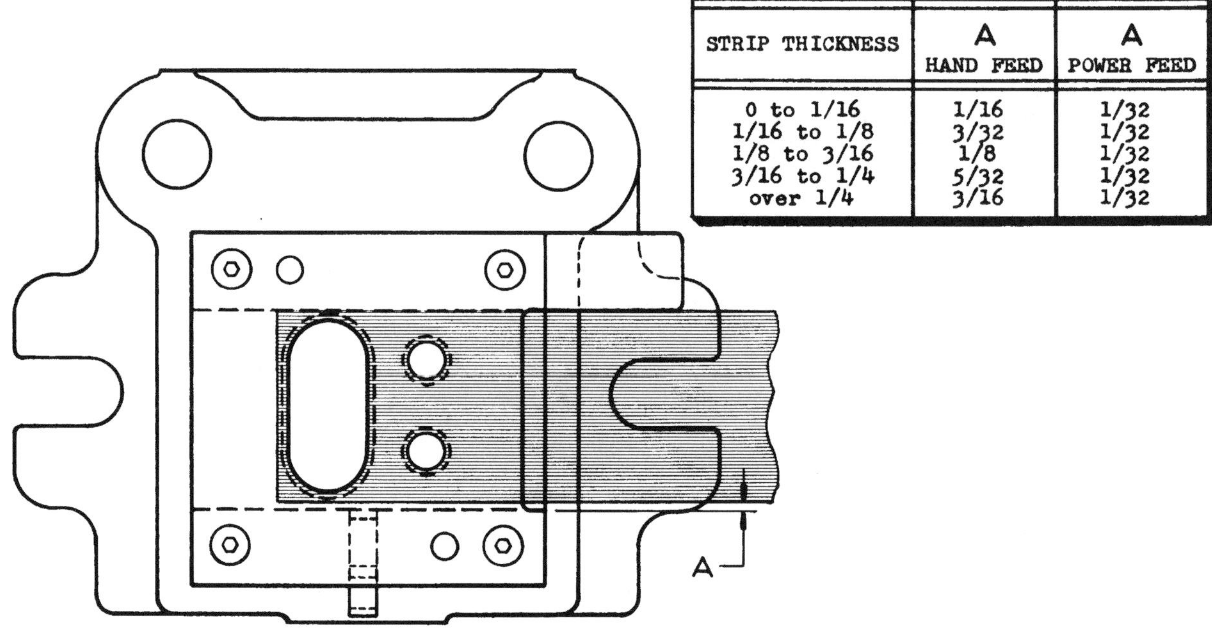

STRIP THICKNESS	A HAND FEED	A POWER FEED
0 to 1/16	1/16	1/32
1/16 to 1/8	3/32	1/32
1/8 to 3/16	1/8	1/32
3/16 to 1/4	5/32	1/32
over 1/4	3/16	1/32

Figure 13.9 Table of recommended **A** distances for both hand and power feeds.

the table in Figure 7.4 for a smooth die hole contour with a dimension of 1.2656 inches (32.14 mm).

2. For 1/32-inch (0.8-mm) thick strip with sharp inside corners in the die hole, the front spacer width, line 1, would be 2 inches and finger stop No. 11 should be used (Figure 13.10).
3. For 1/32-inch (0.8-mm) thick strip, with sharp inside corners in the die hole, the front spacer width, line 1, would be 2 inches and finger stop No. 11 should be used (Figure 13.10).

Of course, slight modifications may be necessary for some dies. When in doubt as to the correct classification of the die hole contour, use the next greater minimum distance. Only the front spacer width and thickness have been mentioned, but it is understood that the back gage is made the same width and thickness.

b) Dimensions of Finger Stops

All dimensions necessary are given in Figure 13.11 for making finger stops ranging from No. 1 to 15. From the table in Figure 13.11, designers can apply dimensions quickly. At the same time, the finger stop can be specified by number when finger stops have been standardized. Note that:

- Stops 1 to 5 are used when smooth contours are present in the die hole
- Stops 6 to 10 are used when the die hole contains inside corners
- Stops 11 to 15 are used when the die hole contains sharp inside corners

c) Front Spacer Thickness and Slot Machining

Recommended front spacer thicknesses are tabulated in Figure 13.12. These allow a minimum of 1/16-inch (1.6-mm) clearance between the top of the strip and the underside of the stripper plate to provide for possible curvature in the strip.

Slot-machining dimensions in a front spacer (Figure 13.13) are tabulated to correspond with the finger stops dimensions in Figure 13.11. Machining the front spacer in this way does not affect its function because it is simply a spacer between the die block and stripper plate.

13.4 SPRING FINGER STOP

Finger stops occupy a good position to act as stock pushers for keeping the strip firmly against the back gage in its travel through the die. Stock

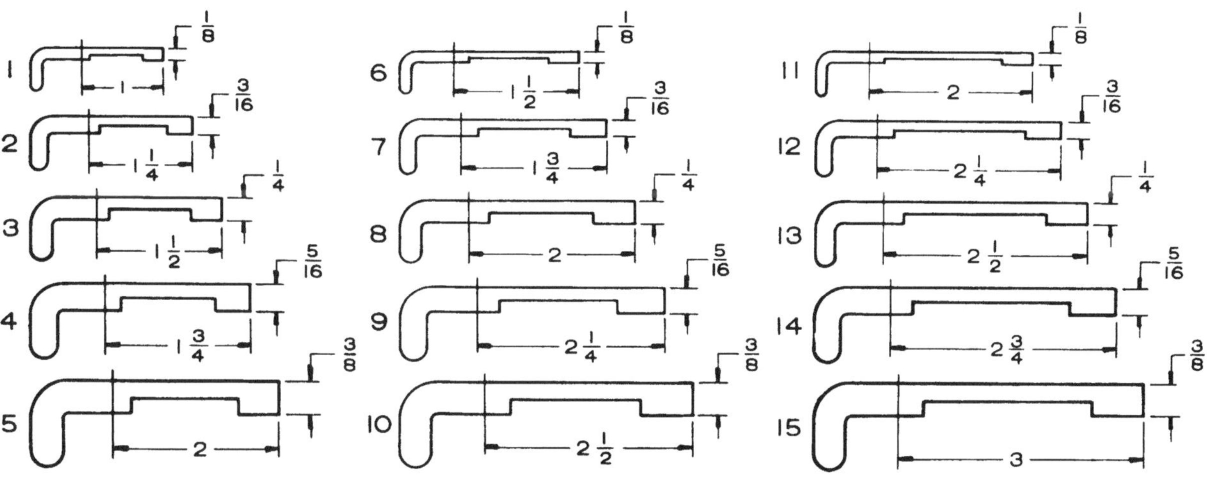

A	1 SMOOTH DIE HOLE CONTOUR		2 INSIDE CORNERS		3 SHARP INSIDE CORNERS	
STRIP THICKNESS	Front Spacer Width	Finger Stop No.	Front Spacer Width	Finger Stop No.	Front Spacer Width	Finger Stop No.
0 to 1/16	1	1	1 1/2	6	2	11
1/16 to 1/8	1 1/4	2	1 3/4	7	2 1/4	12
1/8 to 3/I6	1 1/2	3	2	8	2 1/2	13
3/16 to 1/4	1 3/4	4	2 1/4	9	2 3/4	14
over 1/4	2	5	2 1/2	10	3	15

Figure 13.10 Proportions of 15 finger stops that assure positive stopping of the strip in almost any die design.

NO.	A	B	C	D	E	F
1	.125	1/4	21/32	1 15/32	.062	1.820
2	.187	5/16	13/16	1 3/4	.093	2.259
3	.250	3/8	31/32	2 1/32	.125	2.635
4	.312	7/16	1 1/8	2 5/16	.156	3.012
5	.375	1/2	1 9/32	2 19/32	.187	3.388

NO.	A	B	C	D	E	F
11	.125	3/8	1 7/16	2 1/2	.062	2.914
12	.187	7/16	1 19/32	2 25/32	.093	3.290
13	.250	1/2	1 3/4	3 1/16	.125	3.666
14	.312	9/16	1 29/32	3 11/32	.156	4.043
15	.375	5/8	2 1/I6	3 5/8	.187	4.420

NO.	A	B	C	D	E	F
6	.125	3/8	15/16	2	.062	2.414
7	.187	7/16	1 3/32	2 9/32	.093	2.790
8	.250	1/2	1 1/4	2 9/16	.125	3.166
9	.312	9/16	1 13/32	2 27/32	.156	3.543
I0	.375	5/8	1 9/16	3 1/8	.187	3.920

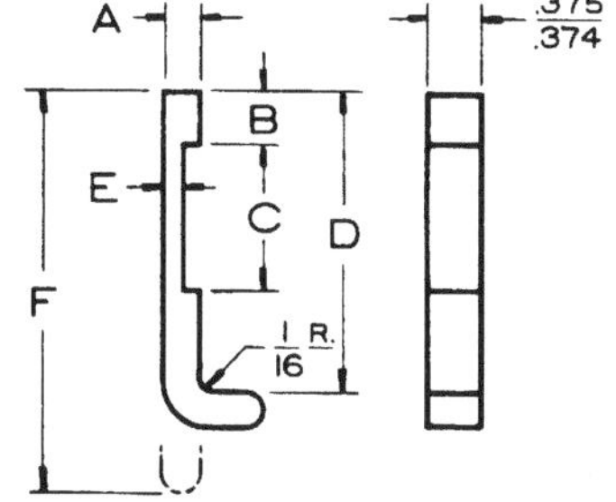

Figure 13.11 Tables of all necessary dimensions for making finger stops.

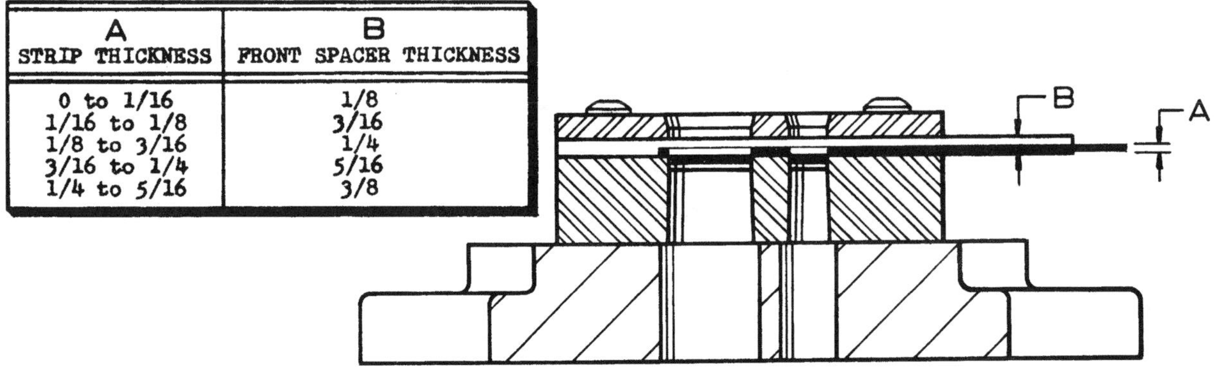

A STRIP THICKNESS	B FRONT SPACER THICKNESS
0 to 1/16	1/8
1/16 to 1/8	3/16
1/8 to 3/16	1/4
3/16 to 1/4	5/16
1/4 to 5/16	3/8

Figure 13.12 Table of recommended front spacer thicknesses.

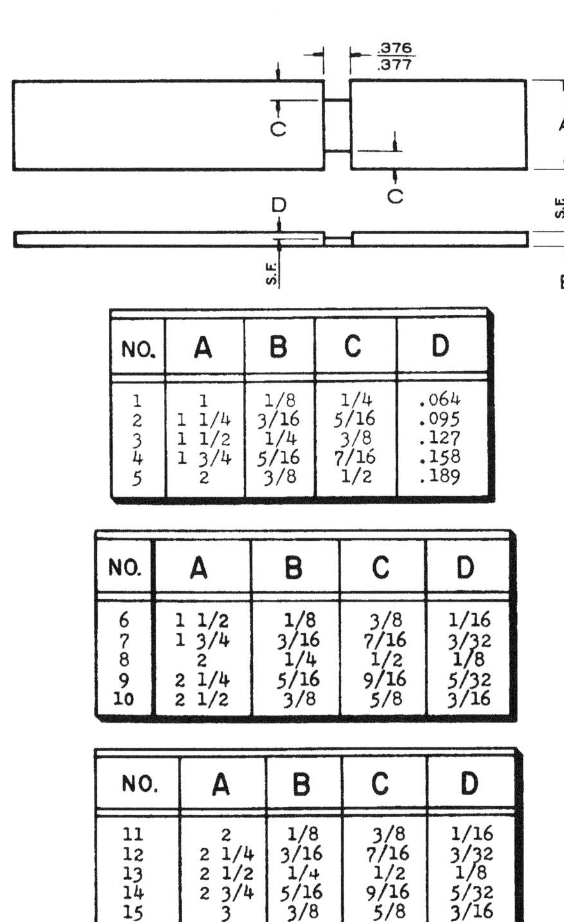

NO.	A	B	C	D
1	1	1/8	1/4	.064
2	1 1/4	3/16	5/16	.095
3	1 1/2	1/4	3/8	.127
4	1 3/4	5/16	7/16	.158
5	2	3/8	1/2	.189

NO.	A	B	C	D
6	1 1/2	1/8	3/8	1/16
7	1 3/4	3/16	7/16	3/32
8	2	1/4	1/2	1/8
9	2 1/4	5/16	9/16	5/32
10	2 1/2	3/8	5/8	3/16

NO.	A	B	C	D
11	2	1/8	3/8	1/16
12	2 1/4	3/16	7/16	3/32
13	2 1/2	1/4	1/2	1/8
14	2 3/4	5/16	9/16	5/32
15	3	3/8	5/8	3/16

Figure 13.13 Tables of dimensions for machining slots in front spacer to accommodate the standard finger stops given in Figure 13.11.

pushers are not ordinarily required for hand feeding because the operator can position the strip against the back gage while advancing it. However, their use eliminates the necessity for side pressure and with them more uniform strip runs are assured.

Stock pushers are indispensible when the strip is to be fed automatically with a roll or hitch feed. Their use makes possible some reduction in scrap bridge allowance with a saving in material cost. For cut-off dies, in which the edges of the strip become edges of the finished blanks, the use of stock pushers results in more accurate parts.

The following illustrations show the method of modifying standard finger stops and of making actuating springs. Applied one way, the spring makes the stop a combination finger stop and stock pusher. Applied another way, the stop becomes a spring-actuated finger stop.

13.4.1 Machining Spring Finger Stops

Conventional finger stop **A** in Figure 13.14 has a slot milled in its underside to retain it in the front spacer and to limit stop travel. Shown at **B**, the stop is modified by end-milling (or spot-facing) radial pockets in both the heel and the toe of the stop to the same depth as the previously machined slot. Dimensions **C** of both stops remain the same.

a) Dimensions of Radial Slots

Figure 13.15 tabulates machining dimensions for applying radial pockets to the line of finger stops described in Figure 13.11 These can be end-milled or spot-faced using a simple drill jig with guide bushings.

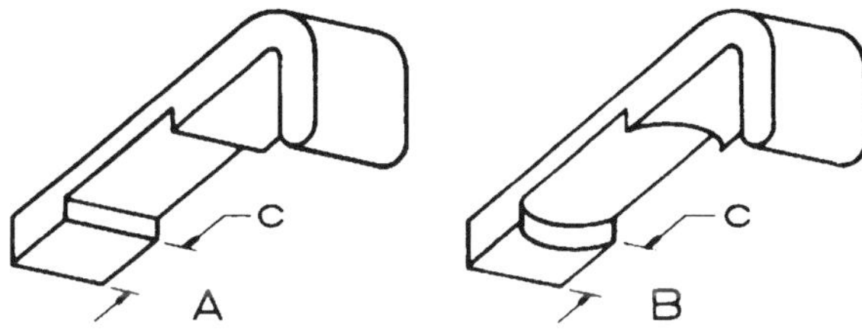

Figure 13.14 Conventional **A** and modified **B** finger stops.

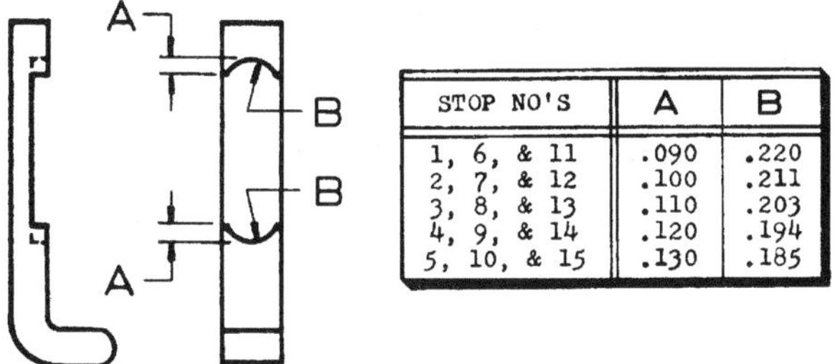

STOP NO'S	A	B
1, 6, & 11	.090	.220
2, 7, & 12	.100	.211
3, 8, & 13	.110	.203
4, 9, & 14	.120	.194
5, 10, & 15	.130	.185

Figure 13.15 Table of machining dimensions for applying radial pockets in the finger stops described in Figure 13.11.

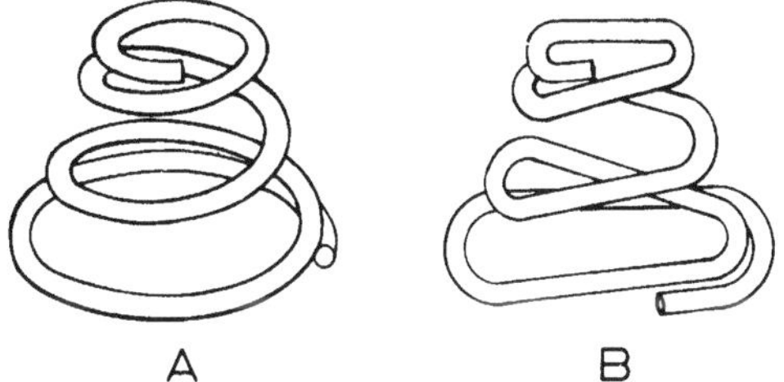

Figure 13.16 Conical **A** and flattened conical **B** springs for use with spring finger stops.

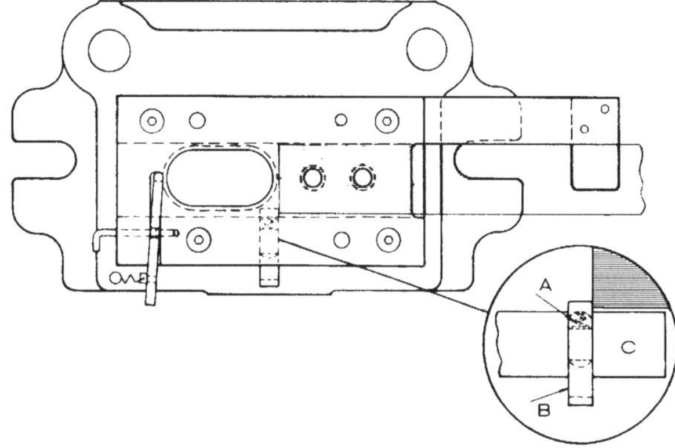

Figure 13.17 Method of applying a conical spring to a finger stop, making it a combination finger stop and stock pusher.

13.4.2 Spring Applications

A conical spring, such as the one shown at **A** in Figure 13.16, can be compressed until it is completely closed, the coils slipping one into the other. A common application of this principle is the flashlight spring. A small conical spring like the one at **A** is used for the heavy finger stops with toe dimensions of 3/8 by 3/8 inch (9.52 by 9.52 mm) Shown at **B** is a flattened conical spring, used for the thinner finger stop ranges.

Figure 13.17 illustrates the application of a conical spring to a finger stop, converting it into a combination finger stop and stock pusher. Spring **A**, acting against the face of the radial pocket in the toe of stop **B**, keeps it in a normally advanced position toward the back gage. Both stop and spring are retained in front spacer **C**. The strip has been brought against the toe of the stop for piercing holes at the first station. The enlarged view shows the stop, spring, and front spacer with the stripper plate removed.

When the same spring is assembled in the back radial pocket of the stop, spring pressure acts to keep the stop away from the strip, as shown in Figure 13.18. In this view, the strip is running through the die with only the automatic stop modifying and limiting its travel.

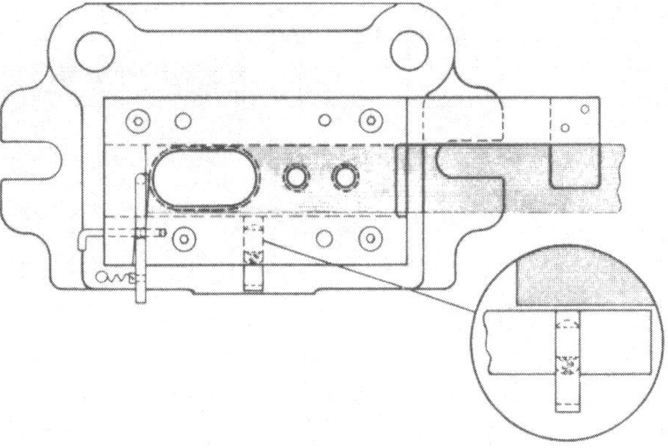

Figure 13.18 Conical spring in back radial pocket keeps stop away from strip.

13.4.3 Stock Pusher and Spring Finger Stop Operations

Stock pusher. The stop (Figure 13.19) has now been retracted by the operator, compressing the spring to allow the strip to advance against the automatic stop. Releasing the stop causes the spring to push the strip firmly against the back gage. It now acts

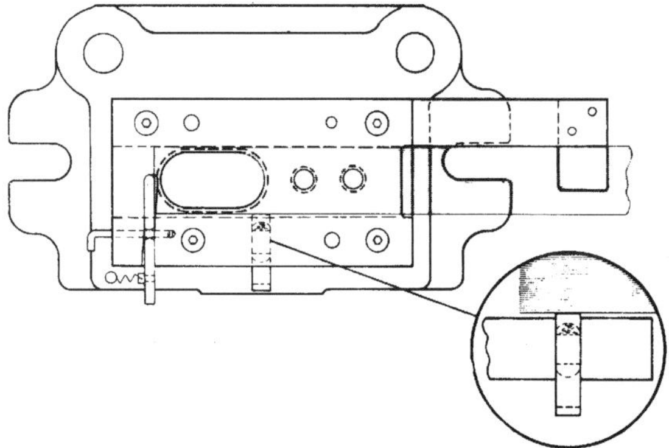

Figure 13.19 Finger stop in use as a stock pusher.

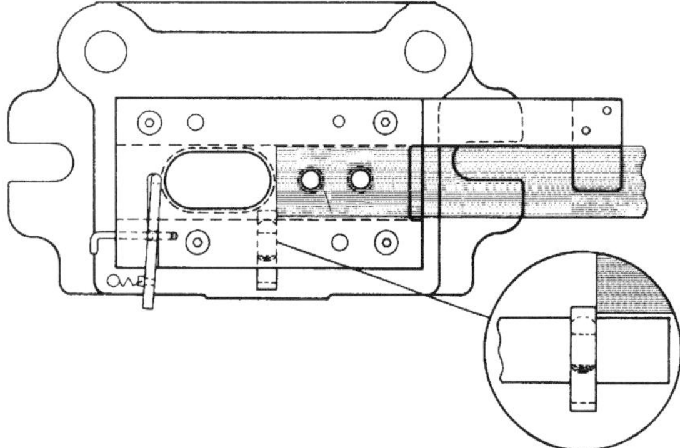

Figure 13.20 The spring finger stop is pushed in by the operator to stop the strip for the first piercing operation.

as a stock pusher until all blanks have been removed from the strip. When a new strip is started, it is used as a finger stop again.

Spring finger stop. To stop the strip for piercing two holes at the first station, the stop is pushed in by the operator, compressing the spring (Figure 13.20). Upon release, it moves back automatically to allow feeding of the strip.

Multiple stops and stock pushers. This practical application is a four-station progressive die for producing pierced and stamped links (Figure 13.21). The two outside finger stops are assembled to act as stock pushers as well. The center stop is assembled as a spring finger stop to reduce excessive drag on the strip. When a number of finger stops are used, usually only the end ones need to be assembled as stock pushers. Thus, the strip is located against the back gage as early as possible in its travel through the die.

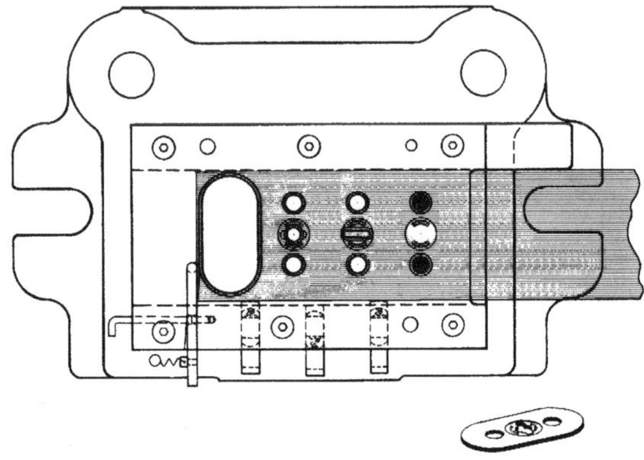

Figure 13.21 A four-station progressive die utilizing two stock pushers and a spring finger stop.

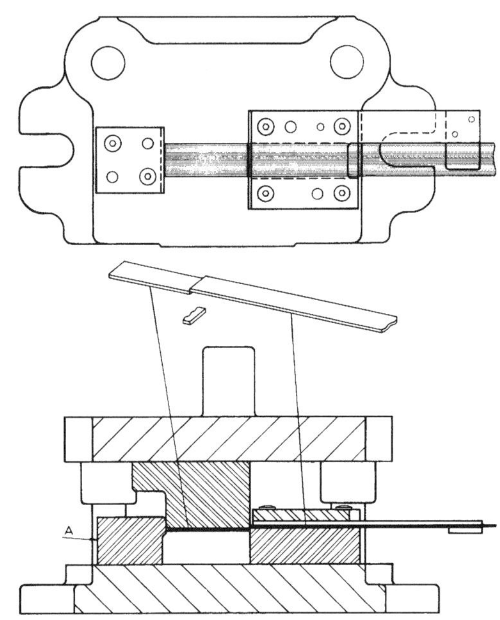

Figure 13.22 Cut-off die in which finger stops are not required.

When are finger stops not needed in a die? For most cut-off dies, finger stops are not required. Figure 13.22 illustrates a simple cut-off die for producing rectangular blanks.

In operation, the end of the strip is advanced until it projects slightly past the cutting edge. Tripping the press trims the end of the strip square. The strip is then advanced against stop **A**, which also backs up the cut-off punch to prevent deflection while cutting. No finger stop is needed because trimming the end of the strip prepares it for register against stop **A**.

HOW TO DESIGN AUTOMATIC STOPS

14.1 INTRODUCTION

Automatic stops, or *trigger stops,* as they are sometimes called, register the strip at the final die station. They differ from finger stops in that they stop the strip automatically, the operator simply keeping the strip pushed against the stop in its travel through the die. For this reason, they are always used when an operator is to feed the strip by hand. Automatic stops can be made of cold-rolled steel or machine steel, cyanide-hardened, but when long runs are anticipated they should be made of tool steel, hardened machining. Design considerations include:

1. Fast, positive action under high-speed, shock conditions.
2. Minimum machining of the stripper plate for strength.
3. Sturdy, gadget-free design; automatic stops perform grueling service in operation, and weak design can be a source of trouble.

This chapter illustrates numerous standardized automatic stops with dimensions tabulated to help designers selects the right one for the conditions encountered. In addition, general rules are outlined to help with the correct application of these die components. This material further explains Step 9 in Chapter 5—Fourteen Steps to Design a Die.

14.2 TYPES OF STOPS

14.2.1 Side-Acting Stop

Taken together, two plan views **A** and **B**, with a common side section-view **C**, illustrate the operation of a conventional side-acting automatic stop (Figure 14.1). At **A**, the strip has been advanced toward the left. Previously blanked strip edge **D** contacts the toe of the automatic stop, moving it to its extreme left position as shown, setting the stop. Descent of the press ram causes the square-head set screw **5** to raise the stop toe to the position shown in section view **C**. Now, the tension spring **3**, acting at an upward angle, turns the stop to the position shown in view **B**. When the press ram goes up, the toe of the stop falls on top of the scrap bridge, allowing the strip to slide under it. After the scrap bridge has passed completely under it, the toe drops to the top surface of the die block, acted upon by spring **3**. Now the stop is ready to be reset upon contact of the next blanked strip edge.

These motions occur at extremely high speeds. On fast runs, the motions cannot be followed by the eye. Six parts make up the stop assembly. They are:

1. Automatic stop
2. Fulcrum pin
3. Tension spring
4. Spring post
5. Square-head set screw, usually 1/4 in. (6.0 mm) diameter
6. Jam nut

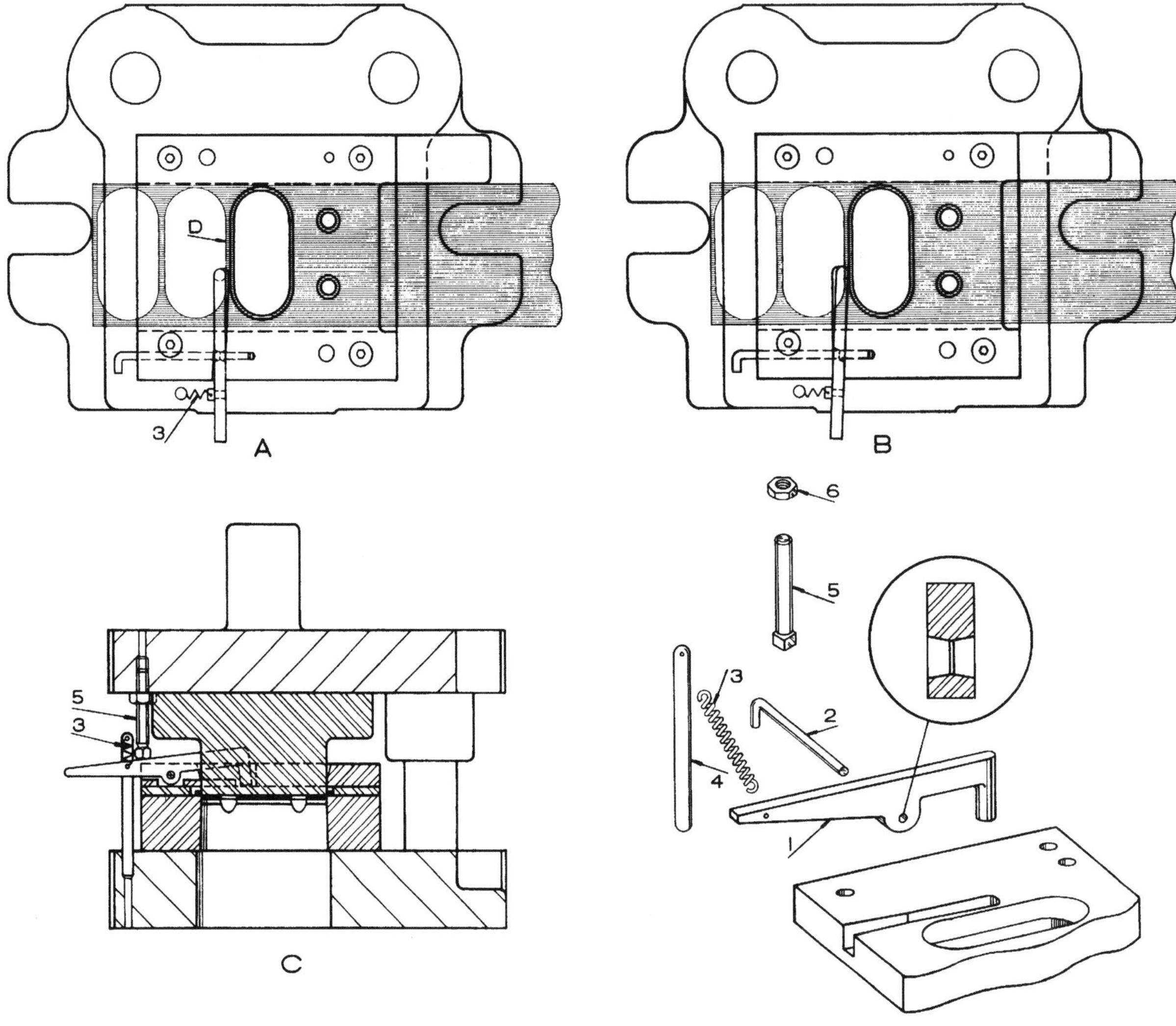

Figure 14.1 Typical side-acting automatic stop.

As shown in the inset of the pictorial view, the hole for the fulcrum pin is taper-reamed from both sides to allow rocking. The slot in the stripper plate is machined angularly to allow stop movement. It is not machined entirely through the stripper plate except at the toe and opposite the fulcrum-hole portion for strength.

a) Stop Holder

Another method of applying a side-acting automatic stop is to confine the stop in a holder (Figure 14.2). Here are the benefits of this construction:

- The holder can be easily hardened to provide long life.
- Very little machining of the stripper plate is required.
- The stop is a self-contained unit, and it can be quickly removed and used in several dies.
- There is no post or exposed spring. Instead, the stop has sturdy parts and simple, clean lines.

View **A** shows the automatic stop set and ready for blanking of the hole. At view **B**, the press has descended, releasing the stop and

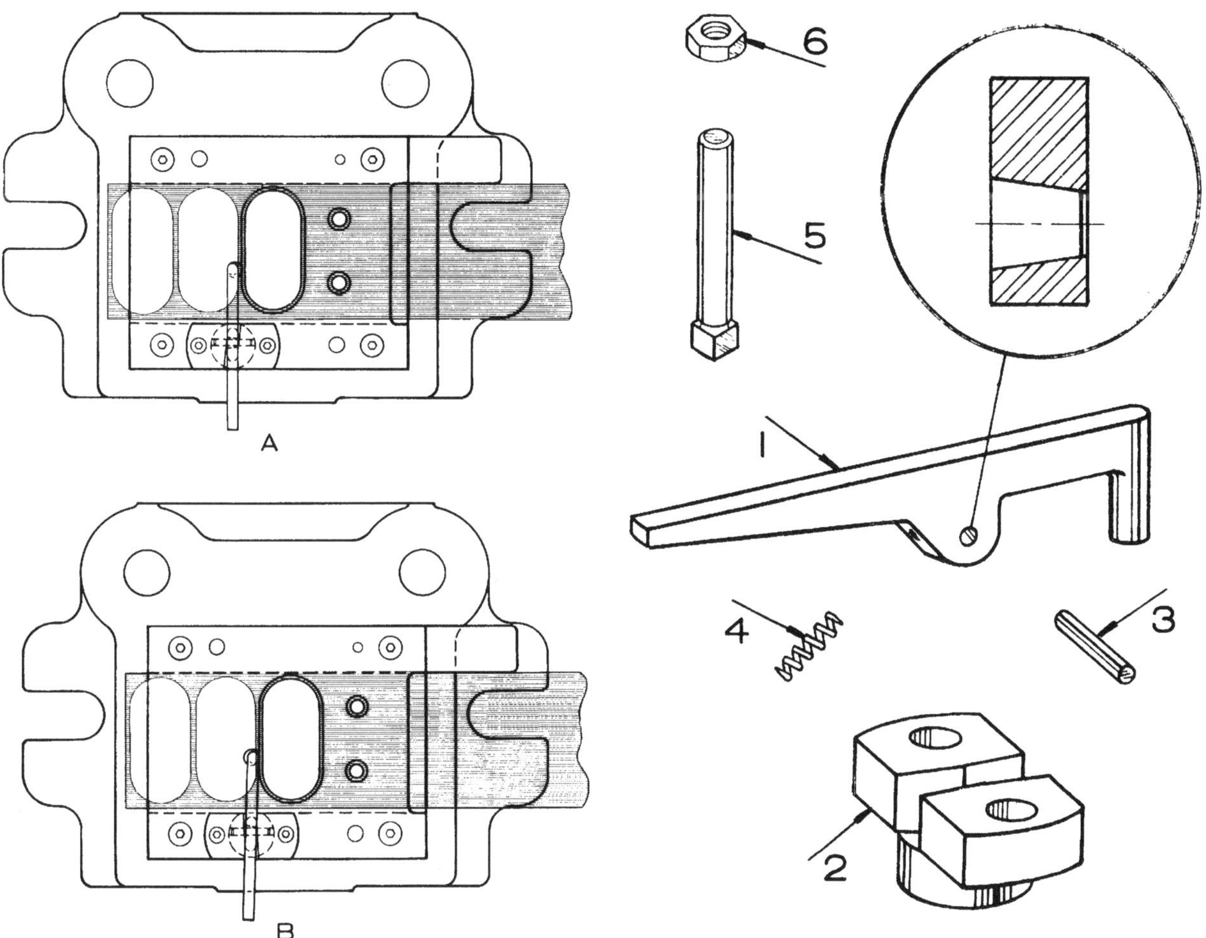

Figure 14.2 Side-acting automatic stop that is incorporated into a holder.

causing it to swing above the scrap bridge. The pictorial view shows stop construction. Stop **1** fits in a tapered slot in holder **2**, held by fulcrum pin **3**. Compression spring **4** provides the required movement. As in Figure 14.1, a square-head set screw **5**, locked by jam nut **6**, is applied to the punch holder for operating the stop. In this case, the fulcrum-pin hole is taper reamed from one side only to allow the spring to provide side thrust, as well as up-and-down movement.

14.2.2 End-Acting Stop

The same type of stop can be designed as an end-acting automatic stop (Figure 14.3). End-acting automatic stops are easier to standardize because they don't have the great range of lengths required in side-acting stops. The primary modification is an end-milled slot instead of a tapered hole to engage the fulcrum pin. Views A and B show the stop set, the strip having pushed it back to its maximum left position. Descent of the press ram causes the square-head set screw to turn the stop to the position shown at view **C**. The spring has now pushed the stop toward the right, the end-milled slot allowing movement. It is obvious from this view that when the upper die goes up, the toe of the stop will drop on top of the scrap bridge, allowing movement of the strip left until the toe is again engaged by the strip. View **A** is a plan view of the die, whereas views **B** and **C** are section views showing operation of the stop.

Dimension **D** is the sum of the following:

1/2 feed

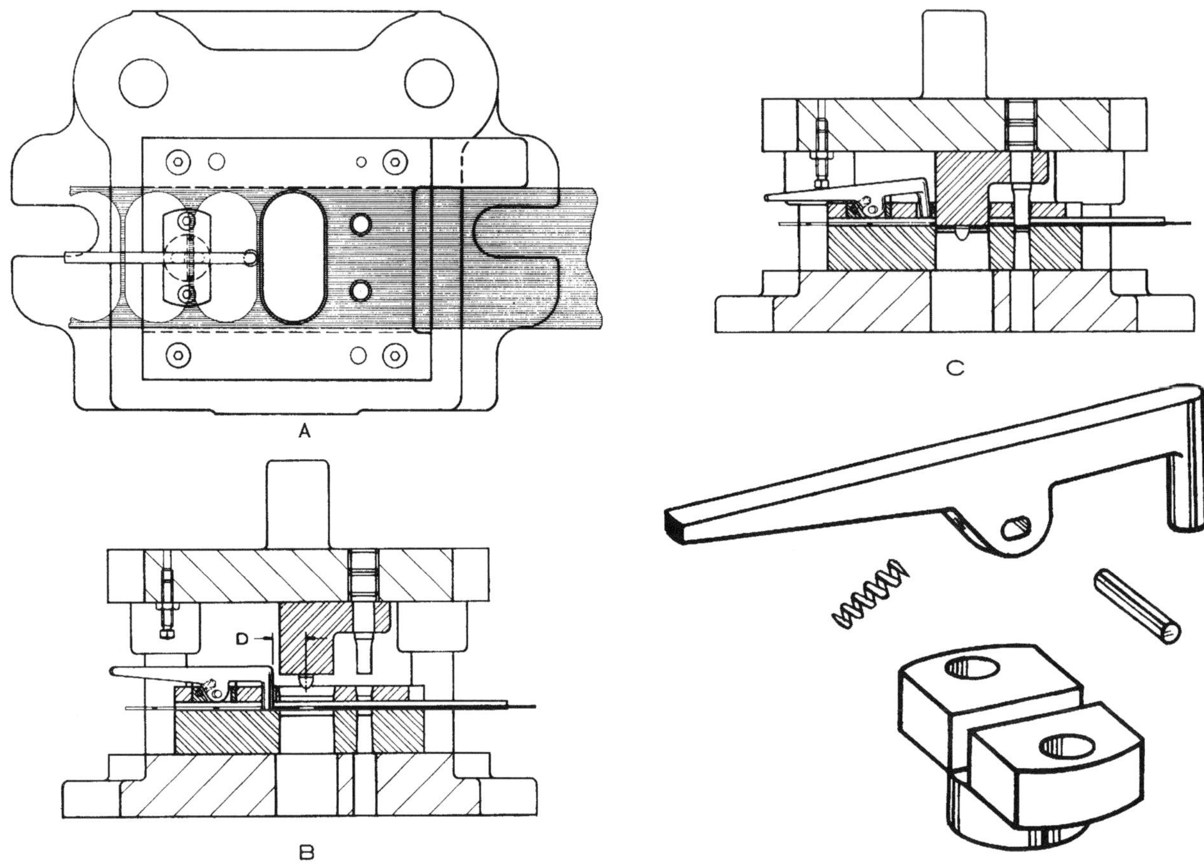

Figure 14.3 Typical end-acting automatic stop.

1 scrap bridge allowance
0.010 inch (0.25 mm).

The allowance causes the strip to be stopped 0.010 inch (0.25 mm) past its required position. Pilots, engaging in previously pierced holes, bring the strip back this amount for final accurate location before blanking occurs.

14.3 ACTUATING AUTOMATIC STOPS

Square-head set screw. Using a square-head set screw for actuating automatic stops is a time-honored practice that usually works well. One disadvantage is that practically every time the die is sharpened, the set screw must be backed up by the amount removed from punch and die members. Many automatic stops have been smashed by the set screw through failure to retract it sufficiently after die sharpening.

Spring plunger. View **A** in Figure 14.4 shows a Vlier spring plunger, available from Vlier Engineering Co., applied as stop actuator. The spring plunger has been backed up by a socket lock screw, although a jam nut would be equally effective. The spring-loaded toe of the Vlier plunger actuates the automatic stop. The heavy series of these standard components should be used to assure positive action. The view at **B** shows the same die in closed position after repeated sharpening. The spring plunger has not been reset in any way. After actuating the stop, the spring-loaded end retracts harmlessly as shown.

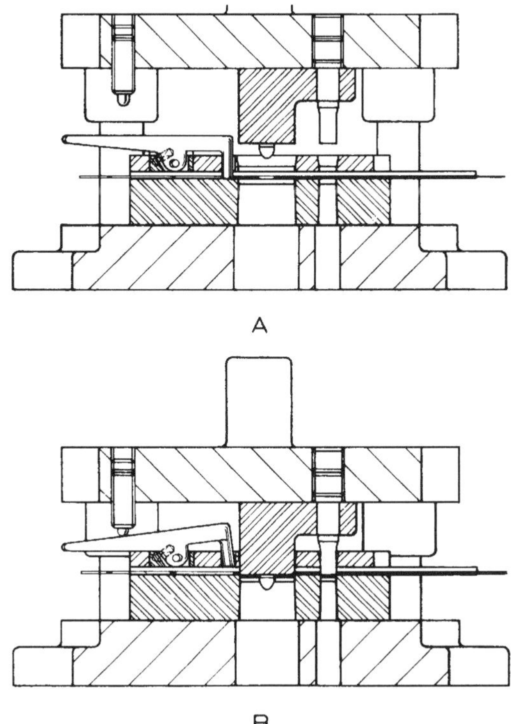

Figure 14.4 Actuating an automatic stop with a spring plunger.

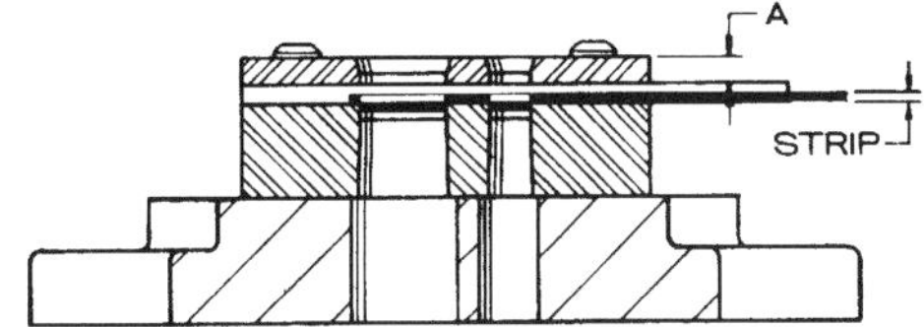

1

STRIP	A
1/16 x 3	.225
1/16 x 6	.325
1/16 x 9	.425
1/16 x 12	.525
1/8 x 3	.350
1/8 x 6	.450
1/8 x 9	.550
1/8 x 12	.650
3/16 x 3	.475
3/16 x 6	.575
3/16 x 9	.675
3/16 x 12	.775
1/4 x 3	.600
1/4 x 6	.700
1/4 x 9	.800
1/4 x 12	.900
5/16 x 3	.725
5/16 x 6	.825
5/16 x 9	.925
5/16 x 12	1.025

2

STRIP	A
1/16 x 3	1/4
1/16 x 6	3/8
1/16 x 9	1/2
1/16 x 12	5/8
1/8 x 3	3/8
1/8 x 6	1/2
1/8 x 9	5/8
1/8 x 12	3/4
3/16 x 3	1/2
3/16 x 6	5/8
3/16 x 9	3/4
3/16 x 12	7/8
1/4 x 3	5/8
1/4 x 6	3/4
1/4 x 9	7/8
1/4 x 12	1
5/16 x 3	3/4
5/16 x 6	7/8
5/16 x 9	1
5/16 x 12	1 1/8

Figure 14.5 Tables of stripper plate thicknesses for (1) representative strip sizes and (2) corresponding commercially available plate thicknesses.

14.4 STANDARDIZED COMPONENTS OF STOPS

Automatic stop proportions can only be determined after we know what stripper plate thickness will be used. The correct value is found by the formula:

$$A = \frac{W}{30} + 2T \qquad (14.1)$$

where:

A= thickness of stripper plate
W= width of strip
T= strip thickness.

Table 1 in Figure 14.5 shows stripper plate thicknesses **A** for representative strip sizes. Table 2 gives recommended commercially available plate thicknesses; for practical purposes, it corresponds with Table 1. From Table 2, the designer can quickly select the correct stripper plate thickness for the specific conditions.

Example

If strip 1/8 inch by 5 inches (3.2 mm by 127 mm) is to be run, select the 1/8 by 6 inches size listed in the table and use a 1/2 inch (12.7 mm) stripper plate thickness. For a strip 3/32 by 4 inches (2.4 mm by 102 mm), use the value given for 1/8 by 3 inches (3.2 by 76.2 mm) in the table. A 3/8 inch (9.52 mm) stripper plate thickness can be used safely.

When in doubt as to the correct corresponding size, use the next thicker plate. Of course, these are minimum thicknesses for adequate strength. Other factors may cause designers to choose a thicker plate. For example, if the fulcrum pin of the automatic stop is to be retained in the stripper plate, the 1/4-inch (6.35 mm) thick plate may be considered too thin and a 3/8-inch (9.52 mm) plate used instead.

Figure 14.6 shows six standardized automatic stops enable ready selections for almost any

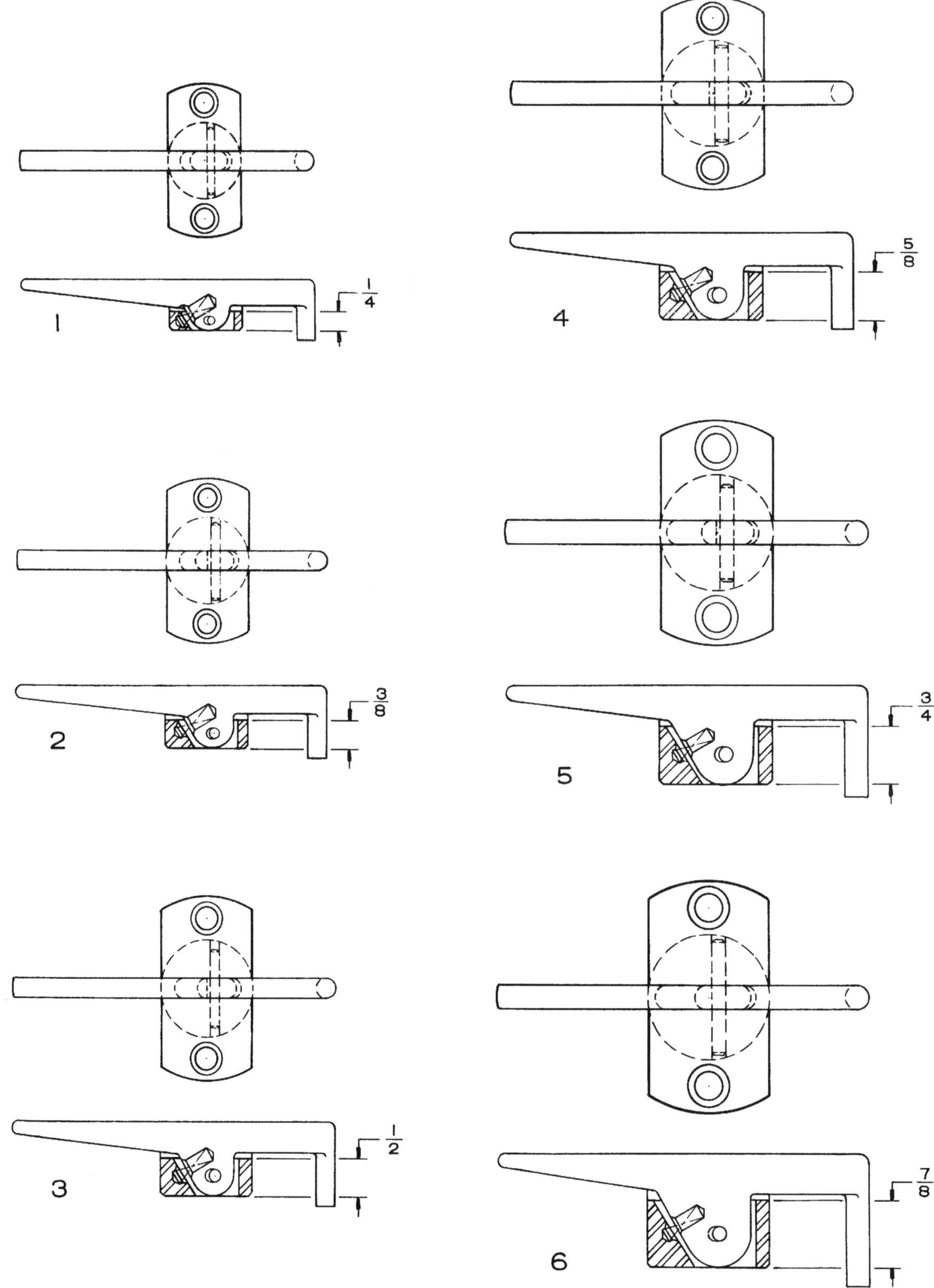

Figure 14.6 Proportions for six standard automatic stops.

die design as they can be used in stripper plates varying in thickness from 1/4 to 7/8 inch (6.35 to 22.23 mm), in increments of 1/8 inch (3.2 mm). Note first that the correct stripper plate thickness is selected from Table 2 of Figure 14.5. Then, from this table, the corresponding stop to use is selected.

a) Dimensions of Stops

Figure 14.7 gives all necessary dimensions for making automatic stops, ranging from No. 1 to No. 6. From this table, the designer can apply dimensions, or the automatic stop can be specified by number. If a side-acting stop like the one shown in Figure 14.1 is to be designed, it can be given the same general proportions.

b) Dimensions of Holders

Figure 14.8 supplies all necessary dimensions for holders to be used with stops No. 1 to No. 6. These holders would be made of machine steel unless unusually severe duty is anticipated, in which case they would be made of tool steel and heat treated.

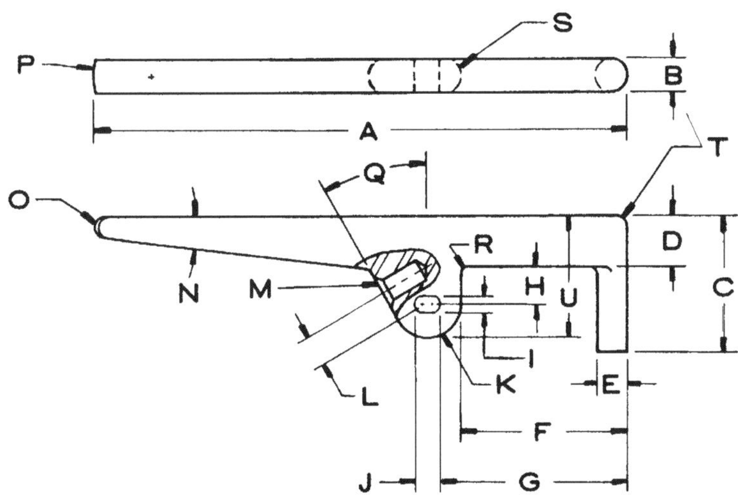

NO.	A	B	C	D	E	F	G	H	I	J	K	L
1	4	1/4	25/32	11/32	1/4	1 3/16	1 3/8	3/16	3/32	1/8	9/32 R.	3/16
2	4 3/16	1/4	15/16	3/8	1/4	1 1/4	1 7/16	1/4	1/8	5/32	9/32 R.	1/4
3	4 3/8	1/4	1 3/32	13/32	1/4	1 3/8	1 9/16	5/16	1/8	5/32	9/32 R.	1/4
4	4 9/16	5/16	1 1/4	7/16	5/16	1 15/32	1 11/16	3/8	3/16	7/32	11/32 R.	9/32
5	4 13/16	5/16	1 15/32	15/32	5/16	1 17/32	1 13/16	7/16	3/16	7/32	13/32 R.	5/16
6	5	5/16	1 11/16	1/2	5/16	1 5/8	1 15/16	1/2	3/16	7/32	7/16 R.	3/8

NO.	M	N	O	P	Q	R	S	T	U
1	3/16 Dr. 3/8 Dp. 45 C/sink 1/16 Dp.	6°	1/16 R.	1/2 R.	30°	3/64 R.	1/8 R.	1/16 R.	21/32
2	3/16 Dr. 3/8 Dp. 45 C/sink 1/16 Dp.	6°	5/64 R.	1/2 R.	30°	3/64 R.	1/8 R.	5/64 R.	13/16
3	3/16 Dr. 3/8 Dp. 45 C/sink 1/16 Dp.	6 1/2°	5/64 R.	1/2 R.	30°	3/64 R.	1/8 R.	3/32 R.	31/32
4	3/16 Dr. 3/8 Dp. 45 C/sink 1/16 Dp.	6 1/2°	3/32 R.	1/2 R.	30°	3/64 R.	5/32 R.	7/64 R.	1 1/8
5	3/16 Dr. 3/8 Dp. 45 C/sink 1/16 Dp.	7°	3/32 R.	1/2 R.	30°	3/64 R.	5/32 R.	1/8 R.	1 9/32
6	3/16 Dr. 3/8 Dp.	7 1/2°	3/32 R.	1/2 R.	30°	3/64 R.	5/32 R.	5/32 R.	1 7/16

Figure 14.7 Tables of dimensions for making the six standard automatic stops given in Figure 14.6.

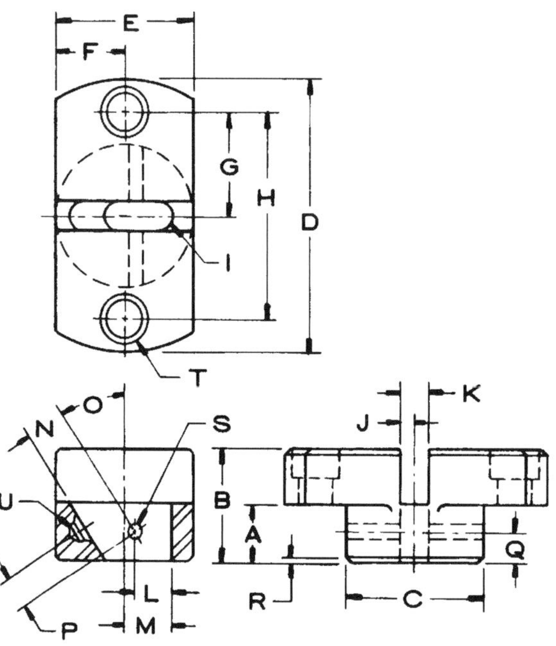

NO.	A	B	C	D	E	F	G	H	I	J	K	L	M	N
1	1/4	21/32	1	2	1	1/2	3/4	1 1/2	1/8	1/8	1/4	5/16	23/64	5/16
2	3/8	13/16	1 1/8	2 1/8	1 1/8	9/16	13/16	1 5/8	1/8	1/8	1/4	11/32	13/32	11/32
3	1/2	31/32	1 1/4	2 3/8	1 1/4	5/8	29/32	1 13/16	1/8	1/8	1/4	11/32	7/16	3/8
4	5/8	1 1/8	1 3/8	2 1/2	1 3/8	11/16	31/32	1 15/16	5/32	5/32	5/16	13/32	1/2	7/16
5	3/4	1 9/32	1 1/2	2 7/8	1 1/2	3/4	1 3/32	2 3/16	5/32	5/32	5/16	15/32	9/16	15/32
6	7/8	1 7/16	1 5/8	3	1 5/8	13/16	1 5/32	2 5/16	5/32	5/32	5/16	1/2	5/8	17/32

NO.	O	P	Q	R	S	T	U
1	33°	7/32	1/8	1/32 x 45°	3/32	17/64 Dr.- 13/32 C'bore 1/4 Dp.	3/16 Dr. 3/32 Dp. 45° C'sink 3/64 Dp.
2	33°	17/64	3/16	1/32 x 45°	1/8	17/64 Dr.- 13/32 C'bore 1/4 Dp.	3/16 Dr. 3/32 Dp. 45° C'sink 3/64 Dp.
3	33°	17/64	1/4	1/32 x 45°	1/8	21/64 Dr.- 15/32 C'bore 5/16 Dp.	3/16 Dr. 3/32 Dp. 45° C'sink 3/64 Dp.
4	33°	5/16	5/16	1/32 x 45°	3/16	21/64 Dr.- 15/32 C'bore 5/16 Dp.	3/16 Dr. 3/32 Dp. 45° C'sink 3/64 Dp.
5	33°	11/32	3/8	1/32 x 45°	3/16	25/64 Dr.- 19/32 C'bore 9/16 Dp.	3/16 Dr. 3/32 Dp. 45° C'sink 3/64 Dp.
6	33°	3/8	7/16	1/32 x 45°	3/16	25/64 Dr.- 19/32 C'bore 9/16 Dp.	3/16 Dr. 3/32 Dp. 45° C'sink 3/64 Dp.

Figure 14.8 Tables of dimensions for holders to be used with the six standard automatic stops.

c) Dimensions of Holes in Stripper Plates

The table in Figure 14.9 gives dimensions for machining the holes in the stripper plate to receive the automatic stop assembly. This type of automatic stop is a self-contained unit. It can be removed quickly from the die by simply removing two screws.

d) Dimension of Set Screw Location

Figure 14.10 tabulates dimensions **A** between the automatic stop and the actuating square-head set screw. By standardizing the position of the actuating screw, more uniform results are assured.

e) Toe Lengths

In Figure 14.11, toe lengths **A** are given for all stops incorporating the variations described in Figure 14.12. We now have 17 standardized automatic stops, many of which vary only in respect to dimension **A**.

Still another factor must be taken into consideration when specifying toe length **C** of

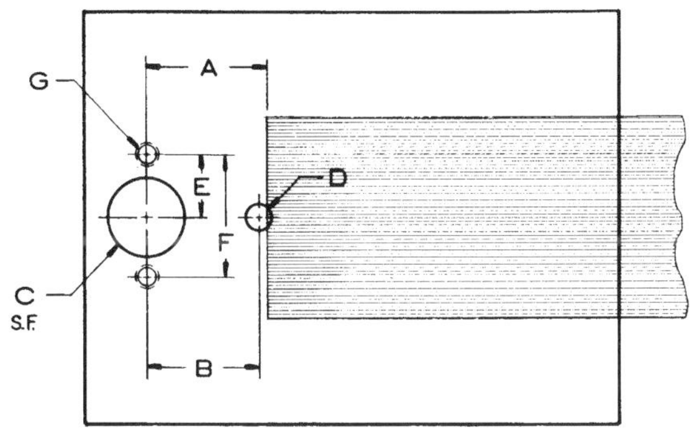

NO.	A	B	C	D	E	F	G
1	1.4687	1.3750	1	5/16	3/4	1 1/2	1/4-20 Tap
2	1.5625	1.4687	1 1/8	5/16	13/16	1 5/8	1/4-20 Tap
3	1.7187	1.6406	1 1/4	11/32	29/32	1 13/16	5/16-18 Tap
4	1.8750	1.7656	1 3/8	13/32	31/32	1 15/16	5/16-18 Tap
5	2.0000	1.8906	1 1/2	13/32	1 3/32	2 3/16	3/8-16 Tap
6	2.1562	2.0625	1 5/8	7/16	1 5/32	2 5/16	3/8-16 Tap

Figure 14.9 Table of dimensions for machining automatic stop assembly holes in the stripper plate.

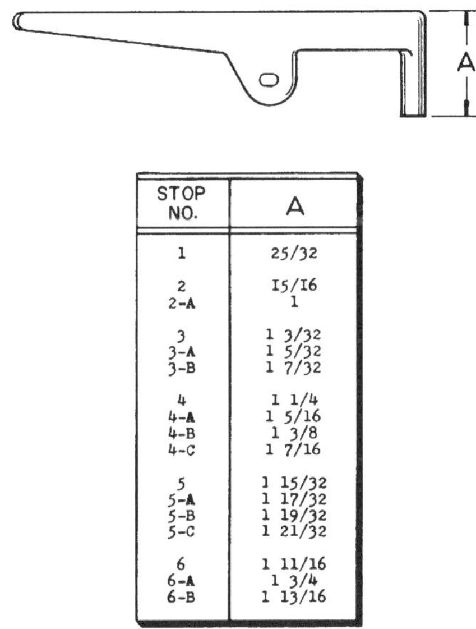

STOP NO.	A
1	25/32
2	15/16
2-A	1
3	1 3/32
3-A	1 5/32
3-B	1 7/32
4	1 1/4
4-A	1 5/16
4-B	1 3/8
4-C	1 7/16
5	1 15/32
5-A	1 17/32
5-B	1 19/32
5-C	1 21/32
6	1 11/16
6-A	1 3/4
6-B	1 13/16

Figure 14.11 Tabulation of toe length **A** dimensions.

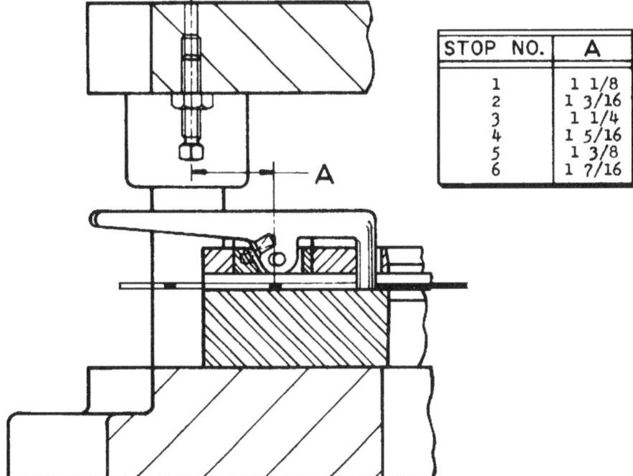

STOP NO.	A
1	1 1/8
2	1 3/16
3	1 1/4
4	1 5/16
5	1 3/8
6	1 7/16

Figure 14.10 Tabulated dimensions **A** of the distance between the automatic stop and a square-head set screw for the six standard automatic stops.

Figure 14.7 and **A** of Figure 14.11. This is the thickness **B** of the back gage and front spacer, because in Table 2 of Figure 14.5, the same stripper plate thickness can be used for more than one strip thickness. For example, a stripper plate 3/8-inch (9.52-mm) thick is used for 1/16 by 6-inch (1.6 by 152.4-mm) strip. But it is also used for an 1/8 by 3-inch (3.2 by 76.2-mm) strip, which would require a thicker back gage and front spacer, raising the stop.

In this table, further automatic stop numbers are applied to allow for these variations. For example, a stripper plate thickness of 1/2 inch (12.7 mm) can have a spacer thickness **B** of 1/8 inch (3.2 mm), which calls for stop No. 3. It may also have a spacer thickness of 3/16 inch (4.8 mm), in which case stop No. 3-A would be used. Still another spacer thickness is 1/4 inch (3.2 mm), with which stop No. 3-B would be used.

f) Fulcrum Pin Dimensions

The table in Figure 14.13 gives dimensions for fulcrum pins for all stops. Fulcrum pins are ordinarily made of cold-rolled steel, but for long runs they would be made of drill rod.

g) Spring Dimensions

Outside dimensions of springs used in these automatic stops are now given in Figure 14.14. All stops use the same size spring, but pressure can be varied by using different wire diameters.

14.5 SIMPLE PIN INSTEAD OF AUTOMATIC STOP

Trigger type automatic stops previously described are used for all but very wide blanks. Run in slow presses, wide blanks can be located by a simple pin stop pressed into the die block

STRIPPER PLATE THICKNESS A											
1/4		3/8		1/2		5/8		3/4		7/8	
B	NO.	B	NO.	B	NO.	B	NO.	B	NO.	B	NO.
1/8	1	1/8	2	1/8	3	1/8	4	3/16	5	1/4	6
		3/16	2-A	3/16	3-A	3/16	4-A	1/4	5-A	5/16	6-A
				1/4	3-B	1/4	4-B	5/16	5-B	3/8	6-B
						5/16	4-C	3/8	5-C		

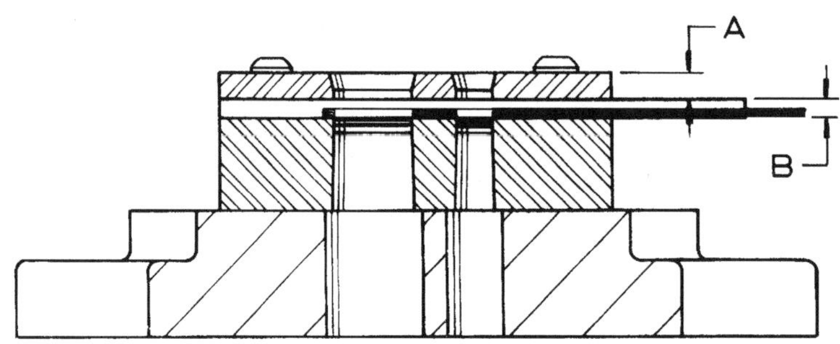

Figure 14.12 Modifications of six standard automatic stop numbers allow for varying thicknesses of back gage and front spacer.

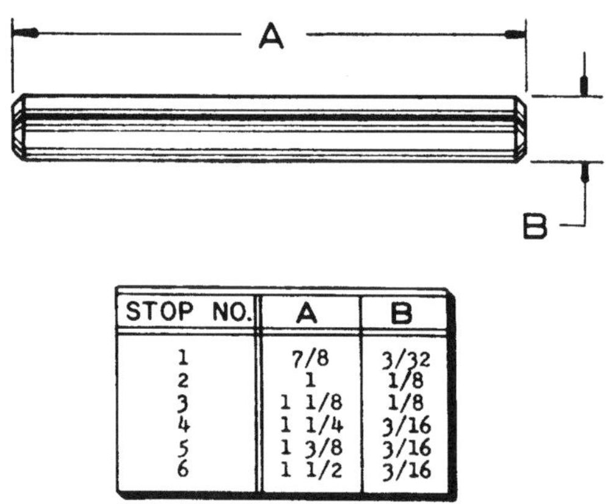

STOP NO.	A	B
1	7/8	3/32
2	1	1/8
3	1 1/8	1/8
4	1 1/4	3/16
5	1 3/8	3/16
6	1 1/2	3/16

Figure 14.13 Table of fulcrum pin dimensions.

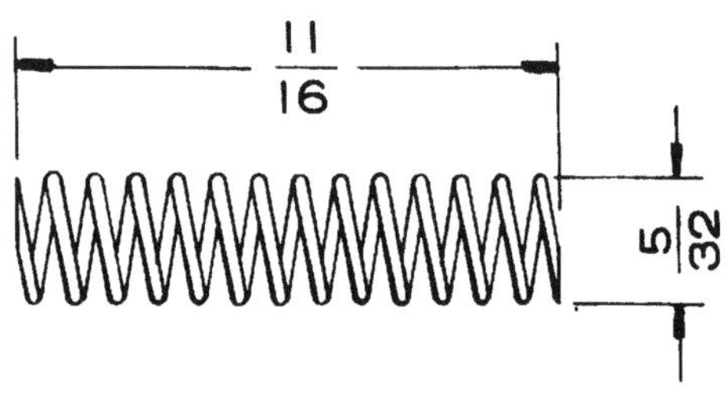

Figure 14.14 Spring dimensions for use in standard automatic stops.

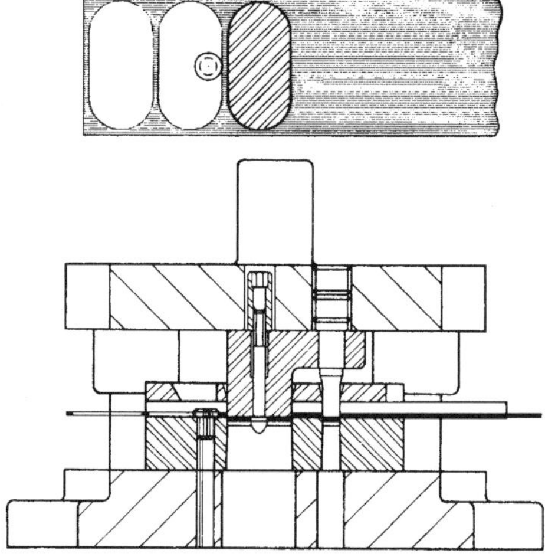

Figure 14.15 Simple pin stop for wide blanks run in slow presses.

(Figure 14.15). In operation, the strip clings around the blanking punch and is removed from it upon contact with the underside of the stripper plate. The operator applies pressure against the strip, tending to move it toward the left so that, as it is released, it falls with the scrap bridge on top of the pin stop to allow feeding. A sight hole should always be applied to the stripper plate so that the operator can see the stop action.

HOW TO DESIGN STRIPPERS

15.1 INTRODUCTION

Stripper plates remove the material strip from around blanking and piercing punches. Severe adhesion of strip to punches is characteristic of the die cutting process. Because of their low cost, solid strippers are the most frequently used type, particularly when running strip stock. Spring strippers, though more complex, should be used when the following conditions are present:

1. When perfectly flat, accurate blanks are required, spring strippers flatten the sheet before cutting begins.
2. When very thin material is to be blanked or pierced, to prevent uneven fracture and rounded blank edges.
3. When parts are to be pressed from waste strip left over from other operations, spring strippers provide good visibility to the operator for gaging purposes.
4. When stripping occurs immediately, small punches are not as subject to breakage.
5. When conducting secondary operations, such as in piercing dies, increased visibility provided by spring strippers allows faster loading of work and increased production.

Stripper plates may be made of cold-rolled steel if they are not to be machined except for holes. When machining must be applied to clear gages, the plates should be made of machine steel, which is not as subject to distortion. This chapter describes numerous methods of applying stripper plates and their components. These methods further explain Step 10 in Chapter 5—14 Steps to Design a Die.

15.2 SOLID STRIPPERS

Figure 15.1 shows the most common method of applying a solid stripper. Plate **A**, machined to receive blanking and piercing punches, is fastened in position on top of the back gage and front spacer with four button-head socket screws. Two dowels accurately locate the stripper plate in relation to the die block and back gage. A small, short dowel locates the other end of the back gage

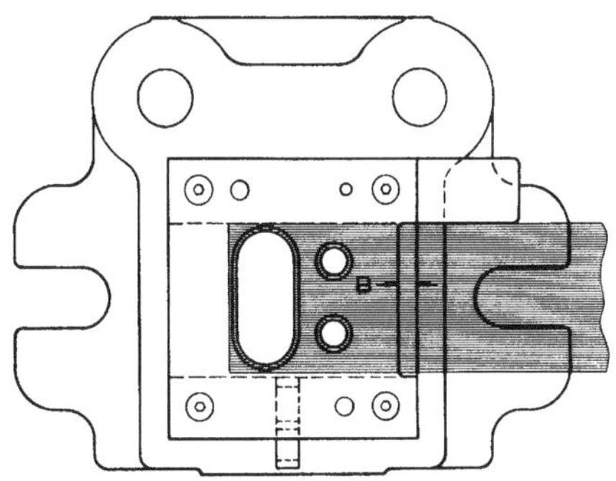

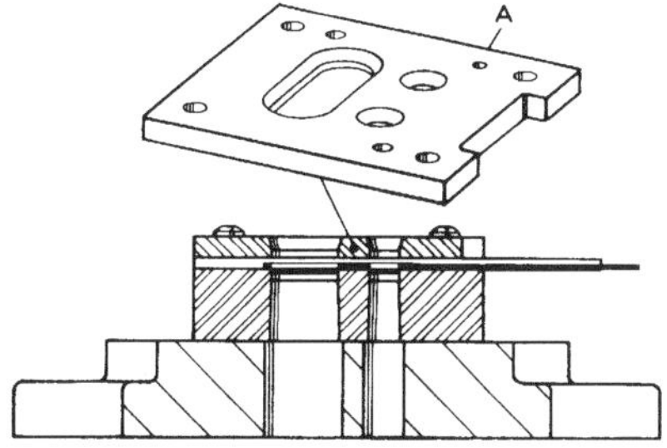

Figure 15.1 Common method of applying a solid stripper.

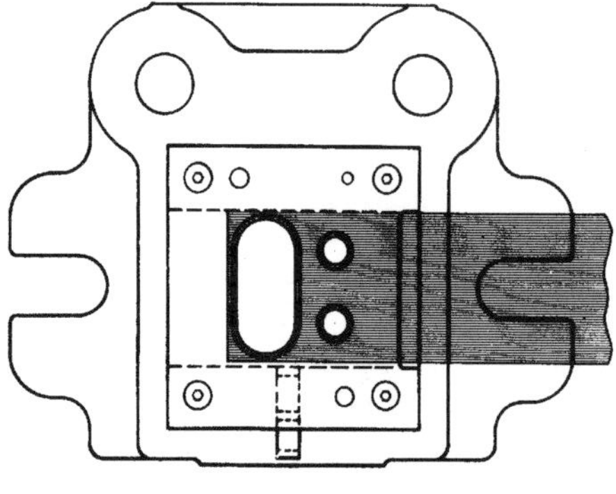

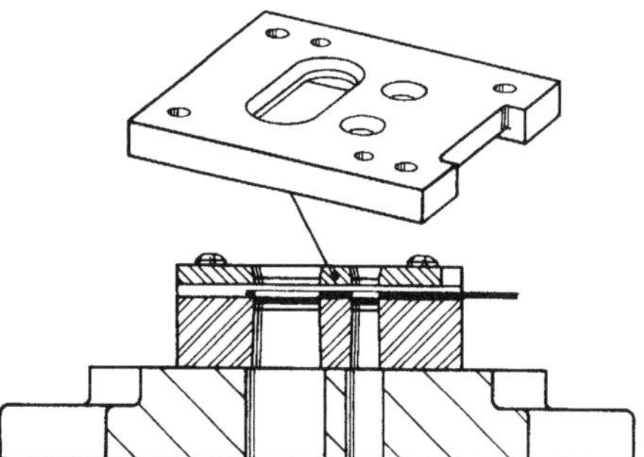

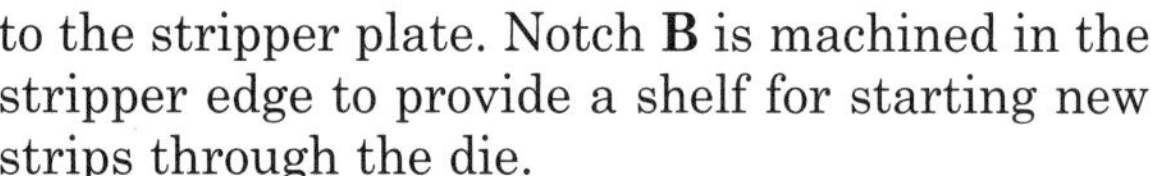

Figure 15.2 Alternate design of a solid stripper, used now only for very small dies.

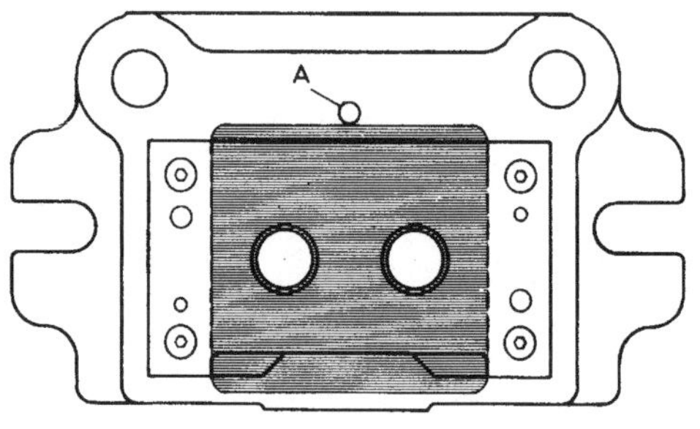

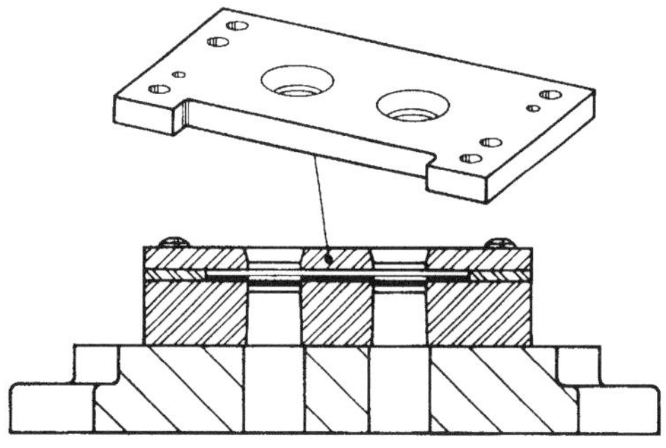

Figure 15.3 Stripper plate for secondary operations.

to the stripper plate. Notch **B** is machined in the stripper edge to provide a shelf for starting new strips through the die.

Alternate method. Figure 15.2 illustrates an alternate method of designing a solid stripper, which is to make it thicker and machine a slot in its underside slightly larger than the strip width. In its travel through the die, the strip is located against the back edge of the slot. Once widely used, this method has become obsolete except for very small dies.

a) Stripper Plate for Secondary Operations

For secondary operation work, stripper plates are fastened on top of two guide rails set at either side (Figure 15.3). This die pierces two holes in a rectangular blank. Gage pin **A** locates the blank endwise, while it is confined sidewise between the guide rails. The stripper plate is relieved at the front to provide a shelf for easy insertion of the blank. The die block is relieved at the front to provide finger room.

15.2.1 Machining Punch Openings

Three methods are used for machining punch openings in the stripper plate (Figure 15.4). At **A**, a 1/8-inch (3.2-mm) straight land is applied from the bottom of the plate. Above the land, angular relief is provided to clear the punch radius when the punch is lowered in die sharpening. At **B**, the cutting diameter of the small punch is kept short for rigidity. The stripper plate is counterdrilled for clearance. At **C**, small punches may be guided in hardened bushings pressed into holes machined in the stripper.

15.3 ELASTIC STRIPPERS

15.3.1 Spring Strippers

Figure 15.5 shows a typical spring stripper. Views A, B, and C illustrate the operation of a spring stripper plate. Springs, arranged around the blanking punch, provide stripping pressure. Four stripper bolts, located at the corners, limit stripper travel. View B shows the die in open position,

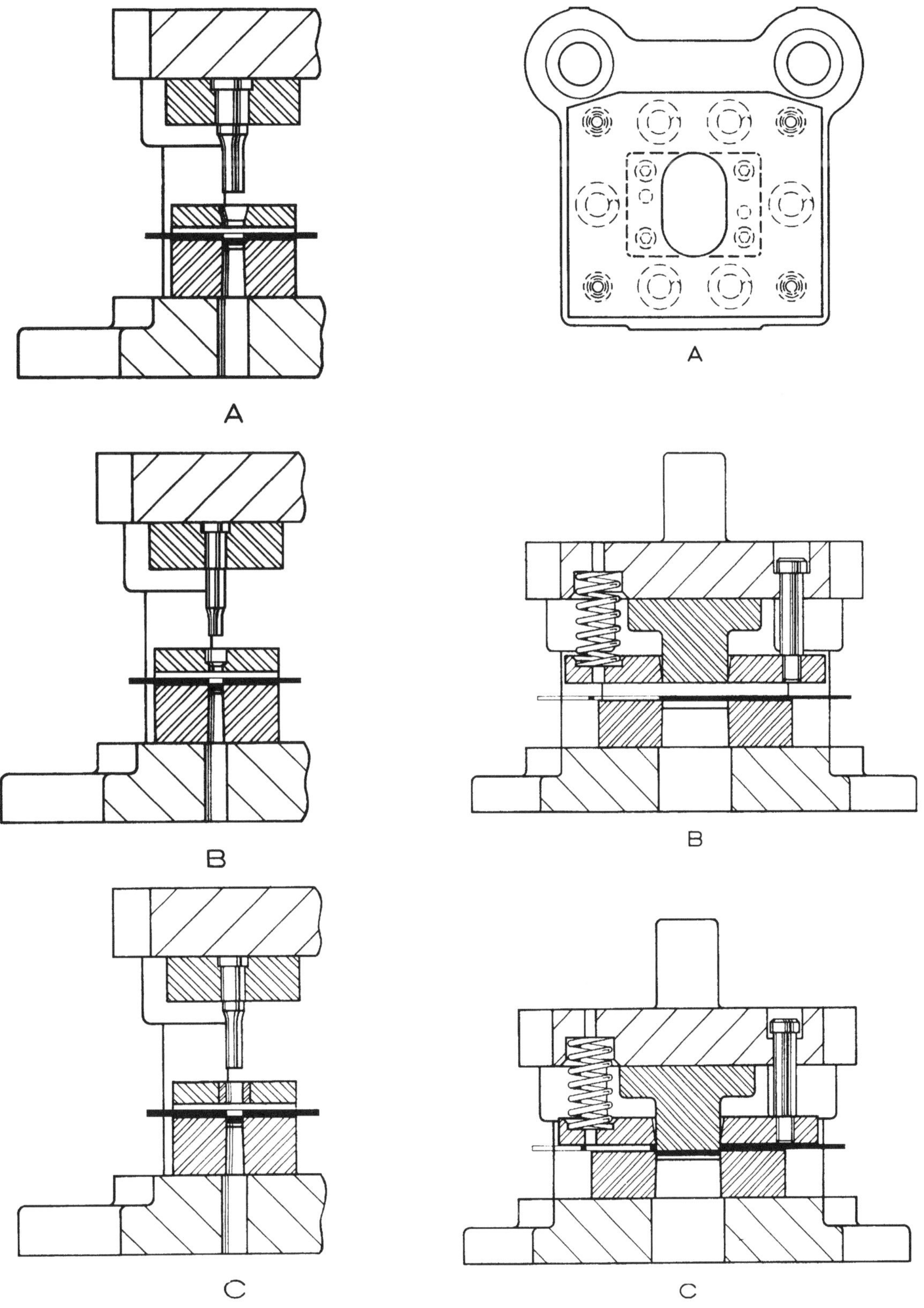

Figure 15.4 Three types of punch openings for strippers.

Figure 15.5 Typical spring stripper.

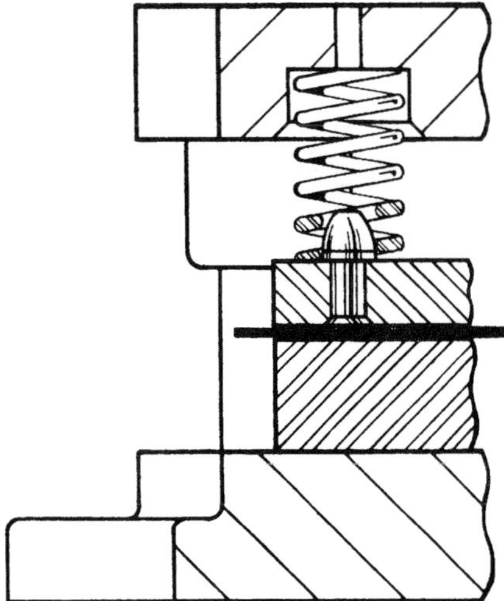

Figure 15.6 Spring pilot retains stripper plate spring.

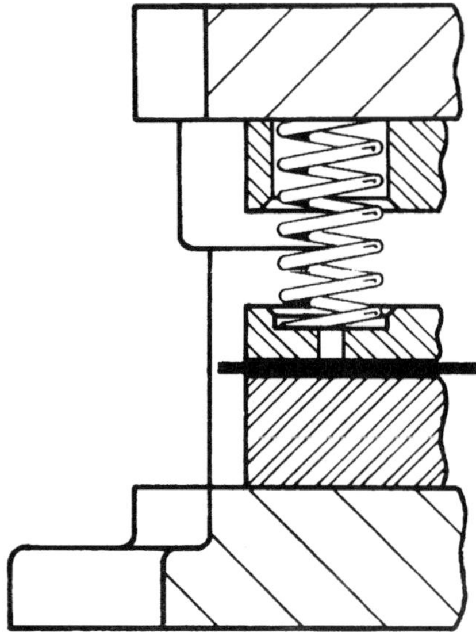

Figure 15.7 A hole bored in the punch plate retains the stripper spring.

whereas view C shows the die as it would appear at the bottom of the press stroke. Springs have been compressed, ready to strip the material from around the punch upon movement of the punches. View A shows the punch holder, blanking punch, and punch plate inverted, as they would appear in the upper right hand view of the die drawing.

For large dies, the cost of the die set becomes important. The stripper plate corners are beveled to clear die posts and guide bushings, thereby allowing use of a smaller die set. Springs are retained in pockets counterbored in the punch holder and in the stripper plate. The holes are countersunk 1/8 inch by 45 degrees (3.2 mm by 45 degrees) in the punch holder and 1/16 inch by 45 degrees (1.6 mm by 45 degrees) in the stripper to guide the spring coils. A small hole is drilled completely through the punch holder and stripper plate while the two are clamped together. This hole provides for engagement of the counterbore pilot.

a) Spring Pilots

Counterboring the stripper plate can be avoided if spring pilots are used to retain springs (Figure 15.6). These can be made to suit conditions, or they can be purchased from commercial sources that offer a standardized line of these components.

b) Machining the Punch Plate

When a punch plate is used to retain piercing punches, holes are bored through the punch plate to hold stripper springs (Figure 15.7). This construction avoids counterboring the punch holder of the die set.

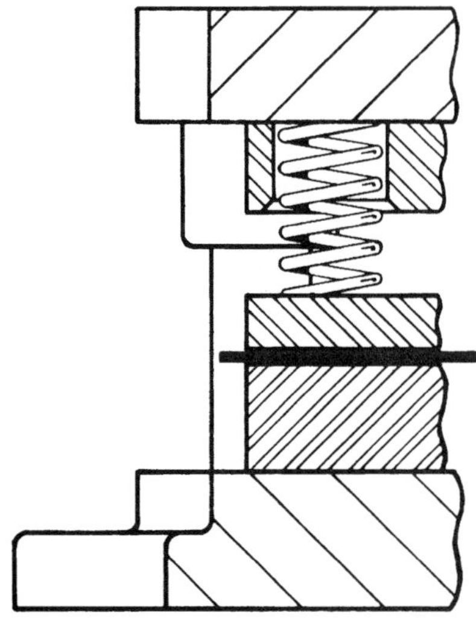

Figure 15.8 A short stripper spring does not require a hole in the stripper for maintaining position.

c) Short Stripper Springs

When stripper springs are short, it is not necessary to counterbore the stripper plate (Figure 15.8). The springs are retained, either by holes in the punch plate or by counterbored holes in the punch holder.

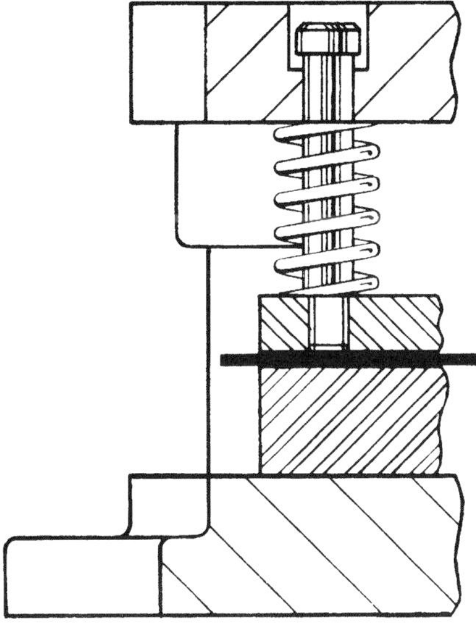

Figure 15.9 The spring retained by a stripper bolt is used in medium-production die.

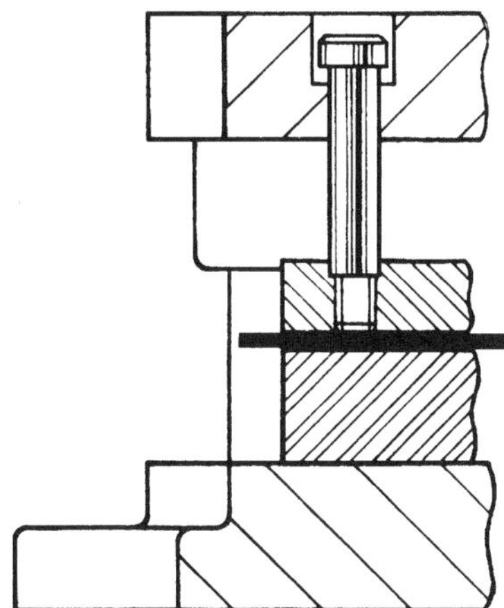

Figure 15.10 Method of applying stripper bolt for severe applications.

d) Springs Retained by Stripper Bolts

For medium-production dies, springs can be applied around stripper bolts, as shown in Figure 15.9. This method is not recommended for long runs because of increased wear on the stripper bolts, and because of the need to remove the bolts in the event of broken springs.

For severe applications, stripper bolts are better applied with their bearing shoulders confined for strength in short counterbored holes in the stripper plate (Figure 15.10). The counterbore is usually made 1/8 inch (3.2 mm) deep and a press fit is specified for the diameter of the hole.

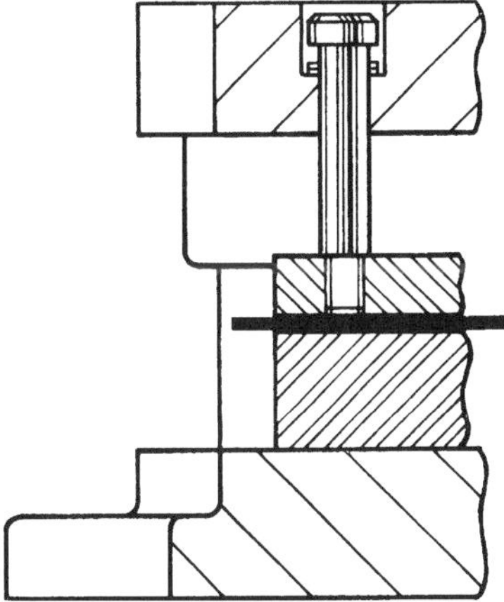

Figure 15.11 Washers are applied under the stripper bolt head when punches are sharpened.

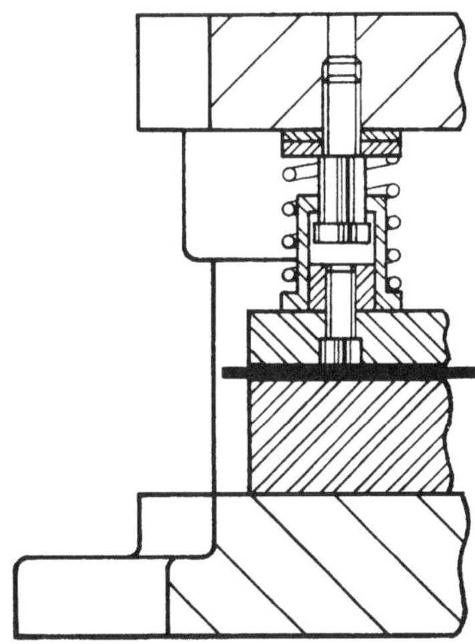

Figure 15.12 Strippit spring units allow removal of stripper without dismantling springs.

When punches are sharpened, spacer washers are applied under stripper bolt heads to maintain the correct relationship between the underside of the stripper plate and the faces of blanking and piercing punches (Figure 15.11). The stripper usually extends 1/32 inch (0.8 mm) past the faces of cutting punches to flatten the strip properly before cutting occurs.

15.3.2 Strippit Spring Units

Strippit spring units (Figure 15.12) are self-contained, commercially available assemblies that allow removal of the stripper plate without dismantling springs. Socket cap screws, or socket flat head screws for thin plates, retain the stripper plate to the Strippit units. Because pressures are

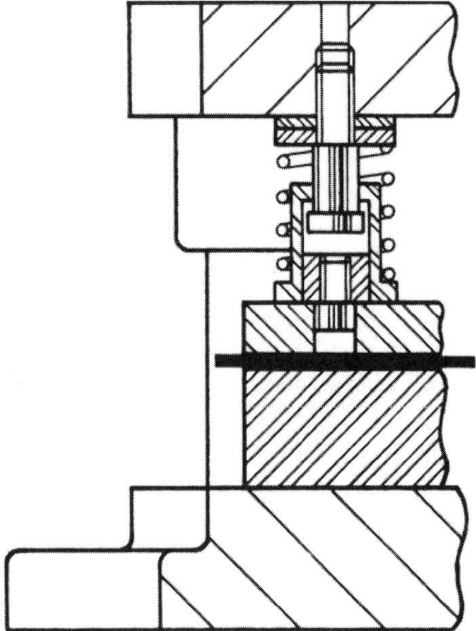

Figure 15.13 Alternate method of applying Strippit spring unit to stripper.

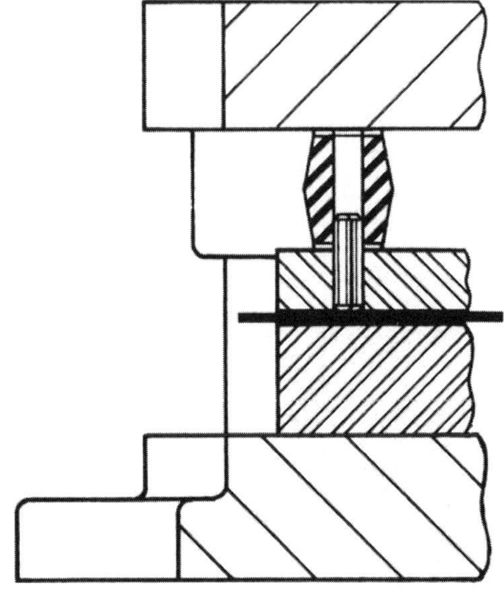

Figure 15.14 Rubber spring for medium- and low production dies.

self-contained, thinner stripper plates can be used, and stripper bolts are not required.

Alternate method. Another method of applying stripper plates to Strippit units is to insert short socket cap screws in the units (Figure 15.13). The cap screw heads engage holes in the stripper plate to prevent lateral movement. Stripper bolts limit travel.

15.3.3 Rubber Springs

For medium- and low-production dies, rubber springs may be used (Figure 15.14). These rubber components are available from commercial sources. Dowel pins, pressed into the stripper, retain the rubber springs in position.

15.3.4 Fluid Springs

Where extremely high stripping pressures are required, fluid springs reduce the number of spring units required in the die, an important factor where space is limited. Fluid springs are filled with an emulsion of silicones in oil. The one illustrated in Figure 15.15 is available from commercial sources.

Another type of fluid spring (Figure 15.16) is made to be interchangeable with some sizes of standard springs. These are also available commercially.

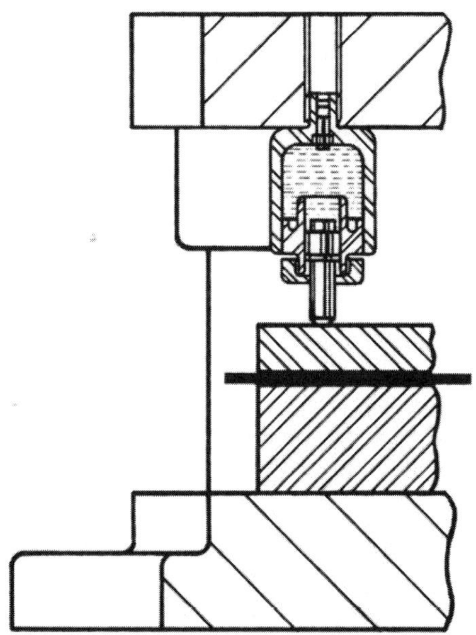

Figure 15.15 Fluid springs are used where extremely high stripping pressures are required.

Both types of fluid springs, as shown in Figures 15.15 and 15.16, are fastened to the die by threading their shanks into holes tapped in the die set.

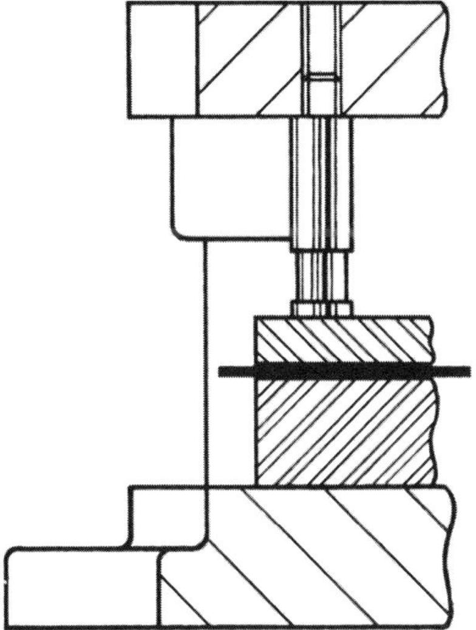

Figure 15.16 Type of fluid spring that is interchangeable with standard springs.

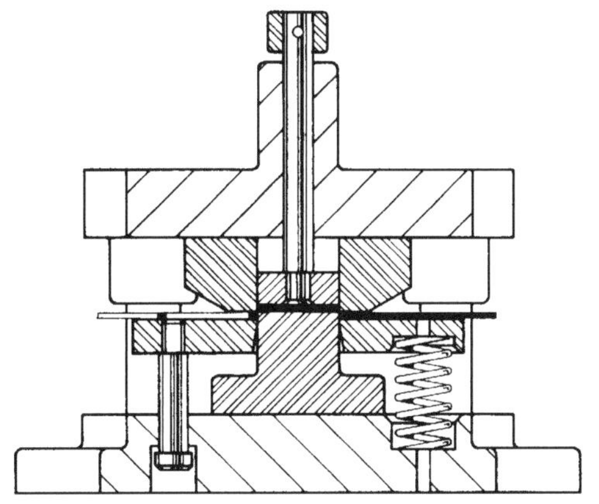

Figure 15.17 Typical inverted die.

15.3.5 Inverted Die

In inverted dies (Figure 15.17), the blanking punch is fastened to the lower die holder of the die set instead of to the punch holder, as in conventional dies. In such applications, the spring stripper is reversed. Reversing the stripper plate does not alter application of the components previously described, except that they are also inverted.

15.4 STRIPPING FORCE

When spring strippers are used, it is necessary to calculate the amount of force required to effect stripping. Stripping force is found by the empirical formula:

$$F = 855 \times P \times T \qquad (15.1)$$

where:

F= stripping force in pounds
P= sum of the perimeters of all perforator faces
T= thickness of material.

This formula has been used for many years in a number of plants; it has been found to be satisfactory for most work. The constant 855 (the reciprocal of 0.00117) simplifies calculations.

But the value of stripping force F should never be reduced in applications where there is any possibility of scrap pieces from other operations being stamped in the die.

After the total stripping force has been determined, the stripping force per spring must be found in order to establish the number and sizes of springs required. Force per spring is listed in manufacturers' catalogs in terms of pounds per 1/8-inch (3.2-mm) deflection. In some catalogs, it is given in pounds per 1/10-inch deflection and must be converted to pounds per 1/8-inch deflection by simple proportion.

Three types of spring deflection must be taken into consideration. View **A** in Figure 15.18 illustrates initial preload of the springs. The compression of the springs is usually 1/4 inch (6.35 mm). View **B** shows a working compression, which shows further deflection of the springs while work is performed. View **C** illustrates grinding compression, usually 1/4 inch (6.35 mm), which is the amount allowed for sharpening punches during the life of the die.

If a spring 1 1/2 inches (38.1 mm) in diameter by 6 inches (152.4 mm) long is tentatively selected, and initial compression is 1/4 inch (6.35 mm), the force for this deflection will be twice the listed amount (as the listed amount is for 1/8-inch (3.2-mm) deflection for that size of spring). To this is added the force for the first 1/8 inch (3.2 mm) of working compression. No more force is added because too high a value would result, and the force would fade away near maximum travel. Should the working compression be less than 1/8 inch (3.2 mm), say 1/16 inch (1.6 mm), only one-half of the listed force would be used in the calculation. For example:

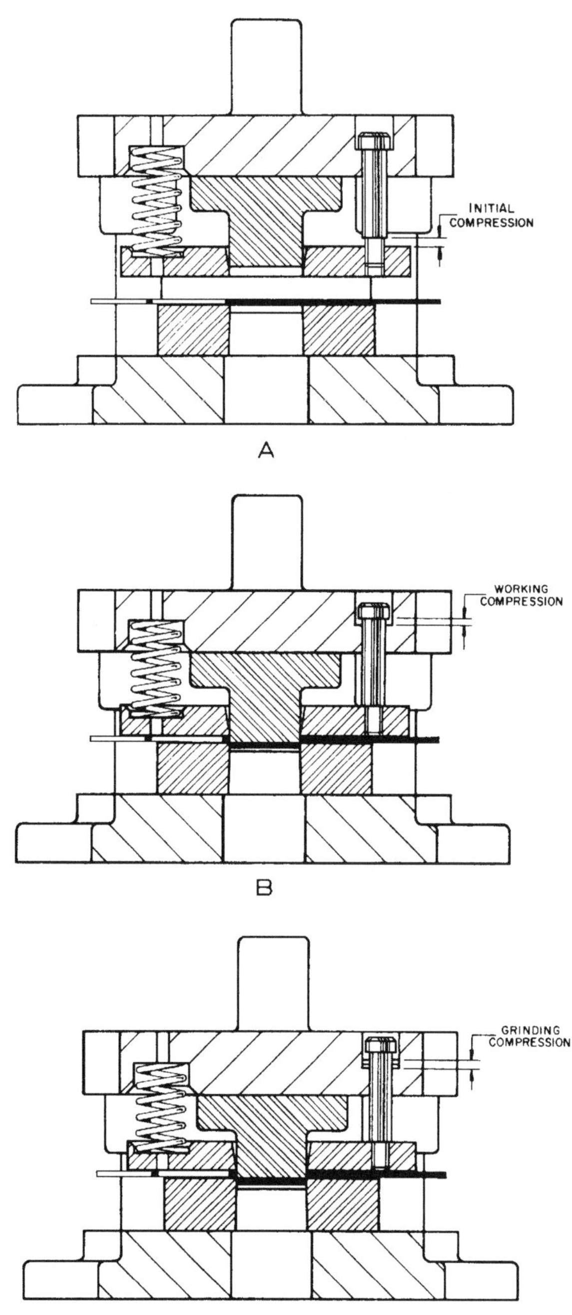

Figure 15.18 Three types of spring deflection that must be considered when determining stripping force.

Listed force per 1/8-inch (3.2-mm) compression:	20 lb (89 N)
1/4-inch (6.35 mm) initial compression: (2 × 20)	40 lb (178 N)
1/16-inch (1.6 mm) working compression: (1/2 × 20)	10 lb (44.5 N)
1/4-inch (6.35 mm) grinding compression: (2 × 20)	40 lb (178 N)
Total force per spring:	90 lb (400.3 N)

The value on the right represents the total amount of force exerted by the compressed spring after 1/4 inch (6.35 mm) has been removed in sharpening. It is the maximum force; the value should be used in any calculation of strength of members.

To determine the number of springs required to effect stripping, we must know the force exerted by each spring before sharpening. Therefore, referring to the example:

Listed force per 1/8-inch (3.2 mm) compression:	20 lb (89 N)
1/4-inch (6.35 mm) initial compression: (2 × 20)	40 lb (178 N)
1/16-inch (1.6 mm) working compression: (1/2 × 20)	10 lb (44.5 N)
Force per spring:	50 lb (222.5 N)

The number of springs required is found by dividing the total stripping force by the force per spring. If the number arrived at is out of proportion to the die, use fewer and larger springs, or perhaps more springs of smaller size. Alternatively, pressure may be increased by applying more initial compression. Manufacturers' catalogs list the maximum amount of compression to apply, and their recommendations should be followed. In general they suggest:

1. For high speed work, both medium pressure and high pressure springs should not be deflected more than 1/4 of their free lengths.
2. For heavy, slow-moving presses, use medium pressure springs with a total deflection not to exceed 3/8 of their free lengths.

15.5 METHODS OF APPLYING KNOCKOUTS

Knockouts remove or strip completed blanks from within die members. They differ from stripper plates in that stripper plates remove the material strip from around punches. There are three types of knockouts:

1. *Positive knockouts.* These eject the blank upon contact of the knockout rod with the knockout bar of the press.
2. *Pneumatic knockouts.* These are actuated by an air cushion applied under the bolster plate of the press.
3. *Spring knockouts.* These employ heavy springs as the thrust source.

The knockout plate or block in contact with the part is usually made of machine steel, but can be made of heat-treated tool steel when it also performs a forming operation.

Knockouts can be applied in two ways: directly and indirectly. In direct knockouts, the force is applied directly from the source. In indirect knockouts, the force is applied through pins arranged to clear other die components, such as piercing punches.

Numerous methods of applying knockouts to dies are illustrated, as well as details of their construction and methods of finding the center of stripping force for correct design.

15.5.1 Operation of a Knockout

Taken together, Figure 15.19 and Figure 15.20 illustrate the operation of a positive knockout. This operation uses an inverted blanking die, with the blanking punch fastened to the lower die holder of the die set, and the die block fastened to the upper punch holder. The knockout assembly

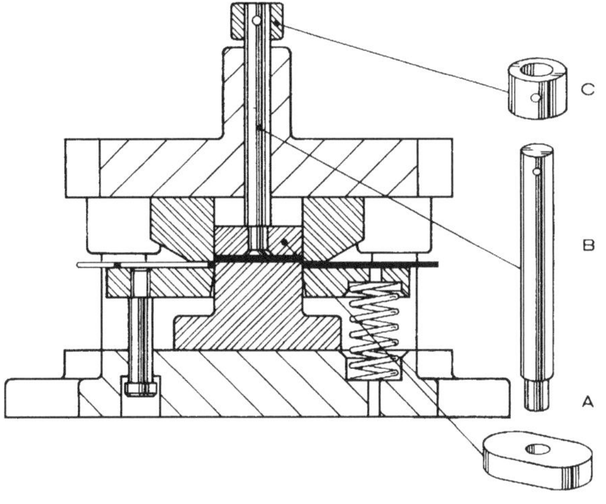

Figure 15.19 Inverted blanking die with positive knockout.

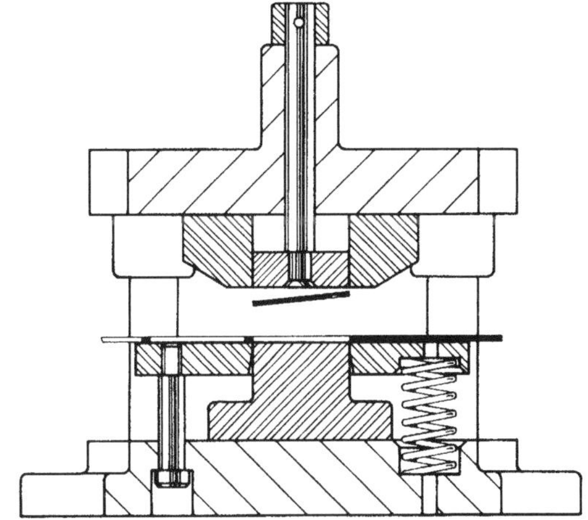

Figure 15.20 A knockout removes the blank near the top of the press stroke.

consists of knockout plate **A**, knockout rod **B**, and stop collar **C**. At the bottom of the press stroke, the blanking punch has removed the blank from the strip and inserted it into the die block, raising the knockout assembly.

15.5.2 Stripping the Blanks

Near the top of the press stroke, the knockout rod contacts the stationary knockout bar of the press. Continued ascent of the upper die causes the knockout to remove the blank from within the die cavity (Figure 15.20). If the job is run in an inclined press, the blank falls to the rear of the press by gravity. Thin blanks may be blown to the rear of the press by air.

Blanks produced from oiled stock tend to cling to the face of the knockout. When this condition is present, a shedder pin applied to one side of the knockout plate will break the adhesion and free the blank. Three types of shedder pins were illustrated in Figure 8.18 in the section Blanking Punches.

15.5.3 Spring Damper

In applying a positive knockout, many designers prefer to provide a relatively weak spring around the knockout rod (Figure 15.21). This spring is not strong enough to strip the part. Its function is to prevent unnecessary travel and bounce of the knockout to reduce wear.

15.5.4 Guiding in the Knockout

Slender piercing punches may be guided in the knockout by providing hardened bushings

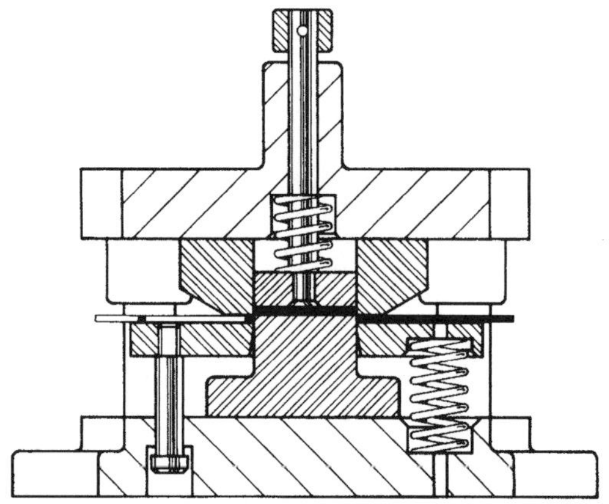

Figure 15.21 Spring around the knockout rod is too weak to strip, but prevents needless travel and bounce.

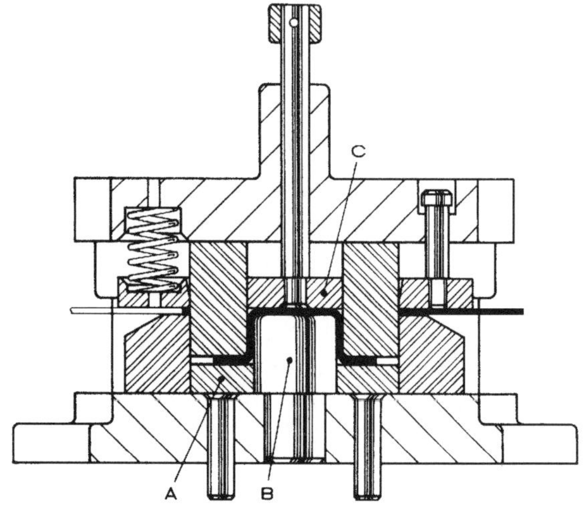

Figure 15.23 Positive knockouts are used to eject drawn shells.

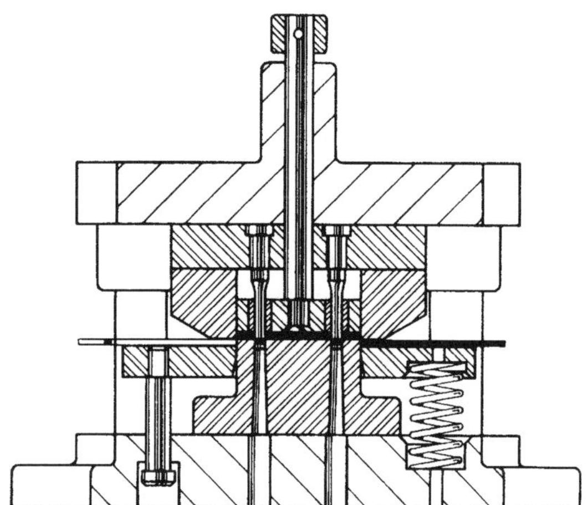

Figure 15.22 Slender piercing punches are guided by hardened bushings in the knockout plate.

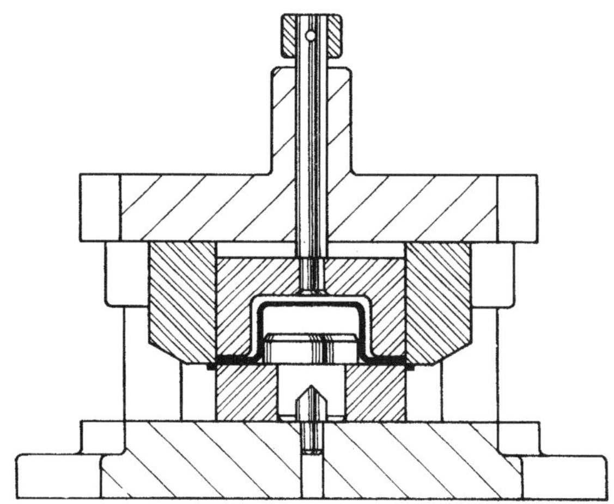

Figure 15.24 Positive knockout is relieved so as to apply stripping force near trimmed edge.

pressed into the knockout plate (Figure 15.22). Close fits between the knockout and other die components must be maintained.

15.5.5 Ejecting Cup and Shells

Positive knockouts may be used to eject drawn shells (Figure 15.23). In this combination die, the upper punch blanks a disk from the strip and, in further descent, draws it into a flanged shell. Lower knockout **A** is raised by pins actuated by the air cushion of the press, and it strips the shell from around draw punch **B**. Positive knockout **C**, actuated near the top of the stroke, removes the shell from within the die and it falls to the rear of the press.

15.5.6 Trimming Operations

Figure 15.24 shows the trimming die for a previously drawn flanged cup. A positive knockout, relieved to clear the shell body, removes the shell from within the die hole. Whenever a knockout is used to remove formed or drawn stampings, it should be relieved so that stripping force is applied only near cutting edges to prevent distortion of the parts.

15.5.7 Forming in the Knockout

For some types of work, the knockout provides the final forming operation (Figure 15.25). This combination die produces a flanged cover by

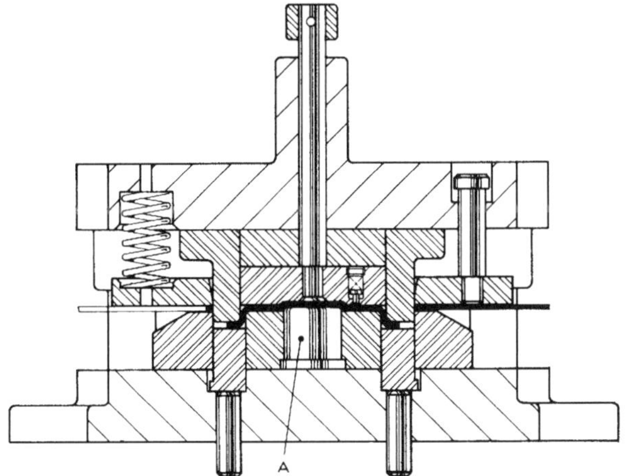

Figure 15.25 Die in which knockout is used for forming or embossing.

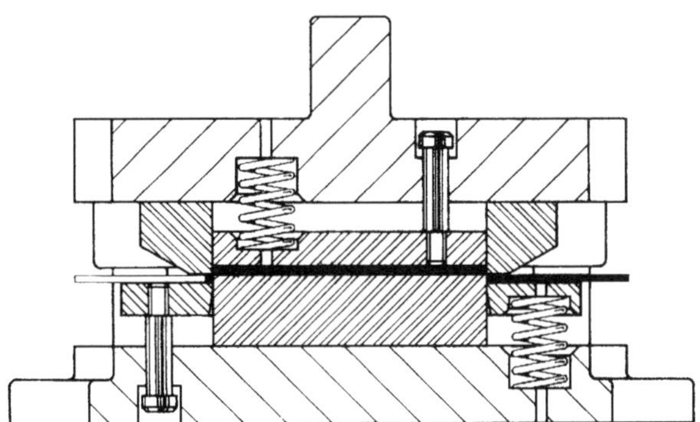

Figure 15.26 Typical spring knockout used in large dies.

blanking a disk and drawing a shallow flanged shell. At the bottom of the stroke, the knockout, in conjunction with the lower embossing punch **A**, embosses a shallow recess in the part. When knockout plates do forming, they are made of hardened tool steel and are backed up by a plate to spank, or set, the form. A shedder pin, backed up by a spring and socket lock screw, prevents adhesion of the part to the knockout face.

15.5.8 Spring Knockouts

Spring knockouts are used in dies too large for the positive types. Stripper bolts limit travel (Figure 15.26). One disadvantage is that the blank is returned into the strip and it must be removed subsequently. In some progressive dies, the blanking station is provided with a spring knockout to return the blank into the strip for further operations.

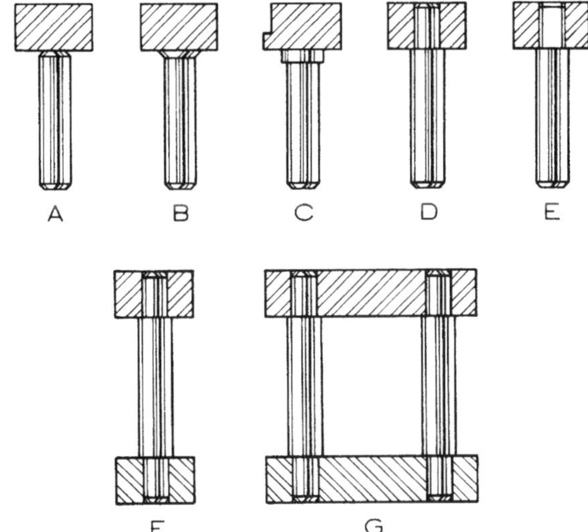

Figure 15.27 Several methods of pinning bottom knockouts.

15.5.9 Bottom Knockout

Bottom knockouts, operated by the press air cushion as in Figures 15.23 and 15.25, have their pins applied in seven ways. In Figure 15.27 at **A**, pins transmit force from the face of the air cushion to the underside of the knockout ring. This method proves suitable for very long press runs. The pins are loose and may become lost if the die is removed often. At **B**, the pins are peened, then machined to form beveled heads to prevent dropping out when the die is removed from the press. At **C**, turned heads retain the pins in the die set. The knockout ring has a flange to retain it in the die. At **D**, the ends of the pins are turned down and pressed into the knockout ring, whereas at **E** the pins are turned down and threaded into the knockout ring.

At **F**, the tops of the pins are turned down and are pressed into the knockout ring. The pins are also turned down at their lower ends and pressed into collars. In the up position, the collar seats against the bottom of the die set to limit travel. It also retains pins and knockout ring to prevent loss. At **G**, long, rectangular knockout plates used in bending and forming dies are retained in the die set by two or more pins with turned-down ends pressed into the knockout bar at the top and also into the bottom retaining bar.

15.5.10 Indirect Knockout

The knockouts described so far have all been the direct-acting types. Indirect knockouts are

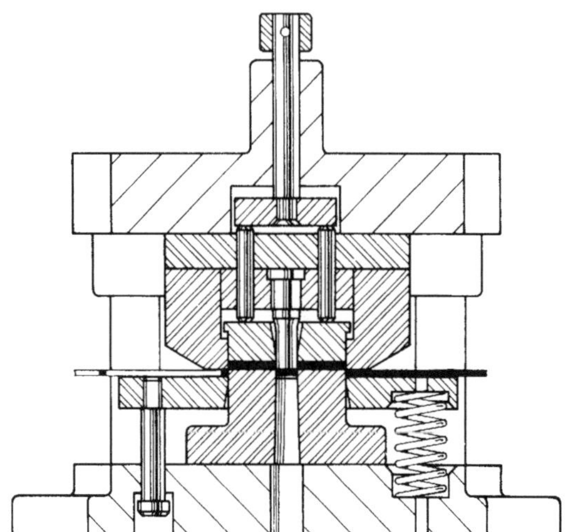

Figure 15.28 Typical indirect knockout.

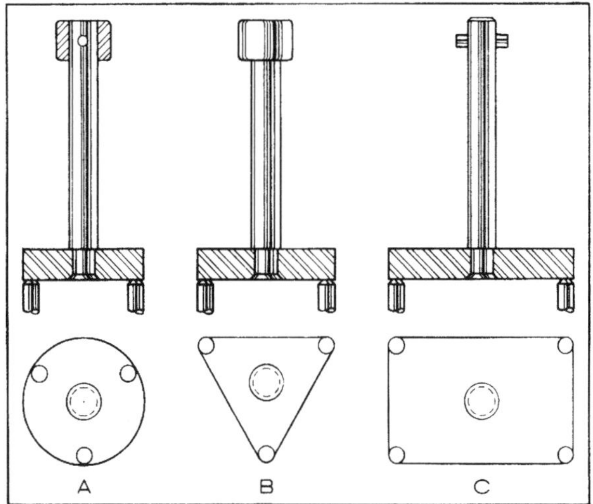

Figure 15.29 Arrangement of pressure pins and stop collars for indirect knockouts.

used when piercing or other types of punches are in line with the knockout rod. Figure 15.28 shows an indirect knockout with force transmitted through pins to clear a central piercing punch. The knockout plate moves in a recess milled in the upper punch holder of the die set. Through the pins, it actuates the knockout block to push the blanks out of the die hole.

15.5.11 Pressure Pins and Stop Collars

The number of pressure pins used and their arrangement depends upon part contour and size. In Figure 15.29 at **A**, knockouts for round blanks have three actuating pins. At **B**, blanks

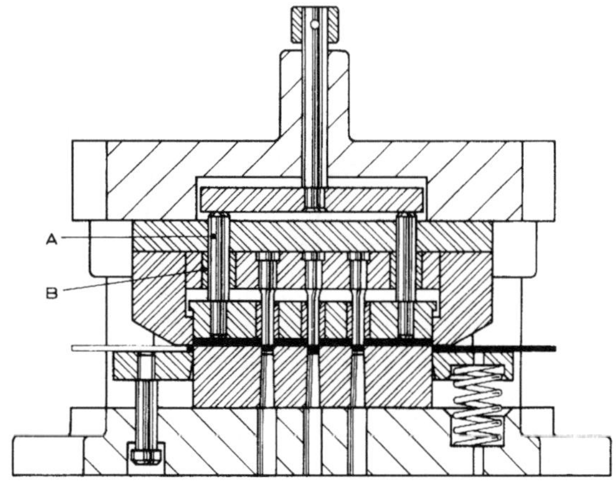

Figure 15.30 Pins **A** guide and transmit force to the knockout block.

with a roughly triangular shape are also provided with three actuating pins. As shown at **C**, blanks that are roughly rectangular or square require the use of four pins.

There are three ways of applying a stop collar to the knockout rod. At **A**, a collar is pressed over the end of the knockout rod and retained by a dowel pressed through both. At **B**, the knockout rod is machined to form a solid shoulder. At **C**, a dowel is pressed through the knockout rod. Although often used, this method is recommended because, should the press be set up improperly, the dowel could be sheared with consequent damage to die components.

15.5.12 Guide Pins

Slender piercing punches may be guided in the knockout block of indirect knockouts (Figure 15.30). Guide pins **A**, pressed into the knockout block, transmit stripping force and also guide the block. They are fitted in hardened bushings **B**. The knockout plate travels in a recess machined in the punch holder of the die set.

15.5.13 Auxiliary Punch Shank

When the press has sufficient opening, an auxiliary punch shank may be fastened to the punch holder of the die set (Figure 15.31). The knock plate travels in a recess machined in the bottom of this auxiliary shank.

15.5.14 Backing Blocks

In another method, the soft steel backing block **A** can be machined to accommodate the

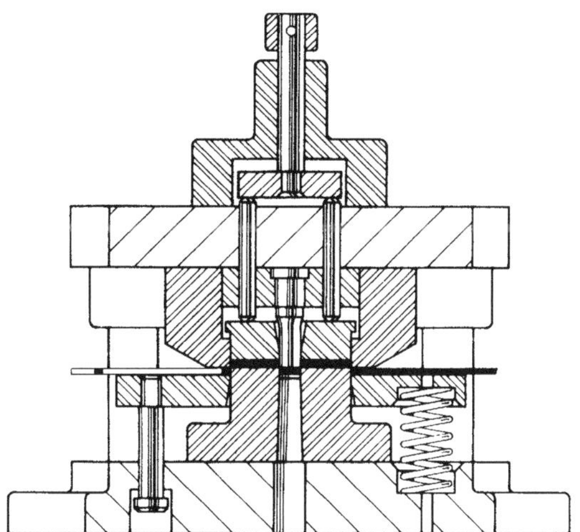

Figure 15.31 Die in which auxiliary punch shank is fastened to the punch holder of the die set.

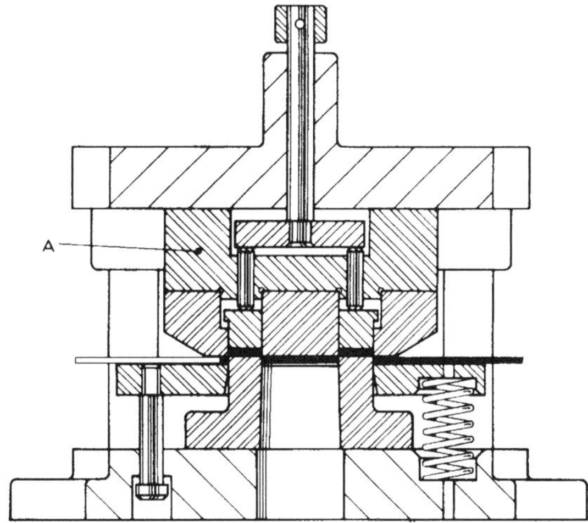

Figure 15.32 Soft steel backing plate **A** machined to accommodate the knockout plate.

knockout plate (Figure 15.32). This die produces ring-shaped parts.

15.5.15 Spider Knockout

When holes are to be pierced in ring-shaped parts, the knockout plate has a spider shape and the punches are retained between the arms of the spider (Figure 15.33). In the drawing, one arm of the spider is drawn at one side and a piercing punch at the other.

15.5.16 Spider Knockout Plates

The number of arms contained in knockout spiders such as the one used for the die in

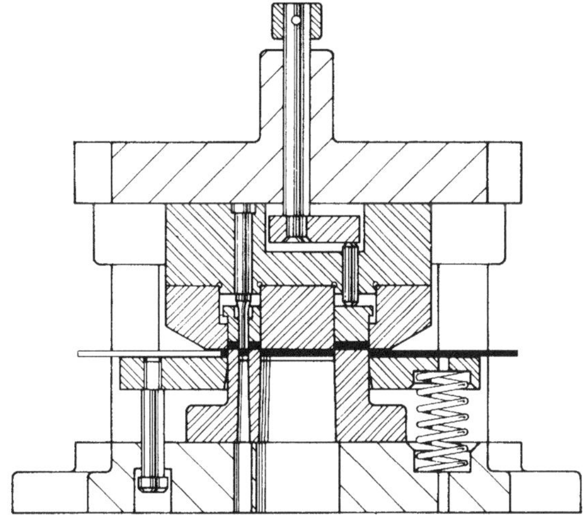

Figure 15.33 The knockout plate is of spider shape to facilitate piercing ring-shaped parts.

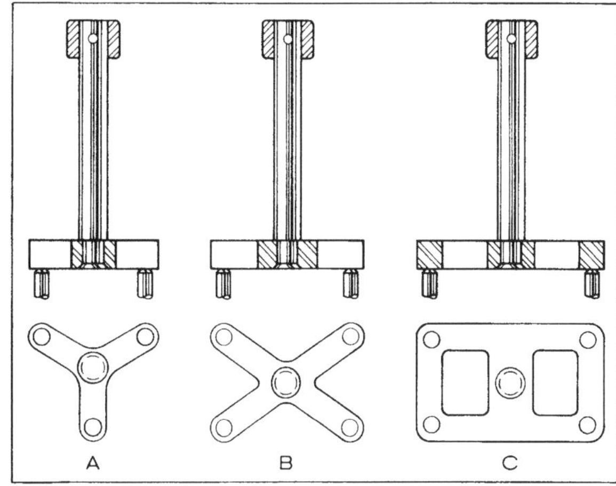

Figure 15.34 Various shapes of spider knockout plates.

Figure 15.34 depends on the number of piercing punches required. When three punches are applied, the spider has three arms to clear the punches. This type is shown at **A**. For four punches, a four-armed spider, at **B**, would be used. Piercing punches, applied at other than radial locations, are cleared by holes in knockout frames, such as the one shown at **C**. The shape of the knockout frame would depend upon the number and arrangement of the punches.

15.5.17 Laser-Cutting Knockouts

One method frequently used to save material is to laser cut (or cut by some other cutting method) the spider knockout entirely through the backing

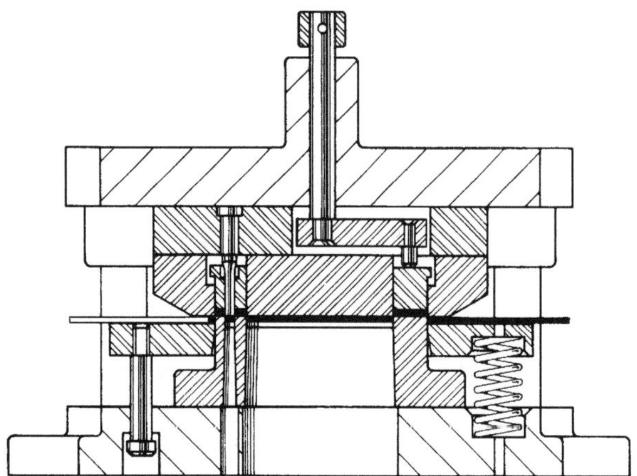

Figure 15.35 Spider knockout plate in this die has been laser-cut.

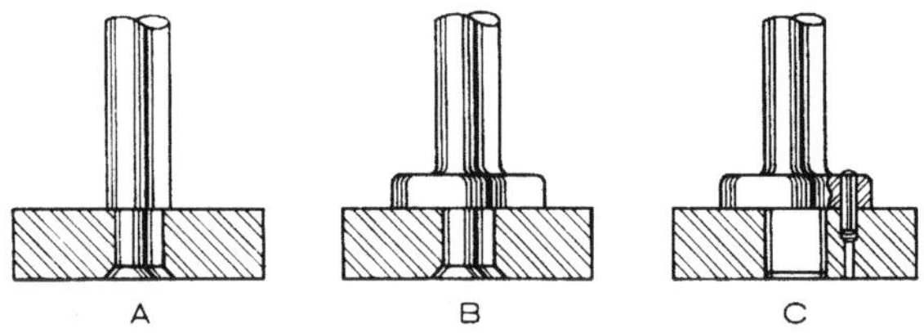

Figure 15.36 Three methods of securing the knockout rod to the knockout plate.

plate. The spider is removed from the plate and then machined thinner to allow up-and-down movement (Figure 15.35). Because the knockout pocket is cut completely through the plate, the pins must be securely fastened to the spider arms, in this case by pressing and riveting their ends.

Still another method of applying large spider knockouts is to laser cut (or cut by another method) the spider through the punch holder of the die set. A steel die set must be specified, and the cutting must be done by the die set manufacturer before stress relieving and final accurate machining.

15.5.18 Knockout Rods

These are the three commonly used methods of securing the knockout rod to the knockout plate (Figure 15.36). At **A**, the rod end is turned down, pressed into the plate, and riveted over. When heavy knockout pressures exist, the knockout rod is provided with a shoulder to prevent it from sinking into the knockout plate, as shown at **B**. In this method as well, the end is turned, pressed in, and peened over as in the previous

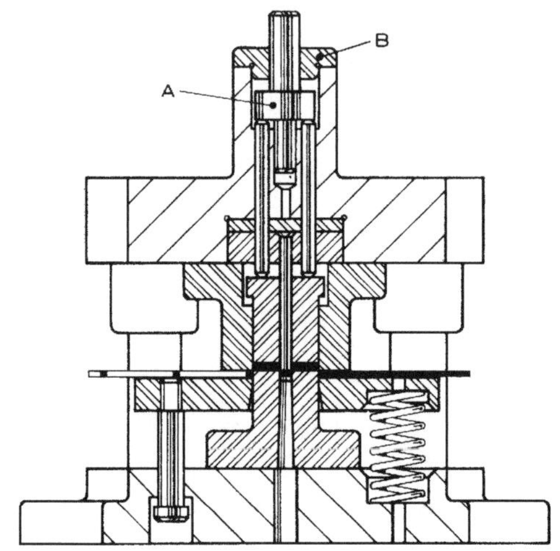

Figure 15.37 Knockout plunger **A**, contained within the punch shank, operates indirect knockout.

example. View **C** shows still another method of fastening heavy knockout rods. The end is provided with a shoulder, threaded into the knockout plate, and prevented from loosening by a dowel pressed through the flange and into the knockout plate. The dowel hole is carried through the plate, but with a smaller diameter to prevent the dowel from working out in operation.

15.5.19 Knockout Plungers

Indirect knockouts for small dies (Figure 15.37) may be operated by a knockout plunger **A**, applied within the punch shank. Cap **B**, fastened to the punch shank with socket cap screws, provides an outboard bearing for the knockout plunger and also keeps it confined within the die.

15.5.20 Spring Knockouts

Small knockouts may be of the spring-operated variety (Figure 15.38). One disadvantage is that the part is returned into the strip and it must be removed as an extra operation. However, the blanking station of some progressive dies is purposely made in the above manner; the part is returned to the strip for other operations, and removed from the strip at the last station.

15.5.21 Spring-Actuated Knockouts

Spring-actuated knockouts may be applied to the lower die in a variety of ways (Figure 15.39). These are used in progressive dies at bending, forming, and embossing stations. At **A**, the

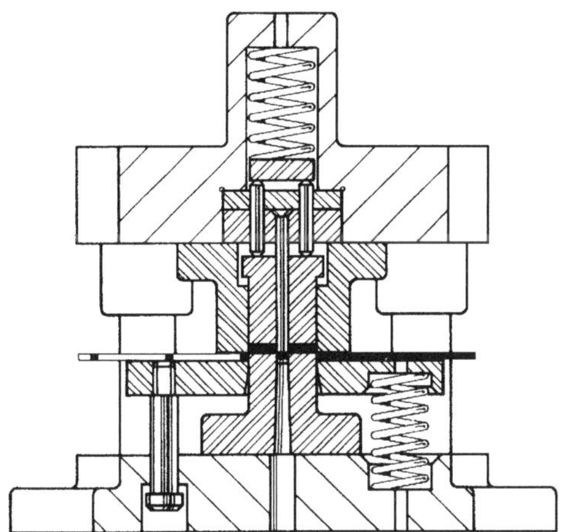

Figure 15.38 Typical small knockout of the spring-operated variety.

knockout is operated by a spring applied in a counterbored hole in the die holder of the die set. As shown at **B**, the hole is carried through the die set and the spring is backed by a plate fastened in the die holder, thereby providing greater travel. At **C**, knockout blocks of irregular shape may be held by one or more stripper bolts, and pressure is applied by a spring retained in a counterbored hole in the die holder. **D** shows one method of applying the spring within a spring housing fastened under the die set. This construction is used when considerable travel is required. At **E**, the spring is held under the die set by a stripper bolt, and pressure is transmitted to the knockout through pins. This design facilitates the use of a very heavy spring when high pressure is required and when the press is not equipped with an air cushion.

15.6 CENTER OF STRIPPING FORCE

Before designing the knockout, the center of the stripping force must be found. If the knockout rod is not applied at the exact center of the stripping force, the rod will be stressed and possibly bent in operation. The stripping pressure should be on a vertical line passing through the specific point that defines the resultant force of the stripping forces. Three methods can be used to determine the center of the stripping force: mathematical, graphical, and experimental. The first

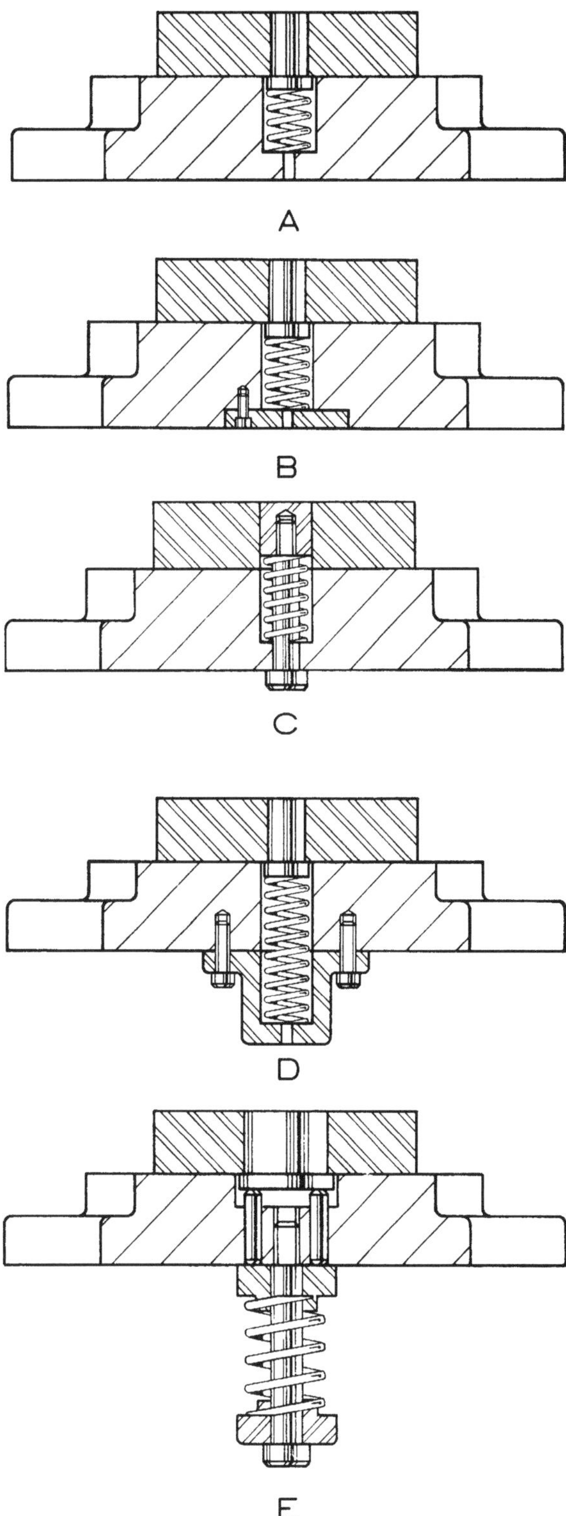

Figure 15.39 Various ways of applying spring-actuated knockouts to the lower die.

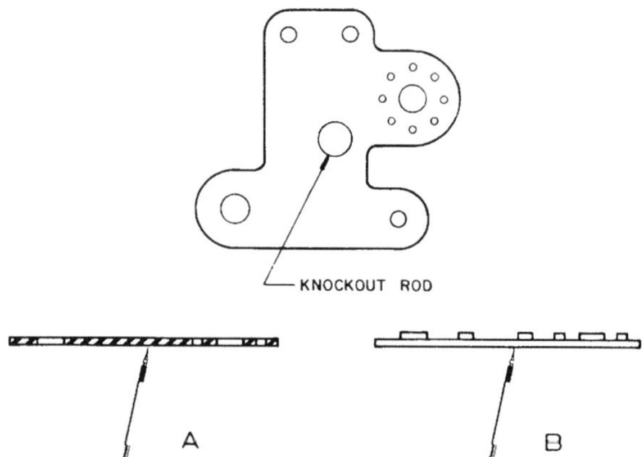

Figure 15.40 Ways of determining the center of the stripping force.

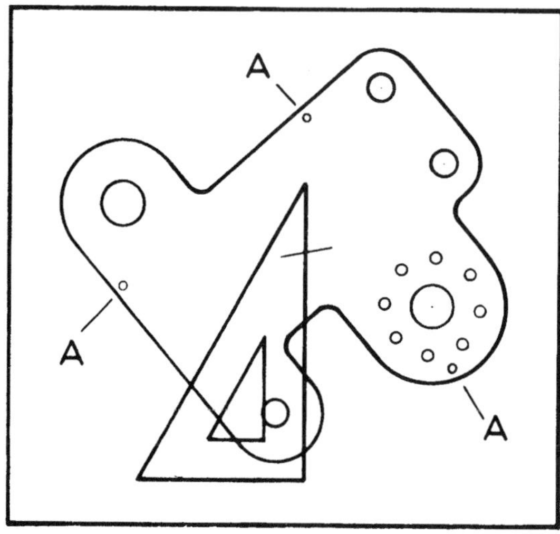

Figure 15.41 Alternate method of determining the center of the stripping force.

two methods are known from statics and they are not explained in this chapter.

However, an experimental method will be described. Symmetrical blanks present no difficulty, but when blank outlines are *irregular,* this method is the one to use (Figure 15.40):

Place a sheet of cardboard (board for show cards is best) under the part drawing. Insert carbon paper between the drawing and cardboard, then trace the part outline, as well as all holes and openings. With scissors cut out a cardboard blank, as shown in the plan view. All holes and openings may be cut out of the blank as shown at **A**. The blank is then balanced on a fulcrum, (a pencil will do). The center of balance is the center of the stripping force.

Cutting out small holes is difficult; here is an alternate method: From the same weight cardboard, cut slugs the shape and size of all holes and openings. Paste these in position on top of the blank, as shown at **B**. Balancing the blank as before will establish the center of the stripping force.

For very large blanks (Figure 15.41), which would be difficult to balance, pierce three small holes **A** close to the edge of the blank. Suspend the blank from a vertical surface, using a pin or thumbtack. With a triangle, draw a vertical line directly under the hole from which the blank is suspended. Turn the blank around and, suspending it from one of the other holes, draw another vertical line crossing the suspension hole. The intersection of the two lines is the center of the stripping force. Suspend the blank from the third hole to provide a check for accuracy.

HOW TO APPLY FASTENERS

16.1 INTRODUCTION

The subject of fasteners is an important one because these components are applied so frequently and employed in such large quantities. Although small, they perform important functions. In the design of tools and dies, fasteners are often the weakest link in the tool. If they are not selected and applied correctly, they can become the cause of failure of the entire tool or die.

In applying fasteners, an experienced designer seems to follow no rules for size, location, or number. But if two experienced designers were to select and apply fasteners for the same job, their choices would be almost identical. In turn, the checker would approve their selections.

Obviously, these professionals have developed a sense of proportion. When designers look over a layout and decide that the screws are too small, too close together, or too close to edges, they are subconsciously being guided by observations of tool components that failed and had the same proportions.

Fortunately, definite rules for application of fasteners can be used by the inexperienced designer or student to avoid many pitfalls. These pitfalls include hardening cracks and breaks, stripped threads, distortion by release of internal stresses, and misalignment of holes. By the use of these rules, the less experienced designer can avoid trouble and quickly develop a correct sense of proportion.

Figure 16.1 shows an exploded view of a typical die for producing blanks from metal strip. All fasteners have been shown removed from the components that they locate and hold. From this

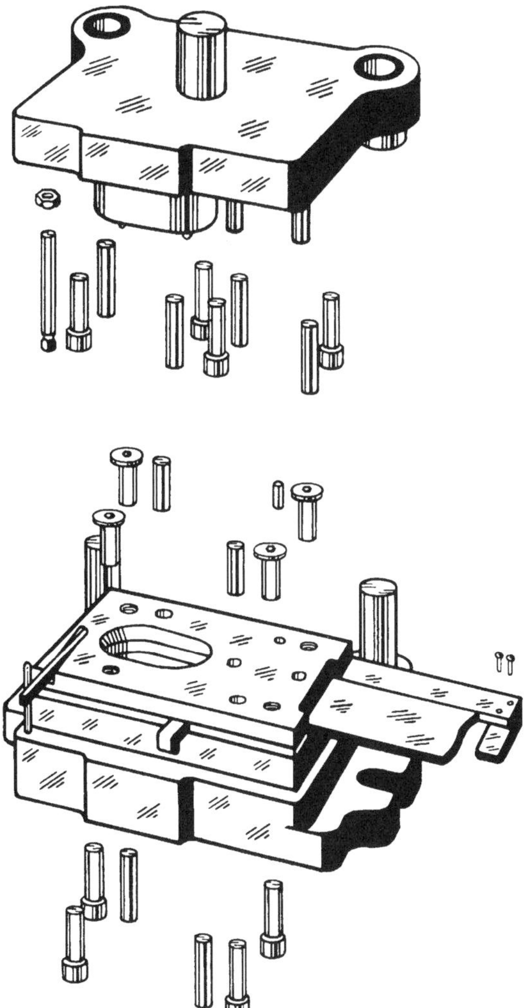

Figure 16.1 Exploded view of die showing number of fasteners used.

drawing, it is apparent that fasteners, although small individually, form a substantial portion of the entire tool when taken together.

16.2 TYPES OF FASTENERS AND METHODS OF FASTENING

Figure 16.2 illustrates the types of fasteners most commonly used in die construction. They are:

1. Socket cap screws
2. Dowels
3. Socket button-head screws
4. Socket flat-head screws
5. Stripper bolts
6. Socket set screws
7. Allen nuts.

Less frequently employed types include the following: hexagon nuts, washers, studs, socket lock screws, socket shoulder screws, and wood screws.

16.2.1 Threaded Fasteners

When components of any mechanical device must be assembled securely and then dismantled occasionally for repair, adjustment, or replacement, threaded fasteners are most effective. Socket cap screws (Figure 16.3) are most often used in tool work, followed by dowels.

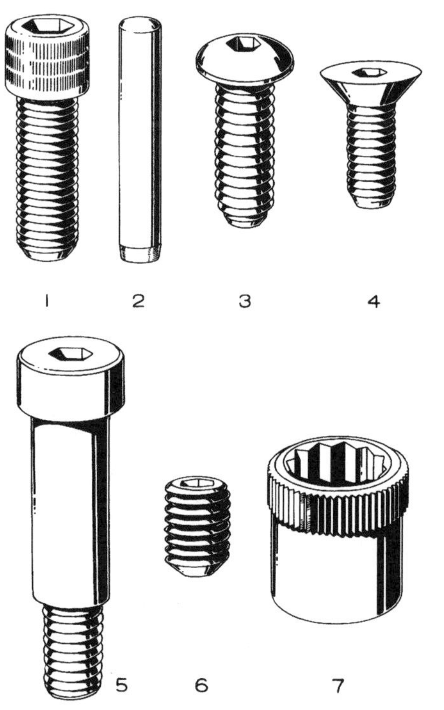

Figure 16.2 Types of fasteners used in die construction.

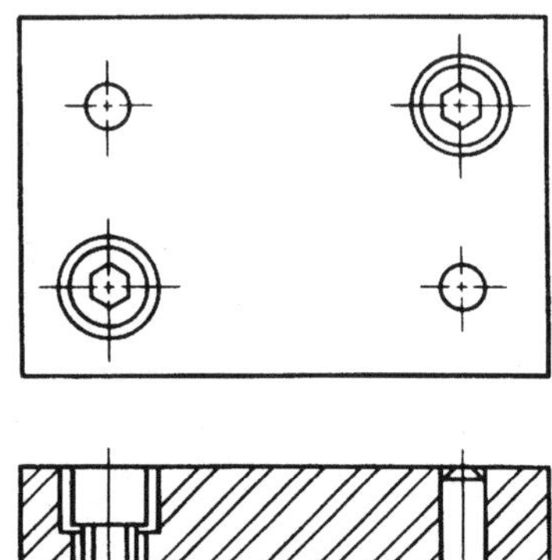

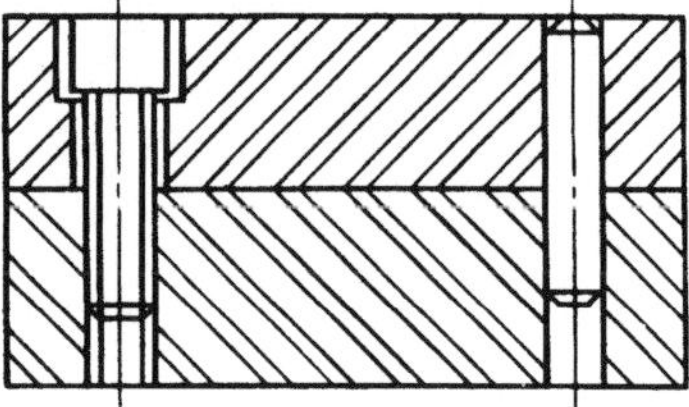

Figure 16.3 Socket cap screws and dowels are used in holding these two blocks together.

Fundamental rules govern the application of screws and dowels. Screws are used to hold components together. They are not intended to locate components sideways relative to each other. Study of the illustration will show why this is so. Two blocks are fastened together with two screws and two dowels. Clearance holes for screws are drilled 1/64 inch (0.4 mm) larger in diameter than the body diameter of the screws; counterbored holes for the screw heads are 1/32 inch (0.8 mm) larger in diameter than the heads. This clearance has been shown slightly exaggerated at the left in the section view.

Such clearances will allow side shift if the screws are loosened. This shift is exactly what can happen when a tool is taken apart for sharpening or repair. A small amount of side movement of this nature may be permissible in many mechanical assemblies, but in precision jigs, fixtures, and dies, it could cause serious damage. For this reason, closely fitted pins called "dowels" are applied to effect accurate relative positioning. Dowels permit no movement between parts. They can be pressed out for repair, then replaced to restore relative positioning to original accuracy.

16.2.2 Applying Fasteners

The first method considered is shown in Figure 16.4 at view **A**. Screws are applied at two diagonal corners and dowels at the other two corners. This method is employed for small and medium-size blocks and plates designed to withstand only small and medium forces. For larger blocks and where larger forces are encountered, the method shown at view **B** should be selected. Screws are applied at all four corners and dowels are offset. When still larger forces are present, a fifth screw is applied to the center of the block, as shown at view **C**.

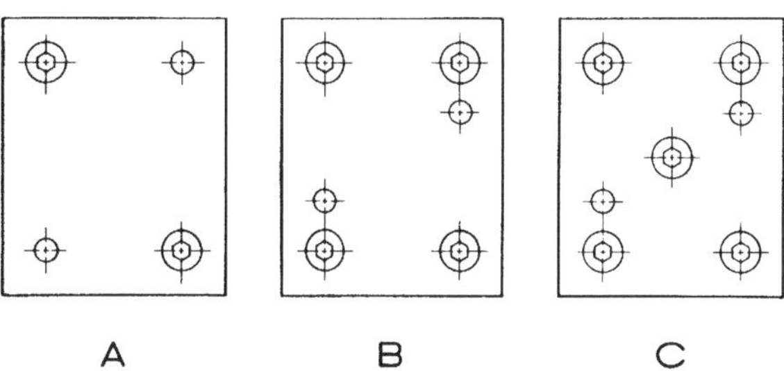

Figure 16.4 The number of fasteners depends on the magnitude of the anticipated force.

In deciding upon the number of screws required, the size of the details, in addition to any forces that will act upon the screws, should be considered. The component being fastened must be pulled down evenly and squarely. Observe these rules:

1. Keep dowels as far apart as possible.
2. Make sure that the screws hold down the part securely flat.

A method of applying fasteners in a die segment is shown in Figure 16.5. In this segment of a sectional die, the construction in view **A** is incorrect because screws should act somewhere along line **C-C** for stability. View **B** shows the correct application, even though the dowels are somewhat closer together.

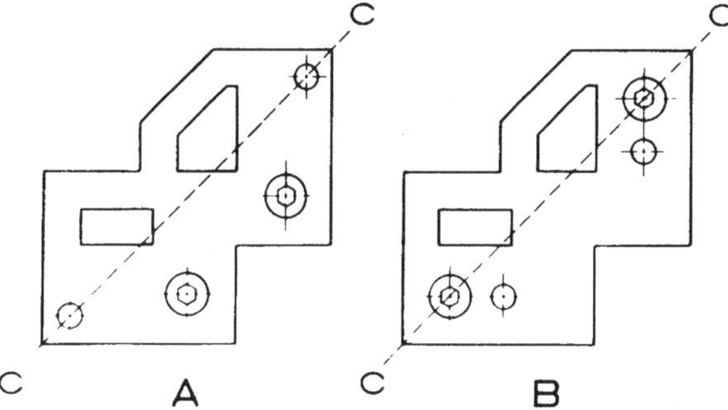

Figure 16.5 Incorrect (view **A**) and correct (view **B**) methods of applying fasteners in a die segment.

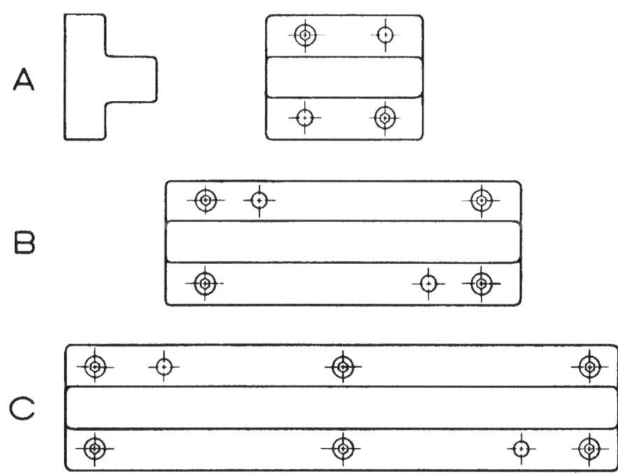

Figure 16.6 Method of fastening flanged punches.

16.2.3 Fastening Flanged Punches

When parts have a projection along the center, as in flanged forming punches (Figure 16.6), two screws and two dowels are applied at opposite corners when punches are short, as in view **A**. For longer punches, four screws are applied at the corners and dowels are offset, as shown in view **B**. Long punches may be fastened with six screws, applied three to each side, as shown in view **C**. Dowels facilitate lateral location and are applied in the same manner as for the punch at **B**.

16.2.4 Spacing of Fasteners

Proper spacing, whether between holes or between holes and edges of parts, is particularly important for tool steel parts to be hardened. If too little space is allowed, there is a strong possibility of the block cracking in the hardening process. But on the other hand, it is often desirable to have screw holes as close to edges as possible, in order to keep dowels far enough apart for accurate positioning.

In Figure 16.7 at view **A**, screws have considerable space between them and outside edges. Also, dowels have adequate space between them and their nearest screw holes. Unfortunately, the dowels are too close together for effective positioning of the component. View **B** shows screws and dowels positioned properly. Space between dowels has increased considerably. We need to know the safe minimum distances **C** and **D**.

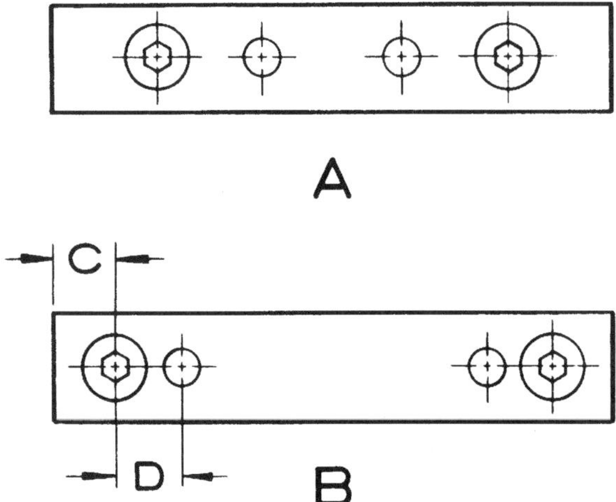

Figure 16.7 Ineffective (**A**) and effective (**B**) spacing of fasteners.

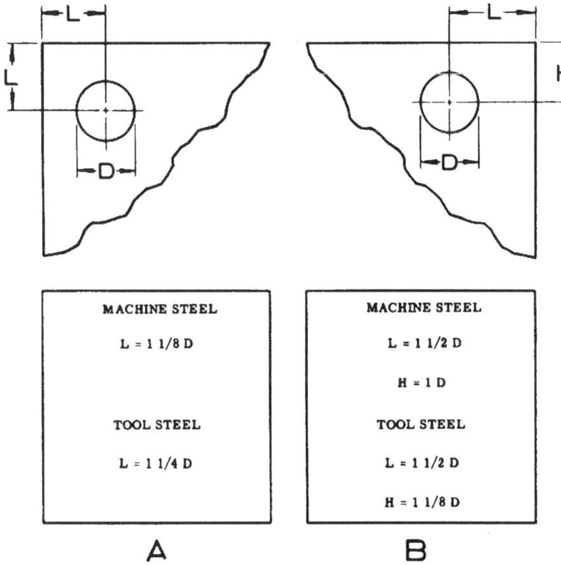

Figure 16.8 Guide that gives minimum proportions for the spacing of holes from corners in machine and tool steel parts.

Figure 16.8 is an illustrated guide that gives correct minimum proportions for applying holes in machine steel and tool steel parts. You will find it invaluable for checking to determine if correct spacing has been applied in your designs. In time, you will come to recognize good proportions instantly. View **A** specifies correct minimum spacing of holes applied at corners. Note that holes may be positioned closer to edges of machine steel parts than they are for parts made of tool steel.

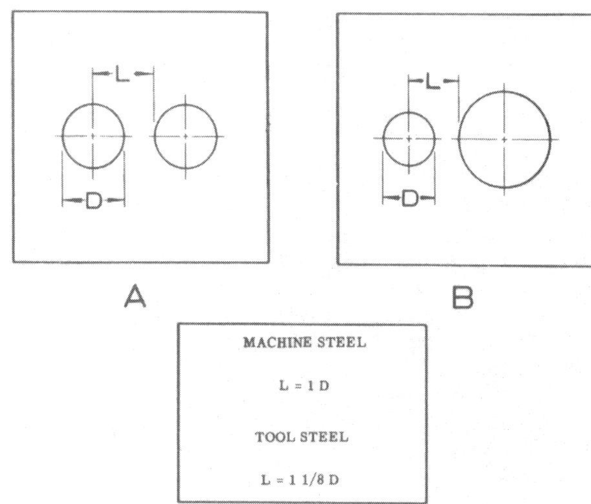

Figure 16.9 Minimum proportions for positioning adjacent holes in machine and tool steel parts.

The extra material applied for tool steel components assures that corners will not crack in the hardening process if it is properly done.

When a hole in tool steel or machine steel is located a greater distance from one edge than in the previous example, it may be positioned closer to the adjacent edge, as shown at view **B**. This condition occurs frequently in the application of screw holes in die blocks and in other tool steel parts.

16.2.5 Distance Between Holes

Minimum proportions for positioning adjacent holes having the same size are given in Figure 16.9 at view **A**. Note that diameter **D** on view **B** applies to the small hole, and that distance **L** is taken from the center of the small hole to the edge of the large hole. This is the common application of drawing holes for a socket cap screw and its adjacent dowel. Under normal conditions, measure distance **L** from the center of the dowel to the edge of the cap screw head. For crowded conditions, measure to the edge of the clearance hole for the screw body.

a) Long, Narrow Parts

Screws and dowels are applied to long, narrow parts in several ways (Figure 16.10). View **A** shows the most frequently used application. Distances **E**, **F**, and **G** are given in Figures 16.8 and 16.9. Components to be taken apart for repairs are foolproofed by offsetting one of the dowels, as in view **B**.

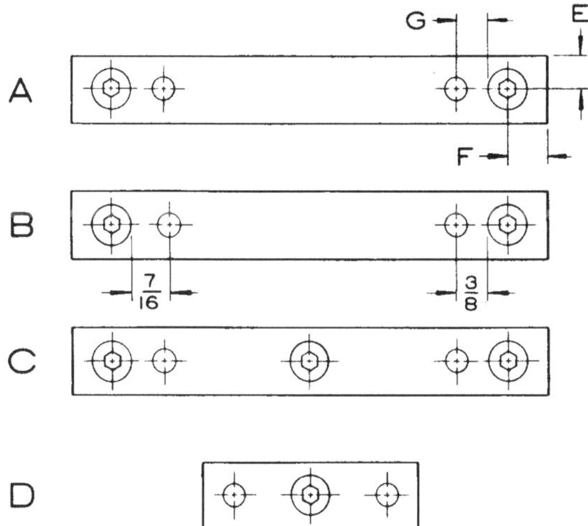

Figure 16.10 Ways of applying screws and dowels in long narrow parts.

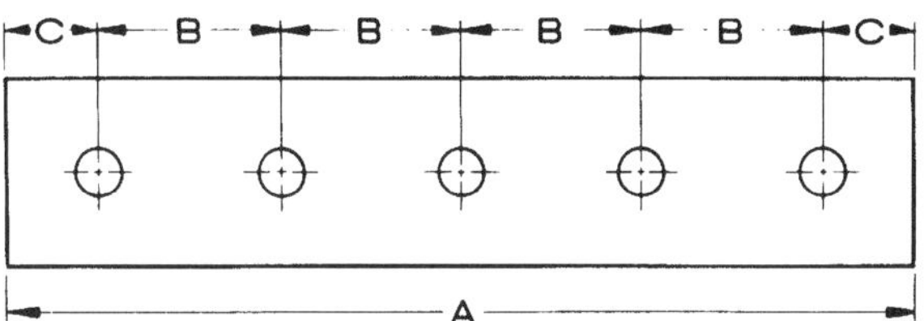

Figure 16.11 Applying a row of holes in a bar.

When greater strength is required, a third screw is applied in the center, as in view **C**. However, this addition does not affect positions of dowels. Short, narrow plates may be fastened with a single screw applied at the center. Dowels at each side provide accurate lateral location. This application is shown in view **D**.

b) Proportions

In applying a row of holes in a bar (Figure 16.11), employ the following steps:

1. Measure length **A**.
2. Decide upon the number of holes required.
3. Divide length **A** by the number of holes.
4. The answer is the distance between holes.

One-half of the amount found in step 4 is the distance between end holes and part edges. Thus, hole locations are properly proportioned. Suppose that 5 holes are required and length **A** is 20 inches (508 mm). Then, 20 divided by 5 is 4. The distance between holes is, therefore, 4 inches (101.6 mm).

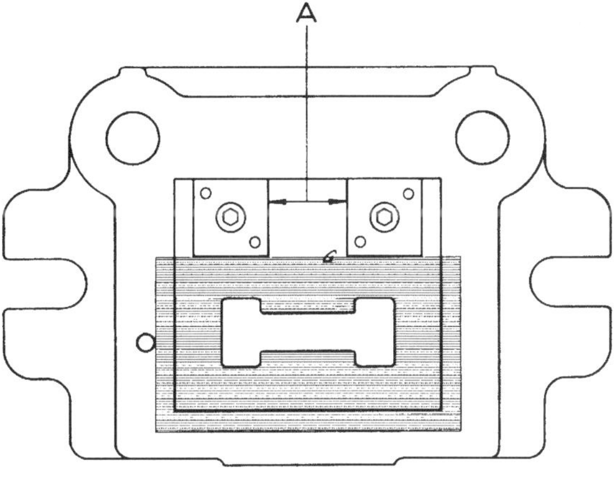

Figure 16.12 Simple method of foolproofing back gages **A** to assure correct positioning upon reassembly.

One-half of 4 is 2. The distance between end holes and part edges is, therefore, 2 inches (50.8 mm).

c) Foolproofing

A simple way to foolproof parts that might be reassembled incorrectly in die repair is to reverse the relative positions of dowels (Figure 16.12). This method is applied to gages **A** in this piercing die. Then, gage positions cannot be reversed accidentally, nor can the gages be assembled upside down because of the holes counterbored for the heads of the screws.

16.3 SCREWS

Figure 16.13 shows screws on a drawing. This is the way screws appear in the plan and front section views of the representative die. Observe that in the flange of the punch, a dowel is drawn at one side and a socket cap screw at the other. The die block is fastened to the die set with long socket cap screws applied from the bottom. The stripper plate, back gage, and front spacer are fastened to the die block with socket button-head screws applied from the top. Double dowels illustrated at the other side of the view provide accurate lateral positioning of components.

Observe that the stripping load against the screws that hold the stripper to the die block is equal to the stripping load against the screws that hold the blanking punch. When only two screws and two dowels are employed to hold and locate the blanking punch, make the screws

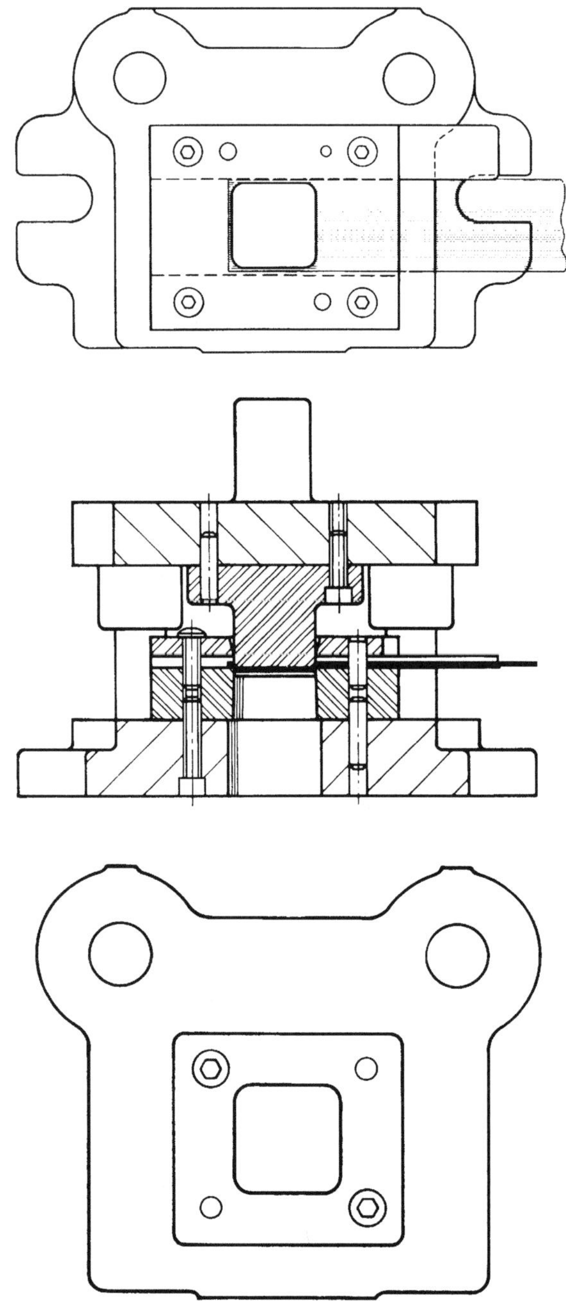

Figure 16.13 Method of application and appearance of fasteners in a typical die drawing.

larger in diameter than those in the die block, where a minimum of four screws is always employed. Compare holding strengths as given in a manufacturer's catalog. In general, screws one size larger will have to be employed to realize an approximately equal holding strength.

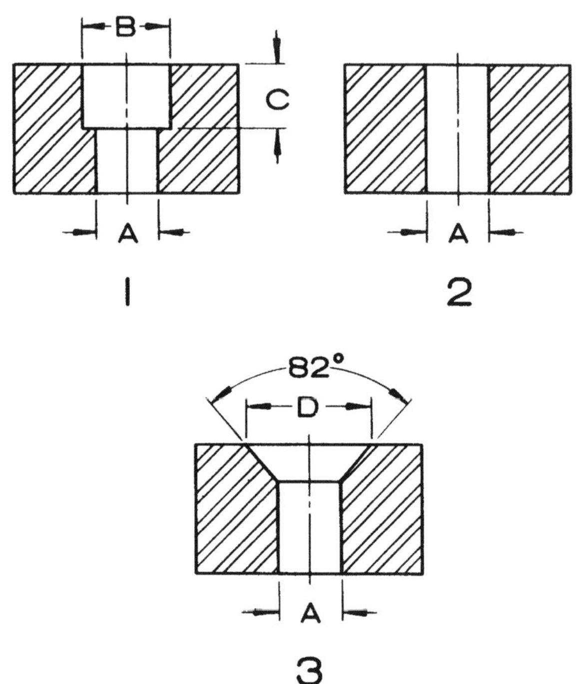

Figure 16.14 Sectional views of holes for various socket screws.

a) Holes for Screws

In specifying holes for socket screws (Figure 16.14), observe the following rules:

1. Holes **A**, which engage the screw bodies, are specified 1/64 inch (0.4 mm) larger than body diameter.
2. Counterbored holes **B** are specified 1/32 inch (0.8 mm) larger than the diameter of the screw head.
3. Counterbore depth **C** is the same as the height of the head of the screw.
4. Countersink diameter **D** is made the same as the head diameter of the flat-head screw.

In the illustration, view **1** shows the section through a hole for a socket cap screw. View **2** shows the hole for a socket button-head screw and view **3** the hole for a socket flat-head screw.

16.3.1 Socket Cap Screws

Socket cap screws are the most frequently used fasteners for tools and dies; they should be specified whenever it is possible to employ them. A socket cap screw is provided with a head within which has been broached a hexagonal socket for driving. Threads of all socket screws in the size range normally used are held to a Class 3 fit, a very accurate fit having precise tolerances. After machining they are heat-treated; a hardening and tempering process is applied to produce a correct combination of toughness without brittleness. The final stage in their manufacture consists of careful inspection for uniformity and accuracy.

Some differences may be noted in socket cap screws obtained from different manufacturers. Some have plain heads; others are knurled for ease in starting screws by hand; some stamp the size on the head; others are provided with two sizes of heads, conventional, and oversize for severe applications. Socket screws are also available in stainless steel for use under corrosive conditions.

Nylock screws are socket-provided screws with a projecting nylon pellet inserted permanently in the body. When the threads are engaged, the nylon is compressed and displaced, providing a locking thrust to prevent loosening under vibration.

a) Thread Engagement

It is important that thread engagement for socket screws be applied correctly. If too little is specified, it is possible to strip the threads in the tapped hole. On the other hand, excessive thread engagement should be avoided because no greater strength is provided and it is difficult to tap deep holes. Also, the deeper a hole is tapped, the greater the possibility is of tap breakage.

The table in Figure 16.15 specifies the minimum amount of thread engagement that should be applied for various materials. The values recommended for steel and cast iron should be memorized because they are employed so frequently.

Rules for thread length. Simple formulas establish thread lengths. Let **L** equal screw length, **I** equal thread length, and **D** equal screw diameter.

- For Coarse-Thread Series (UNC) threads, **I** equals either **2D** plus 1/2 inch (12.7 mm) or 1/2 **L**, whichever is greater.
- For Fine-Thread Series (UNF) threads, **I** equals either 1 1/2 **D** plus 1/2 inch (12.7 mm) or 3/8 **L**, whichever is greater.

b) Proportions of Socket Cap Screws

Socket cap screws (Figure 16.16) are drawn to the following proportions:

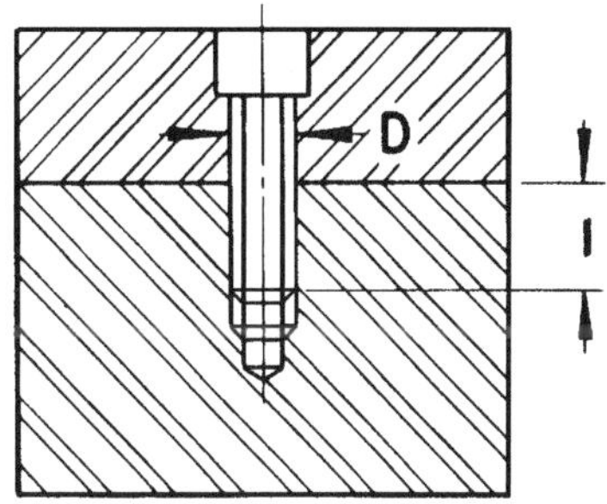

Steel	I = 1 1/2 D
Cast Iron	I = 2 D
Magnesium	I = 2 1/4 D
Aluminum	I = 2 1/2 D
Fiber & Plastics	I = 3 D & UP

Figure 16.15 Table for determining minimum thread engagement for various materials.

A = length of the screw
B = thread length
C = head height
D = the body diameter (note also that that **C** = **D**)
E = root diameter of the thread
F = thread chamfer, drawn 1/32 to 3/32 inch (0.8 to 2.4 mm) wide, depending on screw size
G = angle of chamfer, 45 degrees

When the block to be fastened is made of hardened tool steel, distance **H** should never be less than 1 1/2 **D**. For machine steel, it need only be thick enough for adequate strength. Length of thread engagement **I** is 1 1/2 **D** for steel, with a minimum of 1 **D** under certain conditions, and 2 **D** for cast iron and some nonferrous metals.

Depth of the tapped hole **J** is applied in increments of at least 1/16 inch (1.6 mm), never in increments of 1/32 or 1/64 inch (0.8 or 0.4 mm). Distance **K** between the end of the screw and the bottom of the thread, and between the bottom of the thread and the bottom of the tap drill hole, is normally made 1/8 inch (3.2 mm). Bottoming taps have from 1 to 1 1/2 imperfect threads; this fact must be taken into consideration when drawing blind holes. The thread must extend at

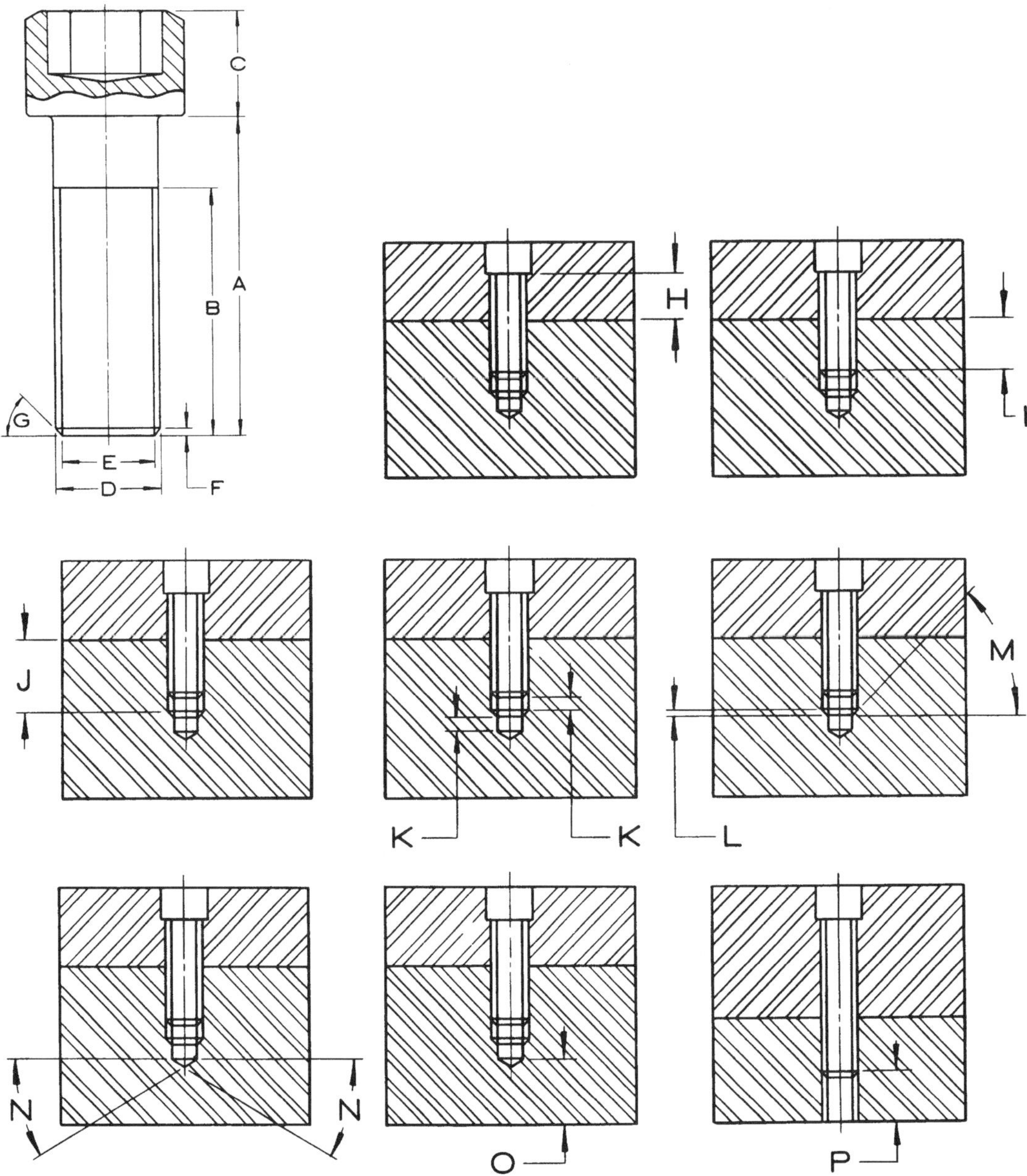

Figure 16.16 Proportions of socket cap screws.

least 1 1/2 times the pitch of the thread past the end of the screw. Length **L** of the thread chamfer is made 1/16 inch (1.6 mm). Chamfer **M** is 45 degrees. At the bottom of the tap drill hole, two lines are drawn at angles of 30 degrees, dimension **N**. They represent the conical depression produced by the end of the tap drill.

For hardened tool steel parts, distance **O** between the end of the tap drill hole and the lower surface of the block must be greater than 1 **D** (one diameter of the screw body). When this distance is less than 1 **D**, show the tap drill running completely through the part. If this is not done, it is possible for the thin, circular section of steel to crack in hardening and fall out as a rough-edged disk.

The hole that a screw engages is tapped all the way through when distance **P** from the end of the screw to the lower surface of the block is 1 **D** (one diameter of the screw), or less.

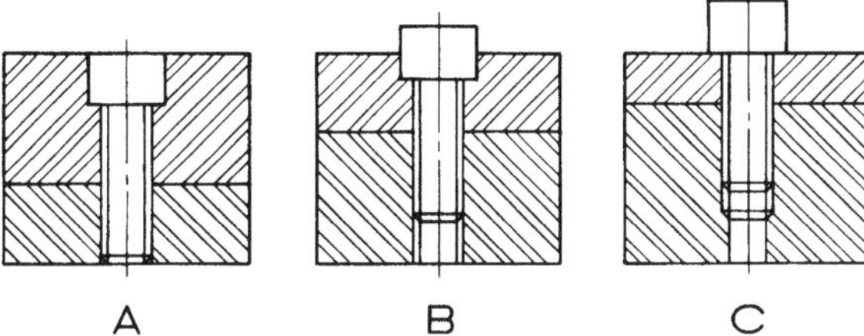

Figure 16.17 Three methods of applying socket cap screws.

c) Applying Socket Cap Screws

There are three methods of applying socket cap screws (Figure 16.17):

- The most commonly used method is to counterbore the hole for the screw, as at view **A**. The screw head engages this counterbored hole and its top comes flush with the surface after assembly.
- In a variation of the foregoing, the hole is not counterbored to full depth, and the head protrudes a certain amount, as at view **B**. This method is used for fastening thinner plates.
- In the third method, the hole is simply drilled and the screw head is left to protrude, as at view **C**. This method would be used for thin parts and in covered places.

16.3.2 Socket Flat-Head Screws

Heads of these screws (Figure 16.18) are machined to an 82-degree included angle, dimension **A**. They are employed for fastening thin plates, which must present a flat, unbroken surface. Avoid their use, when possible, because they cannot be tightened as securely as cap screws. Employ them only for flush applications when plates to be fastened are too thin for counterboring. For tool work, socket flat-head screws are always specified in preference to the slotted type.

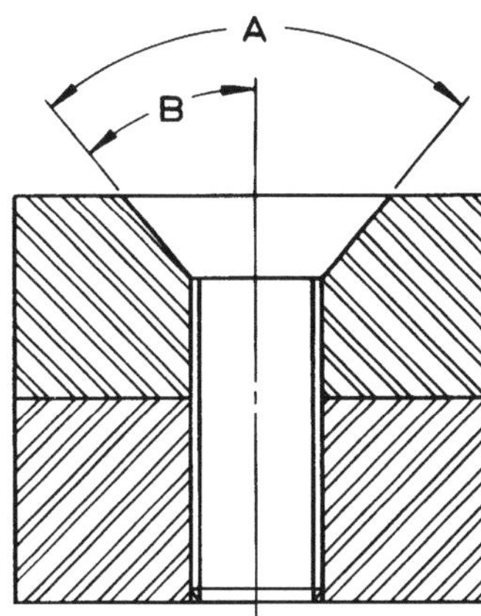

Figure 16.18 Application of a socket flat-head screw.

To draw the side view, first look up the head diameter in a handbook or manufacture's catalog and mark ends of head diameter to the specified dimension. Then the tapered lines of the head are drawn with the protractor set to an angle of 41 degrees from the screw axis, dimension **B**.

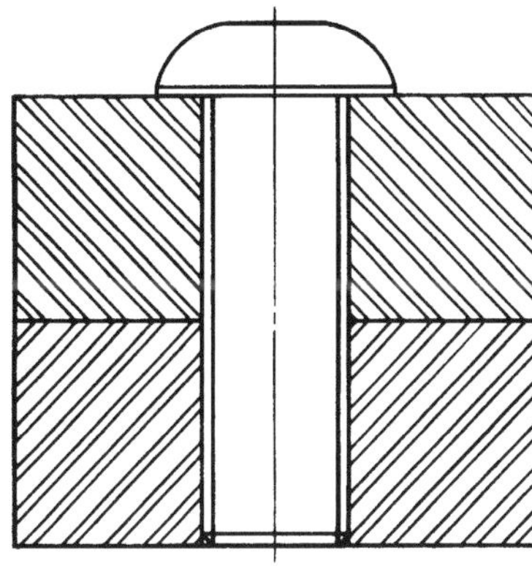

Figure 16.19 Application of a socket button head screw.

16.3.3 Socket Button-Head Screws

These screws (Figure 16.19) are employed for fastening plates in applications where a raised head is not objectionable. They should be used when plates are too thin for counterboring for cap screws and surfaces need not be flush or flat. In the design of small- and medium-sized dies, they fasten the stripper plate, back gage, and front spacer to the die block. Button-head screws project only slightly above the surface. Their rounded contours prevent injuries to operators' hands.

16.3.4 Fastening Plates

When components such as plates are fastened to both sides of a block, the block should be tapped all the way through (Figure 16.20). Screws engage the block from both sides to hold the plates, as shown in view **A**. When the center block is not too thick, it is permissible to hold three parts by threading one and fastening the three with relatively long screws, as in view **B**. Sometimes it is desirable to fasten two details together through a third component. The screw

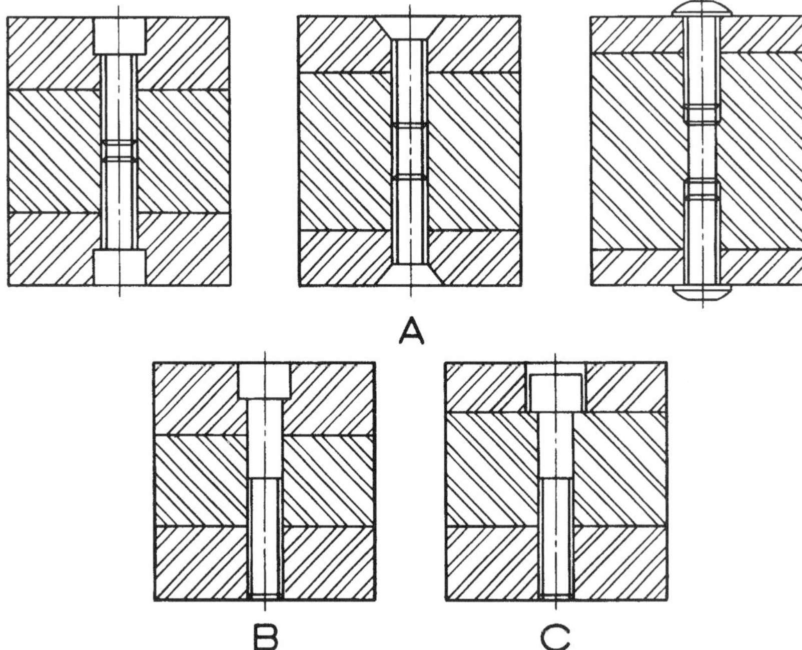

Figure 16.20 Different methods of fastening plates.

head is reached for fastening through a hole, which clears it, as in view **C**.

16.3.5 Set and Lock Screws

a) Socket Set Screws

When first developed, socket set screws were called "safety screws." They were employed to lock rotating machine parts to their shafts. They replaced screws with protruding heads, which were dangerous because clothing could become caught by the whirling projection. A socket set screw (Figure 16.21) is threaded its entire length and is provided with an internal driving socket at one end.

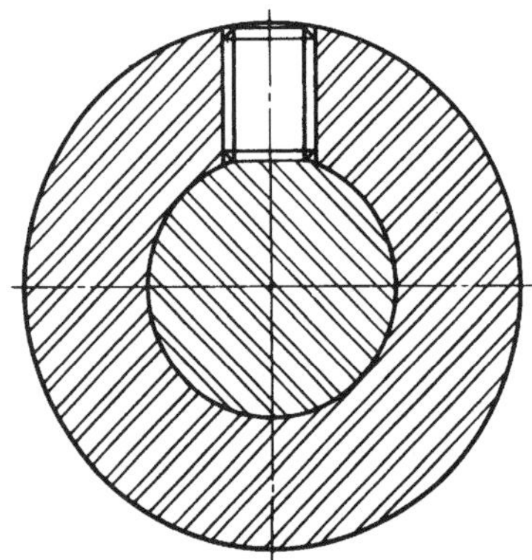

Figure 16.21 Application of a socket set screw.

Point styles for set screws. There are five styles of set screw points (Figure 16.22). Designers must be able to specify the correct style of point for the specific application.

1. The flat point is the most commonly used type for jigs, fixtures, and dies. Flat-point set screws are employed as adjustable stops and clamp screws, as well as for locking hardened shafts.
2. Cup-point set screws are used to lock pulleys, collars, gears, and other parts on soft shafts. The sharp edges cut into the metal of the shaft to effect more or less permanent positioning.
3. Cone-point set screws may be employed for the same applications as cup-point screws. However, a much more positive lock is effected because the shaft is first spotted or drilled to engage the conical point of the screw.
4. Oval-point set screws are employed to lock parts which are to be adjusted, frequently relative to each other. A groove is ordinarily provided of the same general contour as the set screw point that bears against it. This groove may be machined lengthwise, angularly, or in any other direction to the shaft axis.
5. Half dog-point set screws are frequently employed to engage slots milled longitudinally in shafts. They allow lengthwise

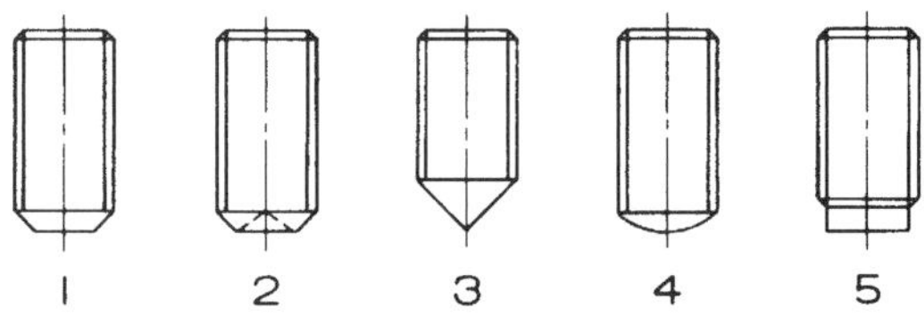

Figure 16.22 Five point styles for set screws.

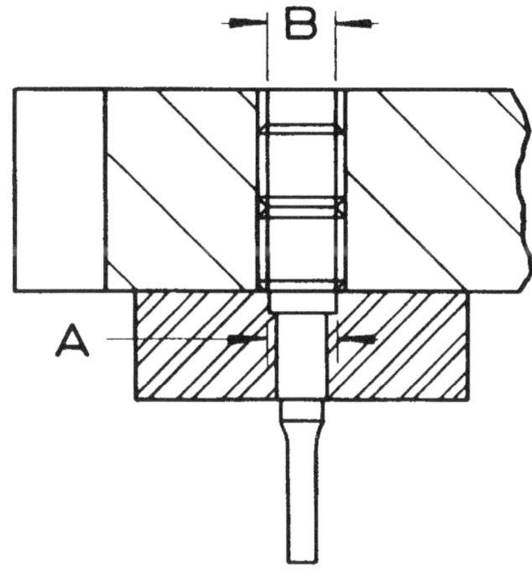

PUNCH HEAD DIA. A	TAP DRILL DIA. B	TAP
3/16	#7 (.201)	1/4 - 20
1/4	#F (.257)	5/16 - 18
5/16	5/16	3/8 - 16
3/8	27/64	1/2 - 13
1/2	17/32	5/8 - 11
5/8	21/32	3/4 - 10
3/4	49/64	7/8 - 9
7/8	7/8	1 - 8

Figure 16.23 Table for selecting top drill diameter and set screw thread size when punch head diameter is known.

movement, but prevent rotation. The point also acts as a stop to limit travel.

b) Punch-Retaining Set Screws

The table in Figure 16.23 facilitates selection of the correct set screw size after punch-head diameter **A** has been established. It will help to eliminate errors because, in every instance, tap drill size **B** is larger than head diameter **A** of the punch. This method provides quick removal and replacement of nicked or broken punches.

c) Square-Head Set Screws

In die design, only one application remains for square-head set screws (Figure 16.24). This is as an actuator for the automatic stop. This right section view of a representative die illustrates the manner in which a square-head set screw **A** is applied to operate an automatic stop. A jam nut **B** locks the screw against rotation. Note that the hole for the set screw is tapped 1/4-inch (6.35 mm) deeper than necessary. This

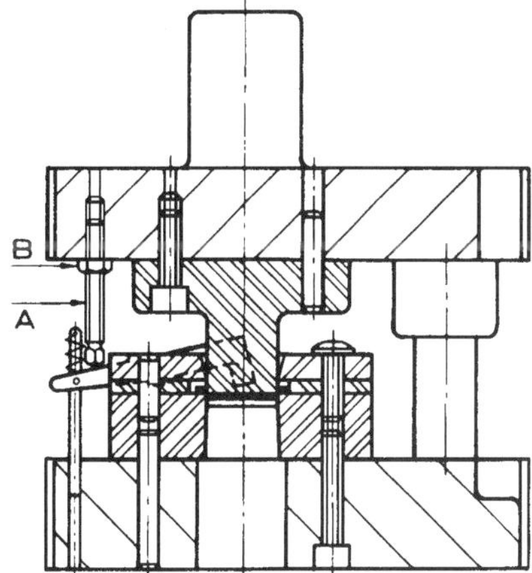

Figure 16.24 Application of a square-head set screw **A** for actuating the automatic stop.

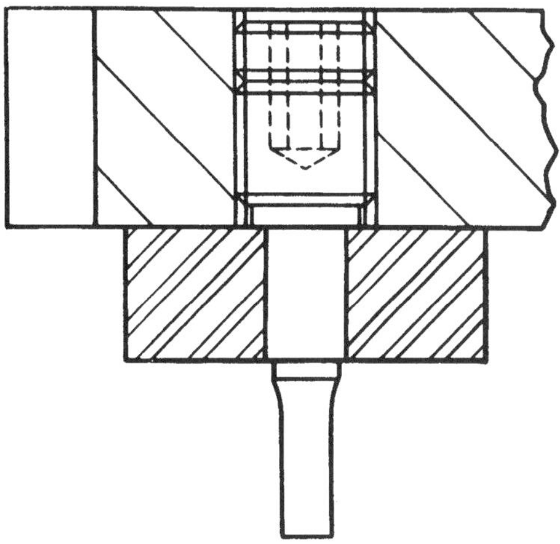

Figure 16.25 Application of a socket lock screw.

permits resetting the screw back as punches are shortened in sharpening.

A square-head set screw is measured from under the head to the end; this is the specification listed in the bill of material.

d) Socket Lock Screws

Lock screws (Figure 16.25) have the same dimensions as socket set screws, except that they are much shorter, and the hexagonal driving socket is broached clear through the screw. The length is one-half the diameter for most sizes. They are employed to lock set screws to prevent their loosening.

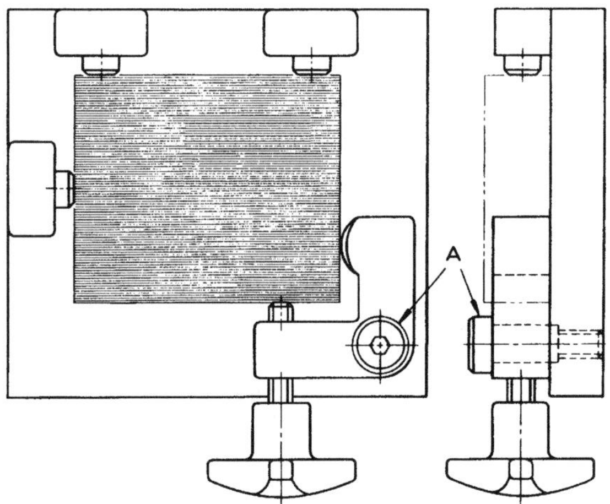

Figure 16.26 Application of a socket shoulder screw.

A big advantage for some applications is that the lock screw need not be completely removed for resetting the set screw. It is simply backed up a turn or so and the set screw socket is reached through the lock screw hole for resetting to a new position.

In the design of dies, lock screws are used to lock punch-retaining set screws under crowded conditions.

16.3.6 Shoulder Screws

Shoulder screws (Figure 16.26) are used to provide pivots for rocking members of jigs and fixtures and also to limit travel of components. A shoulder screw is provided with a body ground to accurate size, a head with an internal driving socket, and, at the other end, a smaller threaded portion.

The illustration shows a corner clamp employed to clamp square and rectangular work pieces against locating buttons for accurate positioning. The corner clamp pivots about a shoulder screw **A** that is threaded into the base.

16.3.7 Determining Strength of Screws

Many factors enter into determining the strength of screws. In designing, we are not concerned about the strength of a screw in its free state. Instead, we must know the amount of strength remaining in it after it has been tightened. A considerable proportion of the strength of a screw is expended in tightening. What is left is the effective load-carrying capacity.

Selecting the right fastener screws from the vast array of materials available can appear to be a daunting task; however, with some basic knowledge and understanding, a well thought out evaluation can be made. The mechanical property most widely associated with standard threaded fasteners is tensile strength.

Tensile strength is the maximum tension-applied load a fastener can support prior to, or coincident with, its fracture (the force or tension required to break the part when pulled in straight tension). Tensile strengths are normally expressed in terms of stress, pounds per square inch (psi) or, in the metric (SI) system, (MPa).

Tensile load strength is determined with the following formula:

$$F = S_t \times A_s \tag{16.1}$$

where:

F = tensile load (lb, N)
S_t = tensile strength (psi, MPa)
A_s = tensile strength area (sq in, sq mm).

For this relationship, a significant consideration must be given to the definition of the tensile strength area A_s. When a standard threaded fastener fails in pure tension, it is designed to fracture in the threaded portion. For this reason, the tensile strength area is calculated through empirical formulas involving the nominal diameter of the screw and the thread pitch: These formulas are:

- Unified National Standard thread tensile stress area:

$$A_s = 0.7854\left[D - \left(\frac{0.9743}{n}\right)\right]^2 \tag{16.2}$$

where:

D = nominal diameter of screw (in.)
n = threads per inch

- Metric (SI) system thread tensile strength area is:

$$A_s = 0.7854[D - (0.9745p)]^2 \tag{16.2a}$$

where:

D = nominal diameter of screw (mm)
p = threads pitch (mm).

Yield strength is the tension-applied load at which a fastener experiences a specified amount of permanent deformation. In other words, the material has entered its plastic zone.

Proof load is a tension-applied load that a fastener must support without evidence of permanent deformation. Proof load is an absolute value, not a maximum or minimum. Proof loads are established at approximately 90 percent of the expected minimum yield strength of the fastener screws.

For fatigue-critical applications, alloy steel socket-head screws, grade 8, should be used. Cap screws can be used in tapped holes or with a nut (like a bolt). Most designer require the cap screw to be tightened to 75 percent of the proof load for a safe working range. But all torque-tension relationships should be viewed with a cautious eye because no one table can indicate the range of conditions a fastener is expected to experience. Torque is only an indirect indication of tension. The torque value to use in an application is best obtained by using a calibrated torque wrench (or transducer) and a Skidmore-Wilhelm type load indicating device to equate actual torque to desired tension. Nearly all of the torque-tension tables which have been developed are based on the following formula:

$$T = k \times D \times F \qquad (16.3)$$

where:

T = torque (in lb)

k = coefficient of friction (0.15 for greased threads, 0.20 for black oxide threads, 0.30 for zinc plated threads, and 0.50 for a dry assembly)

F = axial load (lb)

D = nominal diameter of the thread part (in.)

Example

Lubricated screw D = 1/2-20 pitch, torqued to T = 100 ft lbs (1200 in lbs), can produce 16,000 pounds of clamping force.

$$1200 = 0.15 \times 0.50 \times F$$

$$F = 1200/(0.15 \times 0.50) = 16{,}000 \text{ lbs.}$$

If the screw is not lubricated, a higher present of the torque is consumed by friction. If the above example is assembled dry (k = 0.50), then the clamping force would be only 4,800 pounds.

$$1200 = 0.50 \times 0.50 \times F$$

$$F = 1200/(0.50 \times 0.50) = 4{,}800 \text{ lbs}$$

Shear strength of an alloy steel cap screw loaded in single shear through the threads is about 70 percent of the ultimate tensile strength, so working loads should be lower than this breaking strength. The fatigue strength of cap screws, as measured by the endurance limit, will vary widely depending upon the amplitude, frequency, and type of applied load. Endurance limits of 10–20 percent of the tensile strength have been reported for high strength cap screws.

After the stripping force has been calculated, use the actual manufacturer's date and recommendations for selecting the right fastener screws. When selecting stripper bolts, choose values given for socket cap screws of the same diameter and threads per inch as the threaded portion of the stripper bolt.

16.4 BOLTS

When shoulder screws are employed in dies, they are called *stripper bolts* (Figure 16.27) because they find their most common application in limiting the travel of spring strippers.

This front section view of a blanking die shows an application of stripper bolts **A**, employed to retain a spring stripper and to limit its travel. Four or more stripper bolts are used, but it is common practice to draw only one at one side in order to show one of the springs at the other side of the view.

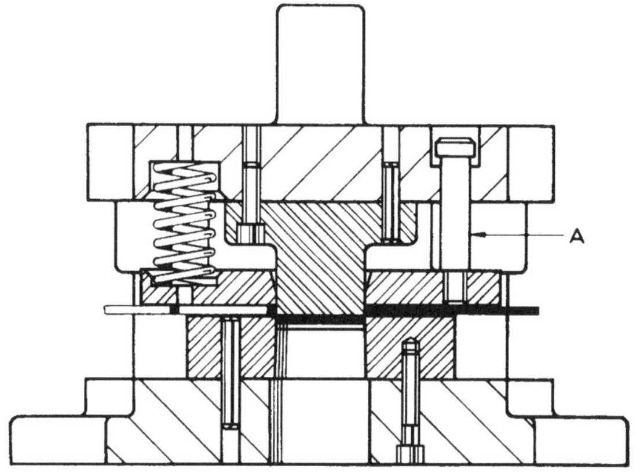

Figure 16.27 Application of a stripper bolt **A** to limit the travel of a spring stripper.

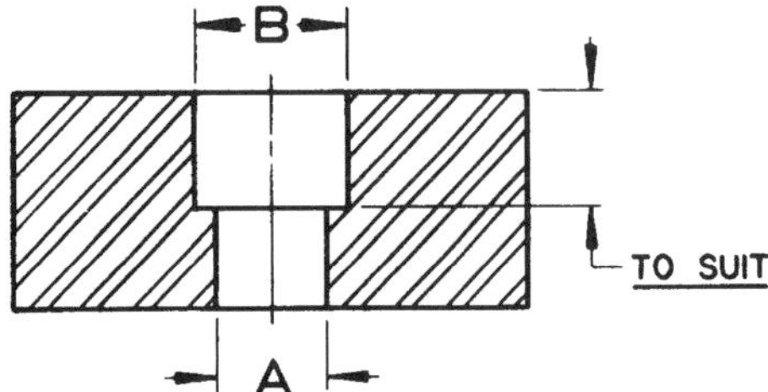

Figure 16.28 Hole for a stripper bolt.

a) Holes for Stripper Bolts

When specifying holes for stripper bolts (Figure 16.28), observe these rules:

1. Holes **A**, which engage stripper bolt bodies, are specified a slip fit or sliding fit by giving nominal diameter, followed by the abbreviation S.F.

Example

1/2 DR. & RM. THRU - S.F.

2. Counterbore diameter **B** is specified 1/32 inch (0.8 mm) larger than the diameter of the head of the screw.

16.4.1 Telescoping Sleeve

When stripper plates or pressure pads require long travel, but stripper bolts cannot extend below the die set because the bolster plate of the press is not provided with a clearance hole, a telescoping sleeve (Figure 16.29) may be employed to increase the amount of travel. At view **A**, the pressure pad is illustrated in raised position. At view **B** it is shown at the bottom of the press stroke with the sleeve telescoped over the stripper bolt.

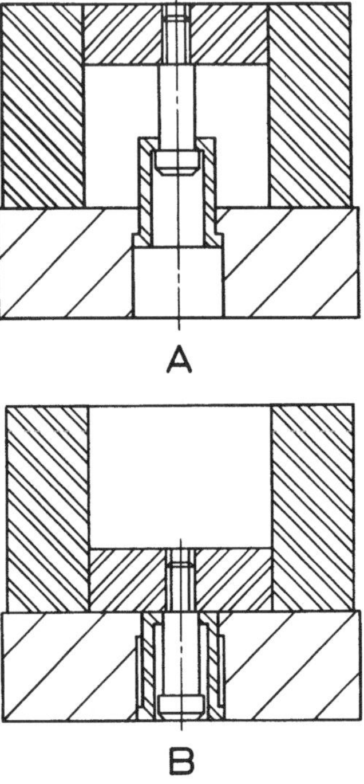

Figure 16.29 Application of a telescoping sleeve provides long travel of a stripper bolt.

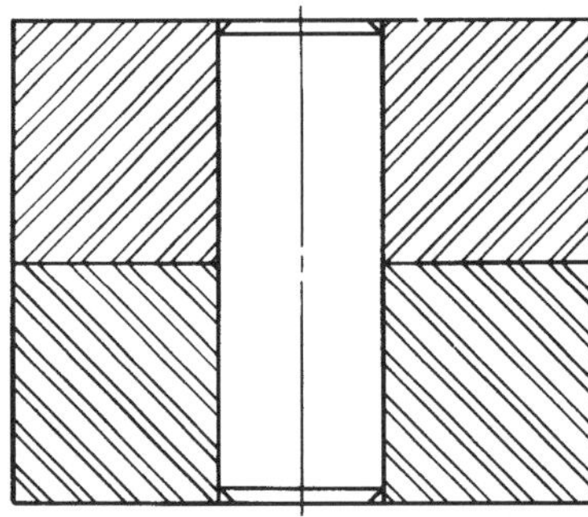

Figure 16.30 Typical application of a dowel.

16.5 DOWELS

Dowels (Figure 16.30) are intended to provide relative alignment of assembled die components. It is important to note that dowels do not provide perfect location. There is nothing perfect about any machine or die. Resistance to shifting is provided by the friction developed between die components by the clamping force of the screws. Thus, the degree of shifting resistance is determined by the coefficient of friction and screw clamping force. Also, dowels facilitate quick disassembly of components and reassembly in their exact former relationship.

Present day dowels are precise tool components, carefully designed and engineered. They are produced with exacting accuracy, both dimensionally and in their physical characteristics. Dowels are made of alloy steel, heat-treated to produce an extremely hard exterior surface with a somewhat softer, tough core. These kinds of dowels are able to resist shear and spread, or mushrooming, when they are driven into tight holes. Surface hardness is Rockwell C 60 to 64; core hardness C 50 to 54.

Depth of the hard exterior case extends from 0.0010 to 0.0020 inch (0.25 to 0.50 mm) depending on size, smaller sizes having a thinner case. Dowels are available in a range of sizes, from 1/8-inch diameter by 3/8 inch long (3.2 mm diameter by 9.5 mm long) to 1-inch diameter by 6 inches in length (25.4 mm diameter by 152 mm in length).

Strength. Single shear strength of alloy steel dowels ranges 0.7 of the ultimate tensile strength or 0.58 of the tensile yield strength.

Oversizes. Dowels are manufactured in two amounts of oversize:

1. Regular dowels, employed for all new jigs, fixtures, and dies, are made 0.0002 inch (0.005 mm) oversize to provide a secure press fit.
2. Oversize dowels are made 0.001 inch (0.025 mm) oversize. They are used for repair work when dowel holes have been enlarged through repeated pressing of dowels in and out, and when holes have been accidentally machined oversize.

Surface finish. Manufacturers precision grind dowel surfaces to a diameter tolerance of plus and minus 0.0001 inch (0.0025 mm). Surface roughness is held to a finish of 4 to 6 microinches average, maximum. This extremely smooth finish reduces the possibility of galling or seizing when dowels are driven in or out of their holes. A rust preventative coating is then applied.

a) Dowel Shape

Standard dowels (Figure 16.31) are given a 5-degree taper at one end for easy starting when

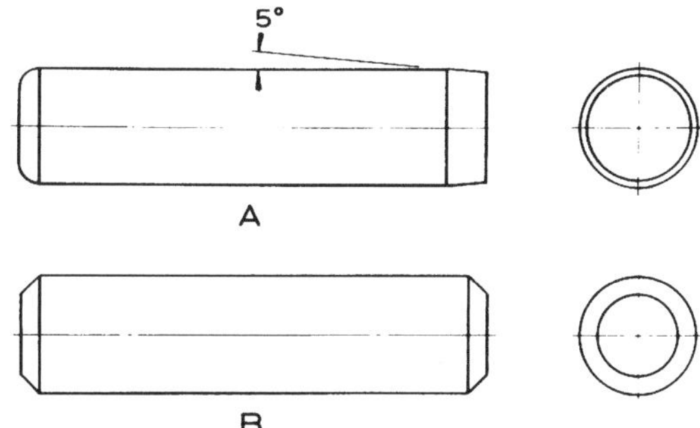

Figure 16.31 Dowel shape in actuality (**A**) and as shown on a drawing (**B**).

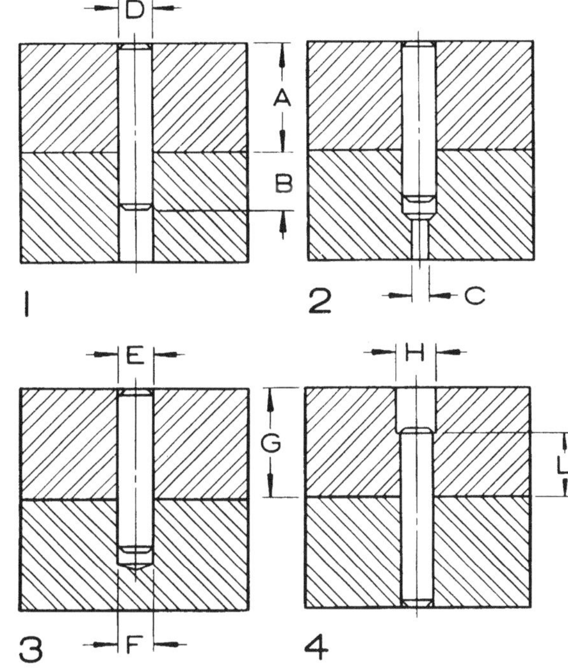

Figure 16.32 Four ways of applying dowels: **(1)** through dowel, (**2**) semi-blind dowel, (**3**) blind dowel, (**4**) relieved dowel.

they are pressed in, as shown at view **A**. A small radius is applied to the other corner. However, to save time, they are drawn conventionally by showing a chamfer at each end, as shown at view **B**.

b) Proportion of Dowels

Conditions under which dowels are employed determine the type of application chosen. These are four (Figure 16.32):

1. *Through dowels.* In this application, the hole is reamed all the way through the components, and dowels can be pressed out from either side. Dimension **A** is two inches or less. Dowel engagement **B** is between 1 1/2 and 2 times diameter **D**.
2. *Semi-blind dowels.* In this application, the dowel hole is drilled and reamed from one side at least 1/8 inch (3.2 mm) deeper than dowel length. A smaller hole is drilled through the block and the dowel can be pressed out from one side only. Dimension **C** of the knockout hole is made 1/2 **D** plus 1/64 inch (0.4 mm), or one-half dowel diameter with clearance for a standard diameter hand punch for pressing out.

3. *Blind dowels.* This dowel is applied in a blind hole, one not drilled and reamed completely through. The application should be avoided whenever possible. Blind dowels are more difficult to fit because of trapped air, and removal can be troublesome. Diameter **E** is made a press fit. Diameter **F** at the blind side should be a slip fit.
4. *Relieved dowels.* When doweling blocks over two inches thick (dimension **G**), standard length dowels are employed and the hole **H** is specified 1/32 inch (0.4 mm) larger than the diameter of the dowel for relief. This relief is applied to that portion of the hole not in actual contact with the dowel surface. Engagement length **L** is 1 1/2 to 2 **D**.

Normally, dowels are press-fitted into both members. Holes are drilled and reamed after the jig, fixture, or die has been completely assembled, and often even after it has been tried out. However, when parts are subject to frequent disassembly—for instance, for sharpening—dowels are press-fitted in one component and made a sliding fit in the other for ease of disassembly.

c) Dowel Diameter

For jigs, fixtures, and gages, always make dowel diameters one size smaller than corresponding screw diameters. For dies, dowels are made the same diameter as the screws because of the high speed and shock conditions present in operation. Table 16.1 lists correct dowel sizes for screws ranging from # 8 to 3/4 inch (19.0 mm) in diameter.

Table 16.1 Screw diameters and corresponding dowel diameters as applied to tools and dies

	DOWEL DIAMETER	
SCREW DIAMETER	TOOLS	DOWELS
# 8	1/8	1/8
# 10	1/8	3/16
1/4	3/16	1/4
5/16	1/4	5/16
3/8	5/16	3/8
7/16	3/8	7/16
1/2	7/16	1/2
5/8	1/2	5/8
3/4	5/8	3/4

Table 16.2 Commonly used reamer drills and corresponding reamers

REAMER DRILL DIAMETER		REAMER DIAMETER	
Nominal Size	Decimal Size	Nominal Size	Decimal Size
5/16	0.307	5/16	0.3125
3/8	0.366	3/8	0.375
7/16	0.427	7/16	0.4375
1/2	0.489	1/2	0.500
5/8	0.616	5/8	0.625
3/4	0.734	3/4	0.750

NOTE: To convert from inches to millimeters, multiply by the conversion factor 25.4.

d) Reaming Dowel Holes

When a hole is to be reamed for press-fitting a dowel, it is first drilled with a reamer drill, then reamed. Reamer drills have the same nominal size as the corresponding reamers, but their diameters are actually smaller. For instance, a 5/16-inch reamer drill has a diameter of 0.307 inch (7.8 mm) and the actual reamer size is 0.3125 inch (7.94 mm), leaving 0.0055 inch (0.14 mm) of metal for the reamer to remove. Table 16.2 lists sizes of commonly used reamer drills and their corresponding reamers.

16.5.1 Dowels for Multiple Parts

In precision tool design, where clearances are held to extreme accuracy, good doweling methods must be employed (Figure 16.33). When more than two parts are to be doweled, the method shown at view **A** may be used when plates are

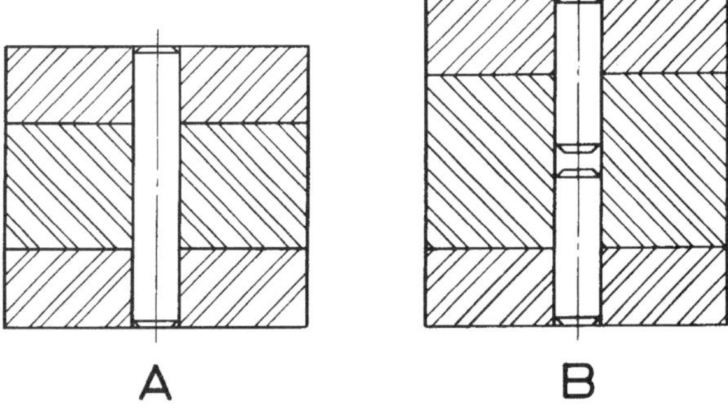

Figure 16.33 Application of dowels for multiple parts.

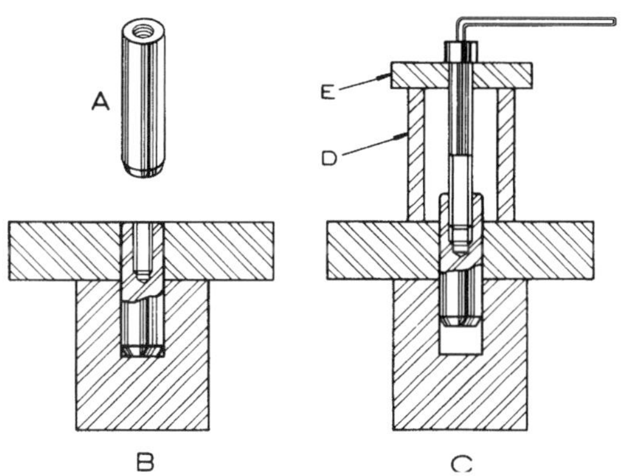

Figure 16.34 One type of removable dowel for use in blind hole applications.

relatively thin. Dowel length is the same as the thickness of the combined plates. For thicker plates and blocks, the method illustrated at view **B** is preferred. Two dowels engage each dowel hole. The outer plates are thus doweled to the center block.

It is common practice to make the length of a dowel four times its diameter, when possible.

Example

If the diameter of a dowel is 3/8 inch (9.5 mm), then the dowel is 1 1/2 inches (38 mm) long.

16.5.2 Removable Dowels

One type of removable dowel is illustrated in Figure 16.34 at **A**. These dowels are used in blind applications where the dowel holes cannot be drilled entirely through the component. A threaded hole is provided in one end as shown. To remove the dowel, a length of pipe or tubing (**B**) is placed over it. Then, rotation of the cap screw acting against a washer (**C**) removes it. Longer lengths of pipe can be used to increase travel.

A different type of commercially available removable dowel is shown in Figure 16.35. Dowel **A** is provided with an axial hole tapped the entire length of the dowel. Threaded into the hole is a long socket set screw **B**. In section view **C**, the dowel is shown pressed into a blind hole. For removal, the set screw is turned, as shown in view **D**, and this jacks the dowel upward out of the hole.

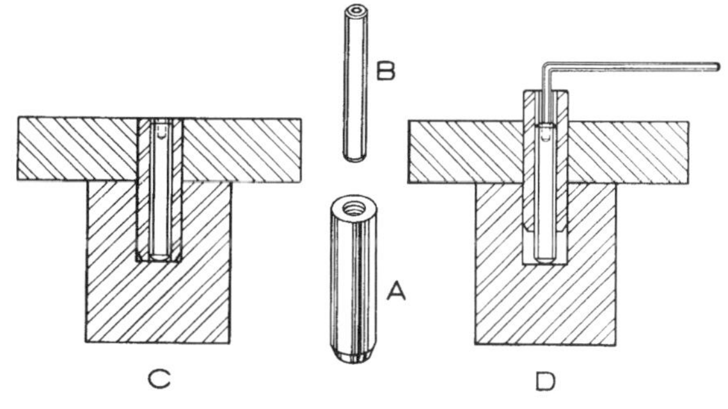

Figure 16.35 Another type of removable dowel for use in blind hole applications.

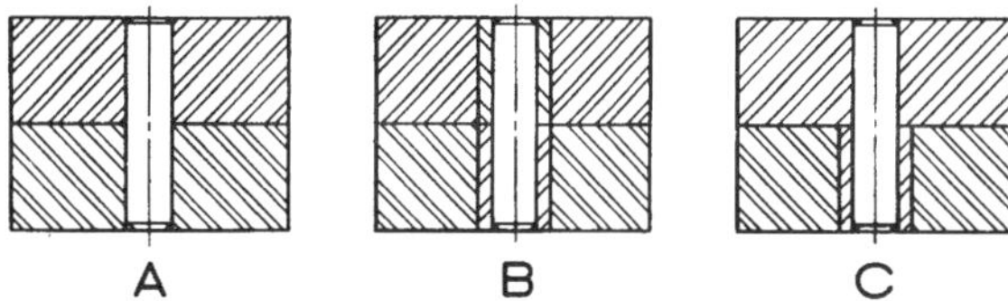

Figure 16.36 Methods of doweling hardened tool steel blocks.

16.5.3 Doweling Hardened Parts

There are three methods of doweling hardened tool steel blocks (Figure 16.36). In the first, shown at view **A**, holes are drilled and reamed before hardening. After hardening, they are either jig-ground or lapped for accurate engagement of dowels. In a second method, shown in view **B**, soft plugs of machine steel are pressed into oversize holes in the hardened blocks. Holes for dowels are drilled and reamed through the soft plugs in a conventional manner.

When doweling a machine steel block to a hardened tool steel block **C**, the hardened block only may be provided with a soft plug pressed into a large hole. Dowel location is transferred from the machine steel plate to the soft plug for accurate location and fit of dowels.

HOW TO SELECT A DIE SET

17.1 INTRODUCTION

After all die details have been designed, a die set of the proper size and style is selected from a manufacturer's catalog and drawn in position. Between 5 and 10 percent of total design time is spent selecting and drawing the views of the die set. This allowance may be increased considerably if the designer does not thoroughly understand the principles underlying die set selection and representation.

Die sets are manufactured in a bewildering variety of sizes and shapes. This chapter will acquaint you with the various styles as well as indicate proper methods of selection and placement.

Advantages realized when die components are retained in a properly selected die set are:

1. Members are kept in proper alignment during the cutting process, even though some looseness may exist in the press ram. Thus, uniform clearances are maintained around cutting edges for producing blanks free of burrs.
2. Die life is increased.
3. Dies can be installed in the press in a minimum amount of time because they are self-contained units.
4. Storage is facilitated. There is no possibility of loss of loose parts.
5. Properly designed dies can be sharpened without removal of cutting members. Standard die sets range from 3 by 3 inches (76-by-76 mm) to 45-by-60 inches (1,143-by-1,524 mm). Die holder and punch holder thicknesses range from 1 to 3 1/2 inches (25.4 to 89 mm).

17.2 PRINCIPLES OF SELECTING THE DIE SET

Ten elements of die set information must be decided before a die set can be ordered:

1. Make or manufacturer
2. Type
3. Size
4. Material
5. Thickness of die holder
6. Thickness of punch holder
7. Type and lengths of bushings
8. Lengths of guide posts
9. Shank diameter
10. Grade of precision.

In selecting a die set from a catalog, first consider the dimensions from front to posts and from side to side. This is the *die set area,* or usable space, to which die details can be fastened. Next in importance is the thickness of the die holder and of the punch holder.

17.2.1 Materials

Selecting the material from which the die set is to be made will depend upon strength requirements. There are three choices:

1. Semi-steel
2. All steel

3. Combination in which the punch holder is semi-steel and the die holder is all steel.

a) Semi-Steel

In manufacturers' catalogs, the material of the die set is listed as either semi-steel or steel. Semi-steel contains only about 7 percent steel in its composition and is considered to be cast iron. Semi-steel die sets are cast to shape and then machined. Some manufacturers may cast punch holders and die holders of Meehanite, which may be considered a high-grade cast iron.

b) Steel Die Sets

When a large hole is to be machined through the die set for blank removal, it is considered good practice to specify a steel die holder. This prevents fracture of the die holder if it is placed over a large hole in the bolster plate, which is done occasionally even in the best press shops. It happens too frequently that a cast iron die holder is actually broken in two because of the weakening effect of a large hole in conjunction with insufficient support under pressure.

Steel die sets are thoroughly stress-relieved by manufacturers before final machining or grinding. Stress-relieving removes any stresses introduced in the material in rolling at the mill, as well as other stresses added during rough machining. If such residual stresses are not removed, they are gradually released with consequent distortion and dimensional change, thereby ruining a precision die.

Obviously, it behooves the designer not to incorporate anything in the design that can introduce stresses in the die set while the die is being built. Welding anything to a die set must be avoided. Rough machining of deep pockets should be done by the die set manufacturer before the stress-relieving operation, and a print showing necessary machining operations should accompany the purchase order. To illustrate the importance of stress relieving in the manufacture of die sets with deep-milled pockets and through holes, here is a representative order of operations actually employed for a larger die holder:

1. Flame or laser-cut holes
2. Stress relief
3. Surface grind
4. Rough machine pockets
5. Stress-relief
6. Surface grind
7. Finish-machine.

17.2.2 Grade of Precision

Die sets are manufactured to two standards of accuracy: *precision* and *commercial*. Punch holder and die holder tolerances are the same for both. The difference between them occurs in the closeness of fit between bushings and guide posts. For precision sets, tolerances between bushings and guide posts are maintained from 0.0002 inch (0.005 mm) to a maximum of 0.0004 inch (0.01 mm). This tolerance assures extremely accurate alignment between punches and corresponding holes in die blocks. For this reason, precision die sets should be specified for all dies that perform cutting operations.

Commercial die sets are given more liberal clearances between bushings and guide posts. These range from 0.0004 inch to 0.0009 inch (0.01 to 0.023 mm). Commercial die sets should be specified only for dies that perform bending, forming, or other non-cutting operations.

17.2.3 Direction of Feed

The direction of the strip's feed will influence selection of the die set. Strip or coil may be fed through a press in any of three directions:

1. *Front to back.* This method may be employed for long runs when strip is fed automatically.
2. *Right to left.* This is the most commonly used feeding direction. It is always employed when the strip is advanced by hand.
3. *Left to right.* This direction is occasionally used when the strip is fed automatically.

Feed direction must always be ascertained before design of a die is started because it will affect the sequence of operations and location of stops.

17.2.4 Press Date

After a die has been tentatively designed or roughed out, the next step is to consider the specifications of the press in which it is to operate. This should be done before the die set is selected because the available space may influence die set size and type. It is very important that a die fit the press for which it was designed; it is the designer's responsibility to ascertain that there be no interferences. The three most important considerations in determining the dimensions of a die set are:

1. *Center post diameter.* Check the press data sheet carefully to make sure that the recess in the press ram will accommodate the punch shank selected. These center posts are called "shanks" in the die set manufacturers' jargon. They cannot be used for clamping the punch holder, but they can be used for aligning the die in the press. Ram (or slide) mounting holes must be provided in the punch holder for mounting.
2. *Shut height of die.* Make certain that the shut height dimension is well within the available die space without the ram adjustment being taken up to its limit. In this connection, make certain that grinding clearance has been taken into consideration.
3. *Distance from the center of the shank to the back of the die.* Make certain that this dimension is at least 1/4 inch (6.35 mm) *less* than the distance from the center of the ram to the frame of the press. The make and model number of the press are marked on the route sheet, and specifications can be found in the manufacturer's catalog or in the company standards book.

Look first for the "Die Space." This is usually specified as the distance from the bed of the press to the underside of the ram with the stroke down and the adjustment up.

Next look for the thickness of the bolster plate. This must be subtracted from the die space to give the shut height from the top of the bolster plate. This is the height of the *tallest* die that will fit into the press. However, some manufacturers give the die space directly as the distance from the top of the bolster plate to the underside of the press ram, and caution should be exercised.

Next, it will be necessary to determine the *shortest* die that the press will accommodate. Look up the dimension for "Adjustment of Slide." The ram of a press is usually provided with an adjustment to accommodate a range of die heights. Subtract the dimension given for adjustment of slide from the shut height. This will be the height of the shortest die that the press will accommodate. However, a die shut height 1/4 inch (6.35 mm) higher must be used to allow for lowering the upper die as punches are sharpened from time to time.

17.3 DIE SET COMPONENTS

Figure 17.1 typifies small- and medium-size die sets made of both cast iron and steel, although different manufacturers may incorporate slight variations. The die set components are:

- Punch holder (**A**)
- Guide bushings (**B**)
- Guide posts (**C**)
- Die holder (**D**).

When the die set is assembled, the lower ends of the guide posts are pressed securely into the die holder. The turned-down portions of the guide bushings are pressed into the punch holder. The bushings engage the guide posts with a close sliding fit to provide accurate alignment.

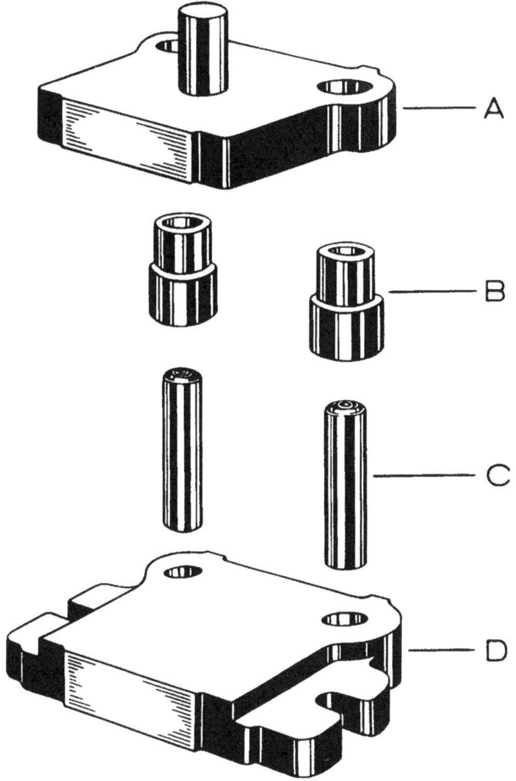

Figure 17.1 Components that make up a die set.

17.3.1 Punch Holders

The upper working member of the die set is called the punch holder (Figure 17.2). The name is easy to remember because of its relationship with the punches, which are normally applied above the strip and fastened to the underside of the punch holder. Surfaces **A** are finished. They are employed by the die maker for squaring and locating punch components of the die. Surfaces **B** are also finished surfaces. The upper one bears against the underside of the press ram. Punch components are fastened to the lower finished surface.

Punch shank. Some punch holders have the punch shank projecting above the punch holder (Figure 17.3); it aligns the center of the die with the center line of the press.

For semi-steel die sets, the punch shank is cast integrally with the body of the punch holder and then machined. For steel die sets, it is electrically welded to the punch holder and then machined.

Punch shanks may also be ordered separately. These are turned down at one end and threaded for engagement in a large tapped hole in the punch holder. Punch shank diameter depends upon the press selected. It is usually determined

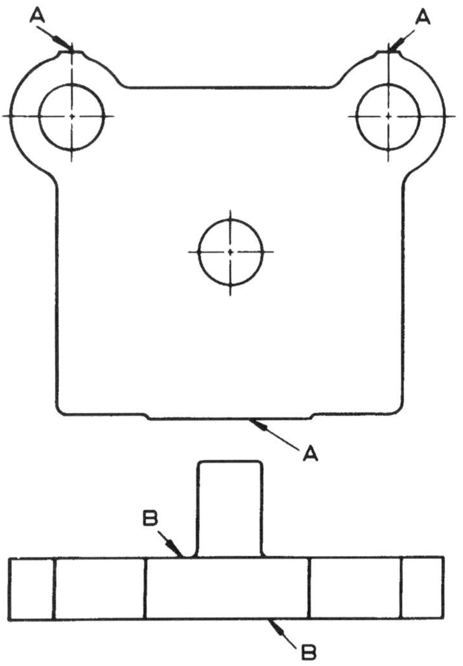

Figure 17.2 A punch holder with shank.

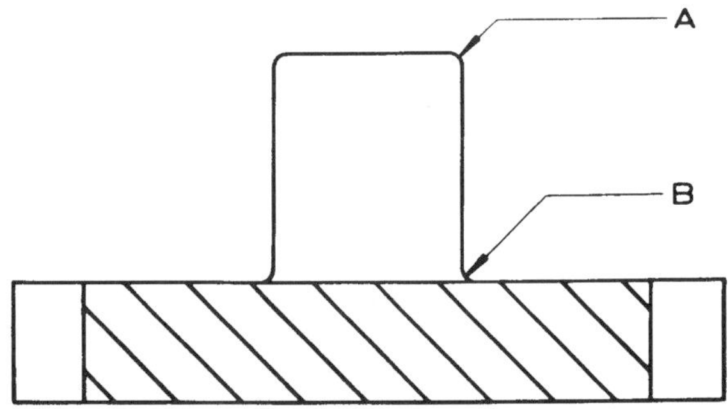

Figure 17.3 Punch shank projecting above punch holder is given a 1/8-inch (3.2-mm) radius at (A) and (B).

from a company standards book and it should be checked carefully for accuracy. After the diameter is known, the length can be found listed in a die set catalog. The round **A** at the top of the punch shank and the fillet **B** at the bottom where it joins the punch holder are given a 1/8-inch (3.2-mm) radius on the drawing.

The shank is employed only for centering the die and not for driving. Instead, such die sets are clamped or bolted to the underside of the ram because of the considerable weight of large punch holders and punch members. The relatively small punch shank would not be a safe method of driving.

To supplement the holding power of the shank, socket cap screws are often inserted upward through the punch holder to engage holes tapped in the press ram. Where this practice is followed, the designers specify and dimension the mounting holes to match the hole pattern in the ram; they must make certain that the holes clear punch components. Dimensions for mounting holes are ordinarily taken from a company standards book.

17.3.2 Die Holders

The die holder (Figure 17.4) is the lower working member of the die set. Its shape corresponds with that of the punch holder except that it is provided with clamping flanges **A** having slots for bolting the die holder to the bolster plate of the press.

Machined surfaces **B** are employed for squaring and locating die components. Surfaces **C** are also finished. The lower one rests on the bolster plate, and the die block and other components are fastened on the upper surface.

Usually, the die holder is made thicker than the punch holder to compensate for the weakening

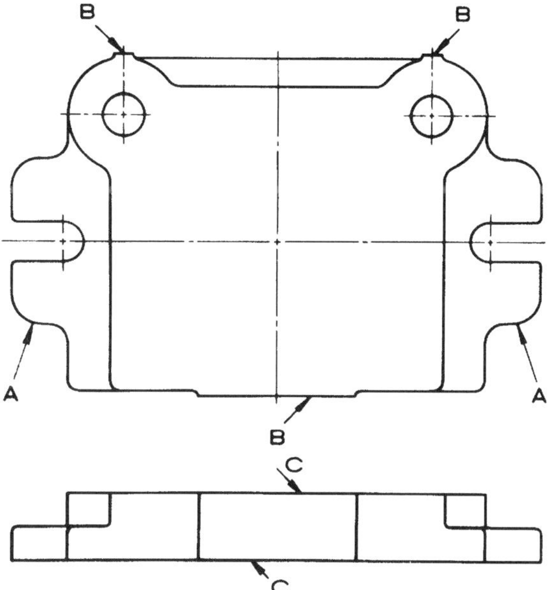

Figure 17.4 Typical die holder with clamping flanges (A) and machined surfaces (B) and (C).

effect of slug and blank holes, which must be machined through it. Common proportions for small- and medium-size dies are:

- Punch holder thickness at 1 1/4 inches (32 mm)
- Die holder thickness at 1 1/2 inches (38 mm)

17.3.3 Guide Posts

Guide posts are precision-ground pins that are press-fitted into accurately bored holes in the die holder. They engage guide bushings to align punch and die components with a high degree of closeness and accuracy. Figure 17.5 illustrates six types of guide posts.

1. Small guide posts are usually hardened and centerless ground, particularly for the commercial die set grades.
2. Larger diameter posts are ground between centers after hardening.
3. Posts may be relieved at what will be the die set surface. This relief is usually applied to precision posts.
4. A non-sticking post end may be incorporated. This provides for quick and easy assembly and disassembly.

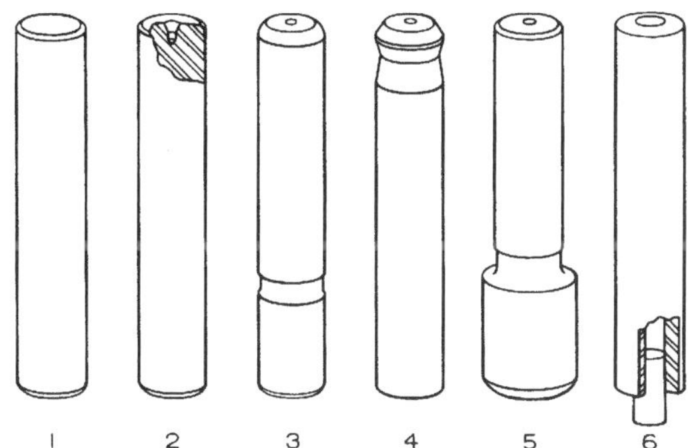

Figure 17.5 Six types of guide posts.

5. Shoulder guide posts are employed in conjunction with shoulder guide post bushings. The large shoulder is the same diameter as the press-fit portion of the guide bushings. In the manufacture of special die sets, the punch holder and die holder are clamped together and holes are bored through both for engagement of bushings and guide posts.
6. Removable guide posts can be easily removed from the die for sharpening. They are employed for large dies and for dies having more than two posts.

Guide posts for precision die sets are hard chromium-plated to provide a high degree of resistance to wear. Also, the addition of a chromium surface reduces friction by more than 50 percent.

For secondary operation dies, guide posts should have sufficient length so that they never leave their bushings during operation. This is a safety feature to prevent possible crushing of fingers accidentally introduced between posts and bushings as the die is operated.

Guide posts are specified at least 1/4 inch (6.35 mm) shorter than the shut height of the die as listed on the drawing. (The shut height is the distance from the bottom surface of the die holder to the top surface of the punch holder, excluding the shank, and measured when the punch holder is in the lowest working position.) This provides a grinding allowance to assure that the top of the posts will not strike the underside of the press ram when the upper die is lowered as punches are sharpened.

a) Non-Sticking Guide Posts

Sticking or jamming in the initial stages of engagement of punch holder and die holder has long been a problem because of the close fits maintained. Sticking occurs until the bushings have engaged the posts sufficiently for complete alignment. Dies must be assembled and disassembled a great number of times in their manufacture, tryout, and sharpening.

The non-sticking guide posts in view **1** of Figure 17.6 are a popular, as well as a commercially available, type of post that features the following characteristics:

- A ground taper **A** guides the bushing over the post.
- A narrow land **B** of the same diameter as the post centers the bushing. The land is narrow enough to allow rocking of the bushing over it.
- This clearance area represents the sticking range **C**. Because metal has been removed, sticking cannot occur.
- This ground lead **D** guides the bushing to engagement with the full diameter of the post.

View **2** shows a post that features a radius at the leading edge to align the bushing.

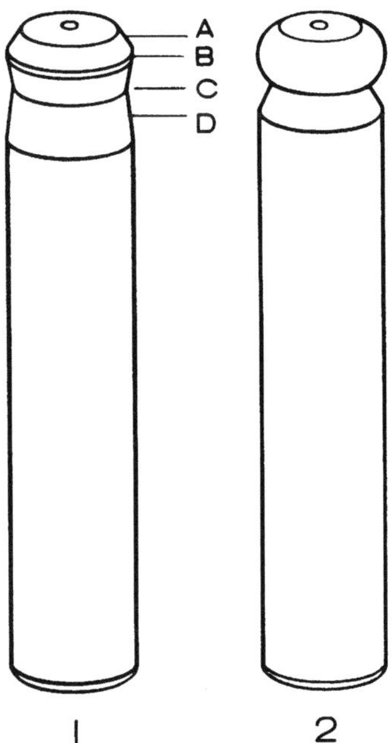

Figure 17.6 Two types of commercially available guide posts developed to overcome sticking or jamming in initial stages of engagement of punch and die holders.

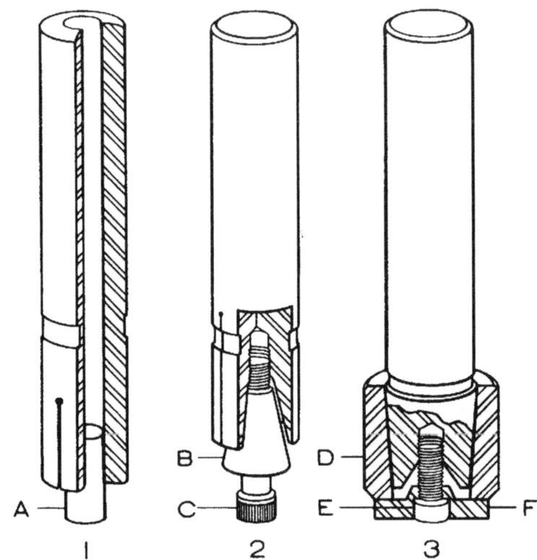

Figure 17.7 Three types of removable guide posts.

b) Removable Guide Post

Guide posts must often be removed for die sharpening, especially in large dies and in dies having more than two posts of the back-post style. Figure 17.7 shows the three types of removable guide posts:

1. The first kind of removable guidepost has an axial hole machined through it and is tapered at one end to engage a taper pin **A**. The post end is slotted. By driving the taper pin, the post is expanded against the walls of the hole in the die holder. To remove the post, a long rod called a *drift* is inserted from the top and the taper pin is pressed out.
2. In the second type of removable guide post, the taper pin **B** is advanced for locking the post by means of a socket cap screw **C**.
3. The third removable post has a taper at the lower end to engage a sleeve for bushing **D**, which is pressed into the die holder. A socket cap screw **E** engages a retaining cap **F** to clamp the post to the bushing. Removal of the socket cap screw allows the post to be lifted up and removed.

c) Offset Post Ends

In another method of assembly employed by a commercial supplier, one of the guide posts is made longer than the other (Figure 17.8). The punch holder engages the long post first and is thus

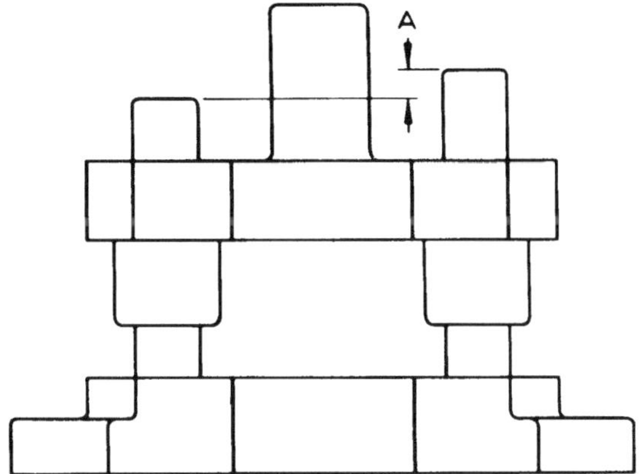

Figure 17.8 Use of guide posts of different lengths facilitates engagement of punch and die holders.

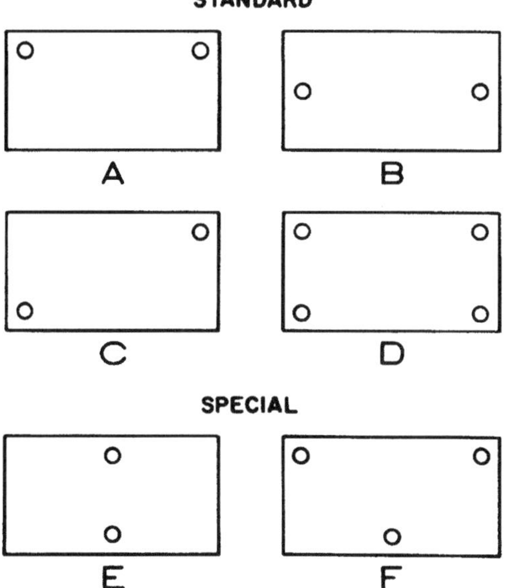

Figure 17.9 Six ways of arranging guide posts in a die set.

aligned before engagement of the other post occurs. Dimension **A** is usually made 1/2 inch (12.7 mm).

d) Post Arrangement

Guide posts may be positioned in any one of six ways (Figure 17.9):

- At **A**, two posts are applied at the back of the die set. This is the most commonly used two-post arrangement.
- At **B**, posts are applied at the sides for feeding strip from front to back.
- At **C**, the posts are positioned diagonally.
- At **D**, four posts are used. The foregoing are standard post arrangements, as listed in die set catalogs.

When rectangular steel die sets are ordered, any post arrangement may be supplied by the die set manufacturer.

- At **E**, for feeding strips sideways in long runs, some designers prefer posts applied at the front and at the back.
- At **F**, others specify three posts for stability as shown.

17.3.4 Guide Bushings

Accurately ground sleeves, or guide bushings, engage guide posts for aligning the punch holder with the die holder. Most bushings are made of tool steel, although they are also available in bronze. There are two types (Figure 17.10):

1. Plain bushings are simple sleeves pressed into the punch holder.
2. Shoulder bushings are turned down at one end; they are pressed into the punch holder against the shoulder thus formed. They are recommended for all dies that perform cutting operations.

Lengths of guide bushings vary, depending upon the manufacturer. In general, we may recognize two different lengths for plain bushings: regular and long. Shoulder bushings are furnished

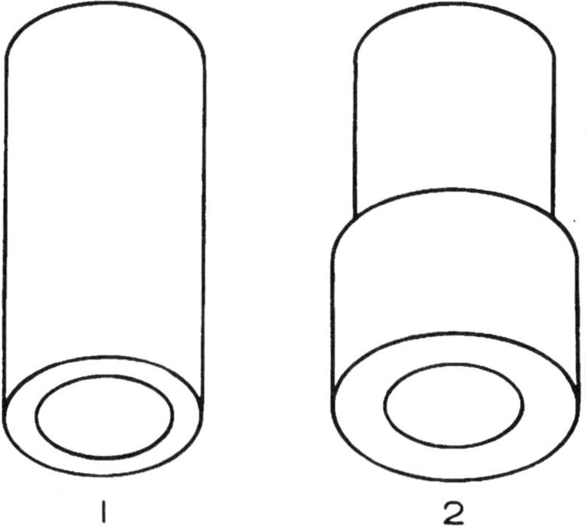

Figure 17.10 Two common types of guide bushings.

in three lengths: regular, long, and extra long. The length selected will depend upon the accuracy requirements of the die. The longer the bushing, the more accurate will be the alignment of punch and die members. This is particularly important in cutting operations, especially for thin stock when clearances between cutting edges are small.

Posts and bushings are assembled by shrink-fitting into holes bored in the punch holder and die holder. The posts and bushings are subjected to deep freezing, thereby reducing their diameters. They are then inserted in the punch holder and die holder. Then upon warming to room temperature, they expand to provide a tight fit between components.

Each guide bushing is provided with a fitting for lubrication. Helical grooves are machined in inside surfaces for retention and distribution of the lubricant.

a) Self-Oiling Guide Bushings

Figure 17.11 shows a guide bushing made of porous powdered alloy steel. Internal pockets are cored in the walls, and these are filled with oil at manufacture. In use, the oil diffuses through the porous walls by capillary action. The stored lubrication is sufficient for the life of the bushing.

b) Demountable Guide Bushings

Figure 17.12 shows two types of *demountable* guide bushings. These are shoulder bushings provided with clamps that engage an annular groove machined in the bushing wall or shoulder. Socket cap screws are threaded into the punch holder to effect clamping. The turned-down portion of the bushing is not a press-fitted into the punch holder. Instead, it is ground to an accurate sliding fit for ease in disassembly.

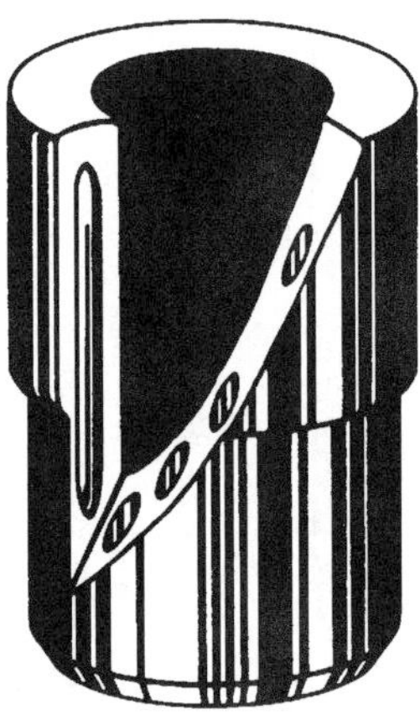

Figure 17.11 A commercially available self-oiling guide bushing made of porous powdered alloy steel.

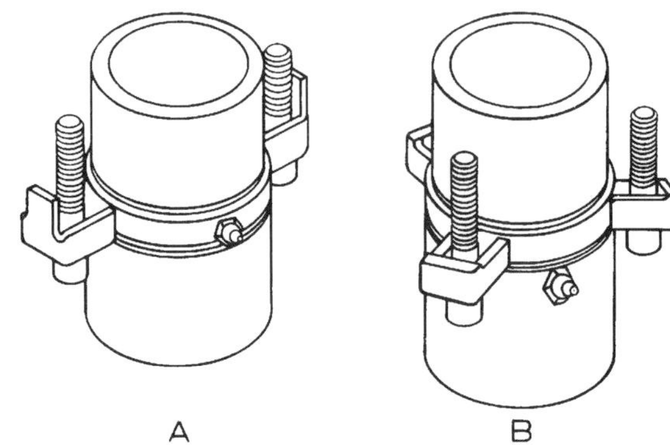

Figure 17.12 Demountable guide bushings are used for long runs.

These bushings are available in both steel or bronze, and they are provided with either two clamps (view **A**) or three clamps (view **B**), depending on size. Demountable guide bushings are specified for long runs when it is anticipated that bushings and posts will require replacement.

c) Boss Bushings

Demountable *bosses* (Figure 17.13) may be employed as guide bushings in large die sets. They are used for heavy duty work when long runs are expected. The bosses are turned down for location in holes in the die set and they are fastened with socket cap screws for easy replacement.

There are three types of boss bushings:

1. *Flange mounted.* The greater portion of the bearing surface is within the die set.

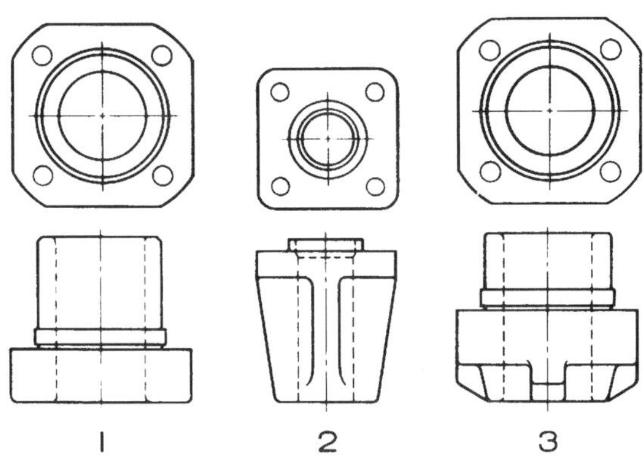

Figure 17.13 Three types of demountable boss bushings for heavy duty work in long runs.

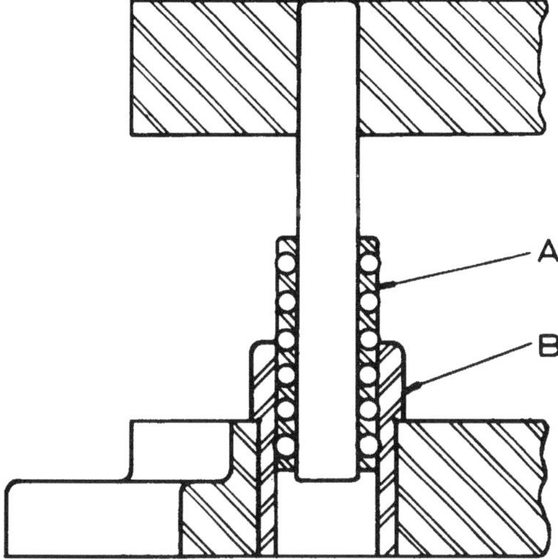

Figure 17.14 Set up in which ball bearings are used instead of a guide bushing.

2. *Demountable.* These are employed as bearings and also as guide post supports.
3. *Long bearing.* These have a bearing surface extending below the mounting flange and supported by ribs.

17.3.5 Ball Bearing Set Up

Some die sets are provided with ball bearings instead of guide bushings (Figure 17.14). Guide posts are pressed into the punch holder and they engage linear ball bearings **A**, which in turn are guided in hardened sleeves **B** pressed into the die holder. The bearings are preloaded to remove looseness or side play. Lubrication is provided by cup grease, which is applied at set-up; this is usually sufficient for the entire run. Ball bearings take up more room than conventional guiding methods and they slightly reduce die space.

17.4 ESTABLISHING THE DIE SET CENTER

Ten design steps are necessary for selecting and applying the views of a die set for a specific die design. Consider the following sequence carefully because it provides a proper working method (Figure 17.15):

1. Determine the diameter of the punch shank. Punch shank data is usually found in a company standards book or manufacturer's catalog, although occasionally it may be necessary to get it directly from the press room. On an appropriate sheet of paper, draw a circle of the required punch shank diameter.
2. On the upper left plan view of the die, draw vertical and horizontal center lines. Their intersection represents the ideal center of the die; the one we should prefer to use if possible. For most dies, the ideal center would be the center of the die block, measured from side to side and from front to back.
3. Transferring measurements, draw corresponding centerlines on the upper right inverted view of the punches.
4. Slip the sheet on which the punch shank circle was drawn under the die drawing and move it to the upper right plan view representing the inverted punches. Apply the center of the circle directly under the intersecting lines representing the ideal center.
5. Inspect punch holder drawing to establish whether or not all holes to be machined through the punch holder will pass outside of the punch shank or inside of it. In the illustration, the punch holder has been turned upside down for viewing the punch shank from the top. Observe that two holes are machined partly in the edge of the punch shank. Machining away portions of the punch edge in this manner is considered poor practice.
6. In this illustration, relative positions of holes and punch holder have been altered so that one of the holes passes entirely within the shank and the other entirely outside of it.
7. Referring back to the drawing, slide the sheet on which the punch shank circle was drawn until it clears holes so that they pass entirely outside or entirely inside of it. The best position found, closest to the ideal center, is shown. Dimension **A**, the wall thickness between the holes and the outside surface, is made 1/8-inch minimum.
8. Transfer the new center position on the die drawing by measurement. Always remember that, because the punch is inverted, horizontal dimensions are reversed.
9. Select a suitable die set and draw its views. Observe that the die block is slightly off

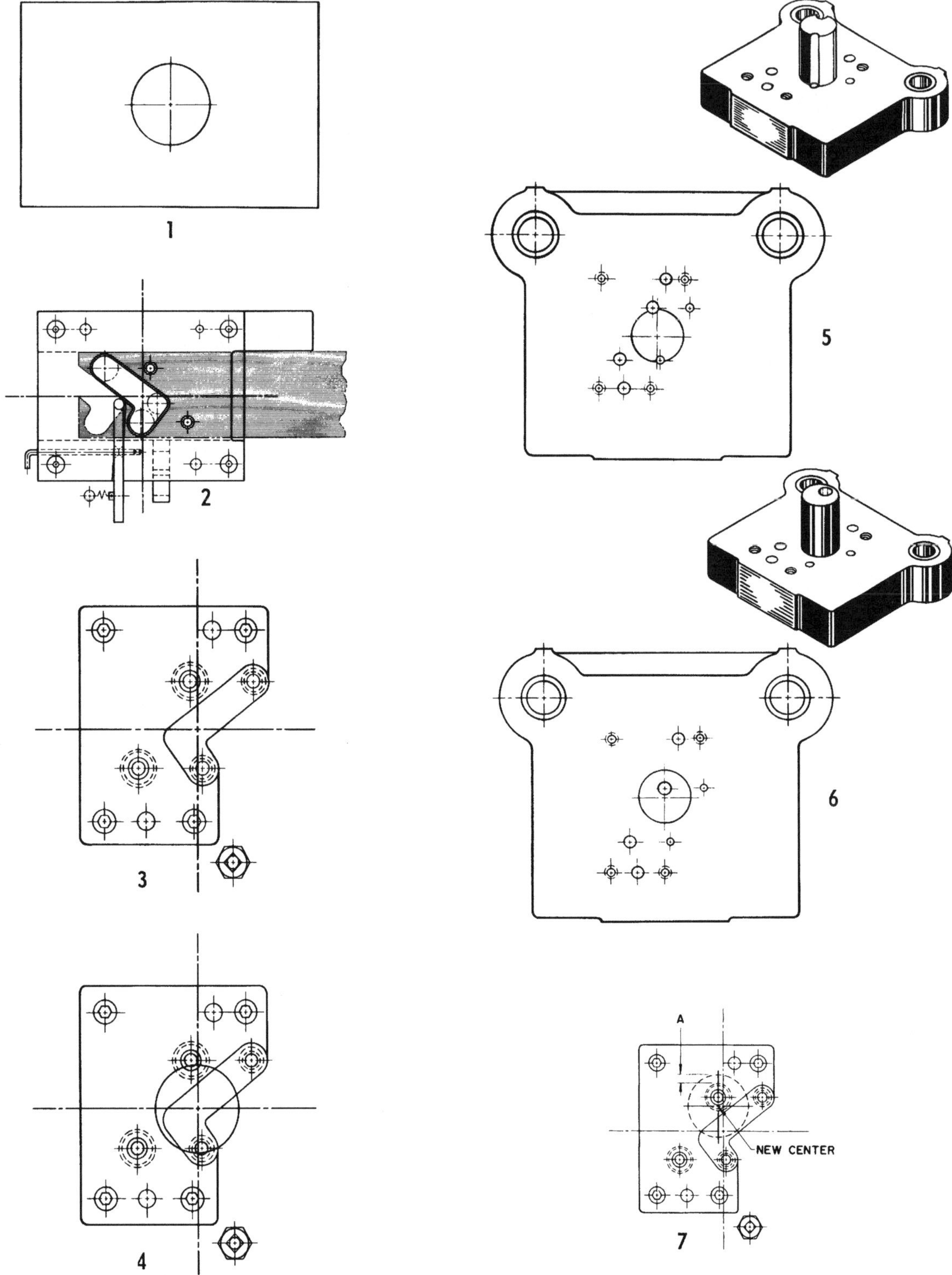

Figure 17.15 Steps necessary for selecting a suitable die set for a specific die design (Steps 1 to 7).

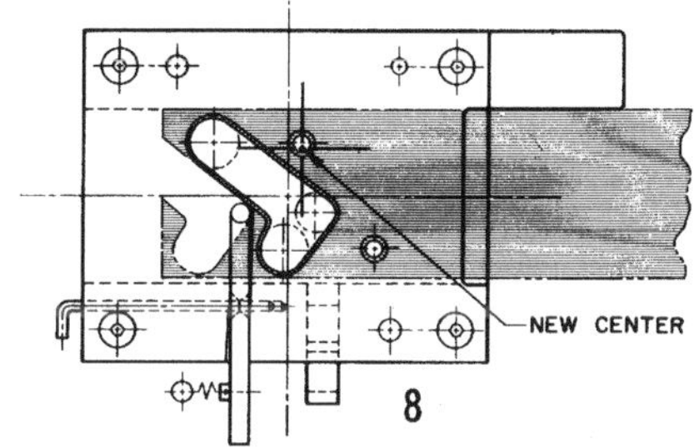

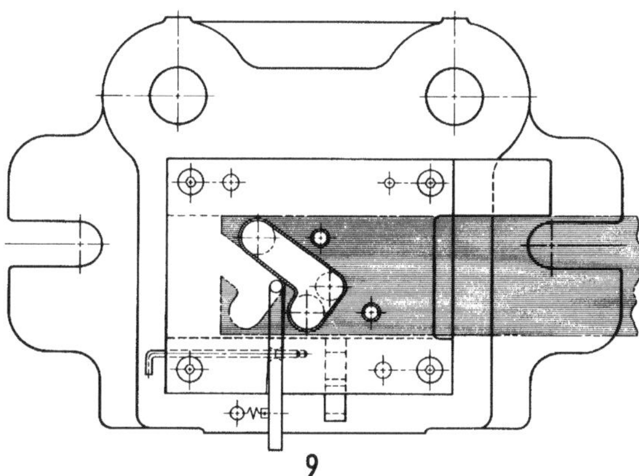

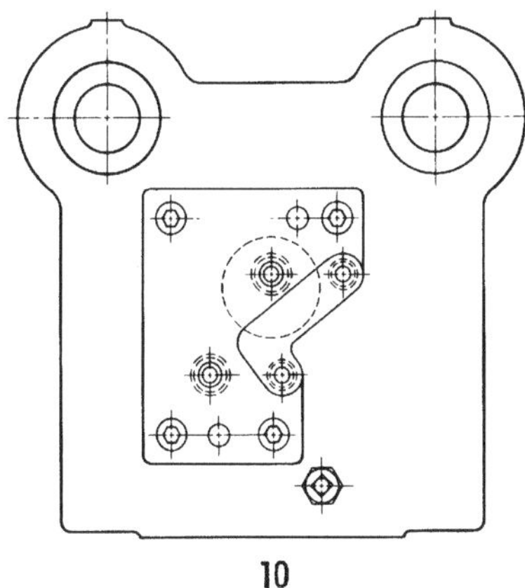

Figure 17.15 (continued) Steps necessary for selecting a suitable die set for a specific die design (Steps 8 to 10).

center, but that it is positioned as close to the ideal center as possible.

10. Complete the view of the punch holder. Front and side section views are then drawn by projection.

In actual practice, the foregoing steps can be performed quickly. They provide the logical method of selecting a suitable die set for a specific die design.

17.5 DRAWING THE DIE SET

Study of any die set catalog will reveal that a number of dimensions are not given. There are three ways of establishing dimensions that are not listed in a catalog:

- Dimensions may be estimated. Those dimensions that are not listed are not critical and may be approximated to proportions printed in the catalog.
- Dimensions may be established with proportional dividers. The point relationship of the divider arms may be set to one of the given dimensions. Points are transferred from the catalog views to the views of the drawing.
- When a component must be drawn at frequent intervals, missing dimensions may be measured and recorded in a standards book for future use.

Figure 17.16 gives all dimensions that are not ordinarily listed in die set catalogs. These dimensions may be used directly when applying the views of the die set to your drawings.

A proper working procedure reduces loss of time in laying out die sets. Seven steps are required (Figure 17.17):

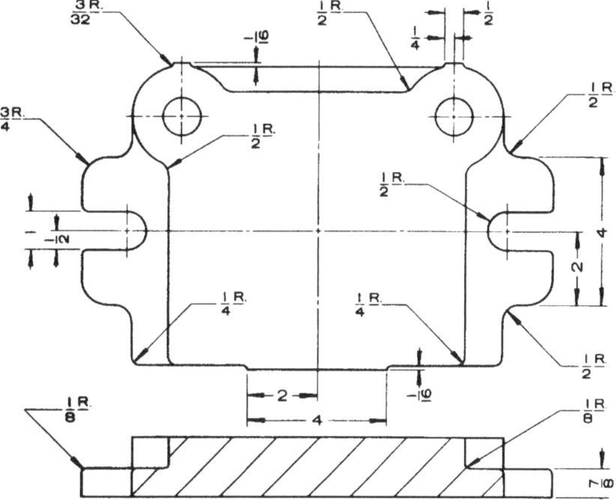

Figure 17.16 Dimensions not ordinarily listed in die-set catalogs.

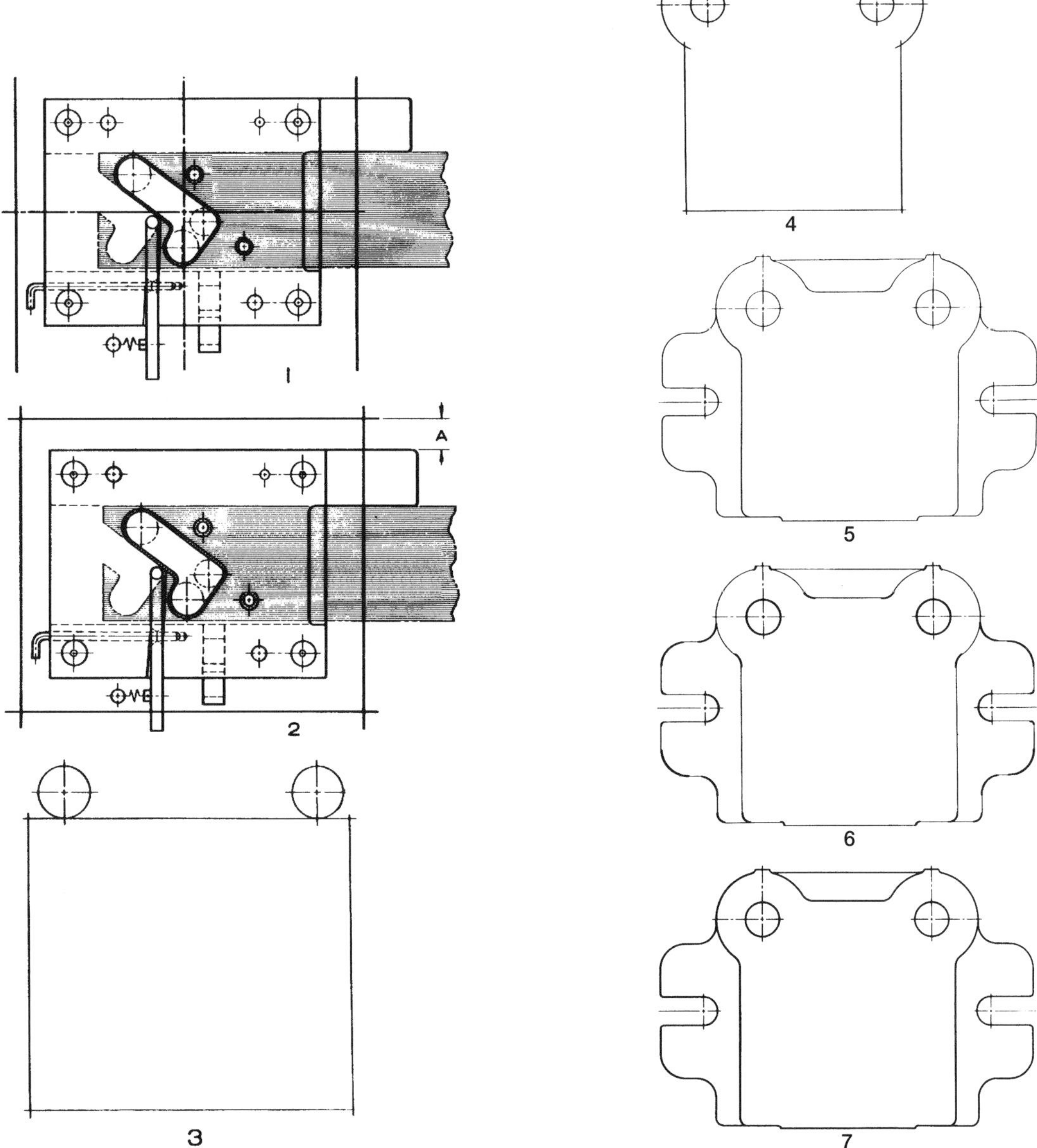

Figure 17.17 Steps taken in laying out a die set.

1. Represent the left and right sides of the die set by vertical lines, as shown in the illustration.
2. In addition, the vertical center of the die set has been established. Preferably the die is centered between the posts and the front of the die set. Look up this dimension in the catalog, then draw horizontal lines to represent the front of the die set. Dimension **A** between posts and die block should be 5/8 inch (15.8 mm) minimum to allow clearance for the grinding wheel nut when the die block is sharpened.
3. Look up the distance between centers of posts and their diameters, then draw circles to represent them.
4. Look up the radii of post bosses and draw them with light lines.

5. Block out the rest of the die set with light lines.
6. Draw all small arcs with heavy lines using a radius template.
7. Darken all remaining lines to final object line width.

The punch holder is drawn in the same manner. Less time will be required because it does not have clamping flanges.

a) Views for Cast Die Sets

Four views are drawn of a back-post die set made of semi-steel or cast iron (Figure 17.18). These are the views ordinarily shown on a die drawing. Observe that the punch holder and die holder are castings; therefore, radii are applied to all corners where the adjacent surface has not been machined. Corners of guide post bosses and of the flange are given large blending radii, a characteristic of this type of die set. The punch holder at the upper right has been inverted and, therefore, the punch shank appears as a dotted circle. Because bushings are viewed directly, they are represented by two solid circles.

b) Steel Die Sets

Steel die sets are drawn somewhat differently. It is necessary to understand the method of manufacture to draw them properly. The first step in producing a steel die holder is to laser- or flame-cut the shape from plate, as in view **A** of Figure 17.19. Note that edges have straight sides all around, and that the plate from which it is cut is the die holder thickness with allowance for machining. In actual practice, a number of blanks would be cut simultaneously in an automatic cutting machine, with the cutting heads guided by a template. In the next operation, the bolting flanges are milled, as in view **B**. A 5-inch (127-mm) diameter cutter is employed, and therefore a 2 1/2-inch (63.5-mm) radius is left at the end of the cut. Corners of the cutter are ground to a 1/8 inch (3.2 mm) by 45-degree chamfer, and a corresponding angle would be left in the corners of the die holder.

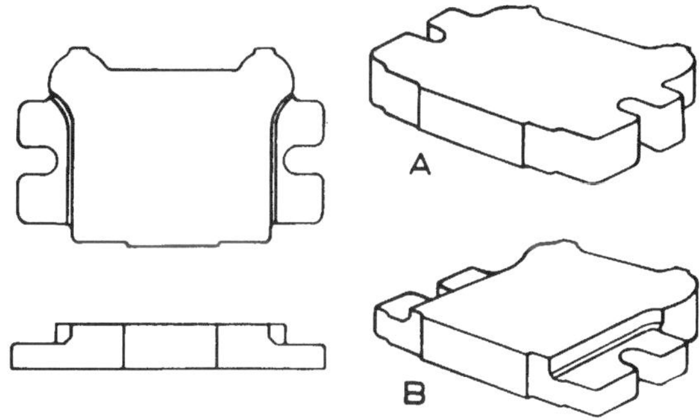

Figure 17.19 Steel die sets are first lasered or flame-cut to shape (A). Then, bolting flanges are milled (B).

Figure 17.20 shows the four views of a steel die set. Comparison with the views of a semi-steel die set will reveal three significant differences. The first is the large radius **A**, and it is 2 1/2 inches (63.5 mm). Another is the absence of a flange at the back of the die holder **B**, and a third is the small angle **C**, applied in the corner of the milled cut. It is drawn 1/8 inch (3.2 mm) by

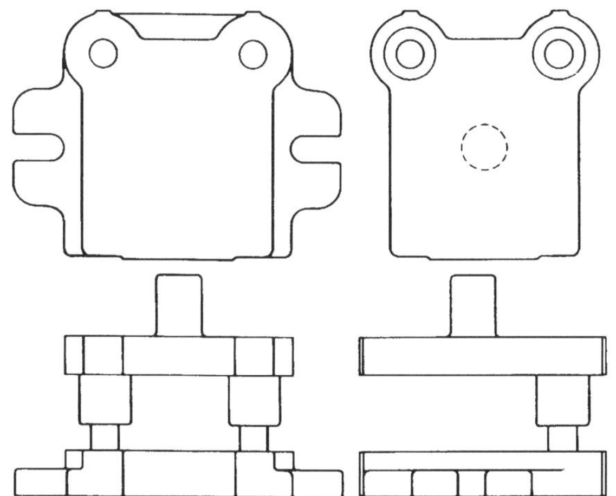
Figure 17.18 How four views of a cast back-post die set are shown on a drawing.

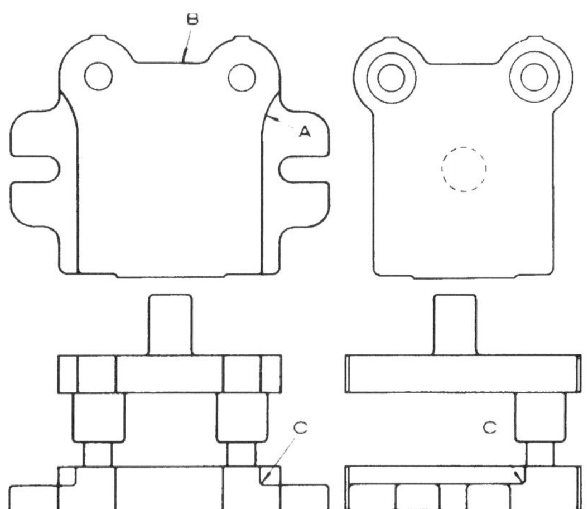

Figure 17.20 How four views of steel back-post die set are shown on a drawing.

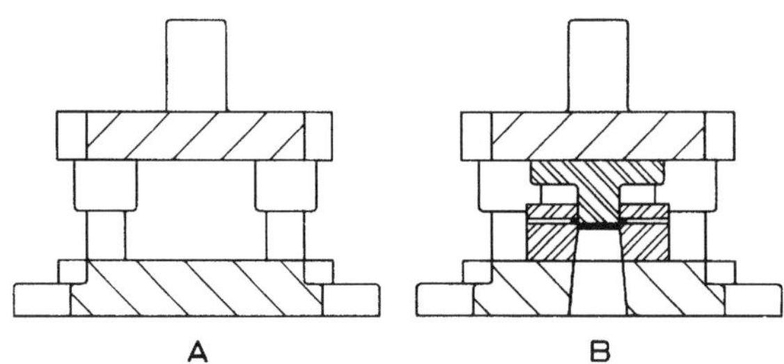

Figure 17.21 How section lines are drawn in a die drawing.

45 degrees. Whenever a steel die set is selected, the views should be drawn in this style and to the proportions given. In addition, sectioning for steel should be applied.

c) Sectioning

The front view and also, usually, the side view of die drawings are sectioned to reveal internal construction (Figure 17.21). Inclination of section lines (hatch) for the punch holder and die holder should be opposite, as shown in view **A**. This "drawing technique" is important for balance. If section lines for both are inclined in the same direction, an optical illusion is introduced and the view will appear to lean.

Section lines for other details are then drawn, alternating their directions from top and bottom toward the strip, as in view **B**.

d) Specifying Die Set Information

Listing the die set. In the bill of material, the die set is always given detail number 1. The information to be specified is too lengthy for the space available in the bill of material and is, therefore, given in the form of a note on the drawing itself, as in Figure 17.22 in view **A**. The box in the bill of

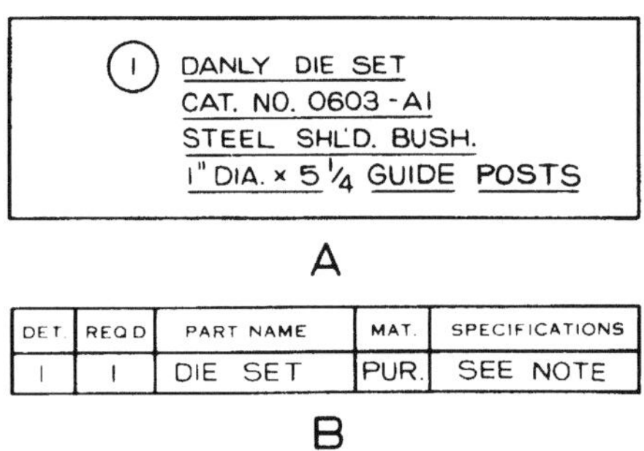

(1) DANLY DIE SET
CAT. NO. 0603 - A1
STEEL SHL'D. BUSH.
1" DIA. × 5 1/4 GUIDE POSTS

A

DET	REQ D	PART NAME	MAT.	SPECIFICATIONS
1	1	DIE SET	PUR.	SEE NOTE

B

Figure 17.22 Die set on a drawing is described in a note (A) and is listed as detail 1 in the bill of material (B).

DIE SET DATA

CAT. NO. OR SERIES, PRECISION 0808 - A1 COMMERCIAL ——
DIE SPACE -- A 8 1/2 (LEFT TO RIGHT) B 8 (FRONT TO BACK)
THICKNESS -- J 1 1/2 (DIE HOLDER) K 1 1/4 (PUNCH HOLDER)
GUIDE POSTS -- L 8 3/4 (OVERALL LENGTH)
BUSHINGS -- TYPE STEEL SHOULDER
SHANK -- DIA. 1 9/16 LENGTH 2 1/8
IF NONE IS REQUIRED STATE NONE

ORDER FROM
DANLY MACHINE SPECIALTIES, INC.
2100 So. LARAMIE AVE., CHICAGO 50, ILLINOIS
(Or Nearest Branch Office)

Figure 17.23 Filled-in rubber stamp print may be used to specify die set data on drawing.

material is filled in, as shown in view **B**, and the reader of the drawing is thus referred to the note.

The die set note should contain the following information:

- The detail number, applied within a detail circle 7/16 inch in diameter.
- Manufacturer's name followed by the words "die set."
- The catalog number.
- The type of bushings.
- Diameter and length of guide posts.

The die set note is applied in a clear space near the center of the die drawing.

Die set stamp. In another method of specifying die set information, a box is printed on the drawing sheet with a rubber stamp, and information is lettered in the spaces provided (Figure 17.23). Stamps are available from die set manufacturers; they vary somewhat in detail. Notice from the example that some dimensions are not absolutely necessary for standard die sets. However, applying extra dimensions such as those for die space tends to reduce errors because a mistake in specifying the catalog number will be discovered.

17.6 MACHINING THE DIE SET

Machining holes for the passage of slugs. There are two methods of applying holes in the die holder for the passage of slugs (Figure 17.24). In view **A**, slug relief holes are machined with straight walls. Dimension **C** is 1/16 inch (1.6 mm) for

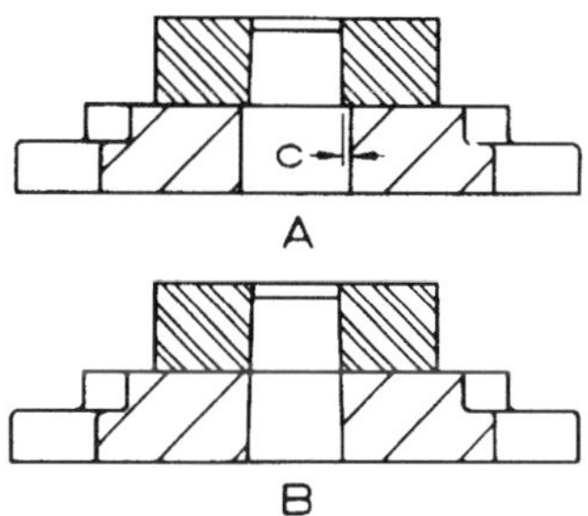

Figure 17.24 Two methods of applying holes for the passage of slugs.

blanks up to No. 16 gage (0.0625 in. or 1.6 mm), and proportionately more for thicker blanks.

The method shown in view **B** is preferred for blanking and piercing soft materials, which have a tendency to stick together and form thick slugs that could jam. The holes in the die set are die-filed to the same angle applied to the die block for relief.

Stripper bolt holes. Counterbored holes for stripper bolts should be proportioned carefully because considerable pounding occurs against the bottoms of the counterbores in operation of the die. Also, a spring stripper may occasionally stick or become jammed. When this happens, a pry bar is used to free it. The stripper then slams up with all the stored energy in the springs. Apply the following minimum proportions (Figure 17.25):

dimension **A** = strip thickness + grinding allowance.
dimension **B** = **D** for a semi-steel die set, and 3/4 **D** for a steel die set.

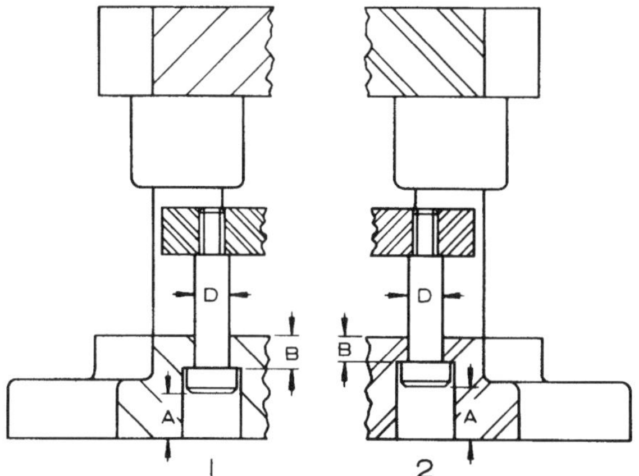

Figure 17.25 Proportions of counterbored stripper bolt holes.

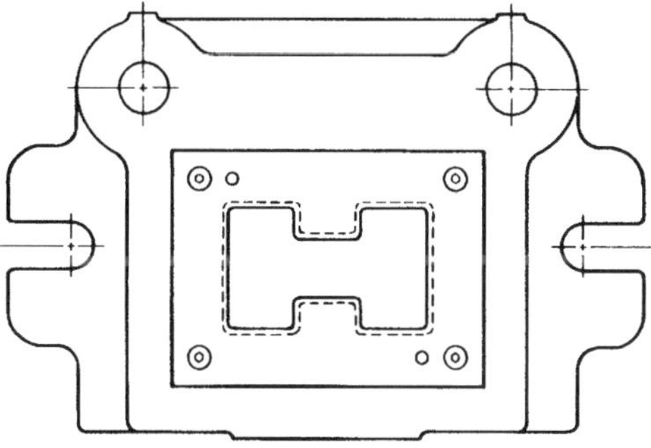

Figure 17.26 All projections of this die block are supported by similar projections in the die holder.

Projections. All projections in a die block (Figure 17.26) must be supported by corresponding projections applied to the die holder. In this illustration, the dotted lines represent the hole machined in the die holder for blank removal. Observe that the die block projections are well supported to prevent overhang and possible breakage under pressure.

Slug slots. When the hole in the bolster plate is smaller than the blank, a slot may be machined in the underside of the die holder (Figure 17.27). At the rear of the die set, it is made slightly wider than blank size. At the front, it need be only wide enough to accommodate a thin bar for pushing blanks to the rear of the press.

Knockouts. When a large spider knockout is incorporated in the design of the die, it can be cut

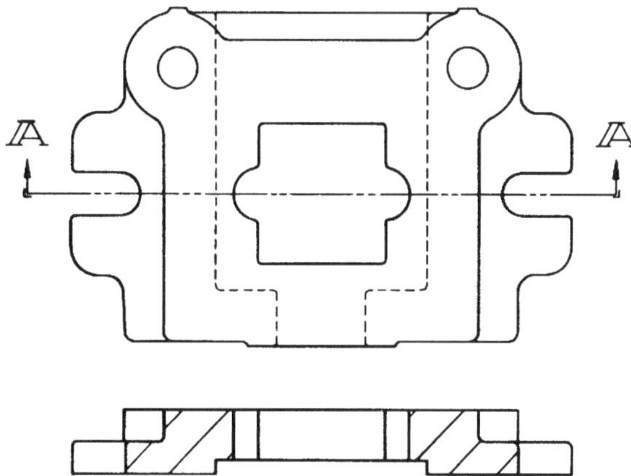

Figure 17.27 A slot machined in the underside of the die holder facilitates removal of blanks when the hole in the bolster plate is too small.

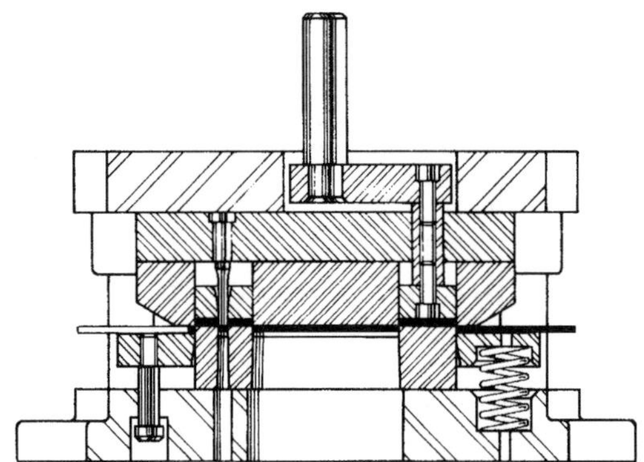

Figure 17.28 Large spider knockout of this steel die set has been cut from the punch holder of the die set.

directly through the punch holder of the die set (Figure 17.28). A steel die set must be specified, and the cutting is done by the die set manufacturer before stress-relieving and accurate machining. After the spider has been cut from the punch holder, it is machined thinner to allow travel; then it is reinserted in the punch holder. The knockout pins are retained to the spider and knockout plate by socket cap screws.

17.7 TYPE OF THE DIE SETS

a) Back-Post Die Sets

Most dies designed for hand feeding of strip are provided with two guide posts applied at the back of the die set. This type gives maximum visibility and accessibility because it is open on three sides. A back-post die set is also used when scrap stock is to be employed for producing small blanks. Scrap material is usually very irregular in shape; other guide post arrangements can interfere with positioning.

There are three distinct types of back-post die sets (Figure 17.29):

1. *Regular.* This type is used for dies with average proportions.
2. *Long.* This type is used for dies that are long and narrow.
3. *Reverse.* This type is used for dies that are relatively longer in measurement from front to back than their measurement from side to side.

Back-post die sets are made in five different styles or shapes (Figure 17.30):

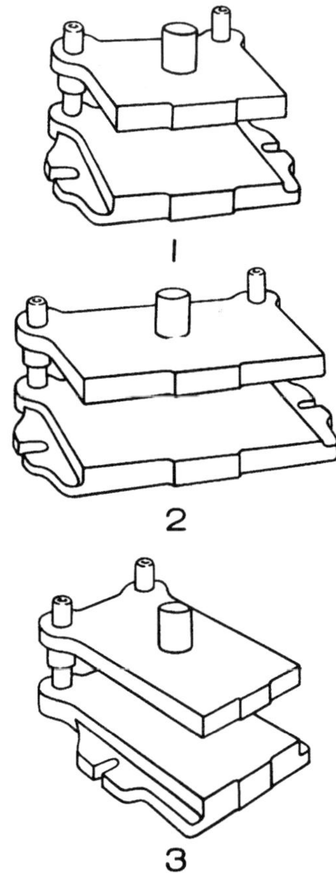

Figure 17.29 Three types of back-post die sets: regular (1), long (2), and reverse (3).

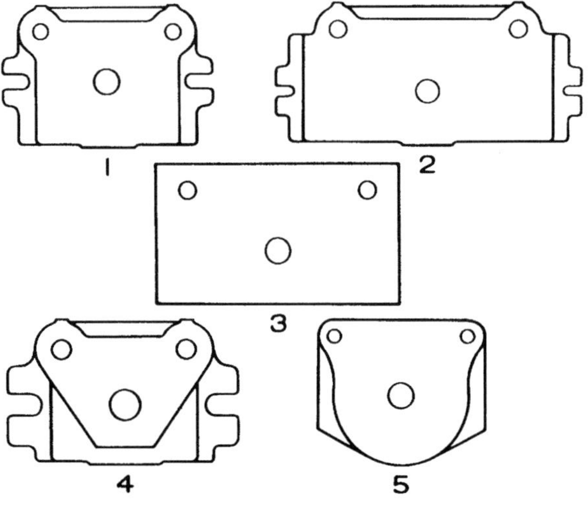

Figure 17.30 Five styles of back-post die sets.

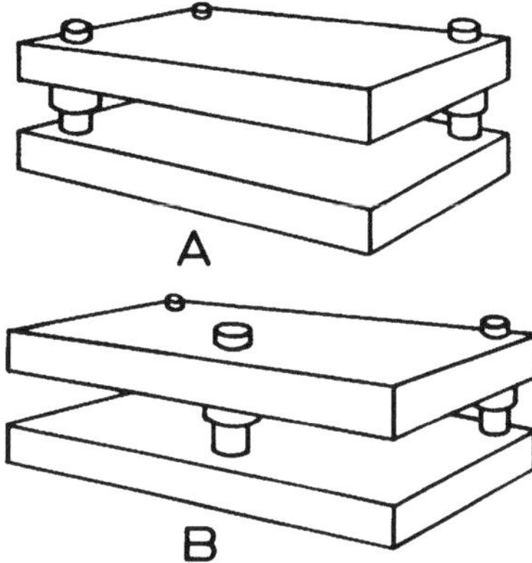

Figure 17.31 Three-post die sets for hand (A) and automatic (B) feeding provide more stability.

1. This style is the most common. It is used for small- and medium-size die sets ranging from 3 by 3 inches (76 by 76 mm) to about 16 by 18 inches (40.6 by 457 mm) in both semi-steel and steel.
2. Large semi-steel die sets ranging from 22 1/2 by 6 inches (571 by 152 mm) to 25 by 14 inches (635 by 355 mm) have sides wider than the distance over the posts, as shown.
3. The larger ranges of steel die sets are made square or rectangular in shape.
4. Many dies have relatively small punch members that occupy little punch holder room. For such dies, die sets with V-shaped punch holders provide a better-proportioned design as well as greater visibility for loading and unloading work.
5. For round punch members a round die set may be used.

Three-post die sets. The addition of a front-post (Figure 17.31) to a back-post die set provides increased stability both for unbalanced cuts and when greater precision is required. These are incorporated only in square or rectangular steel sets. For hand feeding, the extra post is applied at the front, left corner as in view **A**. When the feed is automatic, it is centered, as in view **B**.

b) Four-Post Die Sets

Large die sets are ordinarily provided with four posts. There are two styles (Figure 17.32). View **A** illustrates the conventional type with unbroken edges, used for small- and medium-size four-post sets. View **B** shows the safety type provided with chain slots milled into the sides to facilitate lifting the die with a chain fall. The slots prevent possible slippage as the die is lifted.

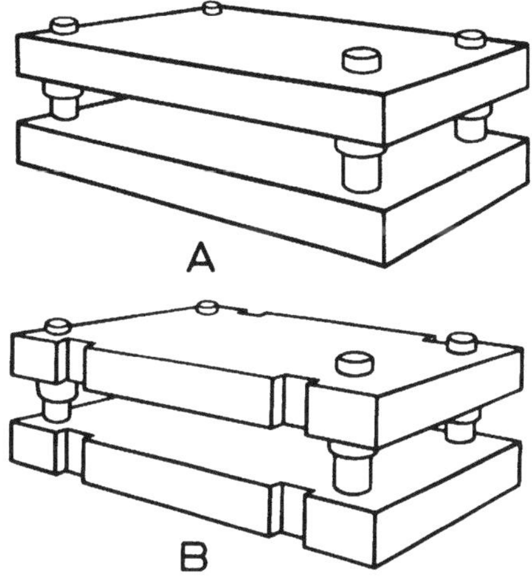

Figure 17.32 Large die sets have four guide posts, some having slots at side (B) to prevent slippage as the die is lifted.

For foolproofing, one post may be offset 1/8 inch towards the edge of the die set, or one post may be made oversize. This prevents accidental reversal in assembly.

Four-post sets are not suitable for hand feeding. They should always be used in conjunction with an automatic feed. They are now widely employed for progressive dies and carbide dies because they provide rigidity.

c) Large, Narrow Die Sets

This type of die set is used to retain dies for cutting, bending, and forming of long, narrow parts. They are back-post sets, and they are available with either two or three guide posts. Two posts (Figure 17.33) are specified for sets ranging from 12 to 72 inches (305 to 1829 mm) in

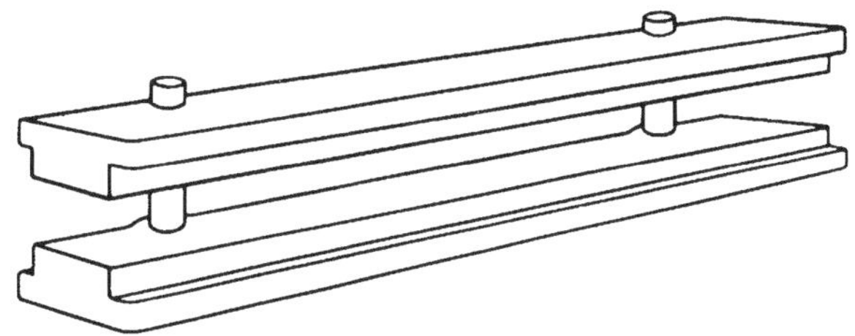

Figure 17.33 Die set for cutting, bending, and forming of long, narrow parts.

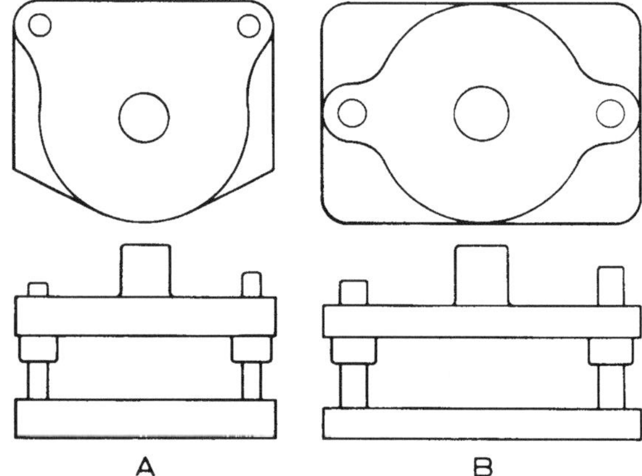

Figure 17.34 Die sets for drawing and trimming round parts.

length, and three posts for sets ranging from 84 to 240 inches (2134 to 6096 mm) in length.

d) Round Die Sets

Figure 17.34 shows two types of round die sets. These die sets are selected for retaining round dies such as drawing dies, trimming dies, and the like. There are two styles: a back-post style as shown in view **A** and a center-post style as shown in view **B**. They are available in diameters ranging from 4 to 48 inches (101 to 1219 mm).

e) Center-Post Die Sets

Die sets in Figure 17.35 are ordinarily employed for secondary operation work such as piercing, coining, and the like. Parts are loaded from the front. The die sets are available in semi-steel in the style illustrated in view **A**, and in steel in the style shown in view **B**. Components may be supplied in combination with a steel die holder used in conjunction with a semi-steel punch holder, as in view **C**.

Another important application for this type of die set is the performing of secondary operations on workpieces having a right and a left side. Parts of one hand may be conveniently loaded from one side. When the other side is to be run, the die set is turned around 180 degrees in the press for ease in loading.

Foolproofing. Diagonal-post die sets (Figure 17.36) are provided with different-diameter posts, dimensions **A** and **B**. Thus, the punch holder cannot be

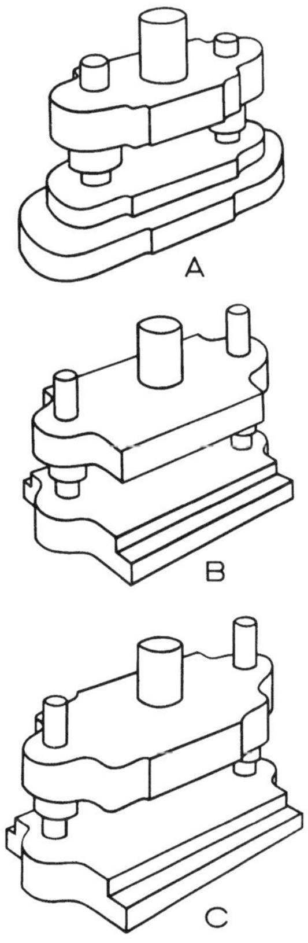

Figure 17.35 Die used for secondary operation work such as piercing and coining.

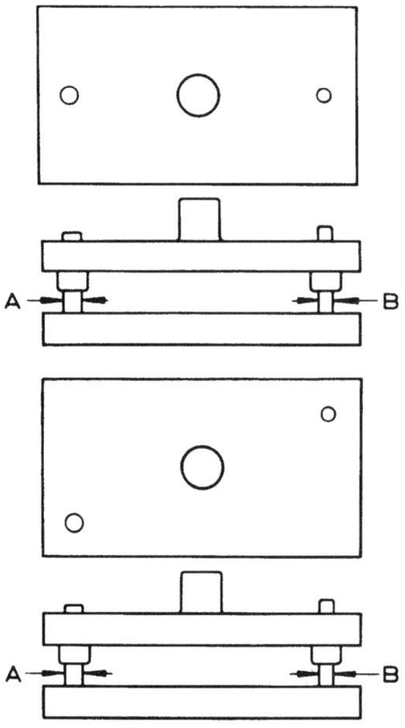

Figure 17.36 Die sets are foolproofed by using different-diameter guide posts.

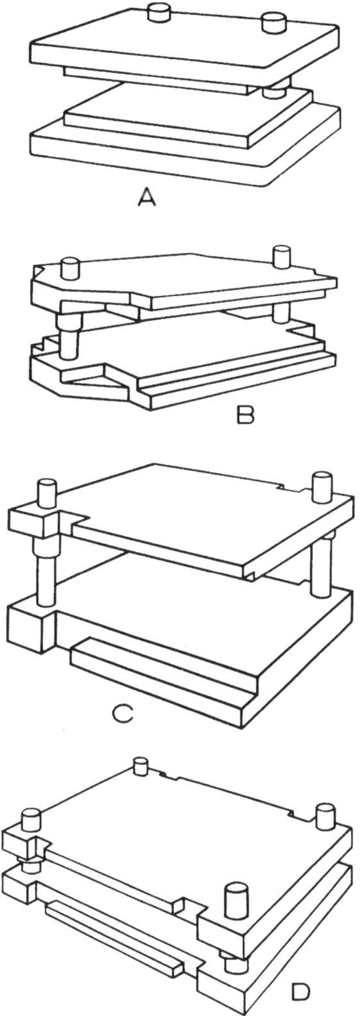

Figure 17.37 Four styles of large semi-steel die sets.

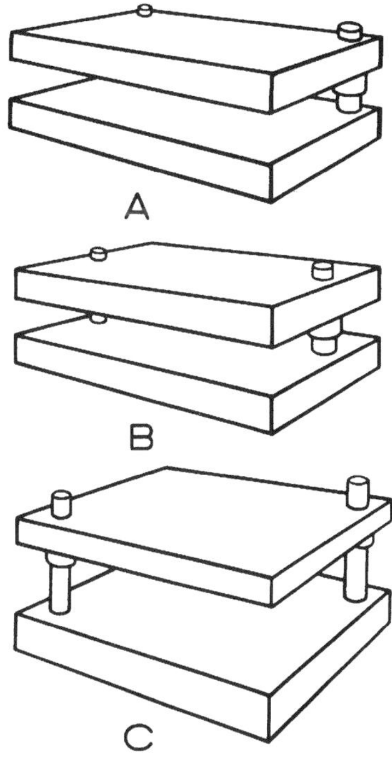

Figure 17.38 The three post arrangements of large two-post steel die sets.

reversed on the die holder. This precaution is important for symmetrical dies.

f) Large Die Sets

Semi-steel. Large semi-steel die sets (Figure 17.37) are available in back-post **A**, center-post **B**, diagonal-post **C**, or four-post **D** styles. All are provided with clamping flanges for mounting in the press. They are assembled with steel shoulder bushings unless otherwise specified.

Steel die sets. Large steel die sets (Figure 17.38) are made of plate. They have ground surfaces and they are square or rectangular in shape. Two-post sets are given one of three post arrangements. View **A** shows the back-post, view **B** the center-post, and view **C** the diagonal-post styles. Sizes are not limited. That is, they may be specified to any length, width, and thickness of punch holder and die holder.

Recommended thicknesses. The die holder and punch holder of a large die set must be given sufficient thickness. Die set catalogs recommend specific thicknesses in relation to side-to-side and front-to-back dimensions; these recommendations should be followed (Figure 17.39). When plates are too thin in relation to their width and length, they will warp, causing misalignment of punch and die members and binding of guide posts in guide bushings. Conversely, when plates are too thick, the overall die cost increases.

The table in Figure 17.39 provides a useful guide to the specification of punch holder and die holder thicknesses. Observe that there are two considerations: the die-space dimensions **A** and **B**, and the force in tons required to perform the work that is to be done by the die. Select the values for **C** and **D** opposite whichever dimension is greater.

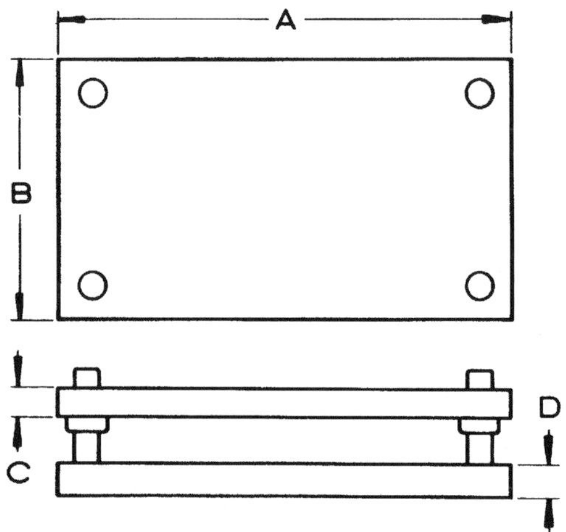

DIE SPACE		PUNCH HOLDER THICKNESS	DIE HOLDER THICKNESS	FORCE IN TONS	
A	B	C	D		
15	10	1 1/4	1 1/2	0	10
30	20	1 3/4	2	10	30
45	30	2	2 1/4	30	50
60	40	2 1/2	3	50	70
75	50	3	3 1/2	70	90
90	60	3 1/2	4	90	110
105	70	4	4 1/2	110	130
120	80	4 1/2	5	130	150
135	90	5	5 1/2	150	200
150	100	5 1/2	6	200	over

Figure 17.39 Table for determining punch and die holder thickness.

Example

If the die set area for a particular die measures 30 by 20 inches (762 by 508 mm) and the force in tons is less than 30, the values of 1 3/4 inches (44 mm) for **C** and 2 inches (50.8 mm) for **D** would be selected. However, if the force in tons were 60, we would use the values opposite force in tons of 50 to 70, and the value for **C** would be 2 1/2 inches (63.5 mm) and for **D** 3 inches (76.2 mm).

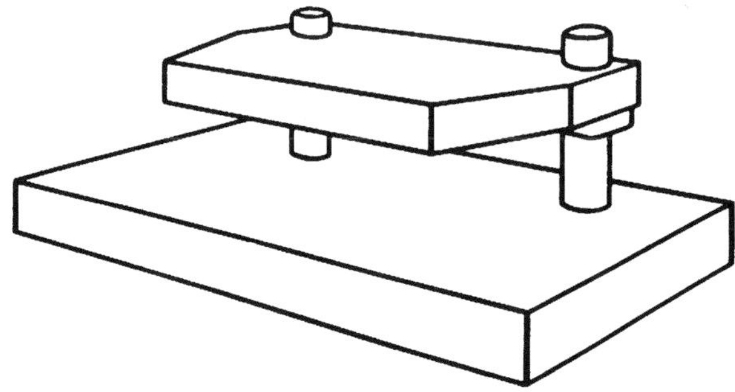

Figure 17.40 Typical heavy duty die set.

g) Heavy Duty Die Sets

This type of die set features an exceptionally large and thick die holder (Figure 17.40); this holder in effect becomes the bolster plate of the press. Heavy duty die sets are particularly useful for long production runs. They are assembled with removable boss bushings to provide adequate alignment between punch holder and die holder.

h) Special Die Sets

Despite the large number of styles and sizes of standard die sets available, it is occasionally necessary to design a special die set for a specific job. This occurs particularly for parts that are exceptionally large or that contain severe offsets. Provide adequate strength by applying ribs or gussets at highly stressed sections. Specific rules cannot be given because of the variety of conditions encountered.

Special die sets are designed by the die designer. A print is then sent to a die set manufacturer where the die set is actually built.

DIMENSIONS AND NOTES

18.1 INTRODUCTION

After the layout of a die has been drawn, dimensions and notes must be applied to complete the information that the die makers will require in order to build the die exactly as the designer planned it. Some dimensions are simple; others must be calculated from considerable background information and experience.

There are two methods of detailing a die. In assembly detailing, all dimensions and notes are given on the assembly drawing. The dimensions can be of two kinds. The first kind includes dimensions that establish relationships between the various components, as needed for assembly of the die. The second kind of dimensions are applied for critical operations performed after assembly to maintain accuracy. Assembly detailing should be used only on designs suited to that method because of the difficulties encountered in building the die if dimensions are vague.

The second method of detailing is to draw and dimension each component individually. Separate detailing is used for more complex dies. It is the best method in most cases because with it, die makers can be furnished with more complete information. Dimensions should not have to be calculated in the tool room because, while the die makers are working on dimensional problems, their expensive equipment is standing idle.

The shut height of the die—the distance from the top of the punch holder to the bottom of the die holder—should always be given on the drawing. It is noted with a fractional dimension followed by the abbreviation S.H. for shut height.

Fractional dimensions on the part print are converted to decimal dimensions on the jig borer layout. This ensures that the die will produce parts well within the required limits. Decimal dimensions are ordinarily given to three places. Often, the digit in the ten thousandths place is left out if it is less than five. When it is five or more, the digit in the third place is increased by one.

18.2 TRIGONOMETRY AS A TOOL FOR SOLVING PROBLEMS

For applying dimensions to die drawings, it is frequently necessary to solve trigonometric problems. Trigonometry involves the determination of lengths of unknown sides or degree of angles of triangles. Our approach to the study of trigonometry will be a practical one. We will employ it as a tool for solving specific problems. Charts will be provided to simplify the subject and to avoid the requirement for memorizing functions or formulas.

a) Right Triangles

Most problems encountered in die design involve the solution of *right triangles* (Figure 18.1). A right triangle has a 90-degree angle. The inclined line, the longest of the three, is called the *hypotenuse.* The side directly across from any given angle is called the *side opposite,* whereas the side nearest to a given angle is called the *side adjacent.*

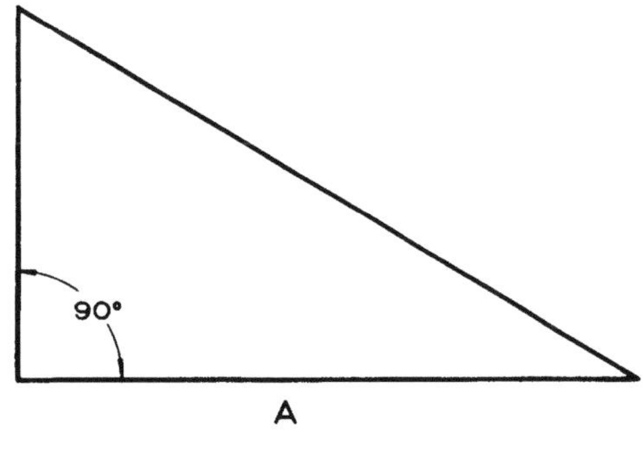

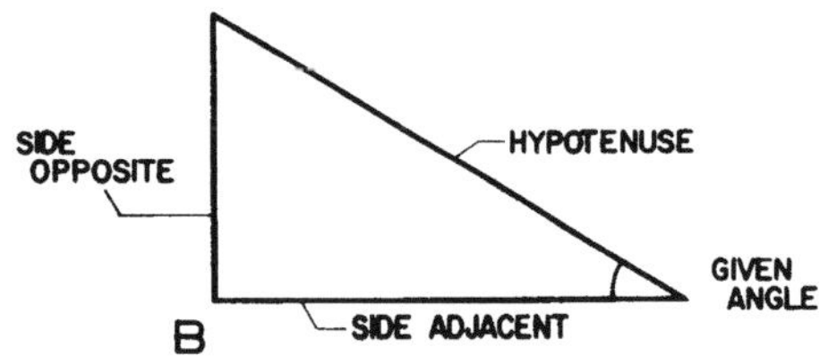

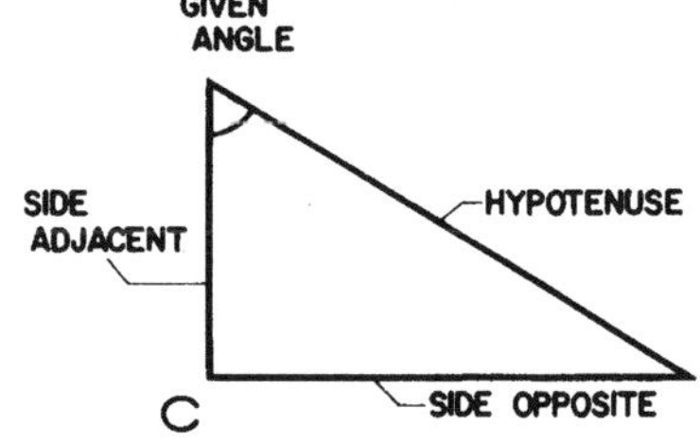

Figure 18.1 Terms relating to right triangles.

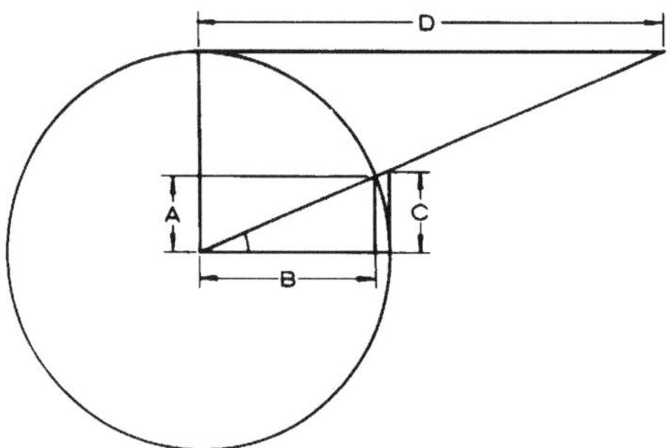

Figure 18.2 Graphic representation of the functions of an angle of a right triangle.

b) Graphic Representation

Figure 18.2 graphically represents the functions of an angle of a right triangle. A circle is drawn with its center at the large apex of a triangle. A vertical line is extended downward from the intersection of the hypotenuse and the circle, line length **A**, and another vertical line from the outside of the circle, line **C**. The following, then, are the functions of the angle (designated with the circular arc) of the right triangle whose hypotenuse is the radius of the circle with a value of 1:

A = sine
B = cosine
C = tangent
D = cotangent

c) Mathematical Relationships

The functions of right triangles (Figure 18.3) may be calculated by the following formulas:

1. The *sine* of an angle is the ratio of the length of the side opposite to the hypotenuse length.

$$\sin A = \frac{a}{c} \tag{18.1}$$

$$\sin B = \frac{b}{c}$$

2. The *cosine* of an angle is the ratio of the length of the side adjacent to the hypotenuse length.

$$\cos A = \frac{b}{c} \tag{18.2}$$

$$\cos B = \frac{a}{c}$$

3. The *tangent* of an angle is the ratio of the length of the side opposite to the length of the side adjacent.

$$\tan A = \frac{a}{b} \tag{18.3}$$

$$\tan B = \frac{b}{a}$$

4. The *cotangent* of an angle is the ratio of the length of the side adjacent to the length of the side opposite.

$$\cot A = \frac{b}{a} \tag{18.4}$$

$$\cot B = \frac{a}{b}$$

In actual practice, functions of triangles are calculated using any calculator with trigonometric functions. They can also be taken directly from a table of trigonometric functions in a handbook. Make sure that you use the table entitled "Natural

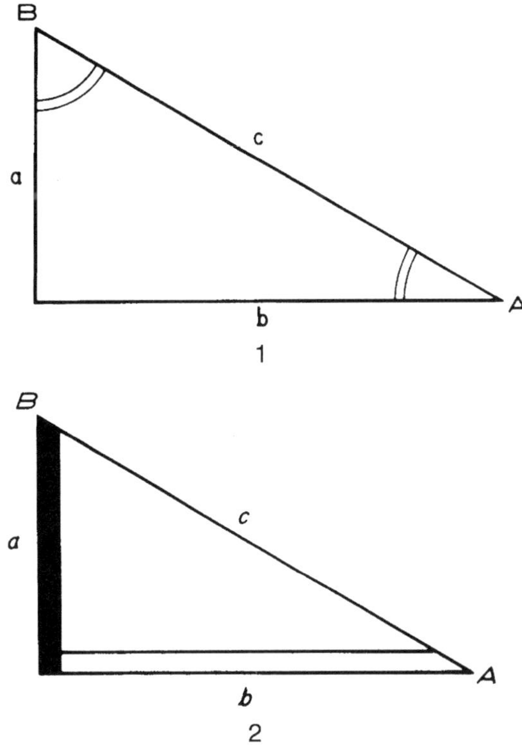

Figure 18.3 (1) Letter designations for use in calculating functions of a right triangle. (2) Solid black side **a** of right triangle is the side to be calculated. Double lines indicate that side **b** is unknown. Hypotenuse **c** and angles **A** and **B** are known.

Trigonometric Functions," and not the one entitled "Logarithms of Trigonometric Functions."

d) Identifying Known and Unknown Sides and Angles

It will be necessary to learn the method of identifying the various sides and angles. In Figure 18.3, observe that capital letters denote angles and lowercase letters denote sides as follows:

A = acute angle
B = acute angle
a = side
b = base
c = hypotenuse.

In our method of identification, a heavy black line denotes the side or angle to be calculated; double lines indicate other unknown sides and angles. In this example, side **a** is the one requiring solution. Side **b** is unknown. Therefore, if we know the values for angle **A**, angle **B**, and hypotenuse **c**, it is a simple matter to solve for the value of side **a**, as will be shown.

e) Trigonometric Equations

Before we begin, however, we must know exactly what information will be required before any triangle can be solved. To compute the length of one side of a triangle, the length of the other two sides must be known or the length of one side and one acute angle must be known. Given this information, we can determine the values of the other sides and angles.

The chart in Figure 18.4 has been prepared to provide a simple method of selecting equations in the solution of trigonometry problems. A careful analysis of its features will enhance its value to you in the development of orderly methods.

Note the five rectangles labeled *side a* at the top of Figure 18.4. Within each box is a triangle and an equation. In every one of these five triangles, side **a** is the unknown side, the one requiring solution. In the first triangle, sides **b** and **c** are known. Angles other than the 90° angle are unknown and they are identified as such by double lines. The formula to employ in solving for **a** when only the base **b** and hypotenuse **c** are known is given above the figure.

In the second triangle, only angle **A** and hypotenuse **c** are known and the formula is given when this condition is present. The same method is followed for the succeeding triangles, and formulas are given for the various combinations of known and unknown sides and angles.

In the next block of triangles, formulas are given for solving for the unknown base **b**, and in the third block for the hypotenuse **c**.

At the top of the next group of triangles, formulas are given for solving for angles **A** and **B**. Underneath this, formulas are given for solving angles and sides when the triangle are not right triangles. In every instance, the known angle—the one used in the formula—is identified on the triangle by a single thin arc.

18.3 DIMENSIONS

a) Part Dimensions

On part drawings for sheet metal workpieces, dimensions are normally applied horizontally and vertically, as shown in view **A** of Figure 18.5 for a simple representative part. Note that a triangle is

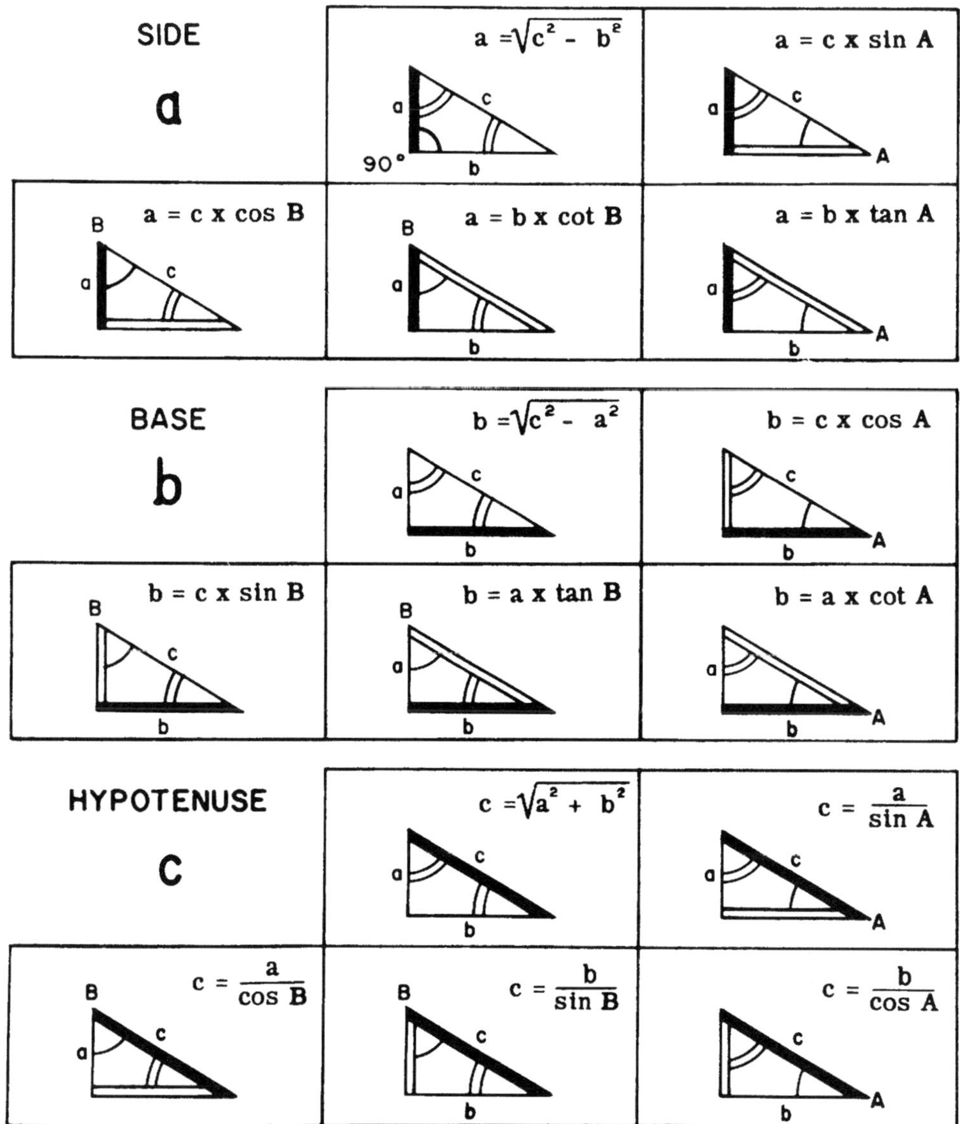

Figure 18.4 Trigonometric equations used in computing sides and angles of triangles.

formed, as illustrated in view **B**. The unknown dimensions of this triangle can be easily calculated by trigonometry because its base and side have been given.

b) Scrap Strip

Figure 18.6 shows a good scrap strip design for the part illustrated in Figure 18.5. The blank is drawn in an inclined position to conserve material. Angular positioning of work pieces is required for most parts except simple round, square, or rectangular blanks, which may be run side by side, positioned either vertically or horizontally.

c) Jig Borer Layout

Holes in die blocks (Figure 18.7) are dimensioned for jig borer layouts. In this type of dimensioning, all horizontal dimensions are given from the line representing the left surface and all vertical dimensions are given from the line representing the upper surface. Grind marks are applied to extension lines to indicate surfaces from which accurate measurements can be taken. In a

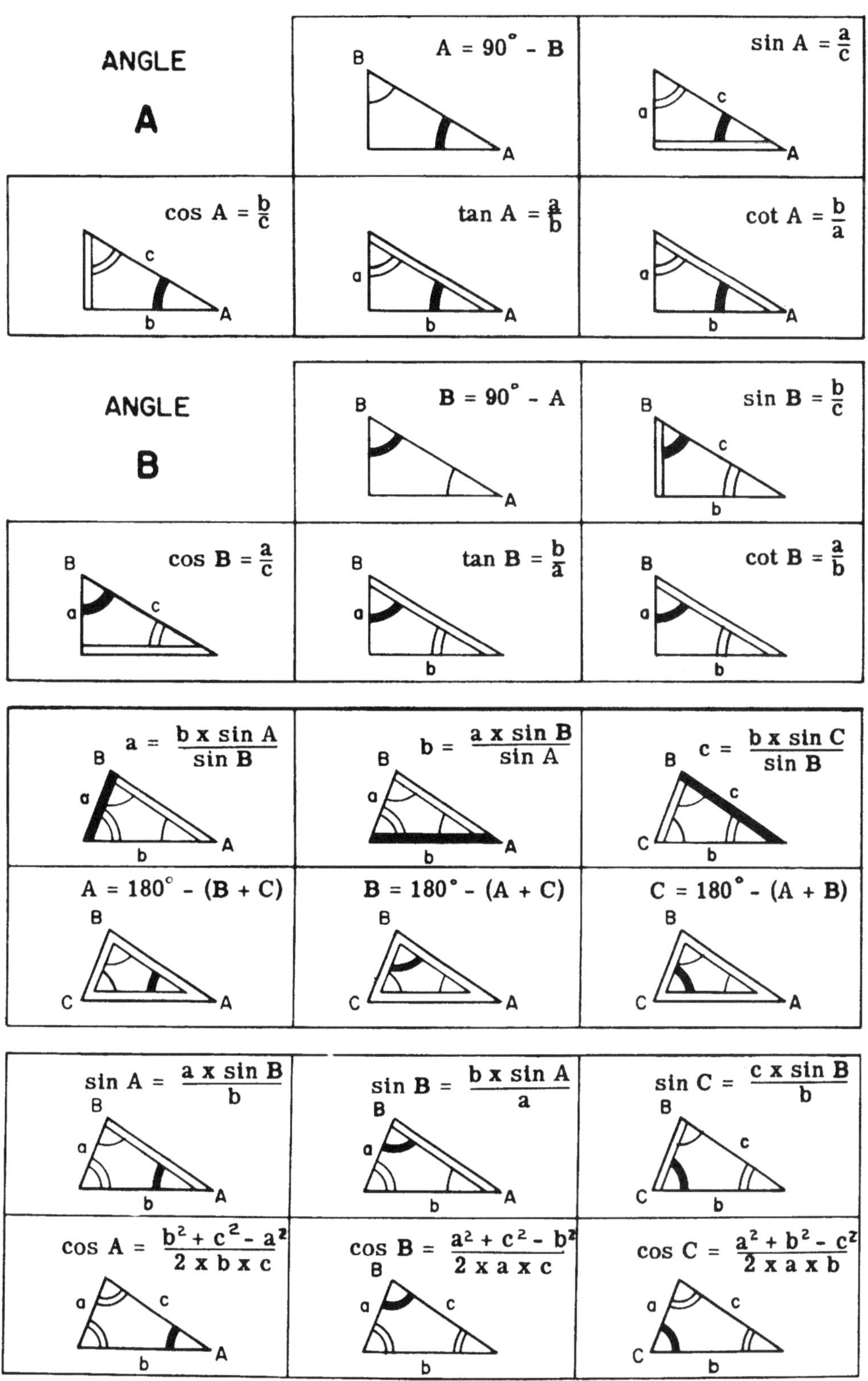

Figure 18.4 (continued) Trigonometric equations used in computing sides and angles of triangles.

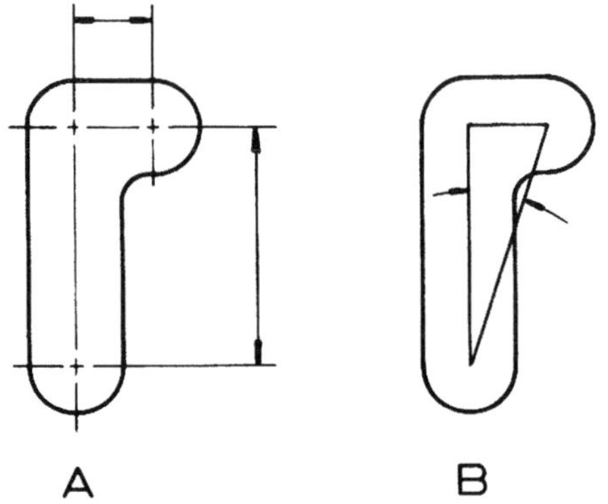

Figure 18.5 Applying trigonometry to part drawing to solve for unknown dimensions.

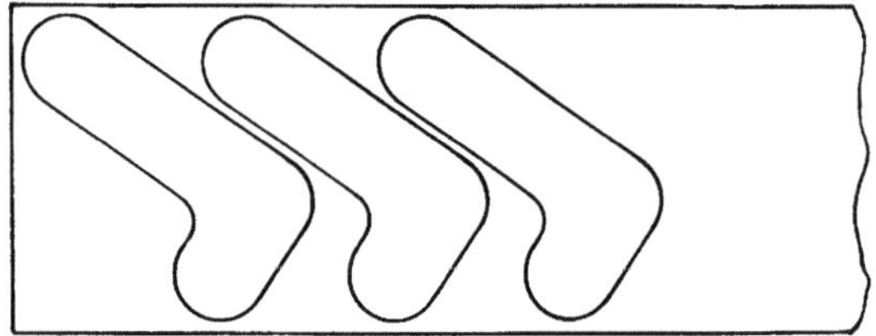

Figure 18.6 Scrap strip design for part shown in Figure 18.5.

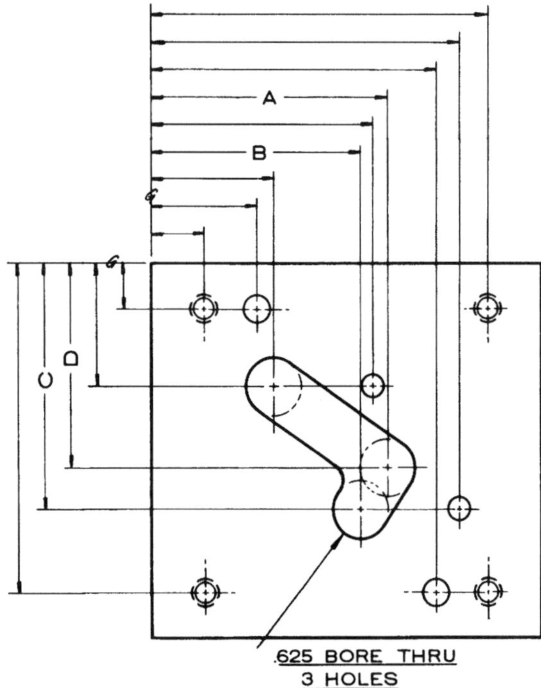

Figure 18.7 Typical dimensioning of die block for jig-boring operation.

variation of the method, the first hole is located with fractional dimensions, both horizontally and vertically. Other holes are then located from it with decimal dimensions in the same manner as shown here.

All internal radii in the die hole contour are completed as full circles with phantom lines. Notes with leaders are employed to indicate that holes are to be jig-bored at those locations. Observe that horizontal dimensions **A** and **B**, and vertical dimensions **C** and **D** cannot be applied directly from dimensions given on the part drawing because of the inclined position of the blank. Their lengths must be calculated. A jig-borer layout guides the jig-borer operator, who machines holes in die blocks and corresponding holes for punches.

In Figure 18.8, a representative layout, note the arrangement of dimensions, all of which are given decimally. All internal radii, those with their

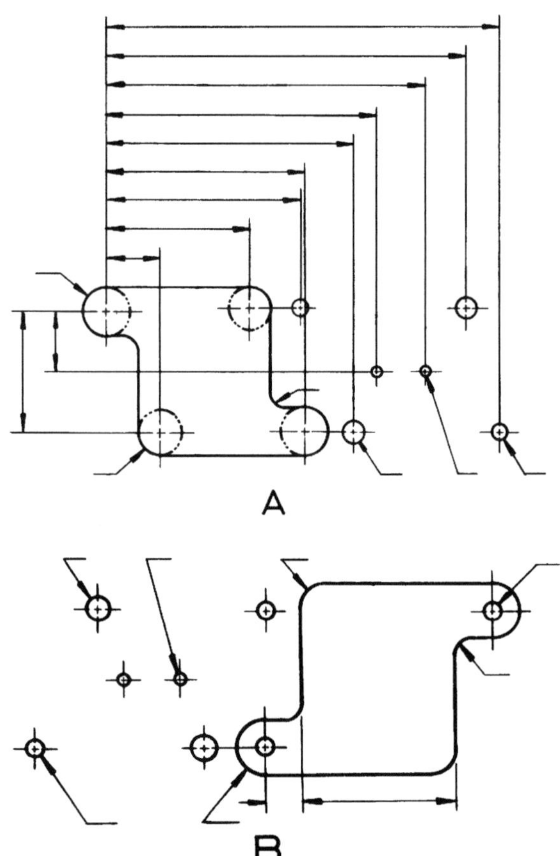

Figure 18.8 Die view (A) and punch view (B) of a typical jig-borer layout.

centers within the large die hole, are completed as full circles with phantom lines; they are specified as jig-bored holes. These accurately positioned holes provide a basis for laying out and machining the irregular die opening. Center-to-center dimensions, which would be identical for the punch view and for the die view, need not be repeated. The punch view **B** gives only sizes of holes to retain piercing punches and actual punch dimensions. Radii are not duplicated on the die view **A**.

18.4 DIE CLEARANCES

Another factor must be considered in applying die dimensions: the amount of clearance between punch and die members. For a blank to part cleanly from the material strip, there must be exactly the correct space between the edge of the punch and the cutting edge of the die. If too little clearance exists, power consumption to operate the press will be excessive. Also, when the punch enters the material strip, the fractures that originate from both sides of the stock—punch side and die side—will not meet, and a ragged edge will be formed on the blank, or on the inside edge of the hole being pierced.

Excessive clearance will dish the blank and produce long, stringy burrs all around the edge. Application of correct clearances will result in a blank free of burrs, and with the burnished portion of its edge extending to the greatest possible depth. This burnished part of the edge should be approximately one-third of the blank thickness.

The proper clearance to apply depends upon the material, its degree of hardness, and its thickness. The accompanying table in Figure 18.9 gives the formula and table of constants by which clearances may be calculated. The formula shows us that the clearance is equal to the stock thickness (in inches) divided by the constant. The values arrived at by the use of this table and formula apply to overall clearances or diameters. When clearance on a side is desired, as when laying out dies that have an irregular contour or for cutting out only portions of a blank, divide the answer by two. For example, if 16-gage (0.0625 inch or 1.6 mm) hard steel is to be punched in the die, look under "Hard Steel" in the table. The constant 14 is given. Divide 0.0625 by 14, and the answer 0.0045 inch (0.11 mm) is applied either to the punch or to the die hole.

Constants for Various Materials

Formula: $\frac{\text{Thickness of Stock}}{\text{Given Constant}}$

MATERIAL	CONSTANT
Copper	21
Brass	20
Soft Steel	17
Medium Steel	16
Phosphor Bronze	16
Hard Steel	14
Boiler Plate (over ¼ in. thick)	10
Soft and Medium Steel (over ¼ in. thick)	10
Aluminum (to ⅛ in. thick)	10
Aluminum (over ⅛ in. thick)	8

PUNCH OR DIE ALLOWANCE TO COMPENSATE FOR PART SIZE CHANGE

STOCK, THICKNESS, GAGE	ALLOWANCE IN.
50 to 22	0.001
22 to 16	0.0015
16 to 10	0.002

Figure 18.9 Table of constants for computing clearances and table of allowances to compensate for part size change.

Below the table of constants in Figure 18.9, another table gives the punch or die allowance needed to compensate for part size change. The use of this information is discussed further in the section Secondary Allowances.

The table in Figure 18.10 shows clearances directly. The values apply to overall clearances or diameters. Stock thicknesses employed by industry are in terms of gages or decimal thicknesses. In either case, the clearance can be interpolated by reference to values given. For clearance per side, divide the given amounts by two.

a) Applying Clearances

For dies of irregular shape, the clearance must be added to the required dimension under certain conditions, whereas at other times it

STANDARD PUNCH AND DIE CLEARANCES

STOCK THICKNESS	SOFT STEEL	MEDIUM STEEL	HARD STEEL	STAINLESS STEEL	PHOSPHOR BRONZE	BRASS	COPPER	ALUMINUM
0.010	0.0006	0.0006	0.0007	0.0008	0.0006	0.0005	0.0005	0.001
0.020	0.0011	0.0012	0.0014	0.0016	0.0012	0.001	0.0009	0.002
0.030	0.0017	0.0018	0.0021	0.0024	0.0018	0.0015	0.0014	0.003
0.040	0.0023	0.0025	0.0028	0.0032	0.0025	0.002	0.0019	0.004
0.050	0.0029	0.0031	0.0035	0.004	0.0031	0.0025	0.0023	0.005
0.060	0.0035	0.0037	0.0043	0.0048	0.0037	0.003	0.0028	0.006
0.070	0.0041	0.0043	0.005	0.0056	0.0043	0.0035	0.0033	0.007
0.080	0.0047	0.005	0.0057	0.0064	0.005	0.004	0.0038	0.008
0.090	0.0052	0.0056	0.0064	0.0072	0.0056	0.0045	0.0042	0.009
0.100	0.0058	0.0062	0.0071	0.008	0.0062	0.005	0.0047	0.010
0.110	0.006	0.0069	0.0078	0.0088	0.0069	0.0055	0.0052	0.011
0.120	0.007	0.0075	0.0085	0.0096	0.0075	0.006	0.0057	0.012
0.130	0.0076	0.0081	0.0093	0.0104	0.0081	0.0065	0.0062	0.0162
0.140	0.0082	0.0087	0.010	0.0112	0.0087	0.007	0.0066	0.0175
0.150	0.0088	0.0093	0.0107	0.012	0.0093	0.0075	0.0071	0.0187
0.160	0.0094	0.010	0.0114	0.0128	0.010	0.008	0.0076	0.020
0.170	0.010	0.0106	0.0121	0.0136	0.0106	0.0085	0.008	0.0212
0.180	0.0105	0.0112	0.0128	0.0144	0.0112	0.009	0.0085	0.0225
0.190	0.0111	0.0118	0.0135	0.0152	0.0118	0.0095	0.009	0.0237
0.200	0.0117	0.0125	0.0142	0.016	0.0125	0.010	0.0095	0.025
0.210	0.0123	0.0131	0.015	0.0168	0.0131	0.0105	0.010	0.0262
0.220	0.0129	0.0137	0.0157	0.0176	0.0137	0.011	0.0104	0.0275
0.230	0.0135	0.0143	0.0164	0.0184	0.0143	0.0115	0.0109	0.0287
0.240	0.0141	0.015	0.0171	0.0192	0.015	0.012	0.0114	0.030
0.250	0.0147	0.0156	0.0178	0.020	0.0156	0.0125	0.0119	0.0312

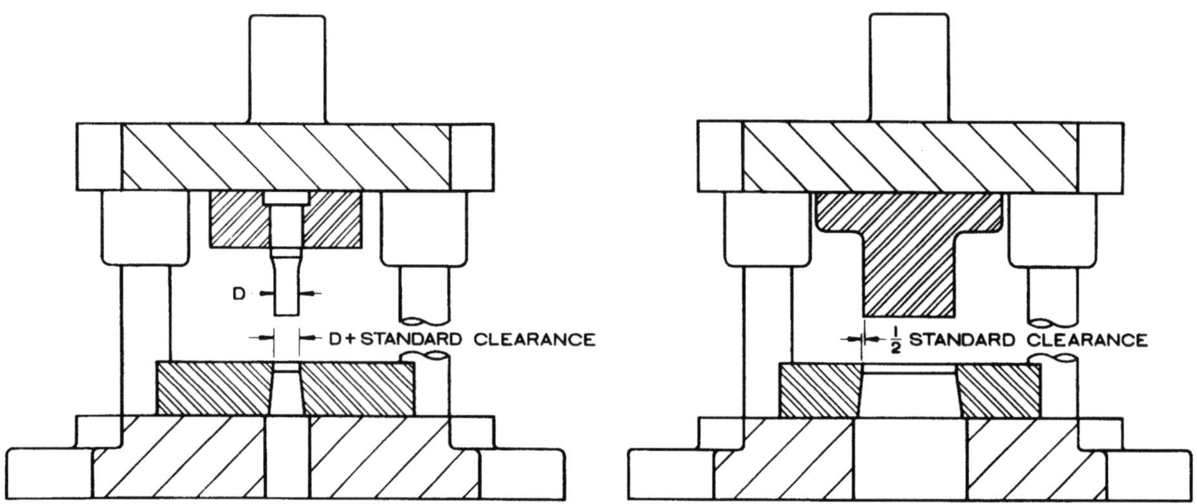

Figure 18.10 Table of standard punch and die clearances, and drawings showing how clearances are applied.

must be subtracted from it. The illustrations in Figure 18.11 show how to apply clearances.

Clearance is applied to either the punch or the die; never to both. Here is the rule to follow: When a slug is produced to be thrown away as scrap, the punch must be to size; in this case, clearance is applied to the die. When a blank is produced that will be kept, and the strip from which it is removed will be thrown away as scrap, the die opening is made to size, and the clearance is applied to the punch. View **1** shows a punch and die hole to which clearances are to be applied. Observe these rules:

1. *When clearance is applied to the punch.* Subtract clearance from all radii with centers

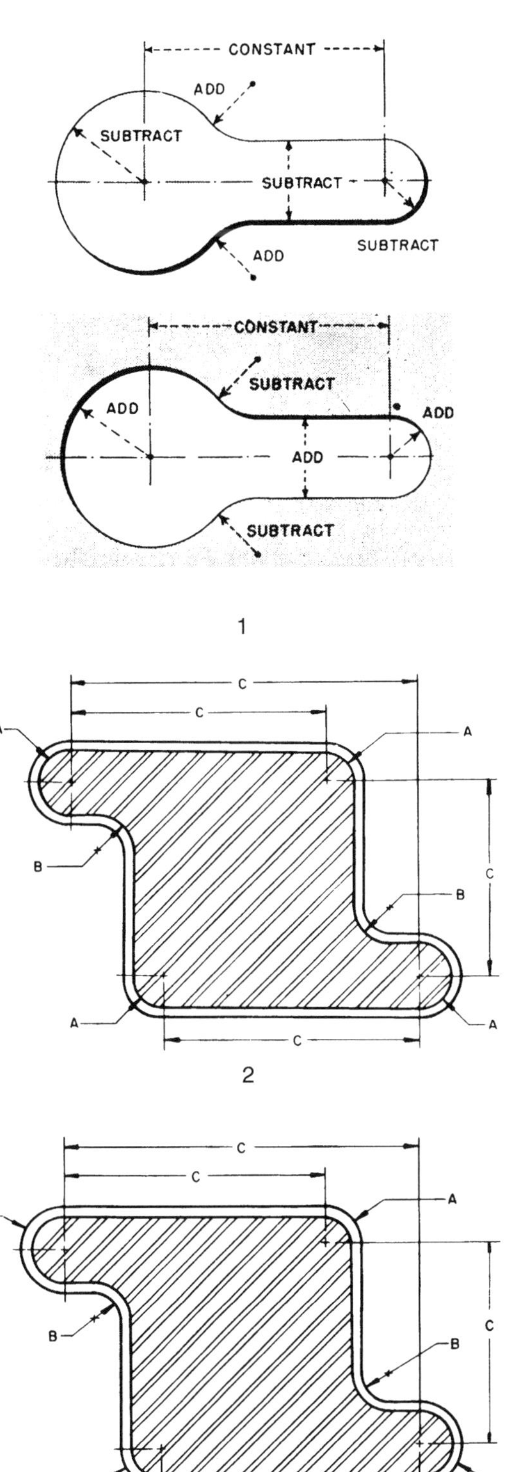

Figure 18.11 Method of applying clearances to punch and die hole of a die.

inside the punch. Add clearance to all radii with centers outside. Subtract from all dimensions between parallel lines. Angles and dimensions between centers remain constant.

2. *When clearance is applied to the die.* Add clearance to all radii with centers inside die. Subtract clearance from all radii with centers outside. Add to all dimensions between parallel lines. Angles and dimensions between centers remain constant.

Let us consider the actual application of clearances to a die drawing. Views **2** and **3** show a blanking punch within a die hole, with clearance greatly exaggerated. The layout in view **2** is a blanking station. Therefore, the die hole must be made to sizes given on the part print. Clearances are applied to the punch and it will be smaller by the amount of clearance. In applying dimensions observe that:

- The amount of clearance is subtracted from each of the radii **A**.
- The amount of clearance is added to each of the radii **B**.
- Distances between centers remain the same for both the punch member and die hole.

The layout in view **3** is a piercing station. Therefore, the punch must be made to sizes given on the part print. Clearances are applied to the die hole and it will be larger by the amount of the clearance. Observe that the order is reversed. For radii at **A**, the amount of clearance is added to the radius. For radii at **B**, the amount of clearance is subtracted from the given radius. Distances **C** remain constant.

Until you acquire facility in applying clearances, it is advisable you make a similar layout, allowing about 1/8 inch (3.2 mm) clearance between lines representing the punch and die members. Dimensioning directly on this layout will reduce the possibility of error because it will be readily apparent whether the amount of clearance should be added or subtracted from any given dimension.

Cutting faces of blanking punches are fitted to the die. Therefore, they do not ordinarily require dimensioning. Also, in the plan view of the punches, do not give locating dimensions. Die makers will know that all punches must fit into their corresponding die openings and will locate them accordingly.

b) Secondary Allowances

After they have been punched out, holes close in a small amount if they are under 1 inch (25.4 mm) in diameter. Blanks under 1 inch in diameter have the reverse characteristic and become larger. A second table in the lower portion of Figure 18.9 lists the amount to either add to the punch or subtract from the die in order to produce an accurate blank.

When the blank is required, the allowance is subtracted from the diameter of the die hole. When accurate piercing is required, the allowances must be added to the punch diameter. For punches and dies of irregular shape, one-half the given value is either added or subtracted all around. Apply the allowance, then add or subtract the clearance.

Example

In piercing a 0.500-inch (12.7-mm) diameter hole in 16-gage (0.0625 inch or 1.6 mm) steel, the allowance (0.002 inch or 0.05 mm) is added to the punch; it is made 0.502 inch (12.75 mm) in diameter. The clearance (0.004 inch or 0.10 mm) is added to this dimension and the hole in the die block is made 0.506 inch (12.85 mm) in diameter.

For a 0.500-inch (12.7-mm) diameter blank, the allowance (0.002 inch or 0.05 mm) is subtracted from 0.500 inch (12.7 mm) and the die hole is made 0.498 inch (12.65 mm) in diameter. The clearance (0.004 inch or 0.10 mm) is subtracted from this dimension and the punch is made 0.494 inch or 12.55 mm in diameter.

Holes or blanks larger than one-inch diameter will not enlarge or shrink appreciably and no allowance need be applied.

When tolerances are liberal, as when hole dimensions are given fractionally, make piercing punch diameters 0.005 inch (0.127 mm) larger than nominal size. This allowance takes care of closing in of holes after piercing and of punch wear. When dimensions are held more accurately, specify punch diameters to the high side of the tolerance.

18.5 BLANK DEVELOPMENT

Blank development entails designing flat blanks from which bent, formed, and drawn workpieces are made. The technique used analyzes the finished workpiece, then in a series of steps transforms it into the flat blank from which it was stamped. The technique is called *blank development*. Designing developed blanks is necessary in order for dimensions to be properly calculated and so that cut-off, blanking, bending, or forming dies can be designed for them. To acquire a thorough understanding of the process of blank development, you will need to learn exactly what occurs when metal is bent. Figure 18.12 illustrates the action. Note the following details:

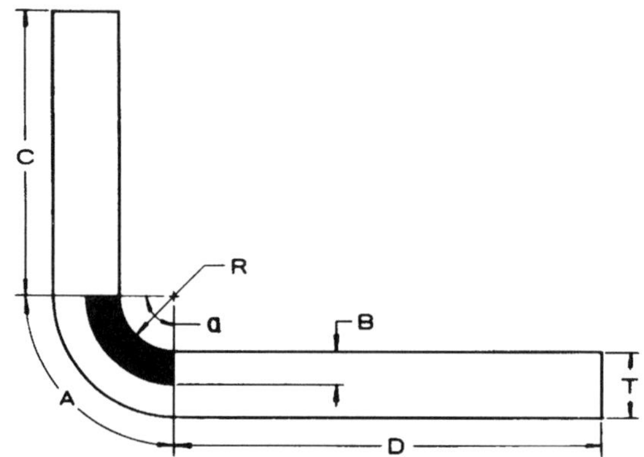

Figure 18.12 Length of blanks is **C** plus **D** plus bend **A** measured along a neutral axis.

1. That portion of the metal on the side of arc **A** has been stretched.
2. That portion of the metal defined by arc **a** and shown solid black has been compressed.
3. Bending may be considered to occur along the inside line **B**. This is the neutral line or axis along which neither stretching nor compression would takes place.

To determine the length of material in the bent arc, it is necessary to compute the length of arc along the neutral line. For calculation purposes, the distance **B** will be considered as either 1/4, 1/3 or 1/2 of thickness **T**, depending on the size of radius **R**. Note that adding **A** (measured along the neutral axis), **C**, and **D** gives the length of the blank.

18.5.1 Developing Bent Parts

To develop a bent or formed blank **A** in an orderly way (Figure 18.13), proceed as follows:

1. Sketch a side view or section of the part, to enlarged scale if necessary, and apply dimension lines as shown.

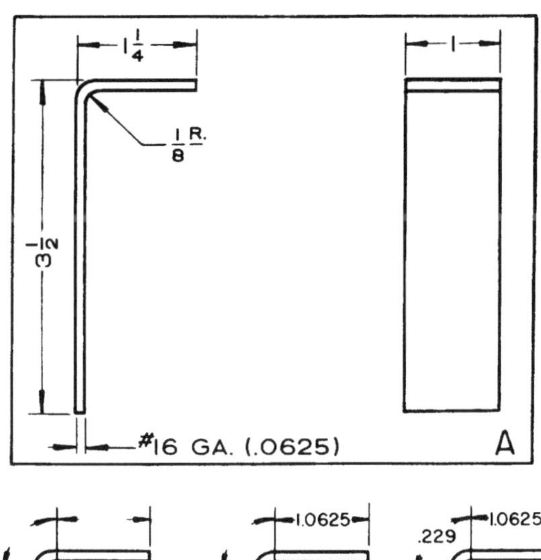

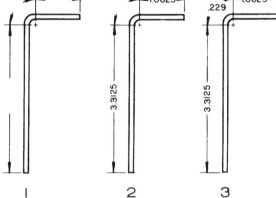

Figure 18.13 Method of determining blank length.

2. From dimensions given on the part drawing, add the thickness of the stock to the given radius; then subtract this from outside dimensions. Fill these in decimally on the sketch.
3. Compute the length of arc and write it in the proper position. Note that adding all dimensions gives the overall length of the part. (See the chart in Figure 18.14.)

For the part under consideration, the formula is used, as the inside radius is 2 times the stock thickness:

$$A = \left(\frac{T}{3} + R\right) \times 1.5708 \qquad (18.5)$$

a) Formulas for Bends

The chart in Figure 18.14 provides all the formulas that you will require for developing corners of bent and formed workpieces. Observe that there are three possible conditions. Each requires the use of a different formula. In the first, the arc has a radius smaller than the stock thickness; appropriate formulas are given below the illustration. In the second, the arc is between one thickness and two thicknesses of stock, and in the third condition, the arc radius is greater than twice the stock thickness. The upper formulas apply for

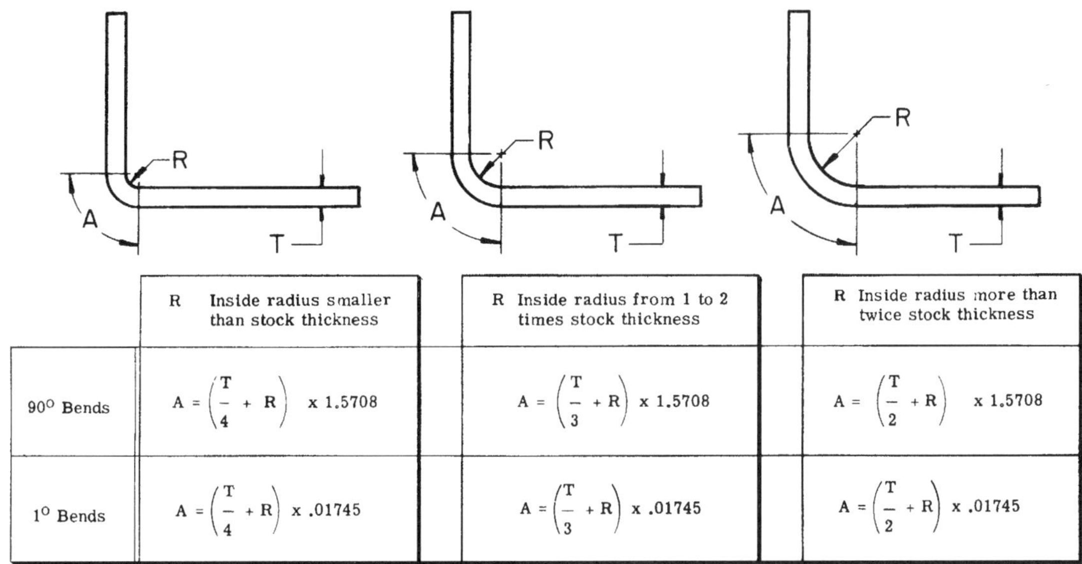

	R Inside radius smaller than stock thickness	R Inside radius from 1 to 2 times stock thickness	R Inside radius more than twice stock thickness
90° Bends	$A = \left(\frac{T}{4} + R\right) \times 1.5708$	$A = \left(\frac{T}{3} + R\right) \times 1.5708$	$A = \left(\frac{T}{2} + R\right) \times 1.5708$
1° Bends	$A = \left(\frac{T}{4} + R\right) \times .01745$	$A = \left(\frac{T}{3} + R\right) \times .01745$	$A = \left(\frac{T}{2} + R\right) \times .01745$

Figure 18.14 Formulas for computing developed lengths of bends.

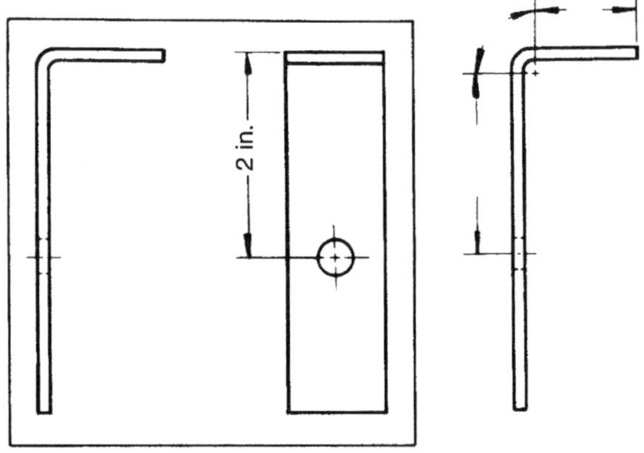

Figure 18.15 Method of determining hole dimension for developed blank.

90-degree bends, which are the most common. The lower formulas apply for bends of 1 degree. These are also used for computing the lengths of bends at angles other than 90 degrees. Multiply the value computed by the actual number of degrees through which the arm of the part is bent.

b) Hole Distance

To establish the developed distance to a hole when it is dimensioned from the surface of the bent edge (Figure 18.15), subtract the sum of the part thickness and the inside radius from the given dimension (in this case, 2). To this value, add the developed arc, taken from the sketch, and the length of the straight arm.

c) Typical Example of Blank Development

The upper portion of the illustration in Figure 18.16 shows a drawing of a part containing a

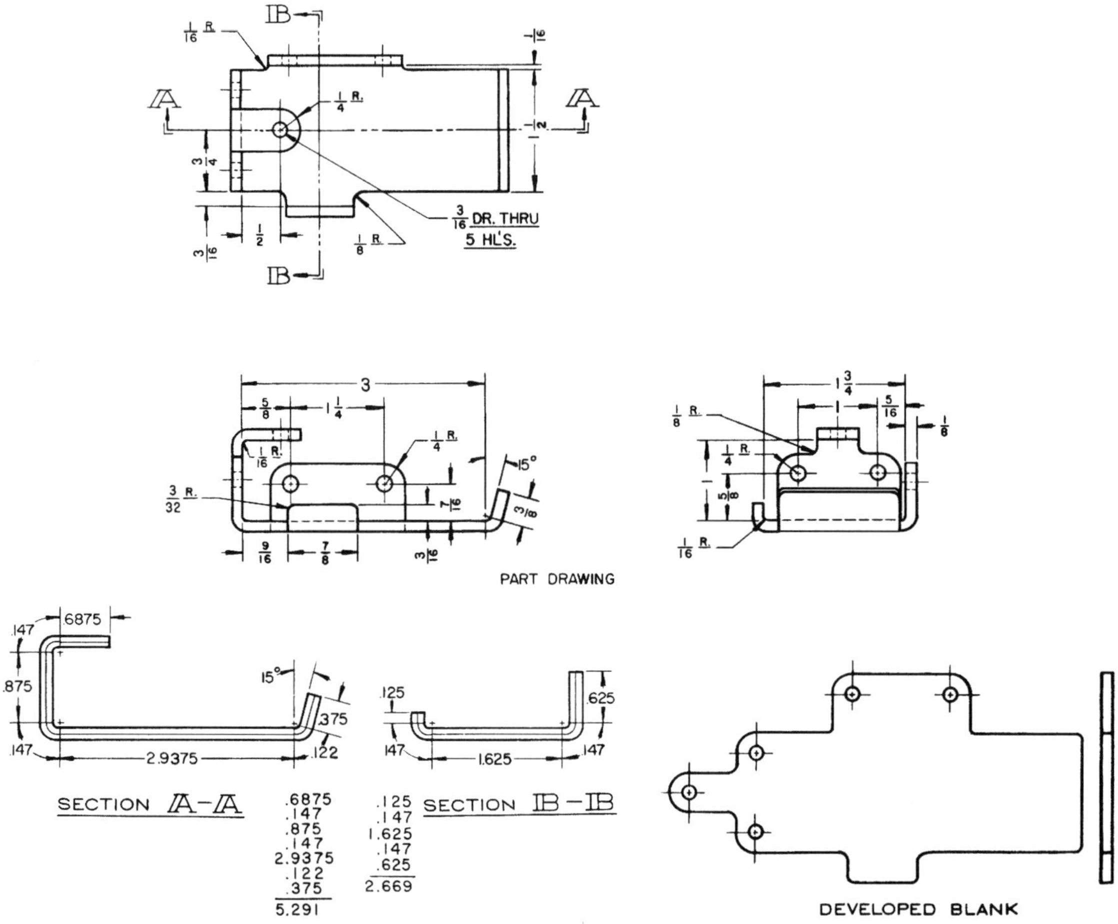

Figure 18.16 Typical example of blank development for bent part.

number of bends. Two sections, a horizontal one marked A-A and a vertical one marked B-B, are taken through the part and drawn below it. Dimensions are applied directly to the sections. Horizontal and vertical dimensions are found by adding and subtracting part drawing dimensions. Dimensions of arcs are found by calculation using the formulas given.

After all dimensions have been applied, the developed flat blank can be easily drawn, as illustrated below the sections. Note that adding all dimensions for both sections gives the length and width of the blank. A rectangle can be drawn that will be the length and width of the finished blank. The drawing is then completed by taking dimensions directly from the sections and the part drawing.

18.5.2 Developing Formed Parts

Although no more difficult, the development of a formed part requires more steps than one that is simply bent. The difference between bending and forming basically occurs in the straightness or curvature of the bend line. In bending, portions of the blank are raised to some angular position and the line of bend is straight. A forming operation is similar, except that the line is curved instead of straight.

A formed workpiece is developed by dividing it into sections, then finding the length of each section by the methods employed for developing a blank for a bent part. Lengths of arcs are determined by employing the formulas given earlier in Figure 18.14.

View **A-A** of Figure 18.17 shows a drawing of a formed workpiece. Wherever the bend occurs in a straight line, it is considered as being bent; wherever the line of bend is curved, it is considered as being formed. Note, however, that any part which contains a form is called a "formed part" regardless of whether or not it also contains bends.

The first step in developing the flat blank is to draw an accurate layout of the top view and front view as shown in view **B-B**. For small parts, this layout is made to an enlarged scale for increased accuracy. Next, draw evenly spaced lines to divide the formed portions of the part into a convenient number of divisions. Extend the lines so they cross both views. These are actually cutting-plane lines. Sections are to be taken through each.

Obviously, the more sections taken, the more accurate will be the development. However, this should not be carried to extremes. Large workpieces may require more divisions, whereas small parts may need fewer divisions to achieve equal accuracy in the development.

On the front view of the layout, draw a line lengthwise to divide the bottom of the part in two. This is shown as line **X** in the illustration.

Now draw the various sections. Observe that the bottoms of the sections are revolved; that is, they are shown as the actual thickness of the part instead of being thicker, as cutting-plane lines in the front and top views would appear to indicate. Dimensions are applied directly to the sections in the same way as for ordinary bends. First, apply dimensions that can be taken from the part drawing. Then scale the layout for other dimensions. Measure carefully along each cutting-plane line, taking a reading from the top of the part to the bottom where it intersects center line **X**. In the section views, this is the distance from the top of the section to the center of the horizontal bottom portion. Lengths of arcs are calculated in the same manner as for regular bends.

Under each section, enter the various lengths and add them together. These will be developed lengths across the workpiece, as measured along each cutting-plane line.

Now determine linear or length dimensions. The length of line **X** must be measured because it is the same as that of the developed blank, as measured along its center line.

The next step is to draw the blank layout **C-C**, proceeding as follows: Refer to the developed lengths of the sections and select the longest dimension. On a sheet, draw a center line and two light horizontal lines the same distance apart. This will be the maximum width of the developed blank. On the layout **B-B**, measure distance **A** and on the blank drawing **C-C**, apply two vertical lines the same distance apart. Next, measure distances **B**, **C**, **D**, etc., and draw vertical lines the same distances apart on the blank drawing. Observe that these distances are measured along the curve of the part and on line **X**, view **B-B**. In blank layout **C-C** they represent distances along the center line. Provide yourself with a die maker's flexible steel scale. Carefully bend the scale to the contour of the line before taking a reading. Values for arcs may be calculated instead.

When you have finished laying out vertical grid lines, mark points on the vertical lines to the respective dimensions of the sections. Dimensions

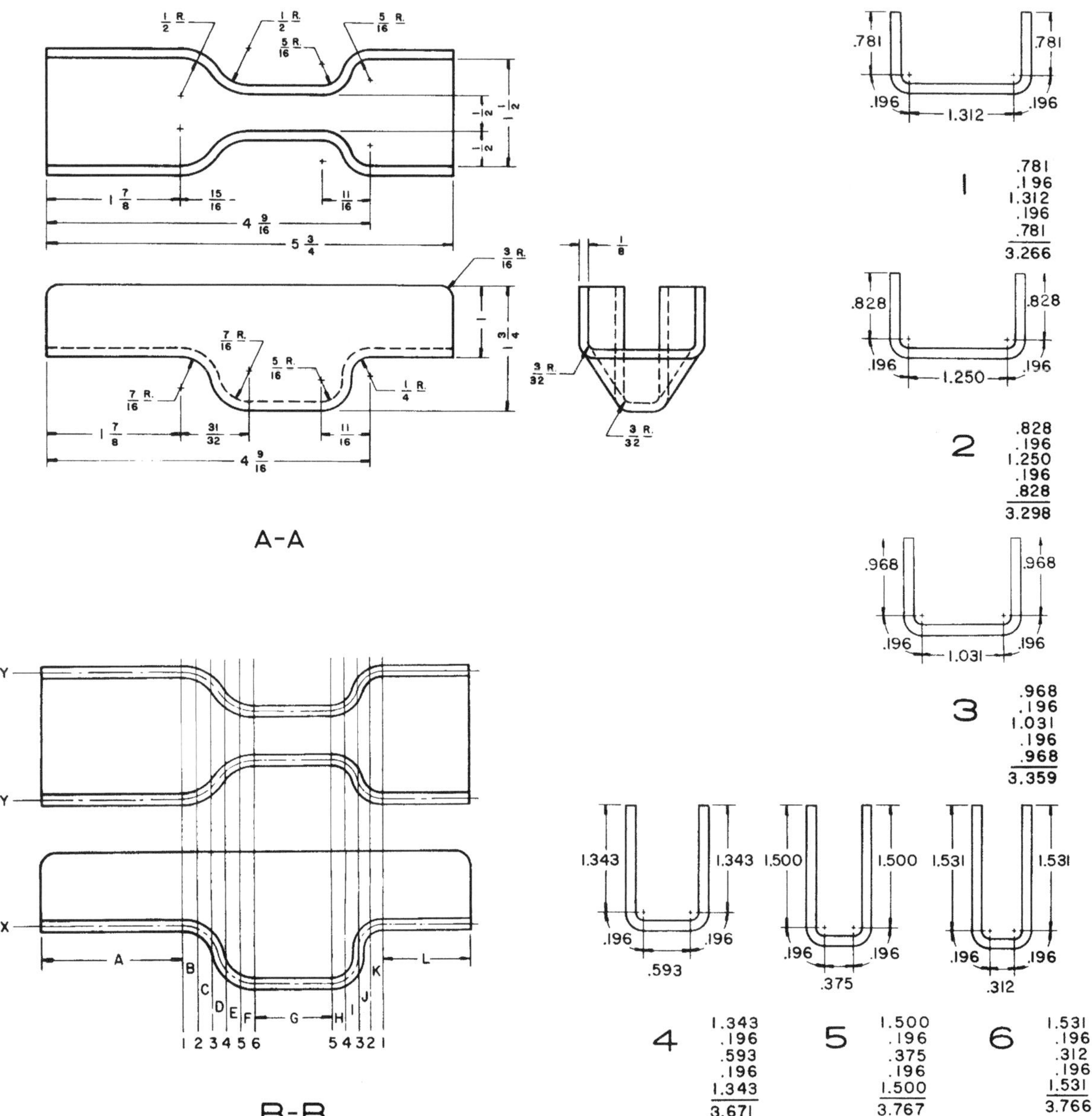

Figure 18.17 Typical example of blank development for formed part.

are marked an equal distance from the center line because the blank is symmetrical.

Several things must be done before an accurate blank layout can be produced. On another sheet, measure the formed portions of the workpiece along line **Y** and draw the divisions as shown on layout **D-D**. The developed layout is then constructed as in layout **E-E**, with the edge distances **A**, **B**, **C**, etc. taken from layout **D-D**, the length of the center line from layout **C-C**, and respective vertical widths from layout **C-C**.

This is necessary because the distances **A**, **B**, **C**, etc. along lines **Y**, which were originally the edges of the blank, are not necessarily the same (in some blanks, they may be the same) as **A**, **B**, **C**, etc., measured along line **X**, which is also the center line of the developed blank. Points are connected with a French curve.

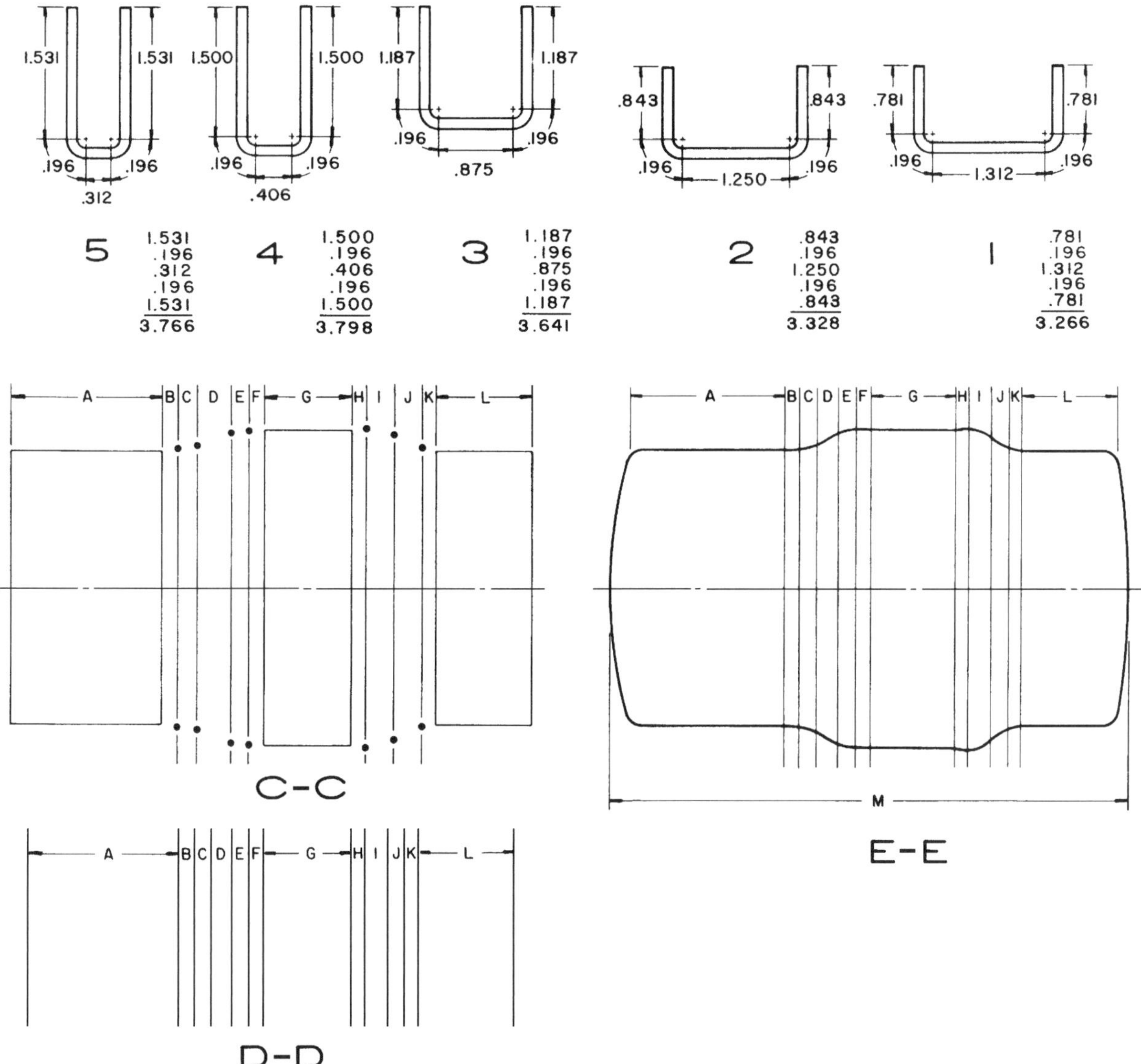

Figure 18.17 (continued) Typical example of blank development for formed part.

Observe that the left and right sides are arcs with large radii instead of straight lines, this is because of the difference in length of the center line as compared with lengths of the upper and lower edges. The small arcs at the corners are 3/16 inch, as taken from the part drawing.

Next, the drawing of the developed blank is fully dimensioned. This drawing is used for tracing or laying out the opening in the blanking die. However, the die opening is not dimensioned on the die drawing. Instead, a print of the blank development is sent to the shop along with the die print. The die maker will lay out and cut two trial blanks to dimensions given. These will be of the same material and gage thickness as the part drawing specifies for finished workpieces.

The forming die is always completed before the blanking die is made; one of the trial blanks would be actually formed in it. Should the formed part be accurate and within the limits established by the part drawing, the die maker then uses the other trial blank as a template for machining the die hole. More frequently, some modification is necessary because of the way in which the metal flows in forming. The die maker will then make two more trial blanks, adding or removing stock in an attempt to correct any irregularity. One of these is then formed and checked for accuracy. This process is continued until an accurate blank is produced.

18.5.3 Developing Drawn Parts

Before a die for producing a drawn cup or shell can be dimensioned, it is necessary to develop the form, that is, to determine the diameter of the flat blank from which the shell is to be drawn. This dimension must be calculated before the design is begun because the diameter to which the first cup can be drawn has a direct relationship with the diameter of the flat blank. Also, the blank diameter must be known before the blanking die can be laid out.

First, you must realize that blanks for drawn parts cannot be developed in the same manner as blanks for bent or formed work. In drawing, metal is gathered or moved from all sides simultaneously; the developed length will be shorter than for corresponding bent or formed workpieces. In Figure 18.18, observe that the developed length of the bent part is 10 inches (254 mm), but that the diameter of a drawn shell having the same proportions is considerably less. Therefore, for drawn shells, simply adding lengths of portions of sections will not provide an accurate dimension for the flat blank. Instead, we must determine the areas of various sections of the shell and compare their sum with the area of the flat blank. The two must be equal.

Two methods are available for determining diameters of flat blanks for drawn shells:

1. They may be calculated mathematically. First the drawn workpiece is separated into its component rings and disks, and the area of each is determined by appropriate formulas
2. Some forms are better developed by a graphical layout. This method is accurate and relatively fast when it is properly understood. We will now undertake the study of both methods.

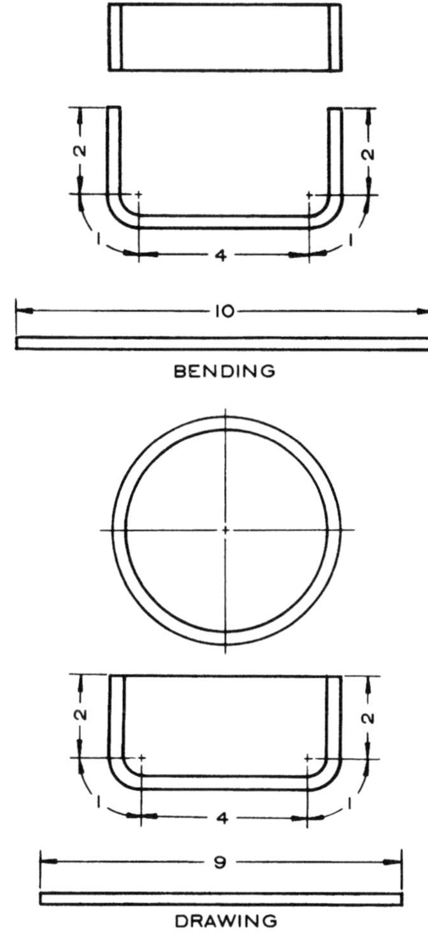

Figure 18.18 Different amounts of material are needed for bending, as compared to drawing.

a) Mathematical Method

The flanged shell in Figure 18.19 at **A** is to be drawn to several diameters. Divide the shell into flat disks and ring-shaped sections. The area of each of these sections must be calculated. Their sum equals the area of the developed blank. The diameter of the blank may be calculated from the formula:

$$D = \sqrt{\frac{4A}{\pi}} = 1.128 \times \sqrt{A} \qquad (18.6)$$

where:

D= diameter of the blank
A= sum of the areas of sections of the shell.

Formulas are given for solving for the areas of each of the nine representative sections. These formulas determine the area of the neutral surface represented by a line in the center of each section and drawn along its length.

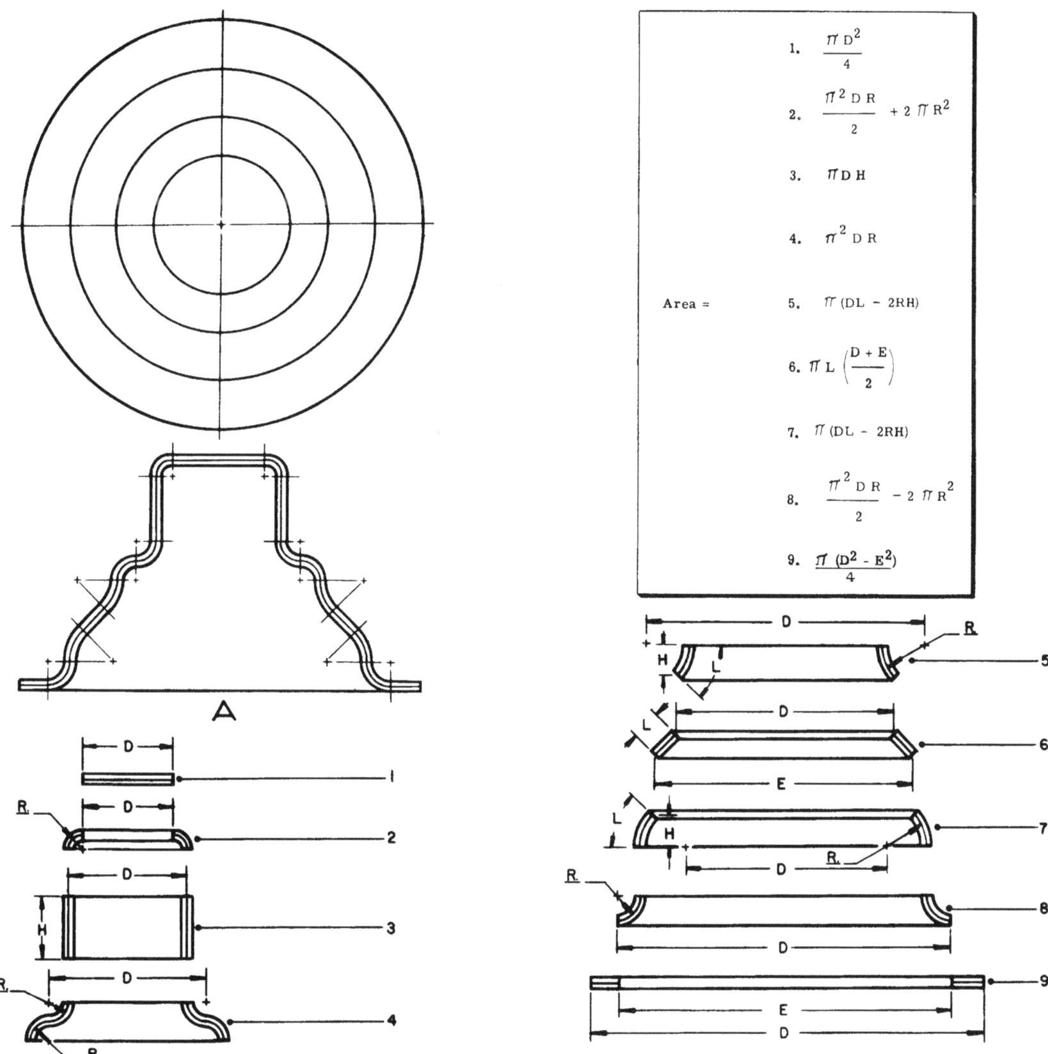

Figure 18.19 Formulas for solving for non-stretched areas of nine representative sections of a drawn shell.

When you are confronted with the problem of developing the flat blank for a round shell, first divide it into elementary forms, as shown. Then it is a simple matter to calculate the areas of each. Their sum will represent the area of the entire shell at the line drawn along the center of its thickness. The sum will also represent the area of the flat blank if no stretching occurs. The corresponding diameter is given on the drawing of the blanking die followed by the word "TRY." The die maker will make a blank by hand and actually try it in the drawing die before beginning to build the blanking die.

NOTE: Equations for blank diameter calculations of many different symmetrical shells may be found in the book *Sheet Metal Forming Processes and Die Design,* Appendix 1, by V. Boljanovic, Ph.D., Industrial Press Inc., 2004.

When a shell has a flat bottom, its division into component parts is represented by a round, flat disk. The formula for calculating its area was given in the previous illustration. When the bottom is spherical in shape (Figure 18.20), different formulas must be employed, depending upon the extent of the spherical surface. This illustration lists appropriate formulas for use.

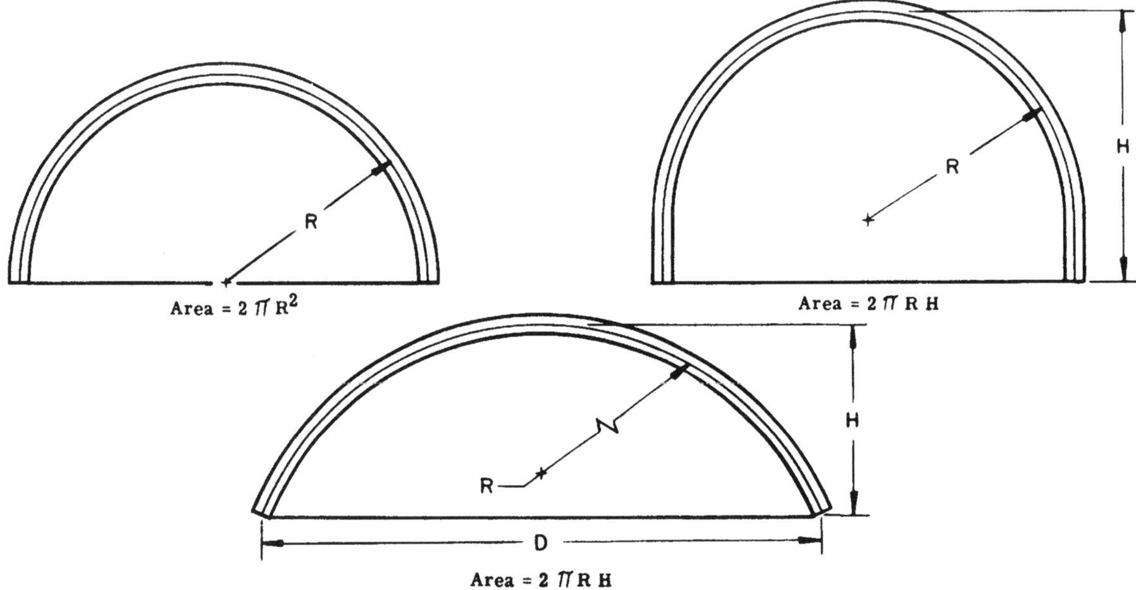

Figure 18.20 Formulas for solving for areas of spherical shape.

b) Graphical Method

Shells may be developed graphically (Figure 18.21), that is, by plotting the diameter of the flat blank with lines instead of computing areas mathematically. The method is an application of the theorem of Pappas and Guldinus, which states that the volume of a body generated by revolving a plane section about an axis in the same plane of the section may be found by the formula:

$$V = 2\pi \times D \times A \qquad (18.7)$$

where:

V= volume of shell
D= distance to the center of gravity of the section from the axis of rotation
A= area of section.

The method is relatively simple. Let us apply it to a flanged shell **A**. Eight steps are taken in developing the blank size, as follows:

1. Draw a section through the left half of the shell and divide it into its component arcs and straight sections, as shown. Also draw a longitudinal line along the center of the thickness. Next apply a center line **Y—V** at the center of the shell. If the shell is small, draw the section to enlarged scale.
2. Number each section. At the right of the view, draw a vertical line and divide it into lengths to correspond with the lengths of the shell sections 1 to 7, measured along the center of the thickness. Solve for lengths of arcs using formulas given in Figure 18.14, or measure them along the center using a flexible die maker's scale.
3. Draw lines **X** and **Y**, each at an angle of 45 degrees.
4. Extend lines from division points on the vertical line to meet the intersections of lines **X** and **Y** as shown, and label them **A**, **B**, **C**, etc.
5. Mark the center of gravity of each section of the shell as shown by heavy dots. Note that the center of gravity of rectangular sections is at the center of each section. The center of gravity of a circular arc is 1/3 of arc thickness **S**, measured from the chord. This is shown as distance **T** in the inset. Extend vertical lines downward from the center of gravity of each section. Below the section view, draw a line **X** at the same angle as line **X** in the diagram at the right.
6. Set the protractor to the angle of line **A** in the diagram. With this setting, draw a line **A** below the section view. Locate it from the intersection of line **X** and the vertical line from section 1, and extend it until it

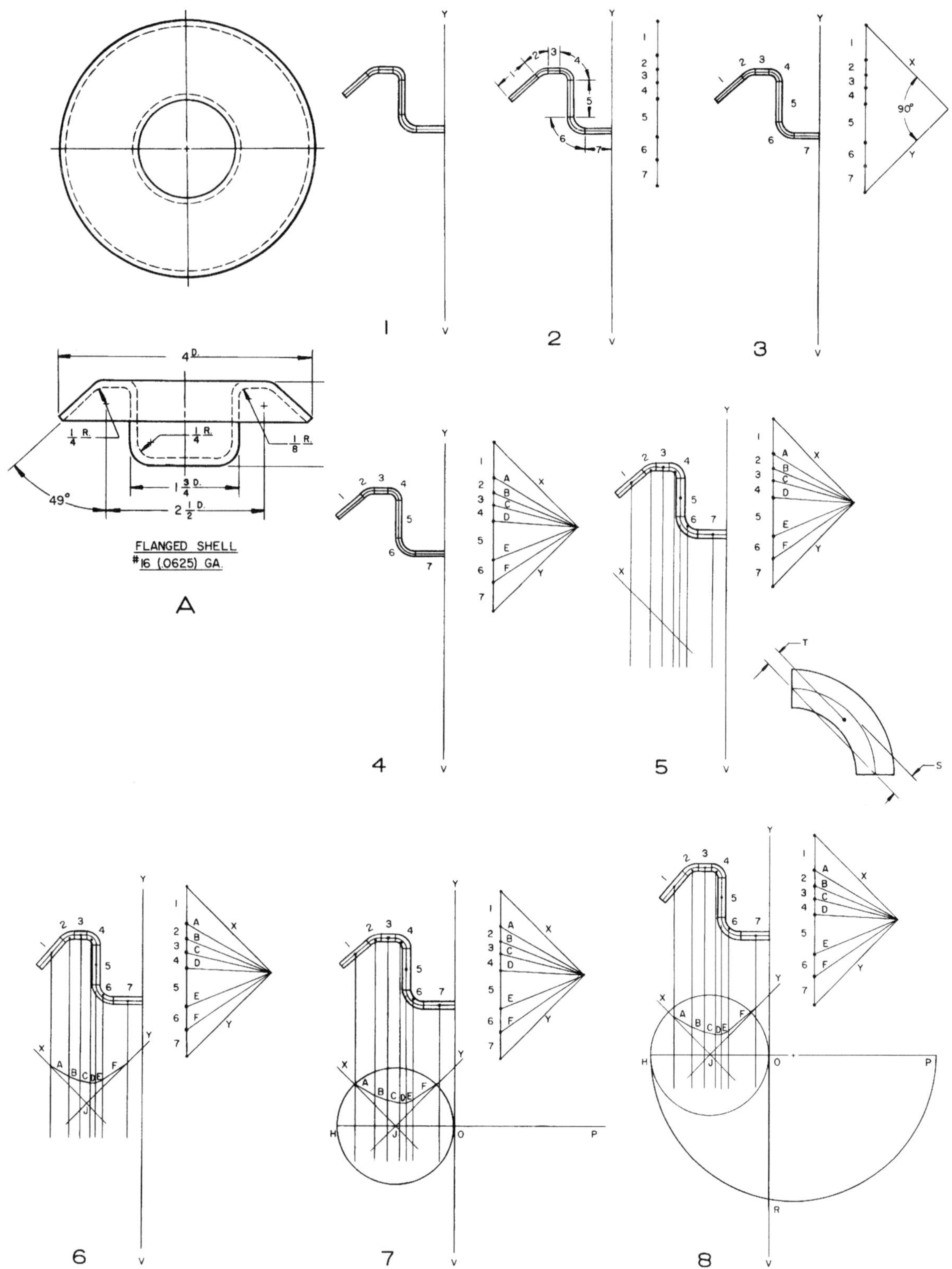

Figure 18.21 Graphical method for obtaining radius of developed blank for a shell.

intersects the vertical line from section 2. Next, set the protractor to the angle of line **B** in the diagram, and on the section view draw line **B** from the intersection of line **A** and section 2 to where it intersects the vertical line from section 3. Continue in this manner until all lines in the diagram have been transferred to below the section view and drawn to the same angular positions. Next, draw line **Y** and continue it until it intersects line **X** at **J**.

7. With **J** as center and with **J-O** as radius, draw a circle **O-H**. Then draw a long, horizontal centerline to cross point **J**.
8. Make line **O-P** the same length as the vertical line 1 to 7 in the diagram. With the compass point on line **H-P**, and with the compass set to one-half its length, draw an arc with the left end tangent to circle **H-O**. The length of the vertical line **O-R** is one-half the diameter of the flat blank required for the shell if the section view was drawn actual size.

c) Square and Rectangular Shells

Square and rectangular shells are developed by a combination of the methods described for developing blanks for formed workpieces and for round drawn shells. The top of Figure 18.22 shows a part drawing for a rectangular drawn shell to be developed to a flat blank. Three steps are required:

1. Sections **A–A** and **B–B** are taken through the long and short sides of the shell and dimensioned using dimensions given in the part drawing and by employing the formulas given previously.
2. Section **C–C**, taken through the corner of the shell to the center of the radius, is drawn and developed by the graphical method. Radius **OR** is called the blank radius.
3. As shown at view **1**, a rectangle is drawn to represent the length and width of the flat blank. Its length equals the sum of all dimensions of the sections **A–A**, and its width equals the sum of all sections at **B–B**. Centered within the rectangle is drawn an inverted top view of the finished shell. As shown at **2**, 45° lines are drawn at the corners to show the corner metal removed. Dimension **D** is made 0.9375 times the blank radius and it is measured from the center of the corner radius of the finished shell. In other words, multiply length **OR** at Section **C–C** by 0.9375 and this figure will be dimension **D**. As shown at **3**, radii are drawn at all corners of the flat blank. Radius **E** of the blank is made the same as distance **D**.

18.6 NOTES

Notes are employed to supplement information given by views and dimensions. Individual companies may have one or more special notes that would apply particularly to the products they manufacture. In your first day in any job, you should be very observant to determine what special methods are followed. In a large plant, you will be supplied with a standards book. Study this carefully for special ways of preparing drawings. You would do well to study a few of their drawings or blueprints in smaller plants.

In addition to conventional notes that apply to tool drawings, die drawings usually have one or more special notes that apply for dies only.

There are three ways of identifying calculated dimensions on drawings. Above the dimension line, and immediately following the dimension, may be lettered one of the following:

1. TRY
2. DEV. (Develop)
3. APPROX. (Approximate)

Select the one in use in the plant for which the die is being designed. Where there is no particular preference, use 'TRY' because it better describes what is actually to be accomplished. Any of the foregoing indicates to the die makers that they are to make a sample blank and actually bend, form, or draw the metal to make sure that dimensions of the formed part are within limits established by the part drawing.

a) Tool Identification Note

This note gives the tool number; it directs that it be stamped or marked on the die (Figure 18.23).

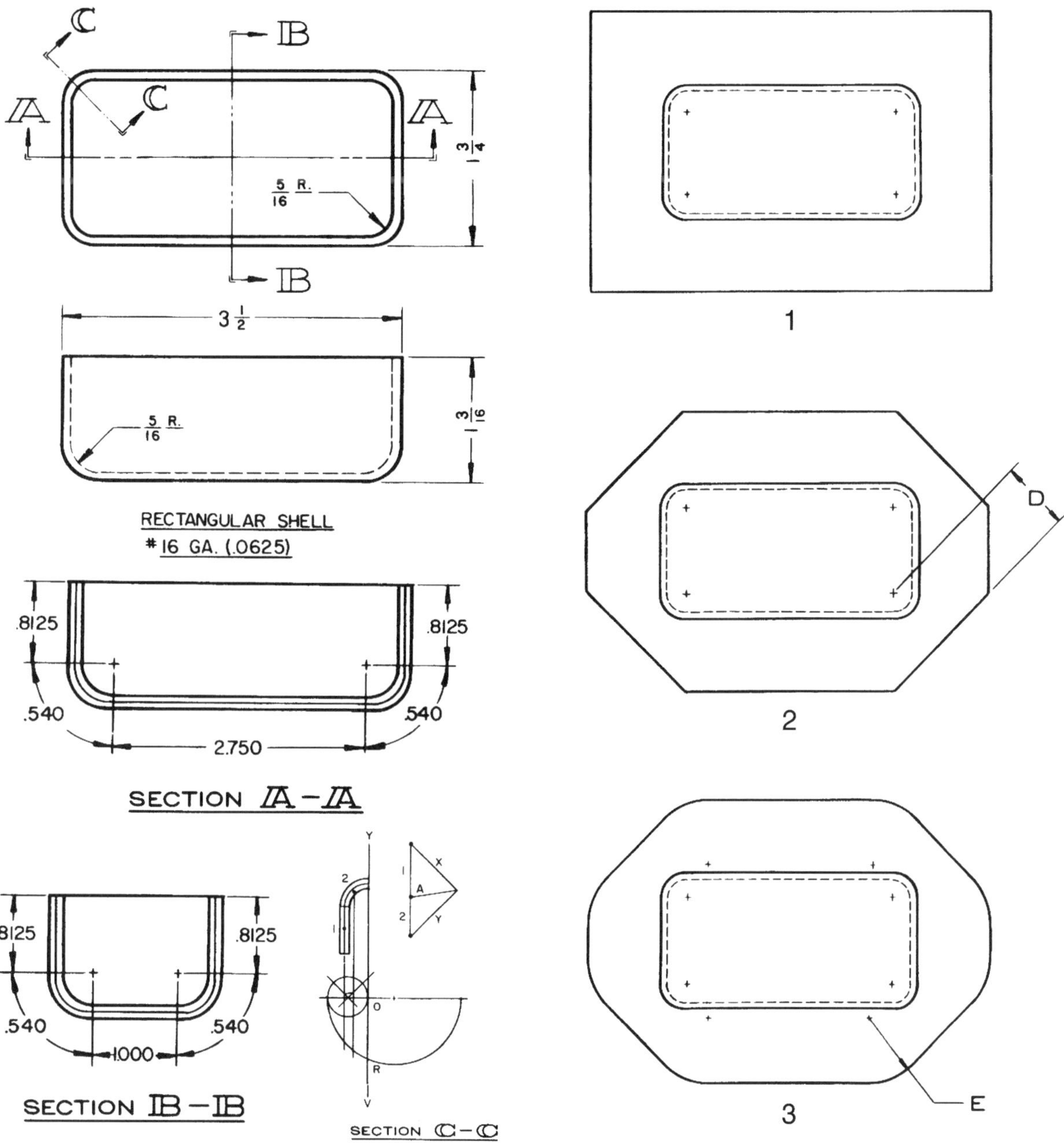

Figure 18.22 Method of developing a blank for a rectangular shell.

In addition, it shows the approximate location where it is to be placed. The note is written by lettering the word "STAMP" or "MARK" and applying the tool number directly underneath. A leader points to the finished pad on the die set where the note is to be applied.

b) Harden and Grind Note

On assembly drawings, the fact that a component is to be hardened and ground is indicated by applying the abbreviation **H. & G.** above the leader of the detail number, as in Figure 18.24, view **A**. When a part is detailed individually, the

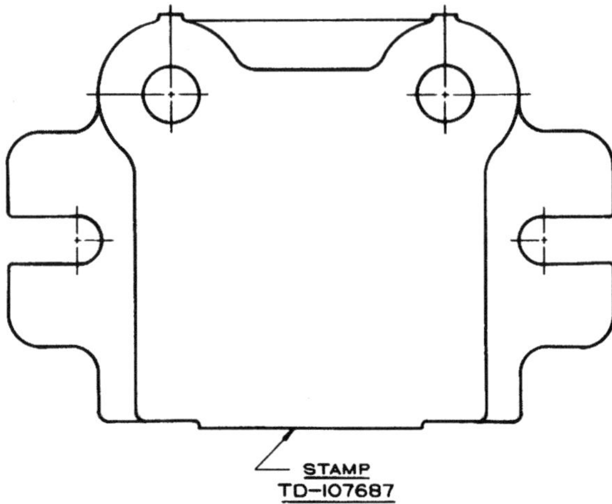

Figure 18.23 Manner in which a tool number note is placed on a drawing.

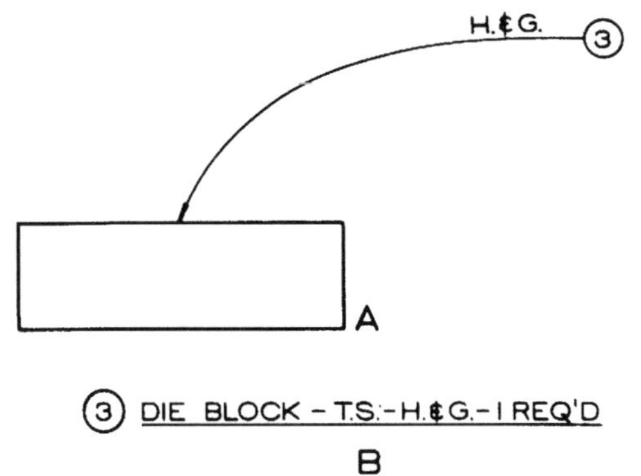

Figure 18.24 Manner in which a harden and grind note is placed on an assembly drawing (A) and below the views of a detail drawing (B).

abbreviation **H. & G.** is lettered immediately following the part and material. It is applied under the views of the detail drawing, as shown in view **B.**

c) Other Notes

Work to part print. This note is applied to drawings that do not contain dimensions for those portions of the punches and die blocks that pertain to the blank or workpiece. The note informs the die makers that they are to obtain all such dimensions directly from the part print.

No allowance has been made for springback. This note is lettered on most bending and forming dies. It informs die makers that springback cannot be tolerated in the workpiece. After they have built the die, they are to try it out and make appropriate corrections. Springback varies between 1/4 degree and 2 degrees; it is the amount that a bent or formed portion of a workpiece deflects backward after bending or forming. For some parts, springback may not be objectionable and the note is then omitted.

Die must produce part to print. This note is applied to drawings of dies that are to form a complete part. It is a clear indication to the die makers that they are to try out the die after it has been built, applying all necessary adjustments so that workpieces will meet tolerances indicated on the part print.

Work to template. When contours of a part are very irregular, this note may be applied on the drawing. A leader is added to indicate the contour against which a template is to be used for checking. Dimensions for making the template are ordinarily taken directly from the part drawing. A template is made from sheet steel. Lines are scribed accurately to given dimensions and unwanted portions are then cut away. As work progresses, components are checked against the template until contours match perfectly.

Develop blank. This note may be applied on layouts of bending, forming, or drawing dies, or a calculated dimension may be given followed by the word “TRY” as previously explained. Both have the same meaning. The bending, forming, or drawing die is to be made first before the blanking die is built. Sample blanks are to be cut by hand and formed to establish final dimensions of the blanking die opening.

Allow for trimming operation. This note is applied on layouts of drawing dies; it means that sufficient material is to be provided for a trimming operation to follow the drawing or forming operation. After blank development, this allowance is added for determining final dimensions of the blanking die opening.

Shave entire contour. This note specifies that a shaving die is to be designed in addition to a blanking die and that suitable extra material must be allowed for the shaving operation.

From the foregoing it should be apparent that, in reading a part drawing, the die designer must pay particular attention to notes because many of them, even short ones, convey far more important information than the amount of space can contain. For instance, the note MUST BE FLAT on a drawing often means that a compound die must be designed instead of the progressive die that might be chosen if the note were absent. The note MUST BE SQUARE applied to the edge of a drawn or formed workpiece means that the edge must be trimmed in a trimming die or by some other means and that material must be provided for this operation. Therefore, all notes should be analyzed before a design decision is made.

THE BILL OF MATERIAL

19.1 INTRODUCTION

Considerable knowledge or experience is necessary for completing the bill of material. The bill of material, or BOM, determines the materials from which the various die members are to be constructed. These decisions can influence success or failure of the completed die in operation. The designer must know the names of the various components that make up the die and which materials have proved successful for similar members in previous work. In addition, the die designer must be familiar with standard and purchased components, the sizes in which they are available, and the conventional methods of listing them. Filling in the bill of material column, therefore, is an important element in the preparation of a die drawing and a die designer must become proficient at it.

The bill of material is a list of the following:

1. Rough sizes of blocks of steel or other materials that will be required for making special components of the die.
2. Standard parts to be taken from stock.
3. Parts to be purchased specifically for the job; these are standard components that are not ordinarily carried in stock.

The bill of material, then, is a complete list of every component in the tool or die.

19.2 DIE DRAWING

The final step in preparing a die drawing is to fill in the bill of material column. On drawings for many die shops, the bill of material is placed at the upper right corner and is read downward, as shown in Figure 19.1. For others, it is located at the lower right corner and is read upward.

The bill of material is usually divided into five columns. When extending the printed lines to complete columns, draw all exactly the same width. If the sheet has one or two horizontal lines printed for the first details, make all succeeding boxes the same height. When the sheet does not have starting lines, space horizontal lines 5/16 inch (8.0 mm) apart. First, extend the left vertical line and mark 5/16 inch (8.0 mm) spaces along all its dividers. Then, draw horizontal lines and complete the columns by extending vertical lines. Draw as many horizontal lines as there are detail numbers in the assembly, and no more.

19.2.1 Columns

Let us now consider the main divisions of information contained in the bill of material. As shown in Figure 19.1, the bill of material is divided into five columns:

1. DET. (Detail)
2. REQ'D (Required)
3. PART NAME
4. MAT. (Material)
5. SPECIFICATIONS

a) Detail Number (Column 1)

In the first column, marked DET., the abbreviation for "detail," write detail numbers, beginning with number one and continuing for the total number of details in the assembly.

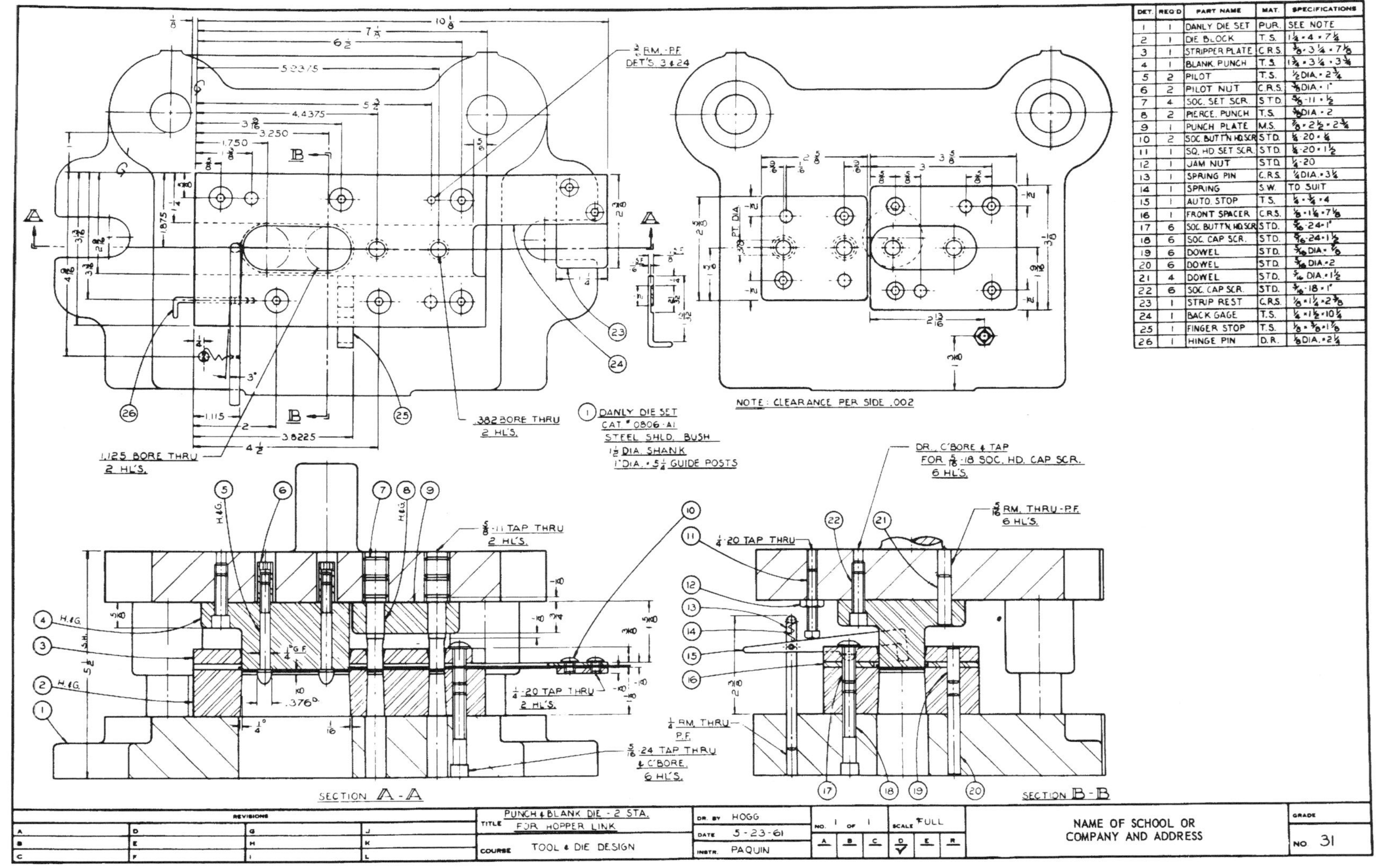
DET. | REQ'D | PART NAME | MAT. | SPECIFICATIONS
1 | 1 | DANLY DIE SET | PUR. | SEE NOTE
2 | 1 | DIE BLOCK | T.S. | 1¼ × 4 × 7¼
3 | 1 | STRIPPER PLATE | C.R.S. | ⅜ × 3¼ × 7⅛
4 | 1 | BLANK PUNCH | T.S. | 1¾ × 3¼ × 3¾
5 | 2 | PILOT | T.S. | ½ DIA. × 2¾
6 | 2 | PILOT NUT | C.R.S. | ⅝ DIA. × 1"
7 | 4 | SOC. SET SCR. | STD. | ⅝-11 × ½
8 | 2 | PIERCE. PUNCH | T.S. | ⅜ DIA. × 2
9 | 1 | PUNCH PLATE | M.S. | ⅞ × 2½ × 2¾
10 | 2 | SOC. BUTT'N HD. SCR. | STD. | ¼-20 × ¼
11 | 1 | SQ. HD. SET SCR. | STD. | ¼-20 × 1½
12 | 1 | JAM NUT | STD. | ¼-20
13 | 1 | SPRING PIN | C.R.S. | ¼ DIA. × 3¼
14 | 1 | SPRING | S.W. | TO SUIT
15 | 1 | AUTO. STOP | T.S. | ¼ × ¾ × 4
16 | 1 | FRONT SPACER | C.R.S. | ⅛ × 1¼ × 7⅛
17 | 6 | SOC. BUTT'N HD. SCR. | STD. | 5/16-24 × 1"
18 | 6 | SOC. CAP SCR. | STD. | 5/16-24 × 1½
19 | 6 | DOWEL | STD. | 5/16 DIA. × ⅞
20 | 6 | DOWEL | STD. | 5/16 DIA. × 2
21 | 4 | DOWEL | STD. | 5/16 DIA. × 1½
22 | 6 | SOC. CAP SCR. | STD. | 5/16-18 × 1"
23 | 1 | STRIP REST | C.R.S. | ⅛ × 1¼ × 2⅜
24 | 1 | BACK GAGE | T.S. | ¼ × 1½ × 10¼
25 | 1 | FINGER STOP | T.S. | ⅛ × ⅜ × 1⅞
26 | 1 | HINGE PIN | D.R. | ⅛ DIA. × 2¼
(1) DANLY DIE SET
CAT. # 0806-A1
STEEL SHLD. BUSH
1½ DIA. SHANK
1" DIA. × 5¼ GUIDE POSTS
NOTE: CLEARANCE PER SIDE .002
.382 BORE THRU 2 HL'S.
1.125 BORE THRU 2 HL'S.
DR., C'BORE & TAP FOR 5/16-18 SOC. HD. CAP SCR. 6 HL'S.
5/16 RM. THRU-P.F. 6 HL'S.
¼-20 TAP THRU
¼ RM. THRU P.F.
5/16-24 TAP THRU & C'BORE, 6 HL'S.
⅝-11 TAP THRU 2 HL'S.
¼-20 TAP THRU 2 HL'S.
SECTION A-A
SECTION B-B
PUNCH & BLANK DIE - 2 STA. FOR HOPPER LINK
COURSE TOOL & DIE DESIGN
DR. BY HOGG
DATE 5-23-61
INSTR. PAQUIN
NO. 1 OF 1
SCALE FULL
NAME OF SCHOOL OR COMPANY AND ADDRESS
NO. 31

Observe these rules:

1. One number identifies all identical parts. For example, all 5/16 dia by 1 1/4-inch (8.0 mm dia by 31.75 mm) long dowels would have the same number. It is lettered only once on the drawing, enclosed within a detail circle with a leader pointing to one of the components.
2. All parts that are different, however slightly, are given individual numbers. For example, a 1/4 dia by 1-inch (6.35 mm dia by 25.4 mm) long dowel would have one number, whereas a 1/4 dia by 1 1/4-inch (6.35 mm dia by 31.75 mm) long dowel would be given a different number.
3. Purchased assemblies, although composed of several parts, are given a single detail number.
4. Weldments are given a single detail number.

b) Number Required (Column 2)

The second column is marked REQ'D, the abbreviation for "required." This lists the number of parts needed for each detail number. Fill in these amounts very carefully because it is easy to miss parts that do not show in every view.

Before specifying the number required, study all the views to determine exactly how many of the particular component are required. In many cases, a single view will not show the total number and other views must be considered as well.

c) Part Name (Column 3)

The third column marked PART NAME lists the name of each component. Learning names of parts is just as important as learning to draw them. They form an important portion of the language of the die designer, which you are now learning to understand.

In listing the name of a part, give the general name first, followed by any word or abbreviation qualifying it. For example, a right hand V block would be listed as V BLOCK - R.H. Note that the name, V block, is listed first, followed by the abbreviation for Right Hand because the hand qualifies or limits the type of V block that is specified.

It is frequently necessary to shorten the part name because of the limited space available. For instance, when specifying a key, we may turn to the Lodding catalog and find a suitable one listed as "Plain Fixture Key." In addition to this rather long name, we must add the word "Lodding," which makes the part name entirely too long for the space provided. The name is better shortened to LODDING KEY. This is sufficient identification when given in conjunction with the catalog number listed in the SPECIFICATIONS column.

d) Materials (Column 4)

In the fourth column, MAT., are listed the materials from which the parts are to be made. For parts available commercially, this space indicates whether they are to be purchased or are available from stock. For dies, the most commonly employed materials are tool steel and machine steel.

Blanks for making cutting members of dies are sawed from bars of tool steel. At least 1/8 inch (3.2 mm) of material should be allowed per side for machining. This is measured from the outside of the bar to final size. Shading in the illustration shows the depth of this layer of decarburized steel and iron oxide, or scale. If not removed previously, the hardening operation will leave soft spots and produce surface cracks.

Direction of grain. Blanks for making punches and small die blocks are preferably cut with the grain of steel at right angles to the cutting faces to reduce hardening distortion. In the illustration, the grain runs in the direction of the parallel lines, always along the length of steel bars. Cutting faces will be at right angles to the grain in finished punches. Of course, positioning cutting faces across the grain is possible only for small components within commercial tool steel sizes.

Large cutting members. When ordering tool steel for large die blocks, consider that distortion is greatest lengthwise with grain direction; stock should be ordered so distortion will be kept to a minimum. Die hole contour will govern the decision. The rule is: Place the longest dimension of a narrow die hole opening across the grain. Distortion in the slot located parallel to the grain will be more extensive than if it is located across the grain.

Avoiding cracks in hardening. Cracks tend to occur along the grain of the material. When sharp corners are present in the die hole, order the stock so the grain will be across the corners and not in line with them. Sharp corners provide a focal point for fractures; correct application of grain direction will help to eliminate this source of trouble.

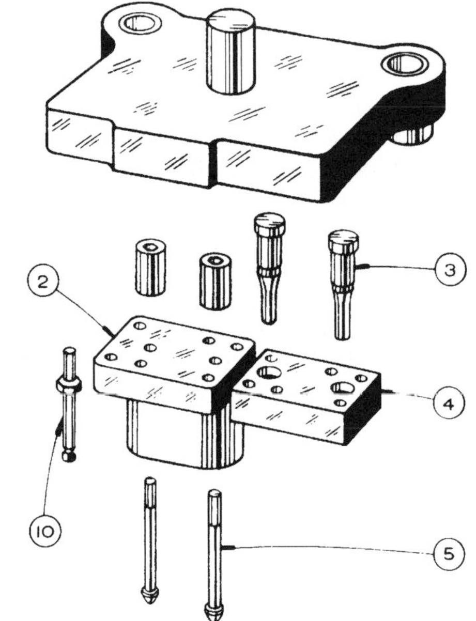

e) Specifications (Column 5)

The final step in filling in the bill of material is to give the specifications of the various components. For square and rectangular bars, this consists of giving the thickness, width, and length. For round bars, the diameter is given, followed by the length. Standard parts such as fasteners are listed by giving the diameter, followed by the number of threads per inch and the length. For purchased components, the catalog number is given. For small springs, the note TO SUIT is applied. When there is not enough space to list the specifications, the words SEE NOTE are lettered in the box in the bill of material and specifications are given in a note on the drawing.

There are two methods of specifying stock for special parts:

1. An allowance is applied for machining wherever it is required.
2. All parts are listed exact size, and the stock cutter applies extra stock where it is required for machining.

About three-quarters of all plants use the first method, and the remaining quarter use the second method. Note, however, that when exact sizes are given, (method number 2), a note is usually printed to that effect near the bill of material column.

19.3 TYPES OF COMPONENTS

Referring once again to the representative die that has formed the basis for these studies, we find that 11 types of components (Figure 19.2) are to be listed in the bill of material, as follows:

1. Die block
2. Blanking punch
3. Piercing punch
4. Punch plate
5. Pilot
6. Back gage
7. Finger stop
8. Automatic stop
9. Stripper plate

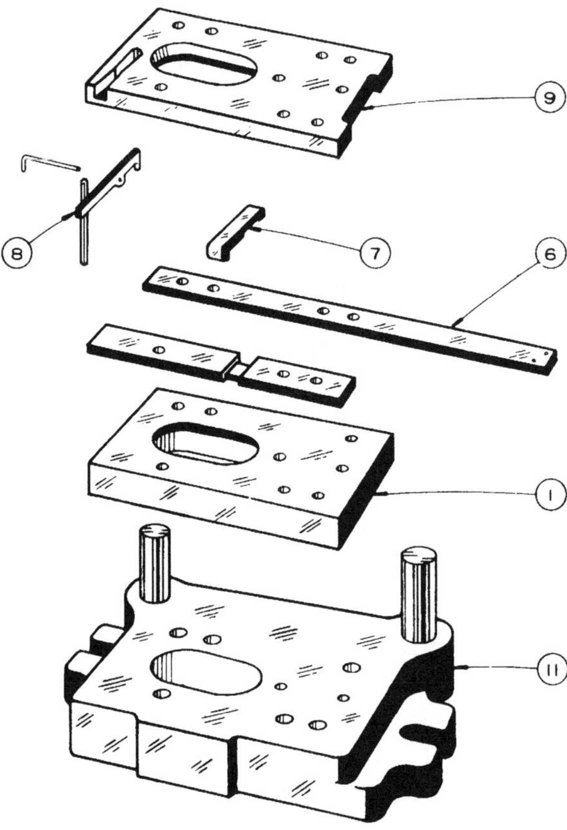

Figure 19.2 Types of components that are listed in the bill of material.

10. Fasteners
11. Die set

Each of these classifications will have other die components associated with it by function. Illustrations that follow will show exactly how each of the 11 basic components should be listed. In addition, listings for related die parts are included so that your understanding will be complete.

a) Die Block

Figure 19.3 shows how the first component, the die block, is listed in the bill of material. In addition the representative bill of material lists those members that could properly be associated or identified with the die block. Observe that many words are abbreviated and compare the abbreviations with the following list, which corresponds to the illustrated list:

1. DIE BLOCK 2. DIE SECTION
3. INSERT 4. COMPOSITE SECTION
5. DIE BUTTON 6. FORMING BLOCK
7. DRAWING RING 8. HORN
9. SPACER

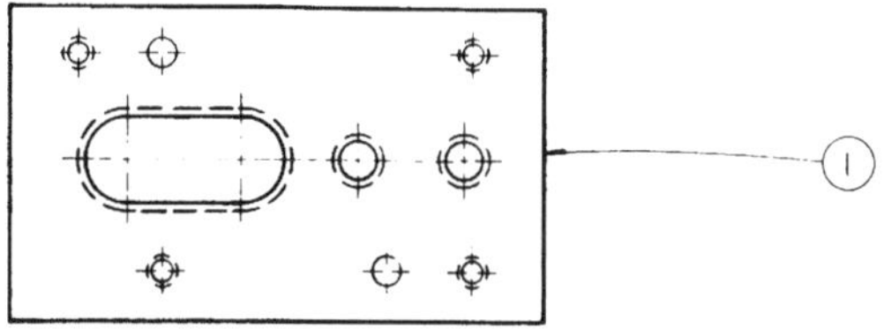

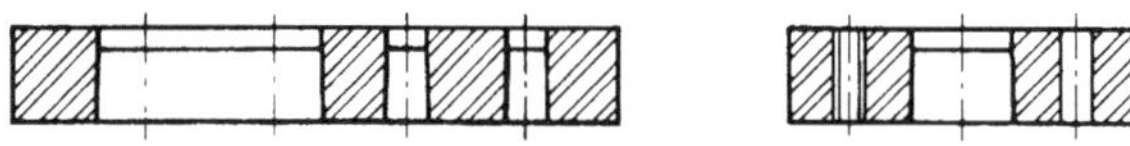

DET.	REQ'D	PART NAME	MAT.	SPECIFICATIONS
1	1	DIE BLOCK	OHTS	1 1/4 X 4 X 4 5/8
2	4	DIE SECTION	T.S.	1 1/4 X 3 X 5
3	2	INSERT	HCHC	1/2 X 1 X 1 1/4
4	6	COMP. SECT.	PUR.	SEE NOTE
5	4	DIE BUTT'N	T.S.	1 1/4 DIA. X 1 1/2
6	1	FORM. BLOCK	T.S.	1 1/2 X 3 X 5 1/4
7	1	DRAW. RING	T.S.	8 DIA. X 2 1/2
8	1	HORN	T.S.	3 DIA. X 6 1/4
9	2	SPACER	M.S.	3/4 X 4 X 7

Figure 19.3 How die block members are listed in the bill of material.

b) Blanking Punch

The punch produces a blank, or portion of a blank, and it is the male member or component of any die. The punch is positioned above the die block in all but a few types of inverted dies. It derives its name from the type of die in which it is applied. The punch of a blanking die is called a "blanking punch" (Figure 19.4). The punch of a trimming die is called a "trimming punch," and so on. Following are listed 17 types of punches to correspond with 17 types of dies. In the illustration, note carefully how the names of punches are abbreviated in the bill of material column.

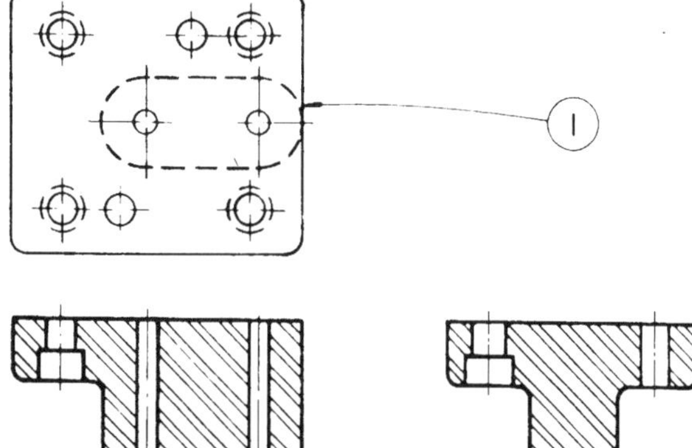

DET.	REQ'D	PART NAME	MAT.	SPECIFICATIONS
1	1	BLANK. PUNCH	OHTS	1 3/4 X 3 X 4 1/2
2	1	C'OFF PUNCH	T.S.	1 3/4 X 4 X 6
3	1	TRIM. PUNCH	T.S.	8 1/2 DIA. X 2
4	1	SHAVE. PUNCH	T.S.	6 DIA. X 1 3/4
5	1	BROACH. PUNCH	T.S.	1 3/4 X 2 X 3
6	1	BEND. PUNCH	T.S.	3 X 4 1/2 X 5
7	1	FORM. PUNCH	T.S.	2 X 4 X 6 3/4
8	1	DRAW. PUNCH	T.S.	9 DIA. X 3
9	1	CURL. PUNCH	T.S.	10 DIA. X 4 1/4
10	1	EXTR. PUNCH	T.S.	1 1/2 DIA. X 2
11	1	SWAGE. PUNCH	T.S.	1 DIA. X 2 1/8
12	1	PART. PUNCH	T.S.	1 1/2 X 3 X 7
13	1	NOTCH. PUNCH	T.S.	1 3/4 X 2 X 3
14	1	LANCE. PUNCH	T.S.	1 DIA. X 3
15	2	STAKE. PUNCH	T.S.	3/4 X 1 X 1 3/4
16	1	COIN. PUNCH	HSS.	1 1/2 DIA. X 2
17	1	BURN. PUNCH	T.S.	3 DIA. X 3 3/8

Figure 19.4 Seventeen types of punches and how each would be listed on a bill of material.

1. BLANKING PUNCH 2 CUT-OFF PUNCH
3. TRIMMING PUNCH 4. SHAVING PUNCH

5. BROACHING PUNCH
6. BENDING PUNCH
7. FORMING PUNCH
8. DRAWING PUNCH
9. CURLING PUNCH
10. EXTRUDING PUNCH
11. SWAGING PUNCH
12. PARTING PUNCH
13. NOTCHING PUNCH
14. LANCING PUNCH
15. STAKING PUNCH
16. COINING PUNCH
17. BURNISHING PUNCH

c) Piercing Punch

A piercing punch pierces holes in blanks or strips. Such holes are ordinarily round, but they may have any contour. Figure 19.5 shows how a piercing punch is listed in the bill of material. In addition, five components that may be associated with the piercing punch are also listed:

1. PIERCING PUNCH
2. PERFORATING PUNCH
3. HOLECUTTING PUNCH
4. COUNTERSINK PUNCH
5. QUILL
6. KEY

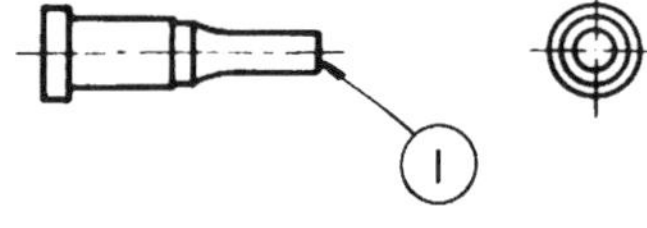

DET.	REQ'D	PART NAME	MAT.	SPECIFICATIONS
1	2	PIERCE. PUNCH	T.S.	3/4 DIA. X 1 3/4
2	4	PERF. PUNCH	D.R.	3/32 DIA. X 1 5/8
3	1	H'L'CUT PUNCH	T.S.	3 1/4 DIA. X 1 3/4
4	1	C'SINK PUNCH	T.S.	1/2 DIA. X 1 3/4
5	2	QUILL	T.S.	3/4 DIA. X 1 3/4
6	1	KEY	M.S.	1/4 X 1/2 X 1

Figure 19.5 How piercing punches and their associated components are listed in a bill of material.

d) Punch Plate

A punch plate (Figure 19.6) is a plate or block used to retain piercing punches. A simple square plate may retain a single punch, whereas a large punch plate may contain hundreds of precision-bored holes for retaining a corresponding number of piercing punches. As the following list indicates, backing plates and spacers are closely associated with punch plates.

1. PUNCH PLATE
2. BACKING PLATE
3. SPACER

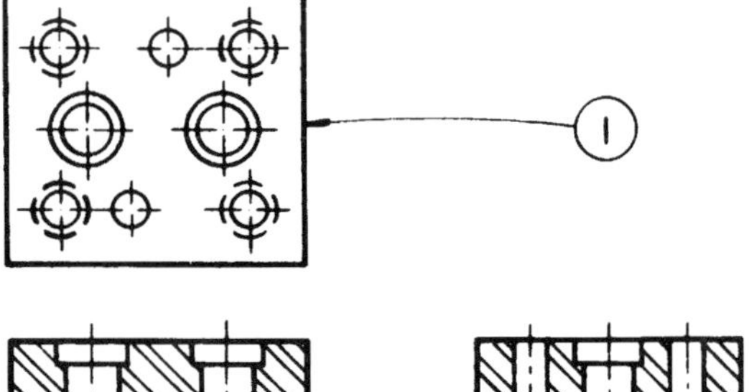

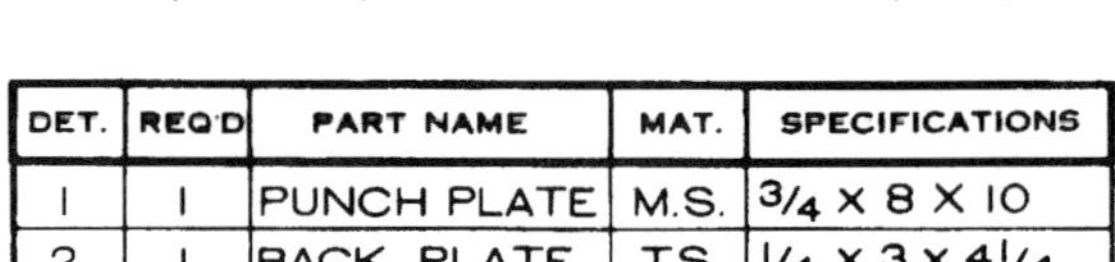

DET.	REQ'D	PART NAME	MAT.	SPECIFICATIONS
1	1	PUNCH PLATE	M.S.	3/4 X 8 X 10
2	1	BACK. PLATE	T.S.	1/4 X 3 X 4 1/4
3	1	SPACER	C.R.S.	1 X 4 X 6 1/2

Figure 19.6 How a punch plate, backing plate, and spacer are listed in a bill of material.

e) Pilots

Pilots (Figure 19.7) provide a method of accurately locating the strip. Associated with a pilot is its pilot nut and, in one type of spring-backed pilot, one or more detents. A locating plug, or locating pin and diamond pin, performs the same function of precision location for some types of secondary operation dies. Following is a list of details related to the pilot:

1. PILOT
2. PILOT NUT
3. DETENT
4. LOCATING PLUG
5. LOCATING PIN
6. DIAMOND PIN

f) Back Gage

In its travel through the die, the strip is positioned against the back gage. Associated with the

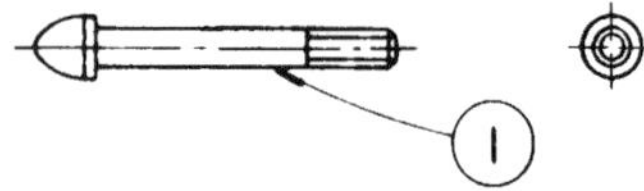

DET.	REQ'D	PART NAME	MAT.	SPECIFICATIONS
1	2	PILOT	T.S.	3/8 DIA. X 1 3/4
2	2	PILOT NUT	C.R.S.	1/2 DIA. X 1
3	1	DETENT	D.R.	3/16 DIA. X 3/8
4	1	LOC. PLUG	T.S.	1 1/2 DIA. X 4 1/8
5	1	LOC. PIN	T.S.	5/8 DIA. X 1 5/8
6	1	DIAMOND PIN	T.S.	5/8 DIA. X 1 5/8

Figure 19.7 How pilot details are listed in a bill of material.

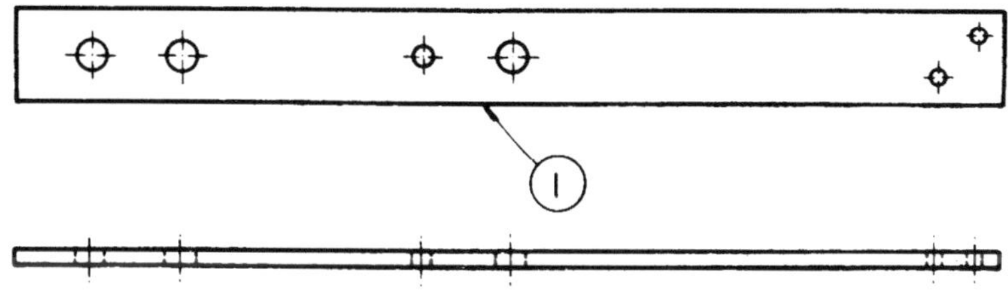

DET.	REQ'D	PART NAME	MAT.	SPECIFICATIONS
1	1	BACK GAGE	T.S.	1/8 × 1 × 8
2	1	FRONT SPACER	CRS	1/8 × 1 × 5 1/4
3	1	STRIP SUPPORT	CRS	1/8 × 1 1/4 × 3
4	1	EQUAL. BAR	CRS	1/4 × 1 × 5
5	2	GAGE	T.S.	1/4 × 1 3/4 × 3
6	1	NEST	T.S.	1/4 × 4 × 6 1/2

Figure 19.8 How details related to the back gage are listed in the bill of material.

back gage (Figure 19.8) are the front spacer and strip support. When the strip must be centered, an equalizing device is designed for positioning it exactly over the center of the die hole. For secondary operations, gages or nests are employed for locating the workpieces. Here is a list of details related to the back gage:

1. BACK GAGE
2. FRONT SPACER
3. STRIP SUPPORT
4. EQUALIZING BAR
5. GAGE
6. NEST

g) Finger Stop

The finger stop (Figure 19.9) positions the strip for the first stroke or operation. The stock pusher, which keeps the strip registered against the back gage, is related to it. Some finger stops serve the dual purposes of stop and stock pusher. A stop block is used in cut off dies.

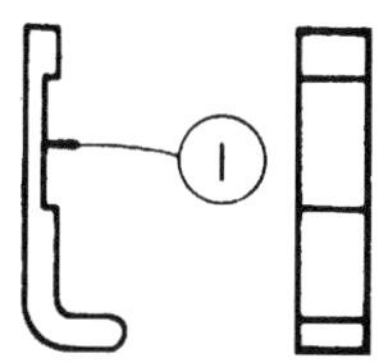

DET.	REQ'D	PART NAME	MAT.	SPECIFICATIONS
1	1	FINGER STOP	CRS	1/8 × 3/8 × 1 7/8
2	1	STOCK PUSH.	T.S.	1/8 × 5/8 × 2
3	1	STOP BLOCK	T.S.	1 1/2 × 2 × 2 1/8

Figure 19.9 How a finger stop, stock pusher, and stop block are listed in a bill of material.

1. FINGER STOP
2. STOCK PUSHER
3. STOP BLOCK

h) Automatic Stop

Dies designed for hand feeding of strip are provided with an automatic stop (Figure 19.10). Large blanking dies run in slow moving presses are provided with a stop pin instead. Seven parts are included in the conventional automatic stop assembly:

1. AUTOMATIC STOP
2. FULCRUM PIN
3. SPRING PIN
4. SPRING
5. SQUARE-HEAD SET SCREW
6. JAM NUT
7. STOP PIN

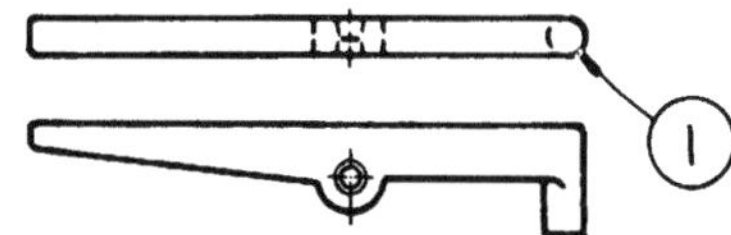

DET.	REQ'D	PART NAME	MAT.	SPECIFICATIONS
1	1	AUTO. STOP	T.S.	1/4 × 7/8 × 4
2	1	FULCRUM PIN	D.R.	1/8 DIA. × 3
3	1	SPRING PIN	CRS	1/4 DIA. × 3 1/4
4	1	SPRING	S.W.	TO SUIT
5	1	SQ.HD.SET SCR.	STD.	1/4-20 × 2 1/4
6	1	JAM NUT	STD.	1/4-20
7	1	STOP PIN	D.R.	1/4 DIA. × 2

Figure 19.10 How parts of an automatic stop assembly and a stop pin are listed in a bill of material.

i) Stripper Plate

A stripper, whether of the solid type or the spring-operated variety (Figure 19.11), is the die member that strips or removes the material from around punches. Ejectors, knockouts, and their accessories are associated with it in function, as shown by the following list:

1. STRIPPER PLATE
2. BLANK HOLDER
3. PRESSURE PAD
4. EJECTOR
5. KNOCKOUT PLATE
6. KNOCKOUT BLOCK
7. KNOCKOUT COLLAR
8. KNOCKOUT ROD

9. PRESSURE PIN
10. GUIDE PIN
11. GUIDE BUSHING
12. DIE SPRING

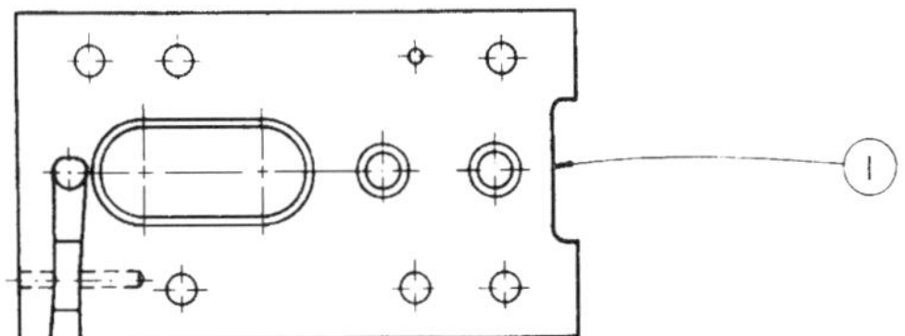

DET.	REQ'D	PART NAME	MAT.	SPECIFICATIONS
1	1	STRIP. PLATE	M.S.	1/2 X 6 X 8
2	1	BLANK HOLD.	T.S.	3/4 X 8 1/2 X 10
3	1	PRESS. PAD	T.S.	1 X 2 1/2 X 4
4	1	EJECTOR	T.S.	3/4 X 1 3/4 X 3
5	1	KN'OUT PLATE	M.S.	3 1/2 DIA. X 5/8
6	1	KN'OUT BLOCK	M.S.	4 DIA. X 3 1/8
7	1	KN'OUT COLLAR	C.R.S.	1" DIA. X 3/4
8	1	KN'OUT ROD	C.R.S.	1/2 DIA. X 7 3/16
9	2	PRESS. PIN	C.R.S.	1/4 DIA. X 2 1/8
10	2	GUIDE PIN	D.R.	1/2 DIA. X 4 1/4
11	2	GUIDE BUSH.	T.S.	1" DIA. X 7/8
12	4	"STD" SPRING	PUR.	#05-YH-10

Figure 19.11 How the stripper and details associated with it are shown in a bill of material.

j) Fasteners

Figure 19.12 shows the method of listing the most frequently used types of fasteners. Observe that fasteners are marked STD., or "Standard," in the Material column. This means that they are carried in stock. In the Specifications column, screw sizes are noted by giving the diameter, followed by the number of threads per inch, and the length. Stripper bolt sizes are noted by giving the body diameter, followed by the length from under the head to the shoulder. Dowels are listed by giving the diameter followed by length. Sizes of nuts are given by listing the diameter of the engaging bolt and the number of threads per inch. Sizes of washers are given by listing the diameter of the engaging bolt. Sizes of rivets are given by listing the diameter followed by length. Compare the following list with the abbreviations in the illustration:

1. SOCKET CAP SCREW
2. SOCKET SET SCREW
3. SOCKET LOCK SCREW
4. SOCKET BUTTON-HEAD SCREW
5. SOCKET FLAT HEAD SCREW
6. STRIPPER BOLT
7. DOWEL
8. HEXAGON NUT
9. JAM NUT
10. WASHER
11. RIVET-FLAT HEAD

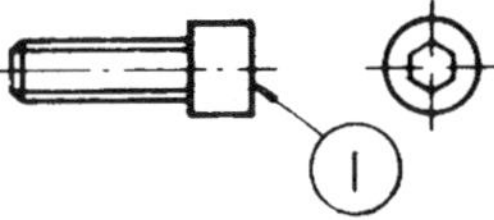

DET.	REQ'D	PART NAME	MAT.	SPECIFICATIONS
1	4	SOC.CAP SCR.	STD.	3/8 - 16 X 2 1/4
2	2	SOC.SET SCR.	STD.	1/2 - 13 X 3/4
3	2	SOC.LOCK SCR.	STD.	1/2 - 13 X 1/4
4	4	SOC. BTT'N. HD.	STD.	5/16 - 18 X 1 1/2
5	2	SOC.FL.HD.SCR.	STD.	3/8 - 16 X 2
6	4	STRIPPER BOLT	STD.	1/2 DIA. X 2 1/2
7	2	DOWEL	STD.	3/8 DIA. X 1 1/2
8	4	HEX. NUT	STD.	1/2 - 13
9	4	JAM NUT	STD.	3/8 - 16
10	2	WASHER	STD.	1/2 DIA.
11	4	RIVET - FL. HD.	STD.	1/4 DIA. X 1/2

Figure 19.12 Listing fasteners in a bill of material.

k) Die Set

Figure 19.13 shows a two-post die set. There is usually not enough room to list all the die set

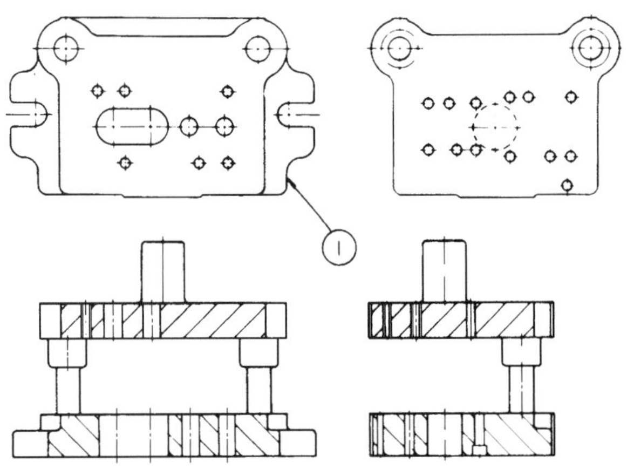

DET.	REQ'D	PART NAME	MAT.	SPECIFICATIONS
1	1	DIE SET	PUR.	SEE NOTE
2	2	SIDE CAM	T.S.	2 X 3 1/2 X 7
3	1	SCRAP CUT.	T.S.	1 3/4 X 2 1/2 X 3

Figure 19.13 How the die set, side cam, and scrap cutter are listed in a bill of material.

information required. The reader of the drawing is therefore directed to a note on the drawing itself, in which specifications of the die set and its component parts are given.

In some types of dies, a side cam or scrap cutter may be fastened to the die set, and these are listed here. Both are usually made of water-hardening tool steel, given simply as the abbreviation, T.S.

1. DIE SET 2. SIDE CAM
3. SCRAP CUTTER

19.4 TOOL STEELS

Tool steel is specified for numerous die components and it may be well to discuss briefly the most commonly used types. They are:

1. Water-hardening tool steel
2. Oil-hardening tool steel
3. Air-hardening tool steel
4. High carbon, high chromium tool steel
5. High speed steel
6. Shock-resisting tool steel
7. Hot work die steel

a) Water-Hardening Tool Steel: "W" series

As its name indicates, water-hardening tool steel is hardened by quenching it in water after it has first been heated to proper hardening temperature. It is employed for parts that can be ground after hardening. Water-hardening tool steel is subject to distortion in the hardening process; it should not be specified for parts with internal contours that must remain accurate and which cannot be ground after hardening.

b) Oil-Hardening Tool Steel: "O" series

Oil-hardening tool steel contains chromium; it is quenched in oil in the hardening process. Warpage or distortion is much less than for corresponding grades of water-hardening steels. When accurate surfaces cannot be ground after hardening, and anticipated production rates are average, oil-hardening tool steel should be specified. The abbreviation is O.H.T.S.

c) Air-Hardening Tool Steel: "A" series

Air-hardening tool steel need not be quenched in either oil or water for hardening to occur. After heating beyond the critical range, it is simply exposed in air until cool. Air-hardening tool steels have minimum Warpage. This is combined with greater toughness and wear resistance than corresponding grades of oil- or water-hardening steels.

d) High Carbon, High Chromium (Deep Hardening) Tool Steel: "D" series

High carbon, high chromium steels have about the same properties as air-hardening steels, except that they possess a greater degree of resistance to wear. High carbon, high chromium steels should be specified for die parts when long production runs are anticipated.

e) High Speed Steel: "T" series

The outstanding quality of high speed steel is its toughness, combined with a high degree of wear resistance. It should be specified for weak die parts such as frail inserts, small diameter punches, and the like. Another excellent application is in dies for cold-working, coining, and upsetting of metal.

f) Shock-Resisting Tool Steel: "S" series

Shock-resisting tool steel contains a smaller amount of carbon; therefore, it is tougher than other types. It is employed for heavy cutting and forming operations where steels with a higher carbon content would be subject to breakage.

g) Hot Work Die Steel: "H" series

These steels are employed in dies designed for forming hot materials because they possess high resistance to softening under heat.

19.4.1 Properties of Tool Steels

Table 19.1 provides a guide for selecting the proper tool steel for any given application. Observe that six types are listed in the first column. The second column gives the non-deforming property for each type. This is an important factor for many die blocks and punches. Other important properties are given in succeeding columns to aid in selection of the type of tool steel with exactly the properties desired.

19.4.2 AISI and SAE Numbering System

The AISI and SAE identification systems are based upon the varying amounts of carbon and

Table 19.1 Guide for the proper selection of tool steels

SAE DESIGNATION	NON-DEFORMING PROPERTIES	SAFETY IN HARDENING	TOUGHNESS	WEAR RESISTANCE	MACHINE ABILITY	RESISTANCE TO SOFTENING UNDER HEAT	DEPTH OF HARDENING
HARDENING TOOL STEELS							
108	POOR	FAIR	GOOD	FAIR	BEST	POOR	SHALLOW
109	POOR	FAIR	GOOD	FAIR	BEST	POOR	SHALLOW
110	POOR	FAIR	GOOD	GOOD	BEST	POOR	SHALLOW
112	POOR	FAIR	GOOD	GOOD	BEST	POOR	SHALLOW
209	POOR	FAIR	GOOD	GOOD	BEST	POOR	SHALLOW
310	POOR	FAIR	GOOD	FAIR	BEST	POOR	SHALLOW
HARDENING TOOL STEELS							
	GOOD	GOOD	FAIR	FAIR	GOOD	POOR	MEDIUM
	GOOD	GOOD	FAIR	FAIR	GOOD	POOR	MEDIUM
	FAIR	GOOD	FAIR	FAIR	BEST	POOR	MEDIUM
HARDENING TOOL STEELS							
	BEST	BEST	FAIR	GOOD	FAIR	FAIR	DEEP
	BEST	BEST	FAIR	GOOD	FAIR	FAIR	DEEP
	BEST	BEST	FAIR	GOOD	FAIR	FAIR	DEEP
	BEST	BEST	FAIR	GOOD	FAIR	FAIR	DEEP
CARBON, HIGH CHROMIUM (DEEP-HARDENING) TOOL STEELS							
	BEST	BEST	FAIR	BEST	POOR	FAIR	DEEP
	GOOD	GOOD	POOR	BEST	POOR	FAIR	DEEP
	BEST	BEST	POOR	BEST	POOR	FAIR	DEEP
RESISTING TOOL STEELS							
	FAIR	GOOD	BEST	FAIR	FAIR	FAIR	MEDIUM
	POOR	FAIR	BEST	FAIR	FAIR	FAIR	MEDIUM
	FAIR	GOOD	BEST	FAIR	FAIR	FAIR	MEDIUM
WORK DIE STEELS							
H11	BEST	GOOD	GOOD	FAIR	FAIR	GOOD	DEEP
H12	BEST	GOOD	GOOD	FAIR	FAIR	GOOD	DEEP

other elements in the steel. The four- or five-digit designation identifies which alloys are present and the percentage of carbon in the steel:

1st number: The first number indicates the major alloying element.

2nd number: The second number indicates the approximate percentage of the major alloying element.

3rd & 4th numbers: The third and fourth numbers indicate the carbon content of the steel.

The AISI and SAE systems of numbering steel were originated by the American Iron and Steel Industry and the Society of Automotive Engineers, respectively. They are widely used. You should be familiar with the system because the material of many stampings, as given on part drawings, is often specified by SAE or AISI numbers. A brief summary follows:

a) Low Carbon Steels AISI 1006 to 1015

This group of steels weld readily, but they do not machine freely, especially when smooth surfaces are required.

b) Low Carbon Steels AISI 1016 to 1030

This group of low carbon steels generally go under the general name of "machine steel." They can be readily welded, easily machined, and have sufficient strength for most tooling applications. When hard surfaces are required, they may be case-hardened, either by carburizing or by cyaniding.

c) Medium Carbon Steels AISI 1033 to 1052

These are sometimes called "low carbon tool steels." They can be hardened a certain amount by heating and quenching in oil. For example, S.A.E. 1040 will harden to about Rockwell C 45. They machine readily and can be welded if allowed to cool slowly to avoid cracking.

d) High Carbon Steels AISI 1055 to 1095

High carbon steels go under the general name "tool steel" and they are specified when good wear characteristics are required. At purchase, tool steels are in the soft state and they machine fairly well. Hardening is accomplished by heating and quenching. After hardening, any further machining must be accomplished by grinding. Tool steel should not be welded.

PRESSES AND QUICK DIE-CHANGE SYSTEMS

20.1 INTRODUCTION

After studying die design, it will be helpful to have an understanding of the construction and operation of the presses, or machines, in which the dies are operated. In this section you will learn about not only the newer and more modern machines and presses, but also about some of the older types that are still operating in many factories today. Several of these older models have been in use for 30 years or more and are still producing parts.

Fundamentally, *stamping presses* are machines with the space to contain and the means to perform dedicated metal-forming tooling with the force, speed, and precision necessary to produce the desired part shape. Mechanical and hydraulic stamping presses are available in several basic designs and a wide range of sizes, tonnage capacities, stroke lengths, and operating speeds.

Both mechanical and hydraulic presses are classified by the type of frame upon which the mowing elements of the press are mounted. The two most common frame types are the *gap-frame* (or *C-frame*) and the *straight side press*. Each has its own advantages and disadvantages.

Press Power

Today's presses get their power from four basic sources. These sources are:

1. *Manual.* These presses are hand-operated or foot-operated.
2. *Mechanical.* These presses are motor-driven and may have a flywheel, single reduction gear, or multiple reduction gearing.
3. *Hydraulic.* These presses may be oil-operated or water-operated.
4. *Pneumatic.* These presses are operated by compressed air.

In this section we will discuss primarily two types of presses—mechanical and hydraulic. Although we are not trying to downplay or ignore either manual or pneumatic presses, an understanding of mechanically and hydraulically driven machines will quickly reveal their similarities, and principles can be applied to all four types.

Press Speed

Although some very high speed, low tonnage presses (15–30 tons) run at up to 1,800 strokes per minute and typically involve light forming, such as electrical connectors, when cutting dies are operated, the speeds usually range from around 20 to 800 strokes per minute. Because drawing and forming dies must be run more slowly to allow time for the metal to flow, their speeds usually range from around 5 to 100 strokes per minute, depending on the part size and the severity of the operation being performed.

20.2 MECHANICAL PRESSES

Mechanical presses typically store energy in a rotating flywheel, which is driven by an electric motor. The flywheel revolves around a crankshaft until engaged by a clutch device. Then, through a series of drive train components, the energy of the rotating flywheel is transmitted to the vertical movement of the slide or ram.

There are three basic drive train variations. In the *direct drive arrangement,* the drive motor, through a belt arrangement, rotates the flywheel. This method provides the highest speeds and is more easily maintained, while allowing less loss of mechanical energy. However, to gain maximum force, which only occurs near the bottom of the ram stroke, the press must always be operated at its maximum as torque is applied to only one end of the crankshaft.

In other drive arrangements using single and double gear reductions along with eccentric gear drives, the misalignment problems are eliminated; clutch and braking systems provide more power for the forming and deep drawing of larger parts.

Although the clutch allows the energy of the revolving flywheel to be transmitted to the crankshaft, the braking system holds the ram in position when the clutch is disengaged.

The stroke of the slide is adjustable within the limits of daylight in a mechanical press. Presses are also classified by the number of slides or ram they have (either single, double, or triple action). *Double-action presses* have two slides moving in the same direction against a fixed bed. *Triple-action presses* have three moving slides, two moving in the same direction with a third moving upward through the fixed bed in an opposite direction. The slide movement in a *single-action press* is similar to that of the inner slide of a double-action press.

Mechanical presses have force that typically ranges from 20 to 5,000 tons. A few special-design, large capacity presses with ratings up to 6,000 tons are in operation.

Strokes range from 0.2 to 20 in. (5 to 508 mm) and speeds from 20 to 1,400 strokes per minute. Mechanical presses are well suited for high speed blanking, shallow drawing, and for making precision parts. The definition of a high-speed mechanical press is generally accepted as being capable of 300 strokes per minute and higher. The press speed for small, high volume parts can be as fast as 1,400 strokes per minute.

Mechanical presses offer high productivity and accuracy and do not require as high a degree of operator skill as do other types of stamping machines.

The two major types of mechanical presses are gap-frame (or C-frame) and straight side presses.

20.2.1 Gap-Frame Presses

Gap-frame presses are available in several design variations:

- Open-back inclinable (OBI)
- Open-back stationary (OBS)
- Two-point gap-frame presses

The gap-frame or C-frame presses allow easy access to three sides of the die area. They require less floor space; in ranges from 20 to 60 tons, they may cost half as much as similar size straight side presses. However, the C-frame press by its design is prone to angular misalignment as the open frame deflects because not enough force is brought to bear on the die. Although misalignment is not always a problem, it is overcome by using heavier and thus more expensive presses.

A popular variation of the gap-frame press is the OBI press. The open-back inclinable press has a C-frame assembled to a base in such a way that the frame can be tilted back to various angles so that the parts fall through the rear opening by gravity. However, OBI presses use a timed blast of air, mechanical devices, or a conveyer system to discharge parts or scrap.

Gap-frame presses are often used in applications where the stock is manually fed. The operators must use both hands to activate the machine. This is a Occupational Safety and Health Act (OSHA) regulation intended to prevent the operator from accidentally cycling the machine while one hand is in the working area. Gap-frame presses produce many thousands of different kinds of parts, ranging in size from tiny instrument components to large appliance, automotive, and space vehicle parts. Some operations performed include blanking, trimming, bending, forming, and drawing. The distinguishing characteristic of a gap-frame press is its open throat. Most gap-frame presses are built in capacities ranging from about 100 to 300 tons of pressure. These presses can be:

- Inclinable
- Fixed (non-inclinable)
- Single-action
- Double-action
- Back-geared.

The frame of an inclinable press can be tilted backward to an angle of around 30°. As the name implies, a fixed-position, or non-inclinable machine does not tilt, and parts and/or slugs must either drop vertically or be removed by some other means.

Single-action presses are constructed with a single ram or slide. These machines are used for blanking, bending, forming, and other operations. They perform a single action with each cycle of the press. A double-action machine is built with dual rams, usually one inner ram sliding within an outer ram. They are used for severe forming and drawing operations. Back-geared presses are provided with gears that mechanically slow the stroke and increase the power of the press.

To enlarge your understanding of the gap-frame press, consider Figure 20.1, which illustrates a typical flywheel-type press. One revolution of the flywheel causes the crankshaft to make one 360° rotation. The rotation of the crankshaft forces the slide to move vertically to the bottom of its stroke distance and return to its original uppermost position, thus completing one cycle of the press.

Small gap-frame presses usually rest on felt or rubber pads. The pads compensate for floor irregularities and reduce vibration transmission to the floor. Usually there are no problems with either the level of the press or the twist or skew in the press bed.

Minster Machine Co.

Figure 20.1 A typical small flywheel gap-frame press.

Minster Machine Co.

Figure 20.2 Large gap-frame press.

Gap-frame presses are manufactured by several different builders, and the outer appearances of each make varies considerably from that of other makers. Figure 20.2 shows a 250-ton large gap-press, two-point (meaning it has two connections) photo press. It has a pressurized, recirculating lubrication system and two pneumatic die cushions in the bed. It should also be noted that this particular machine is a twin-drive machine, which means it is driven by gears on both ends of the crankshaft.

Larger gap-frame and straight side presses have more critical alignment requirements. Twist or skew in the press bed results in misalignment of the machine (changed clearances in gibbing and bearings). Skew or twist that cannot be corrected by leveling the press may be a result from overloading, stress-releasing after machining, or inaccurate machining.

20.2.2 Straight Side Presses

Straight side presses are so named because of the vertical columns or uprights on either side of the machine. This design eliminates the problem of angular deflection. Also, die life and part accuracy are enhanced.

Straight side presses have frames consisting of a crown member, two upright side members, a bed or fundament of the press, and the bolster, which mounts on the press bed and accommodates the die while strengthening the bed. These components are often secured in a preloaded position by four tie rods. They may also be bolted and keyed together or welded into one piece. As a result, straight side presses are stiffer vertically than gap-frame units, and any deflection under load tends to be symmetrical.

Straight side presses are suitable for progressive die and transfer die applications; they cover an enormous range of types, sizes, and speeds. Tonnage capacities range up to 6,000 tons and speeds to 1,500 SPM. Figure 20.3 shows a straight side 80-ton press. This machine is capable of 1,000 strokes per minute with a stroke length of 1 inch (25 mm).

Semi-steel. Large semi-steel die sets (Figure 17.37) are available in back-post **A**, center-post **B**, diagonal-post **C**, or four-post **D** styles. All are provided with clamping flanges for mounting in the press. They are assembled with steel shoulder bushings, unless otherwise specified.

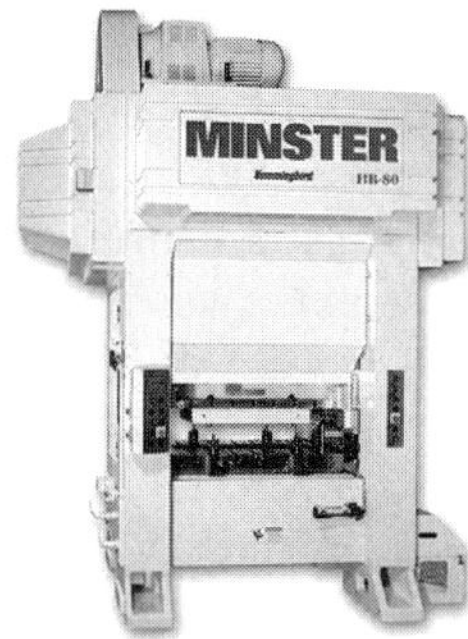

Minster Machine Co.

Figure 20.3 A straight side press.

20.2.3 Screw Presses

Screw presses are not as widely used as mechanical presses, but unique screw press characteristics are driving an increase in their use. As the name suggests, this type of press uses a mechanical screw to translate rotational motion into vertical. Briefly, the ram acts as the nut on a rotating screw shaft, moving up or down depending on screw rotation. Energy is either delivered from a flywheel, which is usually coupled with a torque-limiting (slipping) clutch, or by a direct drive reversing electric motor. The main advantage of screw presses over offset or crank-type mechanical presses is in the final thickness control when the dies impact each other.

Recent developments in mechanical presses are focusing on increased stiffness of the press structure to improve stamping accuracy, automation, and high speed (in terms of die-to-workpiece contact time).

20.2.4 Common Problems of Mechanical Power Presses

There are several common problems with mechanical power presses, which, if they exist, have direct influence on the life of the die and the quality of the stamping parts. Some of the factors considered herein are as follows:

a) Frame

The most common problems are:

- Twist and skew
- Level of the press
- Angular deflection
- Improperly prestressed tie rods
- Welding on the press frame.

Angular deflection occurs on gap-frame presses under load. The vertical deflection should not exceed 0.002 inch per inch (ANSI metric total deflection test). Properly installed tie rods on the front of the press can reduce whatever deflection exists by up to 50 percent.

For leveling the press bed, use only a precision level (accuracy of 0.0005 inch per foot per division or 0.042 mm per meter per division).

b) Ram, Crankshaft, Gears

The most common problems are:

- Drive train gear timing or damaged crankshaft

- Ram level
- Perpendicularity
- Differences in stroke lengths of crankshaft.

Before performing a ram-to-bed parallelism analysis, the press must be level, the bolster plate must sit properly on the press bed, the counterbalance system must be overbalanced, the hydraulic overloads must be pressurized, and the tie rods must be properly prestressed. After ram alignment, gibbing must be readjusted.

c) Clutch and Brake
The most common problems are:

- Wear of the clutch and brake components
- Clutch and/or brake overlap
- Clutch and brake adjustment
- Air leak.

As the clutch and brake surfaces wear, the clutch and brake parts travel a greater distance to engage the clutch and disengage the brake when the air valve pressurizes the system. In this case, a higher volume of pressurized air takes longer to exhaust the stored air to disengage the clutch and engage the brake when the air valve depressurizes the system. The results of an incorrectly adjusted clutch and brake can be longer stopping time, an overheated brake, and excessive wear, if overlap occurs. For clutch and brake plate clearances and air or oil pressure, the values recommended by the manufacturer of the press should always be used.

d) Counterbalance System
The most common problems are:

- Improper air pressure
- Air leak
- Counterbalance alignment.

The ideal counterbalance system provides a constant force for every position of the ram. This is possible only when the volume of the surge tank is infinitely large. The ideal counterbalance system exists only theoretically.

A real counterbalance system provides a force that depends on the position of the ram. If the system is balanced in the middle of the stroke, then it is overbalanced when the ram is in bottom dead center (BDC) and underbalanced when the ram is in top dead center (TDC). When the surge tank is larger, the pressure variation is lower.

e) Vibration
The most common problems are:

- Imbalance of the rotating parts
- Bent shafts
- Misalignment of coupled shafts
- Bearing defects
- Meshing of the gears
- Loose machine components.

Machinery vibration is a summation of many separate mechanical vibrations. The complexity of vibration signals can by displayed in a figure of vibrational amplitude versus time. This format of presentation is ideal for detecting and locating impulsive vibrations (looseness, chipped gear or friction-free bearing defects).

20.3 HYDRAULIC PRESSES

Hydraulic presses are operated by large pistons driven by high pressure hydraulic or hydropneumatic systems used to move one or more rams in a predetermined sequence. They are housed in a variety of types of frames, including C-frames, straight side, H-frames, four-column, and other shapes depending on the application. Figure 20.4 shows a 75-ton open-back hydraulic press.

Hydraulic presses are slow compared with mechanical and screw presses; they squeeze rather

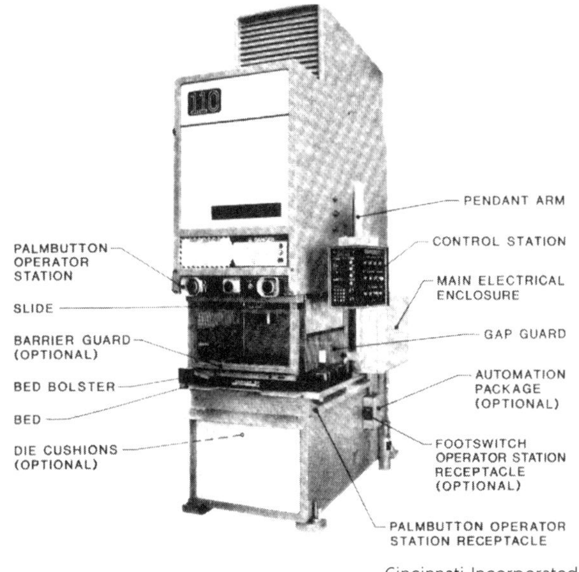

Figure 20.4 A 75-ton open-back hydraulic press.

than impact the work piece. In operation, hydraulic pressure is applied to the top of the piston, moving the ram downward. When the stroke is complete, pressure is applied to the opposite side of the piston to raise the ram.

Speeds and pressures can be closely controlled. In many presses, circuits provide for a compensation control or sequential control, e.g., rapid advance, followed by sequences with two or more pressing speeds. The press can also be regulated to stay at the bottom of the stroke for a predetermined time, raise at a slow release speed, and accelerate until it reaches the original position. When needed, hydraulic press speed can be increased considerably. In many cases, hydraulic presses use microprocessors or computers to control the press operation for parameters such as ram speeds and positions.

Tonnage can vary from 20 tons to 10,000 tons. Strokes can vary from 0.4 in. to 32 inch (10 mm to 800 mm). Hydraulic presses can deliver full power at any point in the stroke; variable tonnage with overload protection; and adjustable stroke and speed.

Hydraulic presses are suitable for deep-drawing, compound die action as in blanking with forming or coining, low speed high tonnage blanking, and force forming rather than displacement forming. Figure 20.5 schematically illustrates the various components that a typical hydraulic press comprises.

In order to assist in the learning process, we will investigate some of those components in a little more detail.

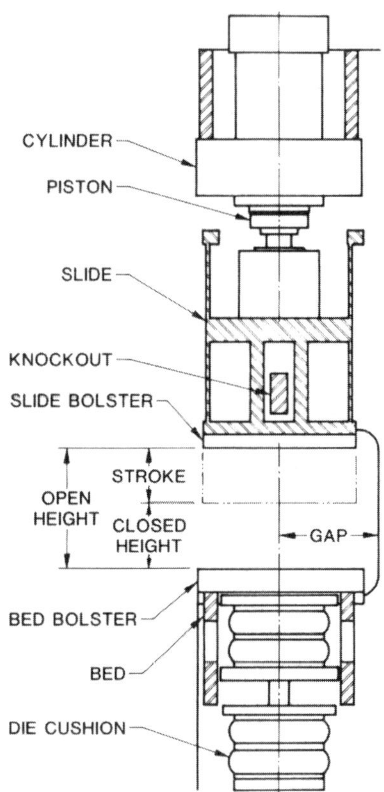

Figure 20.5 A schematic illustration of a hydraulic press.

Hydraulic cylinder. The hydraulic cylinder converts fluid power into linear force and motion. The linear force generated by a hydraulic cylinder is a product of system pressure and the effective area of the piston, minus inefficiencies (losses). When sizing a hydraulic cylinder for a press application, the relationships between pressure, area, displacement volume, flow, speed, and the influence of inefficiencies must be considered.

Slide assembly. The slide assembly is attached to the piston by connecting the holes in the top of the slide assembly to the protruding studs at the bottom of the piston. Then, as pressure is applied to the cylinder, it moves the piston downward, forcing the slide assembly to move in the same direction. Likewise, as the pressure is reversed, the entire assembly moves up, completing one cycle of the machine.

Die cushion. Die cushions consist of two pneumatic actuators connected in series. These actuators are single-acting and exert only an upward force. As the cushions are caused to move downward by the tooling pressure pins, the actuators slightly increase in diameter; the air pressure within the system is increased due to compression of the air within the actuators and surge tanks.

Die cushions are sometimes referred to as "pressure pads." They are usually used for blank-holding purposes during a draw operation, but are also used as a lower knockout mechanism. When the die closes, pins push the actuators down against the air pressure by the amount of knockout travel. When the ram goes up, the actuators push the pins upward to operate the knockout.

Control station. A major area of improvement in today's presses is in the area of controls. The control station allows operators to set up their machines via inputs through the keypad rather than by setting a series of dials inside the cabinet. The setup becomes not only easier, but considerably faster and much more accurate.

The increased use of electronics in this area is also eliminating the need to set up the machine on the shop floor. Microprocessors built into the controls allow the presses to be programmed offline; then, the preprogrammed information is sent to the machine controller electronically. This same theory will allow the programming of robots to load and unload the presses, thereby reducing the risk of injury to operators on the shop floor.

20.3.1 Advantages and Disadvantages of Hydraulic Presses

As in nearly all things, both mechanical and hydraulic presses have advantages and disadvantages. Advantages of the hydraulic press are:

Full power during the stroke. Maximum power is maintained during the entire stroke of a hydraulic press. This allows for rapid movement to a position just above the part, when the stroke is slowed for the working stroke-adjustable slow down speed during forming. The result is more strokes per minute.

Overload protection. Because the pressure is preadjusted, if the pressure exceeds a limit, such as might occur when a part is not properly ejected, the machine shuts down, eliminating catastrophic results to tooling or the machine.

Lower operating costs. Hydraulic presses have fewer operating parts; therefore, there are fewer things to break. Automatic lubrication of moving parts helps eliminate maintenance problems.

Flexibility. Owing primarily to electronic controls and robotics, hydraulic presses fit well into such areas as flexible manufacturing systems (FMS) and factory automation.

No design limitations. The principles of hydraulic force allow for creative engineering. Presses can be designed for traditional down-acting, up-acting, side-acting, and multi-action operation. Power systems can be placed above, below, or remote from the press and force actuators. Large bed presses can be designed for low tonnage applications and small bed presses can be designed for high tonnage requirements.

Unlimited control options. The hydraulic press can be controlled in a variety of ways ranging from basic relays to more sophisticated PLC or PC control systems. Operator interfaces can be added to press systems to facilitate ease of job set by storing individual job parameters for each die. Presses can be controlled for precise pressure and position, including pressure holding, speed control, and dynamic adjustments to real-time operating variances. Ram force and speed can be controlled in any direction with various levels of precision.

Condensed footprint. Hydraulics allow for generation of high pressure over a small surface area. This reduces the overall structure required for support of the force actuators. When compared to mechanical presses, hydraulic presses consume almost 50 percent less space for the same tonnage capability. This size advantage results in lower manufacturing costs and a faster return on investment by requiring less long term overhead expense.

Although hydraulic presses have many advantages, they also suffer from some disadvantages when compared to mechanical presses. Some of the disadvantages are:

Tolerances. Although some hydraulic presses do maintain high tolerances, most of them are limited to approximately 0.020 in. When closer tolerances are required on these machines, it usually is held in the tooling, which increases the cost of the die or other tooling.

Speed. Although high speed is attainable in hydraulic presses, for the most part a mechanical press will produce parts faster. This is especially true when short strokes are used.

Automatic feeding. Because of fewer moving parts, there is less equipment on which to attach automatic feed mechanisms. Therefore, most automatic feed equipment must be integrated into the process via electronics, which increases cost.

20.4 LUBRICATION SYSTEM

The lubrication system of the mechanical and hydraulic presses is very important. If the lubricating system should fail, not only will the press stop, but many of the parts are likely to be damaged beyond repair. Therefore, when lubrication failure occurs, the press can seldom be run again without a major overhaul.

The lubricating system delivers oil to the moving parts of the press to reduce friction and to assist in keeping the parts cool. Most newer and more modern presses are equipped with a pressure recirculating lubricating system that delivers the oil under pressure to the bearings and bushings and other lubricant points.

After selection of the correct lubricant, the next most critical factor to long machine and lubricant life is keeping the lubricant clean and dry. Details such as machine criticality, operating environment, and component clearances as well as lubricant type, viscosity, flow rate, and economic issues, must be carefully considered for optimum lubricant contamination control.

In any press application, the most important aspect for insuring maximum press and lubricant life is the selection of the correct lubricant. This process includes choosing the correct base oil, the correct viscosity, and the correct additives for the application. Next in importance is keeping the oil clean and dry. Particulate and water contamination can have devastating effects on machine and lubricant life.

The primary source of particulate contamination in lubricants is ambient dust and dirt. Although the composition can vary, in general dust and dirt will contain materials such as silicon oxides and aluminum oxides. Elemental indicators of dirt ingression would be silicon (Si), aluminum (Al), and in some cases calcium (Ca) and magnesium (Mg).

If a contaminant particle is larger than the clearance between two slide surfaces, the particle will grind against them, removing metal from the slide surfaces in a process called "abrasive wear." The resultant wear particles can cause a chain reaction by increasing the total number of particulates in the lubricant. Additionally, this newly generated abrasive wear material can get broken into smaller particles and become harder due to the process of work-hardening. These more numerous and harder particles combine with the original solid contaminants to increase the amount of abrasive material that is in the lubricant.

Under conditions of high velocity or high pressure or both, small particles can impinge on a press surface and result in erosive wear. In this case, particles can be much smaller than the machine clearances and still cause extensive damage due to the velocities and pressures involved. Particles generated during erosive wear also add to the overall contamination of the system and further increase machine wear. Erosive wear most commonly occurs in hydraulic systems that contain devices such as servo and proportional. Two most commonly lubricant systems are:

Lost lubricant systems. Lost lubricant systems are used on smaller gap-presses. Older presses are usually lubricated by grease, and newer ones by oil. A hand pump is often used in lost lubricant systems. Major disadvantages of this system are: used lubricant must be cleaned periodically; if a hand pump is used, lubrication of the machine depends on the reliability of the operator.

Recirculating lubricant systems. The oil is pumped from a reservoir, filtered, and then distributed to all lubrication points. The pump is powered by an electric motor or by a press-driven mechanical device. The circulating oil lubricates, cools, and flushes small particles from the bearing and friction surfaces.

If the failure of the lubrication system is not detected early, serious damage to the machine will occur. Thermo-chromatic indicators are often used for early detection of lubrication system failure. By using thermo-chromatic indicators, mechanical or electrical problems can be detected immediately by the operator or maintenance person. The machine can then be inspected and repaired before serious damage to the machine occurs. A few dollars invested in thermo-chromatic indicators can save thousands of dollars on expensive repairs and production losses.

Another sign of lubrication system failure is the presence of bronze particles around the bearing. In this case, it is usually too late for inexpensive repairs because serious damage of the bearing or friction surfaces has already occurred.

20.5 PRESS SELECTION

When selecting a press, compromises must be made in order to use the press for more than just one type of stamping operation because there is not a single universal type of press that provides productive and efficient operation. Such compromises include consideration of the following primary factors:

- Tonnage
- Energy capacity
- Press size and frame design
- Speed
- Control system for press.

Other considerations of the factors can be included, such as the number of operations to be performed, quantities and production rates, size, geometry and accuracy of workpieces, and equipment costs.

A press-rated tonnage for mechanical presses is the maximum force that should be exerted by the slide against the workpiece at a given distance above the bottom of the stroke. The higher the rating is, the greater the torque capacity of its drive members and its capability of delivering more flywheel energy. Presses with flywheel-type drives are basically used for light blanking and piercing operations. The energy requirements of these machines are small and operate at relatively high speeds. Single-geared presses are mostly used for shallow-draw workpieces requiring more energy than flywheel types. Double-geared presses are used for deeper draw operations when a larger amount of energy is needed.

As stamping operations become more automated, the use of computer numerical control (CNC) systems and various electro-mechanical systems to feed material to the press must also be factored in. There are mechanical blank-handing systems where manual handling is not practical due to speed and size. High volume-feeding is done with coil stock, which also requires an investment in additional feeding equipment.

20.6 QUICK DIE-CHANGE SYSTEMS

The metal stamping industry used to enjoy long production runs, high inventory levels, and extended die-change times. But in these modern times of rapidly increasing diversity and smaller batch sizes, the reduction of setup time is of crucial importance for the profitability of many companies. A major factor affecting these bottom line profits is that of equipment, or press, utilization ("up time"). *Press utilization* indicates the amount of time that the equipment is actually producing acceptable parts. The utilization factor should be as high as possible, with the norm between 70 and 85 percent. One way to maximize press utilization is to implement faster, more effective die-changes. Die exchange time can easily range from just a few minutes for a full automatic system to eight hours for a manual system.

A quick die change means that the material for the next part to be produced is already in place, the automation is set up, and the die is located and clamped in position, in the same place, the same way every time, in the shortest time possible.

When setting up a quick die-change system, it is necessary to pay special attention to the following considerations about the press and its die:

- What are the present and long-range production requirements?
- What is the goal for the die-change time?
- How is the press room laid out?
- Which presses are involved?
- How many dies are used in each press?
- What are the minimum and maximum sizes and weights of the dies?

There are many quick die-change systems used, but the "rolling bolster" and "die cart" systems are the most common.

20.6.1 Rolling Bolster Systems

The metal-forming industry uses many types of rolling bolster systems, but two are most common, especially in smaller metal-forming shops. They are single-bolster operation and double-bolster operation systems. Each has some advantages and disadvantages.

a) Single-Bolster Operation

The bolster moves sideways in a straight line into and out of the press. This may be done at either the front, rear, or sides of the press. In this type of system, the die-change sequence begins with the needed die positioned on the moved bolster, usually by the overhead crane, and manually clamped to the bolster. The bolster is then run into the press bed area and automatically clamped into position.

The advantages of this system are that it provides a means to easily install a die at a preset position, ensure uniform clamping, and also ensure uniform positioning in the press. The disadvantage of this system is that sometimes it is necessary the press not be functioning during the entire time of the die exchange.

b) Double-Bolster Operation

This system uses two bolsters which can be either moved side-to-side or front-to-back. The side-to-side movement is preferred. The double-bolster system provides a faster die exchange time. The main advantage of this system is that press down time is reduced and press utilization improved. The disadvantage to this system is that there needs to be more space around the press for the bolster that is not in operation. Also, this type of quick die-change is greatly dependent on the press design, although modifications to the press are frequently possible.

c) Rolling Bolster Locating and Clamping Methods

Locating methods. There are two locating methods commonly employed with rolling bolsters. The first and probably most common is the use of special wedges located on the bed of the presses and on the bolsters. As the bolster is brought into position and lowered, the wedges force the bolster into a preset location. The second method is using index pins on the top of the bolster; locators in the die also assure uniformity. Bolster clamping can be accomplished by manual means, a hydraulically operated swing clamp system, a T-slot clamp system, or a ledge clamp system. The type of clamping system used is mostly dependent on the type of press construction.

Die clamping methods. The upper shoes may be clamped either manually, by inserting hydraulic clamps in the tee slots; by swing clamps located in a fixed location; or by traveling clamps that will adjust automatically to the upper shoe size. Lower die-shoe clamping is normally a manual operation performed when the dies are placed onto the bolster. Hydraulic clamps have been employed to reduce the number of skilled trades and tools required for die setting. Figure 20.6 shows three different die clamps for quick die clamping.

Figure 20.6a shows a mechanical clamp. The main characteristics of this clamp are the following: a safe, quick, and simple way to clamp die; easy insertion into existing T-slots; easy positioning without additional fixing; full pressure that is achieved quickly and easily by turning the low torque power nut 180°; and the ability to self-lock into the clamped position.

Figure 20.6b shows single-acting, sliding hydraulic, block-type clamps with spring return used to clamp a die to a bolster and slide. Each clamp generates 5.2 tons of clamping force, using a preset system pressure of 3,500 PSI (24.1 MPa). A control station is located on the press upright, allowing easy access to clamp/unclamp selection and lifter activation. The quick die-change control is interlocked with the press controls to prevent clamping while the lifters are in the raised position, and to stop the press in the event of loss of electricity or hydraulic pressure. Multiple clamping zones for both the slide and bolster provide reliable operation in the event of a hydraulic pressure loss.

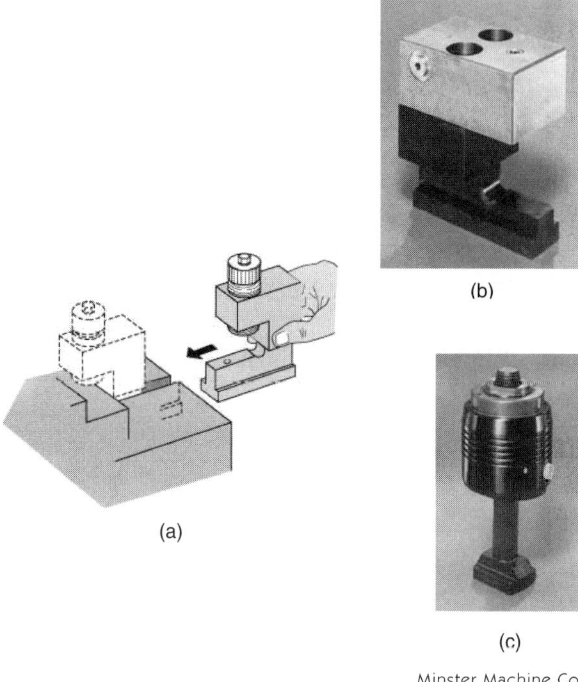

Figure 20.6 Die clamp components: a) mechanical clamp, b) hydraulic clamp, and c) hydraulic piston-type clamp.

Figure 20.6c shows single-acting, sliding hydraulic piston-type clamps with spring return, used to clamp dies to bolster and slide. Each clamp generates seven tons of clamping force, using a pre-set system pressure of 3500 PSI (24.1 MPa). Clamping height from face of bolster or slide is 1 to 3 in (25.4 to 76 mm) whereas the clamps have a maximum stroke of 0.47 in (12 mm). A control station is located on the press upright, allowing easy access to clamp/unclamp selection and lifter activation.

20.6.2 Die Cart Systems

There are many versions of die cart systems, but the most popular die cart systems are:

- Single-stage die cart on rails
- Double-stage die cart on rails
- Two single-stage die carts on rails
- Domino system.

a) Single-Stage Die Cart on Rails

This system incorporates the following elements: one single-stage die cart on rails, one push-pull system (PPS), and two staging tables. The system is dedicated to one press. Because the cart is combined with the die-staging tables, exchange of a die may commence while the press is still running. Figure 20.7 shows a schematic process drawing of a quick die change using this method:

1. The old die is pushed out of the press and onto the die cart. The cart brings it to the table on the right.
2. The cart now moves to the table on the left, picks up the new die, and brings it to the press, where the die loader pulls it into its position on the press.

Normally, this method will be less expensive than the rolling bolster system and will provide the same benefits as the single rolling bolster. Die-exchanging time is no more than five minutes.

b) Double-Stage Die Cart on Rails

A double-stage die cart on rails incorporates the following elements: one double-stage die cart on rails and one PPS. This system is dedicated to one press. Figure 20.8 shows a schematic process drawing of a quick die change by this method:

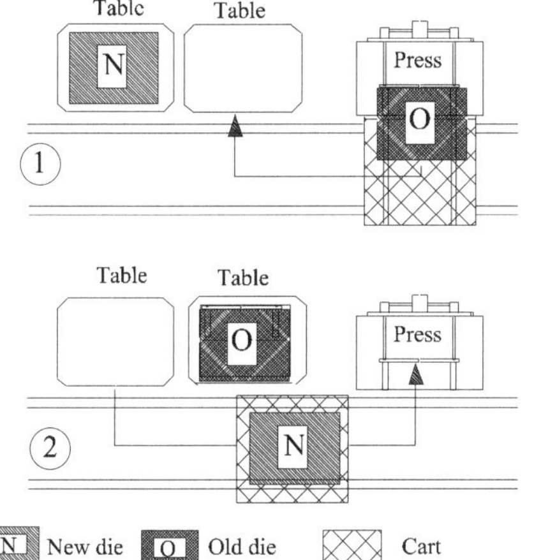

Figure 20.7 Single-stage die cart on rails.

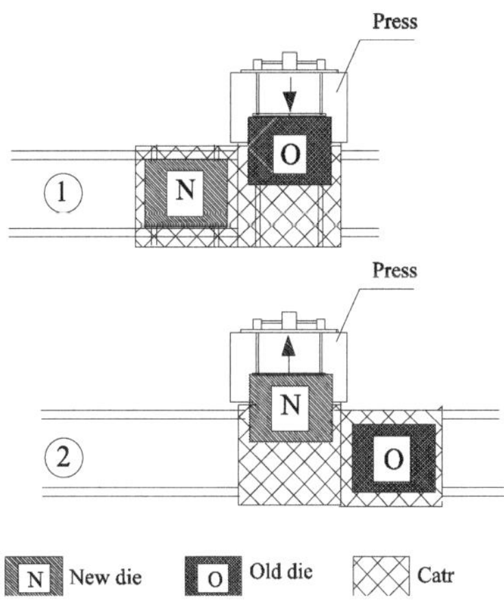

Figure 20.8 Double stage die cart on rails.

1. The old die is pushed out of the press and onto the die cart's empty stage on the right. The cart brings it to the depositing table on the right.
2. The cart now moves to the right to position its other stage, holding a new die in front of the press. Once the loader pulls the die into its operation position on the press the cart may now take the old die to storage.

A double-stage die cart system provides a faster die-exchange time at no more than three minutes. The main advantage of this system is that down time of the press is decreased and the press utilization factor is increased thanks to the very short exchange times, providing an early return on investment.

c) Two Single-Stage Die Carts on Rails

This system incorporates a PPS and a single-stage die cart on rails. This system is usually considered to be very effective in reducing the total die-change time. Figure 20.9 shows a schematic illustration of this method of quick die-change:

1. The cart on the left delivers a new die. It moves on rails, approaching the press from the side. The other cart, equipped with a push-pull unit, gets the old die out of the press. It then moves backwards on its rails, which run straight to the press.

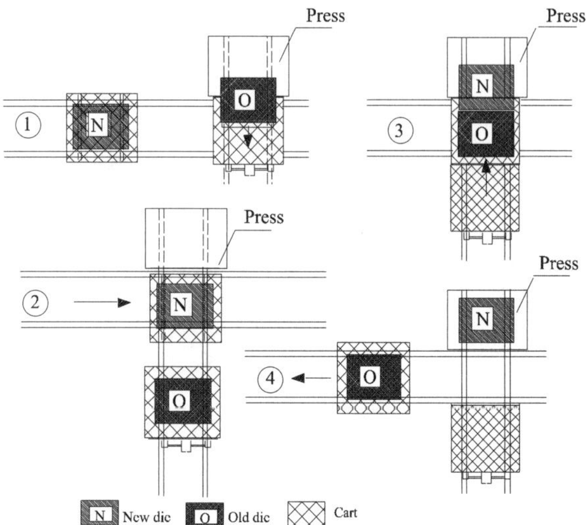

Figure 20.9 Two single-stage die carts on rails.

2. The cart with its new die now moves in front of the press, so that it stands exactly between the press and the cart holding the old die.
3. The push-pull unit on the second cart pushes the new die into the press.
4. The cart that brought the next die now takes the old die back to storage.

This system offers high speed die change and also offers high flexibility. Bolster heights are allowed to vary and the dies may alternately be exchanged. Cost and floor space requirements are major disadvantages of the system.

d) Domino System

A domino system, with several presses in a line, changes dies in each of them simultaneously. Figure 20.10 shows a process schematic of a quick die change for this method:

1. First, the die carts move in front of their respective presses, each one carrying the appropriate new die. In addition, one empty cart moves to the rear of the last press.
2. Carts and press bolsters are perfectly in line. New dies are now pushed into their presses. In turn, old dies are pushed onto the carts at the rear of these presses. This is the domino move.

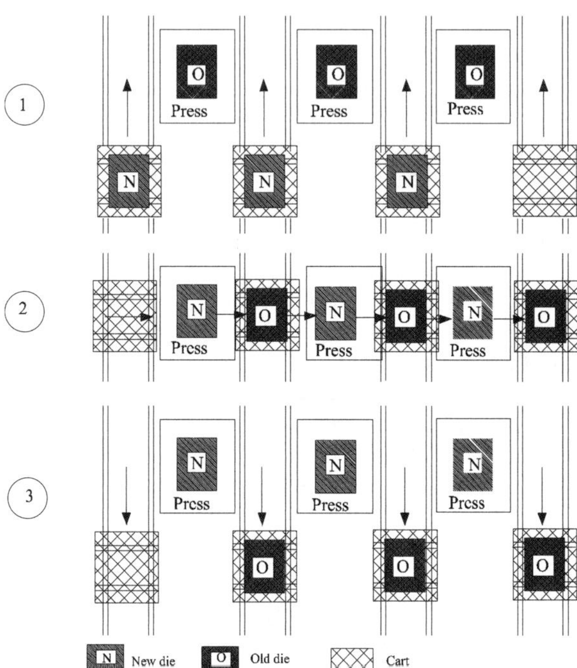

Figure 20.10 Domino system.

3. The carts move back again and take the old dies to storage.

A domino system is the preferred solution when an entire line of automated presses is to be served at the same time. The system is designed and built according to the individual application. Die exchange time is below three minutes.

There are two major disadvantages to all die cart systems. To incorporate automatic capabilities, a common subplate must normally be provided for each die; provisions must be made for rolling these dies onto the bolster or bed of the press. This will increase the cost, depending on the amount of dies needed.

e) Locating and Clamping Methods

Locating method for die carts. The locating method generally used for die carts consists of index pins mounted in the carts, extending into mating sockets in the floor.

Die clamping method. Clamping of the die holder to the press is normally accomplished manually by T-slot hydraulic clamps or ledge clamps. Hydraulic methods of clamping require that a subplate be utilized for each die.

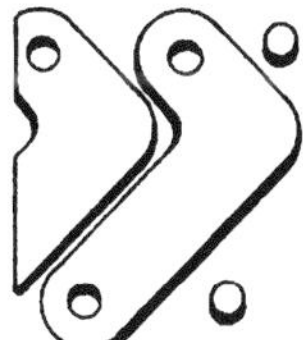

GLOSSARY

ADAPTOR. A block used to mount a die to a press slide.

AERIAL CAM. A cam attached to the punch holder with a driver on the die holder. Also called a flying cam.

AIR BEND DIE. A die in which the blank is bent without it sticking to the bottom of the die. Blank contact with the die is made at only three lines: the nose line of the punch and the two edges of the die opening.

AIR BENDING. Sheet metal-forming operation in which a workpiece is bent without the punch and die closing completely on the workpiece.

AIR CUSHION. Actuated by a large pressurized cylinder, located beneath the bed of a press, used to apply upward pressure to the lower die.

AIR DRAW. A draw operation performed in a single-action press with blank holder pressure supplied by an air cushion.

AIR HARDENING. The process by which an alloy steel is heated to the proper hardening temperature and then allowed to cool in air.

AISI. Acronym for American Iron and Steel Institute.

ALLOY. Pure metal that has been melted together with other chemical elements, forming a new metal structure having specific physical and mechanical properties.

ALUMINUM. Lightweight chemical element of a silver-white metallic color. Aluminum has an atomic number of 13, a melting point of 650°C, and a boiling point of 2450°C. Aluminum is denoted by the symbol Al.

ALUMINUM ALLOY. Pure aluminum that has been melted with one or more other chemical elements into a new metal structure having specific physical and mechanical properties.

ANNEALED. Softest possible solid state of any metal.

ARCS. Partial circles used to describe rounded corners of material and to show bends in material.

ASTM. Acronym for American Society for Testing and Materials.

AUSTENIT STAINLESS STEEL. Non-magnetic stainless steel that cannot be hardened through heat treatment.

AUTOMATIC PRESS STOP. A machine-generated signal for stopping the action of a press by automatically disengaging the clutch mechanism and engaging the brake mechanism.

BALANCING PINS. Pins used in conjunction with pressure pins to distribute and balance the load on a die cushion.

BED. The stationary plate of a press to which the lower die assembly is attached.

BED CUSHION. Commonly required for draw tooling, a bed cushion is a system that applies resistance when pushed upon. This resistance can be dynamic or statically controlled throughout the stroke.

BED HEIGHT. The distance from the bottom of the hydraulic press structure to the working height or top of the bed bolster.

BLANK. An unformed piece of sheet metal before the forming operations; a workpiece resulting from a blanking operation.

BLANK HOLDER. The part that holds the sheet metal in place while forming operations such as deep drawing are being performed.

BLANKING. Die cutting of the outside shape of a work piece.

BOLSTER. The removable plate that serves as the working surface for the bed and ram. A plate is typically bolted to the bed and ram substructures. The bolsters can be machined with a variety of work-holding features such as T-slots, drilled and tapped holes, and lift rails for quick die change.

BOTTOMING. Sheet metal-forming operation in which the punch and the die male complete contact with the workpiece by the use of increased bending force.

BOTTOMING BLOCK. Adjustable block mounted under a pad to determine the proper height of the pad when the die is closed. Sometimes this block is called a stop block.

BOTTOMING THE DIE. Adjusting a press ram so the die is on either the bottom block or the stop block at the bottom of the press stroke.

BOW.

BREAKOFF (OR BREAK-OFF).

BRITTLE. Properties of material in which breakage takes place with little or no bending or other plastic deformation.

BULGING. Expanding the diameter of a tube or other deep-drawn part by pressure from inside.

BURNISH. Smooth, shiny area above the breakout on a sheared edge. Also called shear or cut band.

BURR. A thin, rigid, sharp edge left on sheet metal blanks by cutting operations.

BURR HEIGHT. Height to which burr is raised beyond the surface of the workpiece.

BURRING. Operation in which the rough-cut edges of metal are deburred.

BY POSITION. The press may be designed to "return to position." This method utilizes either a position-sensing device or a limit or proximity switch that can be set to signal that the desired ram extension has been achieved.

CAD. Acronym for Computer-Aided Design.

CAM. Acronym for Computer-Aided Manufacturing.

CAM. A device used to control the sliding motion of the die components during the press stroke.

CAMBER.

CAM DRIVER. A component of the die with an angular surface; transfers vertical motion of the press slide to the mating angular surface on the cam slide.

CAM SLIDE. A device to perform work at an angle to the press stroke.

CARBON STEEL. A steel with up to 1.7% carbon and without substantial amount of other alloying elements. Also termed plain carbon steel.

CARRIER STRIP. The area of a stock strip that ties the workpieces together and carries them through a progressive die until the last operation.

CENTER DRILL. A tool that is a combined drill and countersink. The countersink is at a 60° included angle.

C-FRAME PRESS. A press whose frame has the shape of the letter C. Also called a gap frame press. The C-type press allows easy access to three sides of the die area.

CHAIN DIMENSIONING. A method of dimensioning drawing in which dimensions repeat each other.

CHAIN SLOTS. Machined or cast slots in the punch holder and die holder for handling purposes of die.

CHAMFER. A beveled surface to eliminate an otherwise sharp corner, typically at an angle of about 45°.

CHUTE. A sloping channel or slide for conveying blanks, scraps, or parts from a die or press.

CLEARANCE. Free space between two mating parts. In closed contours, clearance is measured on one side.

CMM. Acronym for Coordinate Measuring Machine. A machine for measuring in three dimensions (X, Y, Z).

CNC. Acronym for Computer Numerical Control.

CNC PUNCH PRESS. Press whose punching operation is controlled by a computer numerical control device.

COIL. A length of metal that has been rolled.

COINING. A compressive sheet metal-forming operation in which all surfaces of a workpiece are in contact with the surface of the punch and the die by the use of increased forming force.

COLD-COINING.

COLD-ROLLING. Passing unheated metal through rollers to reduce thickness. Steel that has been cold-rolled has a smooth surface with slightly increased skin hardness.

COLD-WORKING. Any operation of plastic deformation that is performed at room temperature.

COLUMN PRESS. A four-post single-slide press.

COMPOUND DIE. A die that can perform more than one operation with one stroke of the press.

COMPRESSIVE STRENGTH. The ability of a material to withstand compressive loads without being crushed when the material is in compression.

COMPRESSIVE STRESS. A stress that causes an elastic body to deform in the direction of the applied load.

CONSTRUCTION HOLE. A hole in which the central line is used to dimension other holes or a surface.

COUNTERBORING. Enlarging a hole to a limited depth, producing a flat bottom in the enlargement.

COUNTERSINK. An enlargement at one end of a drilled hole having an 82° included angle to allow the head of a screw to be flat with the surface of a fastened part.

CRANK PRESS. A mechanical press in which the slide is powered by a crankshaft.

CUPPING. Producing a cup-shaped part from a sheet metal blank by using a deep-drawing operation.

CUPPING TEST. A mechanical test used to determine the ductility and stretching properties of sheet metal.

CURLING. An operation for forming an edge of circular cross-section along a sheet workpiece or at the end of a deep-drawn shell or tube.

CUT-OFF DIES.

DATUM. Theoretically exact plane, line, or point from which other features are located on a design drawing.

DAYLIGHT. The distance between the bed bolster and the ram bolster when the ram is fully retracted. This is also commonly known as the open height.

DEBURRING. Removing burrs from metal by various procedures.

DEEP-DRAWING. Sheet metal-forming process to make parts in a die. The blank is drawn into the die cavity by the action of a punch. In this process, the blank diameter is reduced. Deformation is restricted to the flange areas of the blank.

DEFLECTION. A bending or turning aside from a straight course or intended purpose when force is applied to a press member.

DETENT.

DIE. The word "die" is used in two distinct ways. When used in a general sense, it means an entire press tool with all components taken together. When used in a more limited manner, it refers to that component which is machined to receive the blank, as differentiated from the "punch," which is its opposite member.

DIE BLOCK. A block of tool steel prior to its being shaped into a die.

DIE CLEARANCE. The amount of space between the punch and die opening.

DIE CUSHION. Pressurized cylinder which is used to apply upward pressure to the ejector of a part or stripper.

DIE HOLDER. Lower part of die set, on which the die block is mounted, having holes or slots for fastening to the bolster plate of the press.

DIE NIGHT. The distance from top surface of the die holder to the lower surface of the die shoe (holder).

DIE RING. A circular-shaped part of the die that is used in a deep drawing operation to control material flow and wrinkling.

DIE SET. The assembly of the die and punch holders with the guideposts and guide bushings.

DIE STAMPING. The general term for sheet metal-forming parts by a die and a press in one or more operations.

DISHING.

DISTORTION. Any deviation from a desired shape.

DOUBLE-ACTION PRESS. A press that performs two parallel movements independently; the machine may have an interior slide to and from the part and an exterior slide for the blank holder.

DIRECT KNOCKOUT.

DOWEL. A round pin, case-hardened, that fits into a corresponding hole to align two die parts.

DOWEL PULLER. A weight that slides along a rod with a head on one end and threads on the other end; normally used to pull dowels.

DRAW RADIUS. The radius at the edge of a die or punch over which sheet metal is drawn.

DRAWABILITY. The ability of the sheet metal to be formed or drawn into a cupped or cavity shape without cracking or otherwise failing.

DRIVER. A block with one or more angular surfaces; it applies force by the vertical movement of the press to mating angular surfaces on a cam slide.

DUCTILITY. The ability of a material to be stretched under the application of tensile load and still retain the new shape when the load is removed.

EARING. The formation of wavy edge projections around the top edge of a deep drawn part.

ECCENTRIC PRESS. A machine that exerts working pressure using an eccentric shaft.

EJECTOR. A mechanism operated by mechanical, hydraulic, or pneumatic power for removing a workpiece from a die.

EJECTOR ROD. A bar used to push out a formed workpiece.

ELASTIC LIMIT. The maximum stress a material can sustain without any permanent deformation.

ELASTICITY. The property of material that allows it to deform under load and immediately return to its original size and shape after the load is removed.

ELASTOMER. A rubber-like synthetic polymer such as silicon rubber.

ELONGATION. The amount of permanent extension of the material before it fractures.

EQUALIZER PIN. A pin used in conjunction with pressure pins to distribute and balance the load on a die cushion.

EXTRUDING. The drawing out of a flange around a hole which has been punched in a previous operation.

EXTRUSION. A metal-forming process in which a punch compresses a material confined in a container so that material flows through a die in the same direction as the punch.

FEED. The precise linear travel of the stock strip at each press stroke, equal to the interstation distance. Also called pitch.

FIRE CRACKS. An irregular line on the surface of a sheet caused by rolling when the roll is not properly cooled.

FIXTURE. A device to locate and hold a workpiece or component in position.

FLANGE. The formed rim of a part, generally designed for stiffening and fastening.

FLASH. Surplus metal attached to a workpiece after a forming operation.

FLAT.

FLAYING SHARE. A machine that cuts rolled product to length while it is moving.

FLOATING DIE. A die that is mounted so that it can tolerate lateral or vertical movement during the forming process.

FLOW LINES. Structure showing the direction of metal flow during plastic deformation.

FOIL. Very thin sheet metal less than 0.006 inch (0.15 mm).

FOOLPROOFING.

FORMABILITY. The capacity of a material to undergo plastic deformation without fracture.

FORMING LIMIT DIAGRAM (FLD). To assess the formability of sheet metals while forming a workpiece, circle grind analysis is used to construct a forming limit diagram of the material to be used. FLD shows the overall forming pattern of the blank during plastic deformation.

FRACTURE. The surface appearance of a freshly broken material.

FRACTURE STRESS. Nominal stress at fracture.

FRENCH NOTCH. A notch on one or both sides of a strip in a die to control progression of the strip through the die.

GAGE. Nominal thickness of sheet and strip or the diameter of wire.

Galling. Localized damage caused by the occurrence of solid-phase welding between sliding surfaces without local melting.

GANG DIE. A series of die mounted on a die plate.

GAP-FRAME PRESS. A type of press in which the frame is made in the form of a letter C and allows easy access to three sides of the die area.

GAS CYLINDER. A gas-charged cylinder used in die design in place of spring applications. Also called nitro-dyne cylinder.

GRAIN. A single crystalline structure within a microstructure.

GRAIN DIRECTION. Crystalline orientation of material in the direction of mill-rolling.

GRAIN SIZE. The average diameter of grains in a crystalline structure of a metal.

GRAPHITE. Carbon; the hexagonal crystalline structure is layered.

GUIDE BUSHING.

GUIDE POST.

HALF-HARD TEMPER. Low-carbon, cold-rolled steel, produced by cold rolling to a hardness range of 70 to 85 Rockwell on the B scale.

HARDENABILITY. The ability of steel that determines the ease of transformation of austenite to another structure, when cooled from the hardening temperature.

HARDNESS. The ability of material to resist cutting, scratching, and surface abrasion by another hard object.

HEATED PLATES. Plates that have heating capabilities. They can be heated using electric rods, steam, oil, water, or other media. These systems usually require thermal breaks between the heated plates and the hydraulic press structure. Heating controls can be separate from or fully integrated into the press control system.

HEAT TREATMENT. Process of heating and cooling of metal to obtain certain chosen qualities.

HORN. Lower part of the die on which the workpiece nests.

HOT-ROLLED STEEL. Steel that has been roller-formed into a sheet or another shape from a hot plastic state.

HYDRAULIC PRESS. A machine that exerts working pressure by hydraulic means.

HYDRAULIC PRESS BRAKE. A press brake in which the slide is actuated directly by hydraulic cylinders.

INCLINABLE PRESS. A press whose main frame may be inclined backward, usually up to a 45° angle.

INCLINATE CAM. A cam that travels at an angle other than 90° to the press stroke.

INDIRECT KNOCKOUT.

IRONING. A deep-drawing operation accomplished by reducing the wall thickness and outside diameter of a drawn cup.

INSERT. A part of a die or mold made to be removable.

INVERTED DIE. A die in which the die block is fastened to the punch holder and the punch is fastened to the die shoe (the conventional position of die block and punch are reversed).

ISO. Acronym for International Standard Organization.

ISOTROPIC. Having physical and mechanical properties of a material such that they are the same regardless of the direction of measurement.

JIG BORER. A machine to locate and machine holes very precisely.

KERF.

KICKER. A mechanism for freeing a part from a die. Also called knockout.

KNOCKOUTS.

LASER. Acronym for Light Amplification by Stimulated Emission of Radiation.

LASER BEAM CUTTING. A cutting operation that divides material using the energy of a concentrated light beam.

LEVELING. The process of flattening a sheet or strip that has been rolled.

LIFTER. A mechanism for elevating a workpiece in a die to bring it forward to another station, as in a progressive die.

LIMIT DRAWN RATIO (LDR). The greatest ratio of a blank diameter to a punch diameter that can be cup-drawn without cracking or producing other defects in the workpiece.

LOCATE.

LOCATING PIN. Also called a pilot pin.

LUBRICANT. A substance used to reduce friction.

MACROSTRUCTURE. A crystal structure that can be viewed when magnified from 1 to 10X.

MALLEABILITY. The capability of material to be deformed permanently into various shapes under the application of a compressive load without breaking.

MASS PRODUCTION. The production of parts in large quantities, usually over 100,000 per year.

METAL THINNING. The normal reduction in thickness that takes place during a deep-drawing operation.

MILD STEEL. Carbon steel that has no more than 0.25 percent carbon.

MODULUS OF ELASTICITY. The ratio of stress to strain in the elastic domain in tension or compression.

MODULUS OF RIGIDITY. The ratio of shear stress to shear strain in the elastic range.

MULTIPLE DIE. A die for the production of two or more identical parts at one press stroke.

MULTIPLE SLIDE PRESS. A press with an individual slide built into the main slide that can be adjusted to vary the length of stroke.

NATURAL STRAIN. True strain.

NC. Numerical control.

NECKING. Reduction of the cross-section of a localized area of a part that has been subjected to tensile stress during the process of plastic deformation.

NOMINAL STRAIN. Engineering strain.

NOMINAL STRESS. Engineering stress.

NORMALIZING. The process of heating steel to about 50°C above the critical temperature, then cooling it again in still air to room temperature.

OIL QUENCH. Operation in which material is quenched from the hardening temperature with oil as the cooling medium.

PART (WORK OR WORKPIECE).

PARTING LINE. The plane of two mating surfaces of a die or mold.

PATTERN DIRECTION. Orientation of features or surface patterns on sheets and coils.

PENETRATION. Depth of a cutting operation before breakout occurs.

PIERCING. Cutting or punching of openings such as holes and slots in material.

PIERCING DIE. A die that cuts out slag, which is scrap, in sheet material.

PILOT. A pin for locating a workpiece in a die from a previously punched hole.

PITCH NOTCH. A notch usually cut on one side of strip in a progressive die to control stock width and progression of the strip. Also called a French notch.

PLASTIC DEFORMATION. Permanent deformation occurring in forming of metal after elastic limits have been exceeded.

PLASTICITY. The property of a substance that permits it to undergo a permanent change in shape without cracking. A "plastic" material is one that can be deformed without volume change.

PNEUMATIC CYLINDER. A one-way air cylinder.

PRESS ATTACHMENT. A bed-mounted device on a slide forming machine used for punching, piercing, and other press operations.

PRESS BED. The stationary and usually horizontal part of a press that serves as a table to which a bolster plate or lower die assembly is mounted.

PRESS STROKE.

PROGRESSIVE DIE. Die using multiple stations or operations to produce a variety of options. Can incorporate piercing, forming, extruding, and drawing.

PROOF LOAD. A predetermined load to which a specimen or structure is submitted before acceptance for use.

PROOF STRESS. The stress that will cause a specified small permanent set in a material. A specified stress to be applied to a member or structure to indicate its ability to withstand service loads.

PUNCH HOLDER.

QUENCH CRACKING. Appearance of cracking in a metal piece during the process of quenching from a high temperature.

QUENCHING. The quick cooling of a heated workpiece in water or oil.

QUICK-CHANGE INSERT. Tool sections or parts that may be changed without removing the entire tool from the press.

RAKE.

RABBIT EAR. A recess in a die corner to allow for wrinkling or folding of the part.

RAM. A moving member of a press to which the upper part of a die is attached.

REAR CUTOFF. A device on a slide forming machine driven by a cam that is mounted on the rear shaft, allowing the removal of a slug from the strip, thus providing the ability to produce a blank with specialized end shapes.

REDRAWING. The second and successive deep drawing operation, in which a workpiece becomes deeper and is reduced in its cross-sectional dimensions.

REDUCTION. A change in material thickness or cross-section in the process of plastic deformation.

REMOTE POWER SKID. Some press applications may require that the hydraulics be located remotely from the press itself. Other applications may keep the power system from installed at the top around the crown structure of the press. In these cases, the power system and even the controls may be designed into a separate unit capable of being placed either adjacent to the press or away from the press.

RESIDUAL STRESS. The stresses that remain in the workpiece after external forces have been removed.

RESIN. Natural or synthetic polymer that has been neither filled nor reinforced.

RING. The part of a deep-drawing die that holds the blank by pressure against the mating surface of the die to control metal flow and prevent wrinkling. Also called a blank holder.

RUBBER FORMING. A sheet metal-forming operation in which a rubber cushion block is used as a functional male or female die part.

SCRAP STRIP.

SCREW PRESS. A high-speed press in which the ram is activated by a large screw assembly powered by a drive mechanism.

SEGMENT DIE. A die made of parts that can be separated for ready removal of the workpiece. Also known as a split die.

SHAVING. Removal of a thin layer of material with the sharp edge of a die or punch.

SHEAR. Cutting operation in which the material object is cut by straight or rotary blades.

SHEAR MODULUS. Modulus of rigidity.

SHEAR STRENGTH. The ability of a material to withstand offset or transverse load without rupture occurring.

SHEDDER. A pin, ring, or plate operated by spring, air, or a rubber cushion that either ejects blanks, parts, or scrap from a die or releases them from the punch.

SHUT HEIGHT. Clearance in press between ram and bed with ram down and adjustment up.

SINGLE ACTION. A press utilizing only one moving element.

SINGLE-ACTION DIE. A die without blank holder action. It is used with a single action press and does not use a draw cushion.

SLIDE ADJUSTMENT. The distance that a press slide position can be altered to change the shut height of the die space. The adjustment can be made by hand or by a power mechanism.

SPIDER. A plate that bridges two or more transfer pins and distributes force equally, usually used as a positive knockout.

SPRINGBACK ALLOWANCE. The allowance designed into a die for bending metal a greater amount than specified for the finished piece, to compensate for springback.

STOCK. A general term used to refer to a supply of metal in any form and shape.

STOCK GUIDE. A device used to direct a strip or sheet material through the die.

STOCK PUSHER.

STOP PIN. A device used to direct a strip or sheet material through the die.

STRAIGHT LANDS.

STRIPPER. Device applied to the workpiece during the punching operations.

STRIPS. Sheet metal sheared into long, narrow pieces.

STROKE. The amount of possible ram travel. The stroke is the total distance that the ram can travel, from full extension to full retraction.

STROKE CONTROL. The ram travel of the press may be controlled in a variety of ways. Most hydraulic presses are standard with adjustable retract limit switches to limit the retracting distance of the ram. Other adjustable limits may include the following: a slow-down limit; a bottom stop position; and or a bottom stop pressure.

SUB PRESS.

SWAGING.

TEMPERING. Reheating to a temperature below the critical range after a steel has been quenched; tempering is done to relieve quenching stresses or to develop desired strength characteristics.

TENSILE STRENGTH. The greatest longitudinal stress material can sustain without breaking.

TOLERANCE.

TOOLING HOLES. Holes provided in a workpiece for location purposes during production.

TOOL STEEL. Any high carbon or alloy steel capable of being suitably tempered for use in tool-and-die manufacture.

TRANSFER DIE. Progressive die in which the workpiece is transferred from station to station by a mechanical system.

TRANSFER PRESS. A press having an integral mechanism for transfer and control of the workpiece.

TRIPLE ACTION PRESS. A press having three moving slides, two slides moving in the same direction and a third lower slide moving upward through the fixed press bed.

TURRET. Rotary tool holding device in CNC punch press.

TURRET PRESS. Automatic punch press indexing the material and selecting the intended tool out of the rotary tool holding device totally by computer control for piercing, blanking, and forming a workpiece as programmed.

TWIST.

UNBALANCED CUT.

UNSUPPORTED CUT.

UTS. Acronym for Ultimate Tensile Stress.

VENT. An opening, usually quite small, to allow the escape of gases from a die.

WORK HARDENING. Increase in tensile strength of metal resulting from cold-working.

WORKPIECE.

YIELD STRESS. Stress at which a material yields and begins to deform plastically.

YOUNG'S MODULUS. Modulus of elasticity.

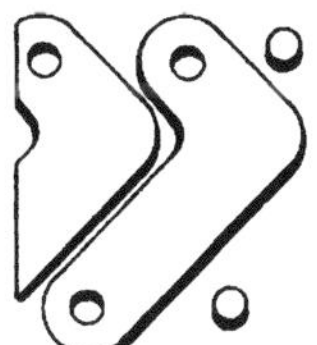

INDEX